Protein Analysis and Purification

Second Edition

Ian M. Rosenberg

Protein Analysis and Purification

Benchtop Techniques

Second Edition

Birkhäuser
Boston • Basel • Berlin

Ian M. Rosenberg
Jackson 715, G.I. Unit
Massachusetts General Hospital
Boston, MA 02114, USA

Library of Congress Cataloging-in-Publication Data
Rosenberg, Ian M., 1949–
Protein analysis and purification / Ian M. Rosenberg.
p. cm.
Originally published: Boston: Birkhäuser, c1996.
Includes bibliographical references and index.
ISBN 0-8176-4340-0 (alk. paper)—ISBN 0-8176-4341-9 (softcover : alk. paper)
1. Proteins—Analysis—Laboratory manuals. 2. Proteins—Purification—Laboratory manuals. I. Title.

QP551.R785 2004
572'.6—dc22 2004049926

Printed on acid-free paper.

Birkhäuser 

ISBN 0-8176-4340-0 (hardcover) ISBN 0-8176-4341-9 (softcover)

Printed in the United States of America. (BS/HAM)

9 8 7 6 5 4 3 2 1 SPIN 10961834 (hardcover) SPIN 10988541 (softcover)

To Mireille, Ella, and Dina

Preface

How one goes about analyzing proteins is a constantly evolving field that is no longer solely the domain of the protein biochemist. Investigators from diverse disciplines find themselves with the unanticipated task of identifying and analyzing a protein and studying its physical properties and biochemical interactions. In most cases, the ultimate goal remains understanding the role(s) that the target protein is playing in cellular physiology.

It was my intention that this manual would make the initial steps in the discovery process less time consuming and less intimidating. This book is not meant to be read from cover to cover. The expanded Table of Contents and the index should help locate what you are seeking. My aim was to provide practically oriented information that will assist the experimentalist in benchtop problem solving.

The appendices are filled with diverse information gleaned from catalogs, handbooks, and manuals that are presented in a distilled fashion designed to save trips to the library and calls to technical service representatives. The user is encouraged to expand on the tables and charts to fit individual experimental situations.

This second edition pays homage to the computer explosion and the various genome projects that have revolutionized how benchtop scientific research is performed. Bioinformatics and *In silico* science are here to stay. However, the second edition still includes recipes for preparing buffers and methods for lysing cells.

Ian M. Rosenberg
Boston, MA

Acknowledgments

I would like to thank Paula Callaghan for giving me the opportunity to correct the first edition and bring this manual into the new millennium. Thanks also go to Barbara Chernow who has guided me through the final steps of this project, and special thanks to my accountant.

Ian M. Rosenberg
Boston, MA

Contents

Protocol uses radioactive substances. Make sure that all institutional rules for the safe handling of radioactivity are followed. Take precautions when disposing of radioactive waste. Refer to Appendix A for further safety tips.

Toxic substances are used. Exercise caution when handling. Refer to Appendix A for further safety tips.

1

An Overview of This Manual

Introduction

This book is designed to be a practical progression of experimental techniques an investigator may follow when embarking on a protein purification or protein analysis project. The protocols may be performed in the order laid out or may be used independently. The aim of the book is to assist a wide range of researchers, from the novice to the frustrated veteran, in the choice and design of experiments that are to be performed to provide answers to specific questions. The manual describes standard techniques that have been shown to work, as well as some newer ones that are beginning to prove important. By following the prominently numbered steps, one can work through any protocol, whether it's a new technique or a task that has been previously performed for which a quick review or updated methodology is needed.

Some topics are introduced with a brief description and a reference. When appropriate, the researcher is encouraged to follow this lead to learn more about a specific technique.

This manual will assist the experimentalist in designing properly controlled experiments. The basic premise remains the same: ask questions and test hypotheses. Sometimes the investigator must make a decision on what conditions to use. Since scientific experimentation is an iterative process, one might have to perform the same experiment three different ways in order to reach a satisfactory conclusion.

There will be no advice for dealing with specific pieces of equipment other than encouragement to read the manual, if you can find it. Throughout all manipulations try to be objective. Be on the lookout for unexpected findings. You will learn the most from surprising results, and they are often the beginning of the next project. It is never possible to record too much in your lab notebook. Treasure the rare moments when everything "works". Do not get discouraged. Remember, things will not always run smoothly.

During the selective survey of methodology, some specific examples are provided. Most methods have many variations, and the one that is detailed has proven to work successfully. There is often a great deal of flexibility in the application of protein methods. Therefore, it should be possible, and it is encouraged, to alter or adapt a technique to your specific needs. Be flexible in finding the optimal conditions and then stick to them.

The evolution of a biomedical research project goes from a foundation of solid understanding of a particular process to a point where novel hypotheses need to be tested. As so often happens in scientific experimentation, the results of an experiment will dictate the follow-up experiment. For instance, one is given an antibody, a cDNA clone, or perhaps a biochemical assay to be used as a means of identifying and characterizing a specific protein that may play a role in a physiological function or disease process. The goal at this stage of the project is to produce enough of the specific protein in a purified form so that functional and biochemical questions can be asked without worrying that the effect might be attributable to a minor contaminant. The first step is to make use of analytical techniques that will give some basic information regarding the protein of interest.

In studying a biological system using a biochemical approach, researchers have traditionally attempted to purify each of the system's components. Using a reductionist approach, each component is then studied in detail with the ultimate aim being to reconstitute the system *in vitro* from the isolated components. The availability of completely sequenced genomes has triggered a shift towards studying the biological system as a whole. Experimental systems are being used that are designed to study all the concurrent physiological processes in a cell.

New methodologies have made brute force protein purification nearly obsolete. Protein science is being transformed by molecular biology. Innovations, the commercialization of reagent production, and simple modern conveniences such as the microwave oven have transformed labs into safer and more efficient places to work. Technology and capitalism have together provided instruments and kits that have replaced laboratory chores that in the 60s and 70s would have taken weeks to months to perform. Most labs no longer distill their own phenol or cast their own polyacrylamide gels. With each breakthrough and technological advance, researchers discover new challenges as they strive to do more in less time with smaller amounts of material. New protein sequences, as they are discovered, are entered into databases, as exemplified by the EMBL Nucleotide Sequence Database, which have grown nearly exponentially. Even with this ever-expanding electronic database, the techniques employed by the protein chemists of the 1960s and 1970s are still worth mastering in the 2000s. Often times, a preliminary step is needed to remove a bothersome contaminant prior to using an affinity method, high performance liquid chromatography (HPLC) or performing polyacrylamide gel electrophoresis (PAGE) to isolate the target protein for microsequencing. Therefore, many basic techniques will be described.

Kits, Cores, and Computers

Most institutions have at least one major resource core. The notion of shared instrumentation, a cost sharing measure and time saver, has evolved into a "core", laboratories that were set up to provide services for analyzing and synthesizing biomolecules. The cores of the late 80s and early 90s such as HPLC and amino acid analysis have expanded into a new generation of cores such as chip technology for genomics, proteomics, mass spectrometry, and real-time PCR reflecting the evolving needs and technologies of the day.

The increased use of pre-made kits is skyrocketing. In some cases it's less expensive to purchase the ready-made reagent than to buy the raw materials and make it in house! Kits make a technique or even a whole field accessible to researchers outside their area of expertise. One does not have to be a molecular biologist to sequence a gene. The downside of the kit generation is that it deprives beginners of the value and the understanding of assembling things from scratch. Young scientists should still know what goes into the test tube and why (DeFrancesco, 2002).

Now, performing scientific research without computers is unimaginable. Computers have become an integral part of lab life, collecting, analyzing and plotting data, writing papers and grants, searching and reading the literature, and emailing, surfing and playing games during long incubations. In fact, with the number and breadth of databases that are available online, in silico scientists can make discoveries without ever going into the lab.

How to Use This Book

This manual is aimed at first-time users and designed to shorten the time to a reliable result for those who are more experienced with a technique. It is comprised of a collection of selected traditional techniques and novel approaches to protein purification and analysis. In addition, newer techniques for purifying recombinant proteins are presented. The analytical chapters contain methods that examine the structure of proteins with emphasis on identifying the target protein in a complex mixture and on the recognition and analysis of posttranslational modifications and how to take advantage of them. In-depth theoretical explanations are not provided. Whenever possible, user-friendly step-by-step protocols for the at-the-bench scientist are provided in a simplified, clear form.

It is hoped that each protocol will be self-contained. Therefore, there will be a fair amount of repetition as many reagents and procedures are referred to repeatedly throughout the manual. Although the protocols have been designed to be complete, information from other protocols may need to be cross-checked. A list of reagents and equipment necessary to successfully complete the protocol will appear immediately prior to the step-by-step instructions.

Names of some solutions might be the same from protocol to protocol, while the composition of the solutions may differ. Be sure that the solution described in a particular protocol is prepared from the correct recipe. Recipes may appear cross-referenced to avoid potential confusion. Use the highest quality water available for the preparation of solutions and reagents. Centrifuge speeds are provided as ×*g* or rpm (revolutions per minute). Consult the nomogram in Appendix F to convert these speeds for your centrifuge and rotor.

Throughout the manual suppliers of certain reagents and equipment are noted where pertinent. You may experiment with substituting your own favorite brand. Useful tips are included as part of the method or directly following it in a "*Comment*."

The Appendices are an essential feature of this manual. They are an eclectic collection of tables and charts, as well as trivial but useful information designed to provide bench-top access to diversified general information found scattered in various catalogs, brochures, manuals, handbooks, and other Methods books.

If you own this book for more than a few weeks and do not bend or smear any of its pages, then you are probably not using it correctly.

Basic Laboratory Equipment

At your work-station you should have access to the following items:

3 pipettes of each of the volumes, 1–20 µl, 20–200 µl, 200–1000 µl
Vortex mixer
Magnetic stirrer
Hot plate (Available for your routine use)
Comfortable chair
Centrifuge, low speed (20,000 RPM)
Refrigerated centrifuge
Ultra-centrifuge (20,000–80,000 RPM)
Microcentrifuge that holds standard 1.5 ml microfuge tubes
Geiger counter
Radiation shielding
Refrigerators and freezers
X-ray film
Cassettes and intensifying screens
Spectrophotometer
pH meter
Scintillation counter
Tissue culture equipment
Balance
Cell disrupter/sonicator
SDS-PAGE supplies
Power supply

Fume hood
Lyophilizer
Microscope
Water purification facility with the ability to produce ultrapure water (double-distilled, deionized, >18Ω) should be used for all reagent preparations
Ice-maker
Computer terminal with internet access and printer
Glassware
Plastic containers of various dimensions

Laboratory Automation

Robotics technologies have revolutionized the way scientific research is being performed in the twenty-first century. Advances from the Human Genome Project have emphasized the concept of high-throughput technologies. To keep pace with the new reality in the lab, tools and resources have changed which change the way that scientific questions are asked and the strategies for answering them. The overall effect of laboratory automation is the increase in the speed at which data is generated. Scientists who once spent a lot of time performing repetitive tasks now spend more time analyzing data and planning experiments.

A computer is a necessity for accessing databases and to store, view and analyze one's own data. The desktop computer has become an integral piece of laboratory equipment. It is routinely used for word processing, storing experimental records and keeping laboratory databases. Computers are used as lab notebook by some investigators. Most importantly, computers allow access to the internet where one can scan catalogs, get technical assistance and order lab supplies. The web can replace lab clutter by cutting down on the volume of catalogs. Electronic journals are readily accessible from the comfort of one's computer, making trips to the library a special event. One can now download that article from an obscure journal found only in a library on the other side of town.

All of the major biological databases including GenBank®, European Molecular Biology Laboratory (EMBL), SWISS-PROT, Protein Information Resource (PIR), MEDLINE, and all of the ACDB species databases managed by the USDA have made information available on the Web. Simple direct access to these databases has changed the way that research is performed. Use frustrating experiences with computers as a chance to learn general computer problem solving skills. Try to know where the best tools for the specific task are located (Brown, 2000).

Automated liquid handling continues to be a major focus of lab automation. Hand-held electronic devices and sophisticated robots will both rapidly and tirelessly perform a range of tedious liquid handling tasks on 96, 384, and 1536 well plates. There are many liquid-handling options available that enable researchers to address their specific needs.

Clearly, not all labs need or require automatic pipetting capabilities or can afford a $75,000 system.

Imaging technology is playing an increasing role in experimental biology. Automation of experiments will require visible results to be recorded and interpreted by computerized imaging systems.

Beyond Protein Analysis and Purification

Much of the methodology presented in this manual is designed to put the researcher in position to characterize the target protein. The ultimate goal is to provide for every protein a database that would include a description of its sequence, its domain structure, sub-cellular localization, posttranslational modifications, variants, homologies to other proteins on the whole protein level and at the domain level, its interacting partners, its cellular function and knock-out and mutant information.

Once a partial amino acid sequence has been obtained, an in-depth computer search for homology with previously characterized proteins is usually performed. Expanding electronic protein databases offer the researcher a growing opportunity to gain clues to the uniqueness and function of a potentially new and important molecule. Eventually, the cDNA will be cloned and antibodies will be produced. The usefulness of this manual does not stop at this point. It provides the tools for answering some of the following biological and biochemically relevant questions pertaining to the target protein and its purification. (1) What are the options? (2) How does a particular method work and what are its strengths and weaknesses? (3) How can the method be exploited while minimizing its limitations? How does it fit with other methods? (4) What is the amount of target protein and the level of synthesis in a particular cell type? (5) What is the subcellular localization (nuclear, cytoskeletal, membrane, membrane receptors, specific organelles) of the target protein? Does the protein translocate in response to specific stimuli? (6) Is the protein posttranslationally modified, and how do these modifications affect its function? (7) Is the protein expressed in a cell-cycle-specific manner? Are there variations in the level of synthesis? (8) Is the level of target protein regulated by hormones, growth factors, heat shock etc.? (9) What is the rate of synthesis in normal and transformed cells? Is the target protein proliferation-sensitive, an oncoprotein, a component of a pathway that controls cell proliferation? (10) What is the function of the target protein? Does the protein belong to a family i.e., phosphoprotein, cytoskeletal protein, kinase or a protein with cyclin-like properties? (11) Is the target protein differentially expressed in various tissues? (12) When performing immunoprecipitation, do other proteins coprecipitate with the target protein? (13) Is the target protein specific to a given disease? How does the protein compare when isolated from healthy and diseased individuals? (14) What is the effect of transfected cDNA in cells that do not express the target protein? In brief, the structural and biochemical information that is obtained will allow the investigator to propose hypotheses and design experiments to test their validity.

References

Brown SM (2000): *Bioinformatics: A Biologist's Guide to Biocomputing and the Internet.* Natick, MA: Eaton Pub
DeFrancesco L (2002): Today's lab. *The Scientist* 16:14–16

2

Protein Structure

Introduction

Proteus, a Greek sea god and keeper of all knowledge, would not give up information easily. Even while being held down, he would struggle mightily, assuming different forms before revealing any of his secrets. Although proteins were not named after Proteus, the description could not be more appropriate.

Proteins are complex macromolecules made up of successive amino acids that are covalently bonded together in a head-to-tail arrangement through substituted amide linkages called peptide bonds. Each protein molecule is composed of an exact sequence of amino acids arranged in a linear, unbranched fashion. Protein molecules have the property of acquiring a distinctive 3-dimensional configuration.

The amino acid backbone may have many posttranslational modifications contributing to the size, charge and function of the mature protein. A functionally active protein, being the product of posttranslational structural modifications, cannot be determined by reference to any single gene. Rather, the active form is often the result of complex biochemical reactions performed on the target protein that are not all controlled by any one gene product.

A. The Amino Acids

A representative amino acid is shown in Figure 2.1. The α-amino acids contain an α-carbon to which an amino group ($—NH_3^+$) and a carboxylate group ($—COO^-$) are attached. The 20 amino acids that make up the building blocks of proteins differ in the structure of their R groups, which may be hydrophilic or hydrophobic, acidic, basic, or neutral. Of the 20 amino acids normally used to build proteins, 19 have the general structure shown in Figure 2.1. Proline, the 20th natural amino acid, is similar but has the side chain bonded to the nitrogen atom to give the amino acid. The chemical composition of the unique R groups of the amino acids is responsible for the most important

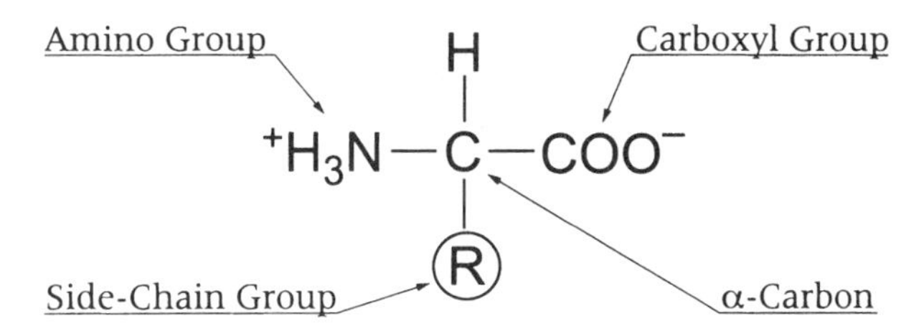

FIGURE 2.1 A schematic version of the general formula of an amino acid. At pH 7.0, both the amino and carboxyl groups are ionized.

characteristics of the amino acids: chemical reactivity, ionic charge, and relative hydrophobicity.

The charged amino acids may be either acidic or basic. At low pH, proteins are positively charged due to the basic groups on lysine and arginine, whereas, at high pH, proteins are negatively charged due to the acidic groups on aspartic and glutamic acids.

The peptide backbone of proteins is composed of amino acids having polar, non-polar, aromatic and charged residues. In their native states, proteins are compact structures with only a small number of amino acids exposed to the surface. The amino acids that are exposed are well suited for their microenvironment. For proteins in an aqueous milieu charged and polar residues are commonly exposed. For proteins that are embedded in membranes, nonpolar, lipophilic amino acids are found in the interface.

The amino acids are presented in Table 2.1 according to their polarity and charge. There are eight amino acids with nonpolar hydrophobic R groups. Five of them have aliphatic hydrocarbon R groups (alanine, leucine, isoleucine, valine and proline), two have aromatic rings (phenylalanine and tryptophan) and one of the nonpolar amino acids contains sulfur (methionine). As a group, these amino acids are less soluble in water than the polar amino acids.

The polar amino acids are more soluble in water because their R groups can form hydrogen bonds with water. The polarity of serine, threonine, and tyrosine is contributed by their hydroxyl groups (—OH); that of asparagine and glutamine by their carboxy-amide groups and cysteine by its sulfhydryl group (—SH). Glycine falls in this group by default. Its R group, a hydrogen atom, is too small to influence the high degree of polarity contributed by the α-amino and carboxyl groups.

Amino acids that carry a net negative charge at pH 6.0–7.0 contain a second carboxyl group. These are the acidic amino acids, aspartic acid, and glutamic acid. The basic amino acids, in which the R groups have a net positive charge at pH 7.0, are lysine, arginine, and histidine which is weakly basic. At pH 6.0 more than 50% of the histidine molecules are positively charged (protonated), but at pH 7.0 less than 10% have a positive charge.

Historically, the amino acids were designated by three-letter abbreviations. Subsequently, to make them more computer-friendly, a set of one-letter symbols has been adopted to facilitate comparative display

TABLE 2.1 Amino Acids

Amino acid	Triple letter code	Single letter code	R group
Amino acids with nonpolar R groups			
Alanine	Ala	A	$-CH_3$
Valine	Val	V	$-CH(CH_3)_2$
Leucine	Leu	L	$-CH_2-CH(CH_3)_2$
Isoleucine	Ile	I	$-CH(CH_3)-CH_2-CH_3$
Proline	Pro	P	$N-C-$ ring with CH_2, CH_2, CH_2
Phenylalanine	Phe	F	$-CH_2-C_6H_5$
Tryptophan	Trp	W	$-CH_2-$ indole (N–H)
Methionine	Met	M	$-CH_2-CH_2-S-CH_3$
Amino acids with uncharged polar R groups			
Glycine	Gly	G	$-H$
Serine	Ser	S	$-CH_2-OH$
Threonine	Thr	T	$-CH(OH)-CH_3$
Cysteine	Cys	C	$-CH_2-SH$
Tyrosine	Tyr	Y	$-CH_2-C_6H_4-OH$
Asparagine	Asn	N	$-CH_2-C(=O)-NH_2$
Glutamine	Gln	Q	$-CH_2-CH_2-C(=O)-NH_2$

TABLE 2.1 *Continued*

Amino acid	Triple letter code	Single letter code	R group
Acidic amino acids (negatively charged)			
Aspartic Acid	Asp	D	$-CH_2-C(=O)O^-$
Glutamic Acid	Glu	E	$-CH_2-CH_2-C(=O)O^-$
Basic amino acids (positively charged)*			
Lysine	Lys	K	$-CH_2-CH_2-CH_2-CH_2-N^+H_3$
Arginine	Arg	R	$-CH_2-CH_2-CH_2-NH-C(=N^+H_2)-NH_2$
Histidine	His	H	$-CH_2-$ (ring: C, N, C, N^+H, C; $C{=}N^+H$)

*at physiological pH.

of amino acid sequences of homologous proteins. General information about the amino acids is presented in Tables 2.1 and 2.2.

B. The Four Levels of Protein Structure

The structure of a protein can be resolved into several different levels, each of which is built upon the one below it in a hierarchical fashion. Proteins are so precisely built that the change of a few atoms in one amino acid can disrupt the structure and have catastrophic consequences.

Primary Structure

Linderstrøm–Lang and his coworkers (1959) were the first to recognize structural levels of organization within a protein. They introduced the

TABLE 2.2 Hydropathy Values, Solubility in Water, and pK Values of the Amino Acids

Amino acid	Hydropathy index*	Solubility (g/100 ml water)	Side Chain pK	α-NH_2 pK	α-COOH pK
Isoleucine	4.5	3.36		9.76	2.32
Valine	4.2	5.6		9.72	2.29
Leucine	3.8	2.37		9.6	2.36
Phenylalanine	2.8	2.7		9.24	2.58
Cysteine	2.5	unlimited	8.33	10.78	1.71
Methionine	1.9	5.14		9.21	2.28
Alanine	1.8	15.8		9.87	2.35
Glycine	–0.4	22.5		9.13	2.17
Threonine	–0.7	unlimited		9.12	2.15
Tryptophan	–0.9	1.06		9.39	2.38
Serine	–0.8	36.2		9.15	2.21
Tyrosine	–1.3	0.038	10.1	9.11	2.2
Proline	–1.6	154		10.6	1.99
Histidine	–3.2	4.19	6.0	8.97	1.78
Glutamic acid	–3.5	0.72	4.25	9.67	2.19
Glutamine	–3.5	2.6		9.13	2.17
Aspartic acid	–3.5	0.42	3.65	9.6	1.88
Asparagine	–3.5	2.4		8.8	2.02
Lysine	–3.9	66.6	10.28	8.9	2.2
Arginine	–4.5	71.8	13.2	9.09	2.18

*The higher the magnitude of the hydropathy index number the more hydrophobic the amino acid. Hydropathy values from Kyte J and Doolittle RF (1982).

terms primary, secondary, and tertiary structure. Primary structure refers to the sequence of amino acids that make up a specific protein. The number, chemical nature, and sequential order of amino acids in a protein chain determine the distinctive structure and confer the characteristics that define its chemical behavior. Although most protein sequences have been established by direct amino acid sequence analysis, the vast majority of primary sequences are being deduced directly from the DNA sequence. Often, one method serves to confirm the other.

The covalent bond that links amino acids together is called a peptide bond. Depicted in Figure 2.2, the peptide bond is formed by a reaction between the α-NH_3^+ group of one amino acid and the α-COO^- group

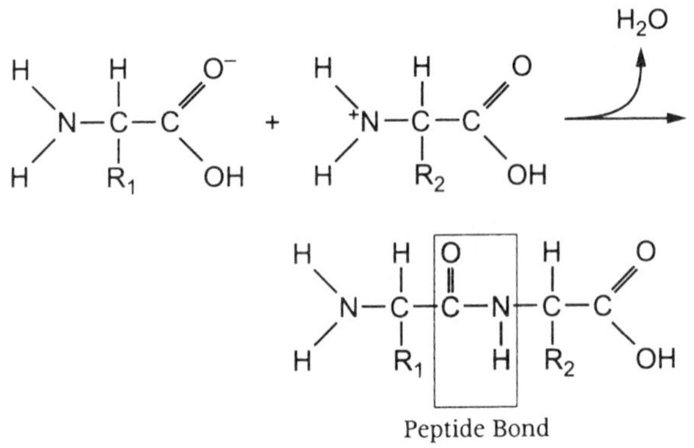

Peptide Bond

FIGURE 2.2 The formation of the peptide bond; a condensation reaction between two amino acids resulting in a loss of water. The peptide bond is highlighted.

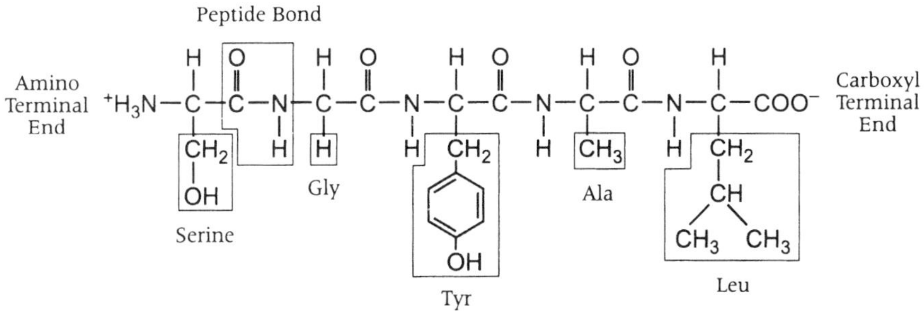

FIGURE 2.3 A schematic linear representation of a pentapeptide.

of another amino acid. One water molecule is removed as each peptide bond forms. The polypeptide backbone is simply a linear ordered array of amino acid units incorporated into a polypeptide chain. All proteins and polypeptides have this fundamental linear order and aside from the modifications to the amino acids, differ only in the number of amino acids linked together in the chain and in the sequence in which the various amino acids occur in the polypeptide chain. Additional covalent bonds, found primarily in secreted and cell-surface proteins, are disulfide bonds (also called S–S bonds) between cysteine residues. These bonds are stable at physiological pH in the absence of reductants and oxidants.

The terminal residue of a polypeptide chain which contains a free α-NH_3^+ group is referred to as the amino terminal or N-terminal residue. The carboxyl-terminal or C-terminal residue has a free COO^- group. In biological systems, proteins are synthesized from NH_3^+ terminal to COO^- terminal, and the accepted convention is to write the amino acid sequence of a polypeptide from left to right starting with the N-terminal residue. A short peptide composed of five amino acids is shown in Figure 2.3.

The term peptide generally refers to a structure with only a small number (typically 2–20) of amino acids linked together. The term polypeptide generally refers to longer chains. The term protein applies to those chains with a specific sequence, length, and folded conformation in their native states. Denatured long chain polypeptides that have lost their quaternary structure are also referred to as proteins.

Secondary Structure

The polypeptide chain of a protein is folded into a specific 3-dimensional structure producing a protein's unique conformation. Secondary structure refers to the regular local structural arrays found in proteins that can be referred to as independent folding units. The secondary structure is determined by the chemical interactions (mainly hydrogen bonding) of amino acid residues with other amino acids in close proximity. The secondary structure is identifiable as substructures, usually α-helices and β-structures. The α-helix, shown in Figure 2.4, is a common secondary structure in proteins. Helices are characterized by their pitch (rise per residue), period (number of residues per turn), handedness (right or left), and diameter. The typical α-helix is right-

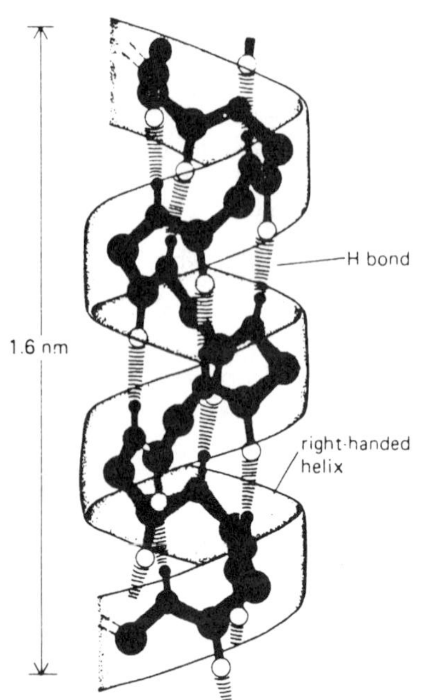

FIGURE 2.4 An α helix. Note the extensive hydrogen bonding. (Reprinted from Alberts et al, 1983 with permission).

handed with a pitch of 1.5 Å and a period of 3.6 residues per turn. All side chains point away from the helix. The length per turn is 5.4 angstroms or 1.5 angstroms per residue. Neglecting the side chains, the α-helix has a diameter of about 6 angstroms. The diameter of the α-helix is sufficiently small that the structure is more like a filled cylinder than an open spring (Cohen, 1993). The lengths of α-helices vary in proteins, but on average, there are ~10 residues per helical segment. Theoretically, helices can be right- or left-handed. However, α-helices comprised of L-amino acids are never left-handed. If a right-handed helix were a spiral staircase and you were climbing up, the banister would be on your right-hand side. Some amino acids do not accommodate themselves to helices whereas others lead to helix formation.

There are variations on the α-helix in which the chain is either more tightly or more loosely coiled. Hydrogen bonds between corresponding groups that are closer or further apart in the primary structure by one residue are designated the 3_{10} helix or the π helix, respectively. One or two turns of the 3_{10} helix is a common occurrence at the ends of helical portions of proteins. In the 3_{10} helix, every n and n + 3 amino acid is linked by hydrogen bonds. The name arises because there are 10 atoms in the ring closed by the hydrogen bond, and there are only three residues per turn. The 3_{10} helix is only about 5% as abundant as the α-helix.

Two types of β pleated sheet or β structure (Figure 2.5) commonly occur in proteins. Antiparallel pleated sheets are formed by extended, adjacent segments of polypeptide chain, whose sequences with respect to the direction from $—NH_3^+$ to $—COO^-$ run in the opposite direction. Antiparallel sheets may be formed by a single chain folding back on itself or by two or more segments of chain from remote parts of the molecule. Chains running in the same direction form the parallel pleated sheet.

Along a single β strand, the amino acid side chains are positioned alternately up and down. When β strands are assembled into a sheet, the side chains are aligned in rows with all the side chains in a row pointing up or down. This is true for both parallel and antiparallel β sheets.

The polypeptide chains in many proteins often fold sharply back upon themselves, giving rise to secondary structures called β-bends, which are stabilized by hydrogen bonds. The amino acids occurring most frequently at the bend are glycine, proline, aspartic acid, asparagine, and tryptophan. An example of a polypeptide chain spontaneously folding to a more favorable conformation is shown in Figure 2.6.

Proteins are generally built up from combinations of secondary structure elements, α-helices and β–sheets, connected by loop regions of various lengths. Secondary structural elements combine in ways that result in the formation of a stable hydrophobic core. They are often arranged in motifs, simple, commonly occurring super-secondary elements with characteristic geometric arrangements. Secondary structure elements and motifs combine to form domains. A domain is part of the polypeptide chain that can fold independently into a stable

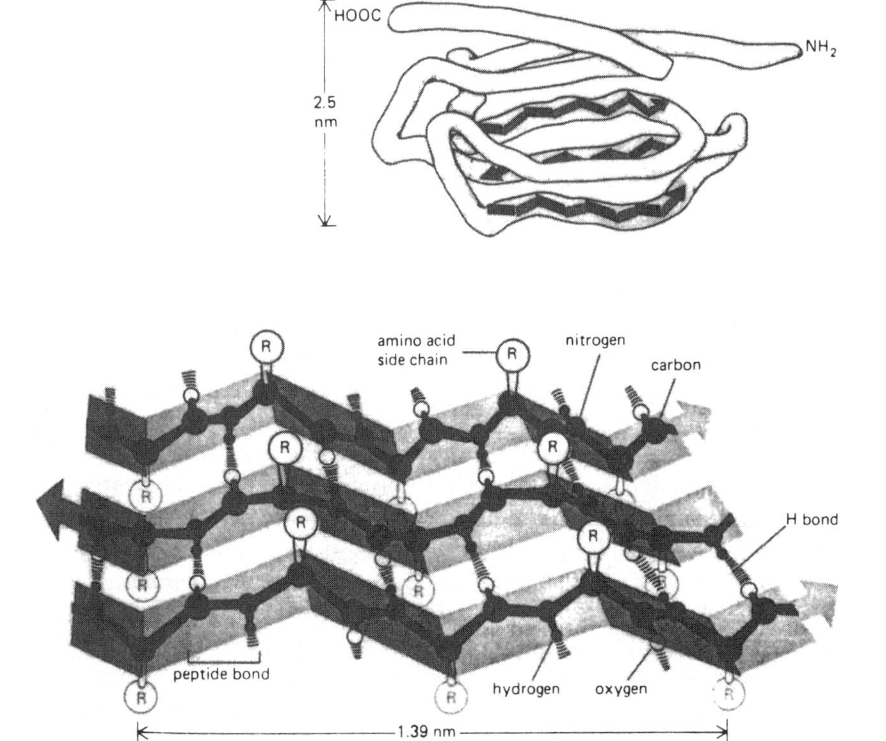

FIGURE 2.5 An idealized antiparallel β sheet is shown in detail. Note that every peptide bond is hydrogen-bonded to a neighboring peptide bond. (Reprinted from Alberts et al, 1983 with permission).

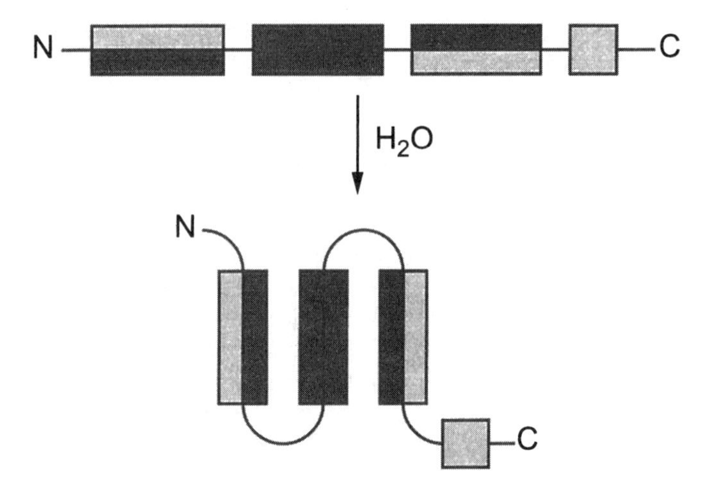

FIGURE 2.6 Folding of a protein chain in water. The hydrophobic segments (shaded areas) rearrange to form a compact core surrounded by the hydrophilic portions. (Reprinted from Fletterick et al, 1985 with permission).

tertiary structure. A protein can consist of a single domain or multiple domains.

In the evolution of proteins, domains are important units that have been shuffled, duplicated and fused to create larger, more complex proteins. Although the possibilities of distinct amino acid sequences are unlimited, the number of different folding patterns for the domains is not. Extrapolation based on existing databases of protein sequence and structure indicates that most of the natural domain sequences assume one of a few thousand folds (Govindarajan et al, 1999), of which ~1,000 are already known (Orengo et al, 2002).

Tertiary Structure

Unraveling the pathway by which unstructured proteins spontaneously fold to their native functional form has been a central goal of protein chemists since before Anfinsen's landmark paper (1973). The sequences of existing proteins have been selected through evolution not only to adopt a functional 3-dimensional structure after folding but also to optimize the protein folding process both temporally and spatially, given the constraints of the cellular context. An attempt to understand the complicated structure of a polypeptide is greatly simplified by realizing that much of the complex 3-dimensional (tertiary) structure can be described as an assembly of regular secondary structural elements. Tertiary structure is the intramolecular arrangement of the secondary structure independent folding units with respect to each other. The 3-dimensional organization of several hundred polypeptide chains has been revealed by crystallography and nuclear magnetic resonance spectroscopy. This level of organization is determined by the noncovalent interactions between helices and β-structures together with the side-chain and backbone interactions unique to a given protein. The tertiary structure can be inferred from an analysis of the packing of these secondary structural elements. Essential for the tertiary structures is a delicate balance between many noncovalent interactions: hydrophobic regions formed by nonpolar R groups of the amino acids; ionic; van der Waals interactions; and hydrogen bonds. Disulfide bonds are a major force as they most likely stabilize the conformation after folding occurs. These bonds form spontaneously when the appropriate thiol groups are brought into juxtaposition as the result of cooperative interactions of the R groups that lead to correct folding.

Quaternary Structure

Many proteins in solution exist as aggregates of two or more polypeptide chains, either identical or different. Polypeptide chains are designated by letters, an example being hemoglobin, referred to as $\alpha_2\beta_2$. Many proteins are dimers, trimers, tetramers, or even higher-order aggregates of identical polypeptide chains. Quaternary structure refers to the stoichiometry and spatial arrangement of the subunits and is determined by the amino acid sequences of the subunits. The subunit stoichiometry of a protein is the number of each type of polypeptide that has combined to produce the specific structure. Each subunit interacts with the other subunits through hydrophobic and polar interactions suggesting that the interacting surfaces must be highly complementary. Dissociated subunits can recombine to give a functionally competent native protein. Not only do the specific amino acids confer secondary and tertiary structure but the quaternary structure is also dictated by the amino acid sequence of the polypeptide chains.

Two fundamental types of interactions can take place between identical subunits, isologous and heterologous, as demonstrated in Figure 2.7A,B. In an isologous interaction, the interacting surfaces are identical, giving rise to a closed, dimeric structure (Morgan et al, 1979). In the heterologous association, the interfaces that interact are not identical. The surfaces must be complementary but need not be symmetric (Degani and Degani, 1980). This association is potentially open-ended unless the geometry of the interaction is such as to produce a closed cyclical structure as shown in Figure 2.7C.

A multimeric protein is composed of more than one folded polypeptide. Each of the polypeptides composing such a protein folds to form a tertiary structure unique to the amino acid sequence of that polypeptide. These folded polypeptides will then associate with each other to form the multimeric protein. The arrangement in space of these folded polypeptides in the protein is its quaternary structure.

A protein quaternary structure database exists that contains ~10,000 structurally defined proteins of presumed biological importance (http://pqs.ebi.ac.uk). Each assembly consists of at least two protein chains (Sali et al, 2003).

C. Chemical Characteristics of Proteins

Proteins have ionic and hydrophobic sites both internally (within the folds of the tertiary structure) and on the surface where the primary structure comes in contact with the environment. The ionic sites of a

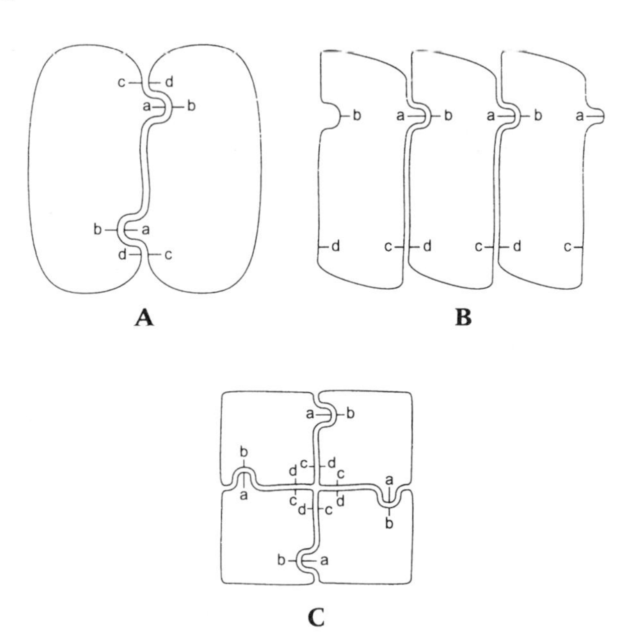

FIGURE 2.7 Schematic illustration of isologous and heterologous association between protein subunits. (A) Isologous association to form a dimer. (B) Heterologous association leading to an infinitely long polymer. (C) Heterologous association to form a closed, finite structure, in this case a tetramer. (Adapted from Monod et al, 1965 with permission).

protein are provided by the charged amino acids at physiological pH and by covalently attached modifying groups (e.g., carbohydrates and phosphate).

Proteins are polyelectrolytes containing positively and negatively charged groups. The net charge on a protein is contributed by the free α-amino group of the N-terminal residue, the free α-carboxyl group of the C-terminal residue, those R groups capable of ionization, and by the unique array of modifications attached to the protein. In proteins, the ionizing R groups greatly out-number the single ionizing groups at the two terminal residues. At physiological pH the $\alpha\text{-COO}^-$ and $\alpha\text{-NH}_3^+$ groups are ionized, with the deprotonated carboxyl group bearing a negative charge and the protonated amino group a positive charge. Therefore, an amino acid in its dipolar state is called a zwitterion.

At a certain pH, referred to as the isoelectric point (pI), the numbers of positive and negative charges on a protein are equal and the protein is electrically neutral. A protein has a net positive charge at pH values below its pI and a net negative charge above the pI. If a protein has a high pI (e.g., 10), this implies that it is basic and that the excess positive charges contributed by arginine and lysine residues are rendered neutral at a high pH. Conversely, a protein with a low pI is rendered neutral when the charges are neutralized at low pH, which occurs when the protein is in an environment with many free H^+ ions which will protonate the negative charges on aspartic and glutamic acid residues.

Hydrophobicity

The hydrophobic character of a protein is due to nonpolar amino acids (e.g., valine and phenylalanine) and covalently attached hydrophobic groups (e.g., lipids and fatty acids). The hydrophobicity of a protein can be changed by partially denaturing the protein and exposing the protein's interior primary structure which contains most of the hydrophobic amino acids.

The final folded structure of a protein is a thermodynamic compromise between allowing hydrophilic side-chains access to the aqueous solvent and minimizing contact between hydrophobic side-chains and water. It follows that the interiors of water-soluble proteins are predominantly composed of hydrophobic amino acids, while the hydrophilic side-chains are on the exterior where they can interact with water.

A hydropathy (strong feeling about water) index, presented in Table 2.2, has been devised where each amino acid is assigned a value reflecting its relative hydrophilicity or hydrophobicity. An appropriate evaluation of a given amino acid sequence should be able to predict whether a given peptide segment is sufficiently hydrophobic to interact with or reside within the interior of the membrane which itself is hydrophobic. A computer program has been devised that uses a moving-segment approach that continuously determines the average hydropathy within a segment of predetermined length as it advances through the sequence (Kyte and Doolittle, 1982). The consecutive scores are plotted from the amino to the carboxy terminus. The procedure gives a graphic visualization of the hydropathic character of the chain from one end to the other. A midpoint line is printed that corresponds to the grand average of the hydropathy of the amino acid compositions found in most sequenced proteins, a universal midline. For most proteins there is an excellent agreement between the interior portions of the protein and the region appearing on the hydrophobic side of the midpoint line, as well as the exterior regions appearing on the hydrophilic side of the line. Potential membrane spanning segments can be identified by this procedure.

Figure 2.8 shows the hydropathy profile of erythrocyte glycophorin, illustrating the concept of hydropathic indices. Glycophorin has an easily recognizable membrane-spanning segment in the region of residues 75–94. As can be seen from Figure 2.8, a positive hydropathic stretch is indicative of hydrophobic amino acids.

Consensus Sequences

It is clearly not sufficient to determine the primary structure of a protein or deduce it from the DNA sequence and expect that this will explain all the properties of a protein. Elucidation of the complete covalent structure of a protein includes the knowledge of the primary structure, as described above, and the chemical nature and positions of all the modifications to the protein that take place in the cell and are necessary for its correct function, regulation, and antigenicity.

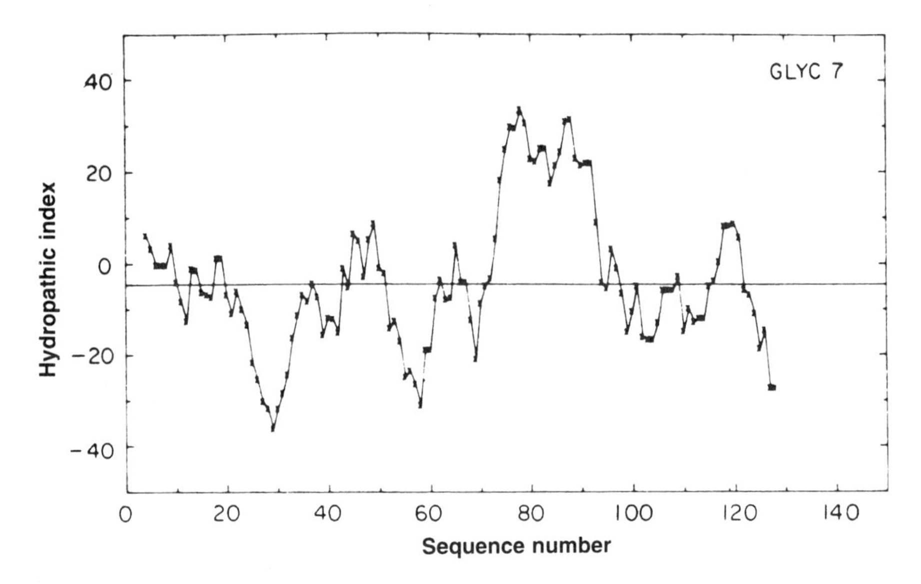

FIGURE 2.8 Hydropathy profile of erythrocyte glycophorin which has an easily recognizable membrane-spanning segment in the region of residues 75 to 94. (Reprinted from Kyte and Doolittle, 1982 with permission).

When an investigator has succeeded in defining the amino acid sequence of the target protein or translates a cDNA sequence, a careful analysis of the primary amino acid structure may yield a great deal of structural and functional information about the protein. Consensus sequences have been defined for many posttranslational modifications and motifs which may suggest a function for a previously undefined protein.

The term "consensus sequence" refers to those sequence elements containing and surrounding the site to be modified. These sequence elements are considered essential for its recognition. It generally takes the form of a short linear sequence of amino acids indicating the identity of the minimum set of amino acids comprising such a site (Kennelly and Krebs, 1991).

The sequences of many proteins contain short, conserved motifs which are involved in recognition and targeting. These sequences, which appear in the primary, linear structure of the protein, are often separate from other functional properties of the molecule in which they are found. They do not include elements from different polypeptide chains or from widely scattered portions of a single polypeptide chain. Therefore, they are not the result of distant segments being brought together as the protein assumes its native conformation. The conservation of these motifs varies; some are highly conserved while others permit substitutions that retain only a certain pattern of charge across the motif.

The usefulness of consensus sequence analysis is based on its simplicity. Summarizing the complexities of the substrate recognition sequence as sets of short sequences has facilitated the evaluation of a large body of observations. A word of caution: the existence of an apparent consensus sequence does not assure that a protein is modified. Consensus sequence information functions best as a guide whose implications must be confirmed or refuted experimentally. A partial list of consensus sequences for a variety of posttranslational modifications and cell functions is presented in Appendix E.

Proteomics

A View of the Proteome

The annotation of the human genome indicates that the number of genes is between 30,000 and 40,000. However, the estimated number of proteins encoded by these genes is two to three orders of magnitude higher. Therefore, organism complexity is generated more by a complex proteome than by a complex genome.

Proteomics is many things to many investigators. The proteome is defined as the time and cell specific protein complement of the genome. This encompasses all proteins that are expressed in a cell at one time, including isoforms and protein modifications. Therefore, expression proteomics aims to measure up-and-down-regulation of protein levels. Whereas the genome is constant for one cell, being largely identical for all cells of an organism, the proteome is very dynamic with time and in response to external factors, and differs markedly between cell types (Rappsilber and Mann, 2002).

Every cell type or tissue has its own proteome, which represents only part of the genome. There are many ways that the same polypeptide backbone can be posttranslationally altered making a famous biological dogma, one gene one enzyme, put forth by Beadle and Tatum, no longer tenable.

Functional proteomics attempts to characterize cellular components, multiprotein complexes, and signaling pathways. Proteomics attempts to provide a snapshot of protein synthesis rate, degradation rate, posttranslational modification, subcellular distribution and physical interactions with other cell components.

Several diverse mechanisms can result in the expression of many protein variants from the same gene locus: single nucleotide polymorphisms (SNPS), gene splicing, alternative splicing of pre-mRNA, proteolytic cleavage, and co-and posttranslational modification of amino acids. Even though the difference between two proteins can be very small, a single amino acid change or the modification of a single amino acid, this difference could be crucial for the function of the protein, classically demonstrated by hemoglobin where a single amino acid substitution results in sickle cell anemia.

Protein abundance cannot be predicted from mRNA abundance, and posttranslational modifications cannot be predicted from deduced amino acid sequence. Simply predicting a polypeptide sequence can even be problematic because of alternative splicing of mRNAs or frameshifts (Gygi et al, 1999). The identity of a protein must therefore be defined by its structural formula (the connectivity of all atoms), not just the amino acid sequence, but any and all of the modifications.

Proteomics would not be possible without the previous achievements of genomics that provided the "blueprint" for all possible gene products (Tyers and Mann, 2003). However, unlike DNA analysis, which spawned technological advances like automated sequencing and the polymerase chain reaction, proteomics must deal with a set of problems for which technological shortcuts do not as yet exist. The biological protein analyst must deal with the problems of limited and variable sample material, degradation and posttranslational modifications due to developmental and temporal conditions.

One fundamental challenge to proteomics is the accurate detection of proteins that are present in vanishingly small amounts. The abundance of individual proteins differs by as much as four orders of magnitude. Low abundance proteins such as transcription factors and kinases that are present at 1–2000 copies per cell represent molecules that perform important regulatory functions (Hucho and Buchner, 1997). The protein diversity that can result from a single gene demands a more precise analysis.

An organized effort has begun to develop an infrastructure in proteomics that is aimed at unraveling the complexity of the proteome in health and disease. To this end the Human proteome Organization (HUPO, http://www.hupo.org) was founded. HUPO's mission is threefold: to consolidate proteome organizations into a worldwide organization, to encourage the spread of proteomics technologies; and to coordinate proteome initiatives aimed at characterizing specific cell and tissue proteomes (Hanash, 2003).

Developing Proteomic Technologies

Some of the major technology platforms that have developed along with proteomics include 2-D gel electrophoresis and mass spectrometry. Historically, 2-D electrophoresis provided the technology that illustrated the potential for cataloging expressed proteins in databases (Celis et al, 1996). After obtaining the protein fraction, the method of choice for proteomic studies is one- or 2-D gel electrophoresis. The advantage of one dimensional SDS-PAGE is that virtually all proteins are soluble in SDS. Also, the range of relative molecular mass from 10 kD–300 kD is covered, and extremely acidic and basic proteins are easily visualized.

Because even the best 2-D gels can routinely resolve no more than 1,000 proteins, it is clear that only the most abundant proteins can be visualized if a crude protein mixture is used. Proteins present in low abundance, like regulatory proteins, will probably go undetected if a total cell lysate is used. Applying larger amounts of sample is usually not an option since the resolution will breakdown if the gel is overloaded. One solution is to enrich for the target protein by using a purification step prior to electrophoresis. For example, a 66 kD protein that is present at about 1,000 copies per cell, a protein of medium abundance, less than 2 picomoles (100 ng) in one liter of cell culture would not be visualized from whole cell extracts. But if it could be affinity enriched it could probably be detected by silver stain. For 2-D gel electrophoresis-based proteomics the protein mixture is fractionated by 2-D electrophoresis. 2-D image analysis software is commercially available. Typically, the software packages can digitize images from 2-D gels, detect, quantify, and identify spots of differential intensity between gels.

References

Alberts B, Bray D, Lewis J, Raff M, Roberts K, Watson JD, eds. (1983): *Molecular Biology of the Cell.* New York: Garland Publishing, Inc.

Anfinsen CB (1973): Principles that govern the folding of protein chains. *Science* 181:223–230

Celis JE, Gromov P, Ostergaard M, Madsen P, Honore B, Vandekerckhove J, Rasmussen HH (1996): Human 2-D PAGE databases for proteome analysis in health and disease. *FEBS Lett* 398:120–134

Cohen FE (1993): The parallel β helix of pectate lyase C: Something to sneeze at. *Science* 260:1444–1445

Degani Y, Degani C (1980): Enzymes with asymmetrically arranged subunits. *Trends Biochem Sci* 5:337–341

Fletterick RJ, Schroer T, Matela RJ, Staples J, eds. (1985): *Molecular Structure: Macro molecules in Three Dimensions.* Boston: Blackwell Scientific Publications

Govindarajan S, Recabarren R, Goldstein RA (1999): Estimating the total number of protein folds. *Proteins* 35:408–414

Gygi SP, Rochon Y, Franza BR, Aebersold R (1999): Correlation between protein and mRNA abundance in yeast. *Mol Cell Biol* 19:1720–1730

Hanash S (2003): Disease proteomics. *Nature* 422:226–232

Hucho F, Buchner K (1997): Signal transduction and protein kinases: the long way from the plasma membrane into the nucleus. *Naturwissenschaften* 84:281–290

Kennelly PJ, Krebs EG (1991): Consensus sequences as substrate specificity determinants for protein kinases and protein phosphatases. *J Biol Chem* 266:15555–15558

Kyte J, Doolittle RIF (1982): A simple method for displaying the hydropathic character of a protein. *J Mol Biol* 157:105–132

Linderstrøm–Lang KU, Schellman JA (1959): Protein structure and enzyme activity. In: *The Enzymes, Vol. 1.* Boyer PD, Lardy H, Myrbäck K, eds. New York: Academic Press

Monod J, Wyman J, Changeux J-P (1965): On the nature of allosteric transitions: A plausible model. *J Mol Biol* 12:88–118

Morgan RS, Miller SL, McAdon JM (1979): The symmetry of self-complementary surfaces. *J Mol Biol* 127:31–39

Orengo CA, Bray JE, Buchan DW, Harrison A, Lee D, Pearl FM, Sillitoe I, Todd AE, Thornton JM (2002): The CATH protein family database: a resource for structural and functional annotation of genomes. *Proteomics* 2:11–21

Rappsilber J, Mann M (2002): What does it mean to identify a protein in proteomics? *Trends Biochem Sci* 27:74–78

Sali A, Glaeser R, Earnest T, Baumeister W (2003): From words to literature in structural proteomics. *Nature* 422:2126–225

Tyers M, Mann M (2003): From genomics to proteomics. *Nature* 422:193–197

General References

Creighton TE (1984): *Protein Structures and Molecular Properties.* New York: W H Freeman Publishing Co.

Rossmann MG, Argos P (1981): Protein folding. *Ann Rev Biochem* 50:497–532

3

Tracking the Target Protein

Introduction

Protein analysis and purification is aimed at elucidating the structure and function of proteins. In order to analyze a protein, you must have a reliable detection system that unambiguously enables you to follow the target protein. This is especially true when the target molecule is in crude or even semipurified form. Often, analytical results will lay the groundwork for a successful purification scheme.

The purification of a bioactive molecule is frequently accomplished by using a definitive assay designed to recognize a property of the target protein. The system you choose should be sensitive, accurate, reproducible, reliable, and capable of being performed with ease on a large number of samples. The reagents and equipment necessary should be readily available at a reasonable cost.

The sensitivity of the assay used for detecting the target protein is often the limiting factor as the protein is present in extremely small amounts. The assay should be as specific as possible to avoid wasting time and possibly producing an artifactual result. Typically, biochemical assays are designed to monitor either the disappearance of a substrate or the appearance of a product. This is routinely achieved by using optical absorption or fluorescent methods, or with a radioactive substrate that when modified in a specific manner by the target molecule can be readily quantified.

As shown in Figure 3.1, the bioassay will enable the investigator to ultimately obtain enough of the target protein in purified form so that other valuable reagents, like specific antibody and a cDNA clone can be produced. Figure 3.1 also suggests experimental directions that may be pursued as a project develops.

The protocols to be followed depend upon the starting material and on the analytical tools that are available. If you have a specific antibody, be it monoclonal or polyclonal, the procedures and analytical strategies you choose will differ from those that would be followed if the project is based on either a cloned gene or a bioassay. In most cases, the ultimate goal is the same regardless of what you start with: to

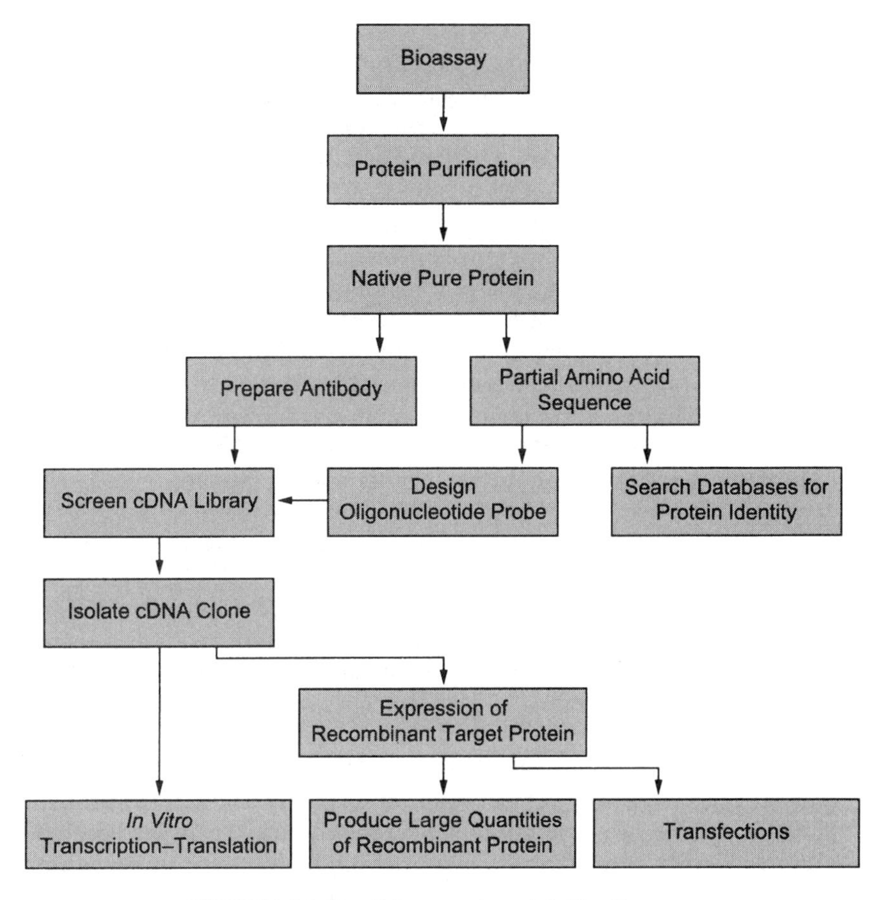

FIGURE 3.1 Possible experimental directions.

determine the structure and function of your molecule. However, your unique situation will determine the strategies to be used to reach the objective.

In most cases, the availability of an antibody that recognizes the target protein will lead to a successful outcome. Many of the techniques that are described rely on a specific antibody to identify the target protein. Using the antibody, analytical reactions can be performed on a complex mixture and the effect on the target protein can be specifically followed. The antibody can also be utilized to screen an expression library, identify the bacterial colony expressing the target protein, and ultimately clone the gene.

If you possess a clone of a gene whose protein product you wish to analyze, one approach to take would be to synthesize the target protein in a biological system by introducing the gene into a suitable expression vector. This methodology is presented in Chapter 11.

In the early stages of analysis when a biological activity is being followed, the active molecular agent may be nucleic acid, protein, carbohydrate, lipid, any combination, or any other molecule such as a prostaglandin, leukotriene, or polyamine. In order to characterize the molecular nature of the active factor, one useful approach is to attempt to destroy the activity by treating it with an enzyme or subjecting the factor to a simple extraction procedure. For proteins, the extract is routinely treated with broad specificity proteolytic enzymes such as proteinase K, pepsin, pronase, and trypsin (see Appendix G). It is important to include a suitable control to be sure that the treatment itself does not have an effect in the assay system being used. If the control is valid and the activity is destroyed, this constitutes strong evidence that the active fraction is proteinacious. However, one must proceed cautiously. A negative result should not be viewed as absolute proof that the active factor is not a polypeptide.

Proteins are often denatured by heating, resulting in the loss of bioactivity. If a solution of the active material is exposed to 56°, 65°C and, in the extreme, to boiling for times ranging from 3 min to 1 h and the bioactivity is destroyed, this is an indication that the active molecule is proteinacious. Again, negative results should be interpreted with caution. Denaturing treatments such as heat or incubation with the anionic detergent sodium dodecyl sulfate (SDS) may cause the protein to partially lose its tertiary structure and open up, making it more susceptible to protease degradation. Inactivation of a biomolecule in a crude preparation by 65°C heat treatment may reflect an increased susceptibility to proteases more than heat lability. This should be kept in mind when performing early characterization experiments on the target molecule.

There are many molecules that retain their bioactivity after boiling and protease treatment yet are made up partially or totally of protein. Some examples are glycoproteins, proteoglycans, and glycosaminoglycans. Crude preparations may contain the active protein which is protected by noncovalent associations with irrelevant molecules. Another possibility may be that the crude extract contains an enzyme that can destroy the active molecule which itself is nonproteinacious.

The bioactive molecule might be lipid in nature. In this case perform a chloroform/methanol extraction and assay both organic and aqueous phases. If the activity partitions into the organic phase, then the active fraction may be lipid.

A. Labeling Cells and Proteins

Sensitive and convenient analytical reactions have been devised to characterize proteins present as minor components of a complex mixture without going through a long, multistep purification scheme. The availability of a specific antibody, either polyclonal or monoclonal, enables the enrichment and isolation of the target protein by immunoprecipitation. Biochemical reactions designed to analyze the target protein can be performed on the protein as part of the immunoprecipitated complex. The protein is then separated electrophoretically and analyzed. Alternatively, the target protein could be transferred and immobilized on a matrix such as nitrocellulose where additional analytical procedures may be performed.

This chapter describes methods to tag the target protein, that is usually present as a minor component of a cell lysate, and then specifically follow it. General methods are presented for metabolically labeling live cells, and for labeling proteins located on the cell surface. In addition, the Chloramine T method of labeling a pure protein in solution is presented. Other specific labeling protocols are presented where applicable throughout the manual.

There are two main approaches to radiolabeling cellular proteins. Each has certain advantages and limitations. In one approach, proteins present on the cell surface are labeled post-synthetically. Highest specific activities are achieved with radioiodination techniques. The other approach is the biosynthetic incorporation of radioamino acids into the nascent polypeptide chain. Since the radiolabel is distributed throughout the polypeptide and all biosynthetic forms are labeled, this approach facilitates subsequent analysis of protein structure, such as peptide mapping, amino acid sequencing and 2-dimensional electrophoresis. **A word of caution:** *Do not assume that a molecule that is radiolabeled efficiently by one technique will label equally well with another method.* Do not hesitate to modify a given technique to suit your specific experimental needs. Perform the controls to eliminate any artifactual results.

Metabolic Labeling Cells in Culture

Cells can be grown in tissue culture in the presence of different precursor molecules that are specifically incorporated either directly into the protein's primary structure or as part of a modifying group. Many posttranslational modifications have been well studied, and methodologies exist for their analysis. Several of these are listed below and are described in Chapter 9:

- Phosphorylation
- *O*-linked glycosylation
- *N*-linked glycosylation
- Myristoylation
- Palmitylation
- Isoprenylation
- Glypiation
- Sulfation

To successfully perform metabolic labeling experiments, you must be skilled in the fundamentals of tissue culture and sterile technique.

In most metabolic labeling experiments, [^{35}S]methionine is the radioactive amino acid of choice. The [^{35}S] signal is stronger than [^{3}H] and [^{14}C] and is available commercially at high specific activities. In addition, the intracellular pool of methionine is smaller than most amino acids, making it possible for exogenously added [^{35}S]methionine to be incorporated rapidly into proteins.

Proteins can be metabolically labeled with any amino acid using this simple protocol by adjusting the formulation of the tissue culture medium and substituting a radioactive amino acid for its non-radioactive isoform. Special media which lack one or more amino acids are commercially available (GIBCO/BRL).

PROTOCOL 3.1 Metabolic Labeling Adherent Cells

Materials

CO_2 incubator and laminar flow hood

Methionine minus medium (Met-): DMEM Medium lacking methionine and lacking serum, or media containing 5% dialyzed serum (Appendix Protocol G.2)

Phosphate buffered saline (PBS): See Appendix C

[^{35}S]methionine: cell labeling grade (1000Ci/mmol)

1. The cells should be healthy and growing rapidly. Choose dishes that are ~75% confluent. Remove the medium and wash the cells with PBS or Met-medium.
2. Starve the cells by incubating them at 37°C for 30–60min in Met-media. For a 10cm dish, use 3ml of Met-. For other size dishes or wells, use enough medium to just cover the cell layer.
3. Add the appropriate amount of [^{35}S]methionine, ~100μCi/ml, 3ml for a 10cm dish, directly to the dish and gently swirl.
4. Incubate the cells at 37°C in a CO_2 incubator for the desired length of time. As a starting point try 2–4h.
5. For steady-state labeling incubate the cells for 16–20h in HEPES-buffered tissue culture medium containing 1/10 the normal concentration of methionine supplemented with 5% dialyzed serum (Appendix Protocol G.2).
6. When the labeling period is complete, remove the radioactive medium and wash the cells twice with 10ml PBS to remove free label. Treat the washes as liquid radioactive waste. The cells are ready to be lysed.

Comments: If the target protein is thought to be a secreted protein, save the labeling medium for analysis. Otherwise, *dispose of the radioactive medium according to your institution's guidelines for radioactive waste.*

PROTOCOL 3.2 Metabolic Labeling Cells Growing in Suspension

Materials

Low speed centrifuge, bench-top model

PBS: Appendix C

Methionine minus medium (Met-): DMEM Medium lacking methionine and lacking serum or 5% dialyzed serum (Appendix Protocol G.2)

15 and 50ml plastic disposable centrifuge tubes

[^{35}S]methionine: cell labeling grade (1000Ci/mmol)

Pasteur pipet

1. Harvest the cells in a 50 ml centrifuge tube by a 10 min centrifugation at ~200–500× *g*.
2. Resuspend the cells in PBS and wash them twice with PBS.
3. Resuspend the cells in Met-medium (10^7 cells/ml) and starve them for 30–60 min in your 37°C incubator.
4. Add the radioactive amino acid (~100 µCi/10^7 cells) and incubate at 37°C for the desired labeling interval.
5. To remove the labeling media, transfer the cell suspension to a 15 ml disposable plastic conical centrifuge tube and pellet the cells by centrifugation at ~200–500× *g*. Carefully remove the labeling medium. If you are interested in secreted proteins, save the medium; otherwise *dispose of it according to your institution's radioactive waste disposal guidelines.*
6. Resuspend and wash the cells once in 10 ml of PBS. Remove the PBS from the cell pellet and treat it as liquid radioactive waste. If no additional reaction is desired at this stage, the cells are ready to be lysed (Protocol 3.9).

Comments: An alternative to [^{35}S]methionine is Expre^{35}S^{35}S-label sold by Perkin Elmer. This contains both [^{35}S]methionine and [^{35}S]cysteine. As mentioned above, usually 100 µCi/10 cm plate is sufficient. However, if the target protein is not abundant, try using higher concentrations of label and extend the labeling period.

In some experimental systems the labeling solution can be recycled, especially if it contains a high concentration of radiolabel and the same cell type is to be labeled. Save the high concentration labeling media by storing it at –20°C. After the starvation step, discard the media and replace it with the recycled labeling media containing the radioactive amino acid.

A specific protein may label poorly if it contains few, or lacks the amino acid used for labeling, or because it has a low rate of synthesis. In these cases, try labeling the cells for longer times and changing the radioactive amino acid. Two or more radioactive amino acids can be mixed together for metabolic labeling.

PROTOCOL 3.3 Pulse-Chase Labeling

Pulse-chase experiments are designed to analyze posttranslational processing events that may be modifying the protein of interest. By using short labeling periods (the pulse step), followed by the removal of the radioactive amino acid and the addition of saturating amounts of cold amino acids (the chase step), the target protein can be followed as it is modified in a time dependent fashion. Other valuable information, such as the turnover rate of the protein, can be readily assessed using pulse-chase protocols. Following pulsing and chasing, the cells are ready to be lysed.

Materials

See Protocols 3.1 and 3.2.

Chase medium: Complete tissue culture medium. For methionine chase add 1 mM methionine, for cysteine chase add 0.5 mM cysteine.

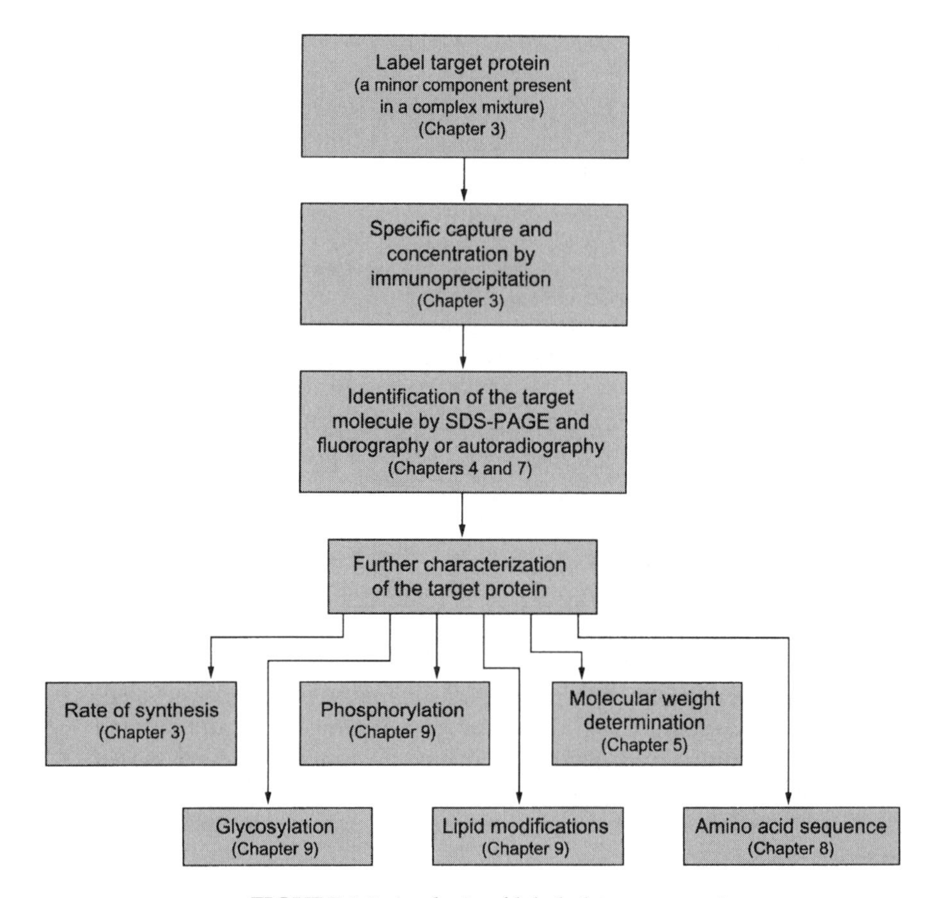

FIGURE 3.2 Analysis of labeled target protein.

1. For metabolic labeling, use either Protocol 3.1, for adherent cells, or 3.2 for cells in suspension. Pulse labeling is usually performed for 2–30 min periods.
2. Immediately after the designated labeling period, wash the cells with PBS then incubate them for various time intervals in chase medium (the chase step).
3. Harvest the cells and prepare a total cell lysate (Protocol 3.9 for cells in suspension, Protocol 3.10 for adherent cells). If you are interested in secreted proteins, save the chase medium from every time point.

Comments: Potential reactions for the identification and analysis of the target protein are outlined in the flow chart shown in Figure 3.2.

Labeling Proteins Present at the Plasma Membrane

If you suspect that the target protein is located on the outer surface of the plasma membrane, it should be possible to tag it by surface labeling. Radioiodinating a protein involves the introduction of radioactive iodine into certain amino acids, usually tyrosines. Histidine residues are also iodinated by some methods. Radioactive ^{125}I can be incorporated into proteins either enzymatically or by chemical oxidation. For the majority of proteins, iodination does not interfere with the protein's function. However, if the iodinated protein becomes functionally compromised, try a different method. Three labeling techniques are described; others are mentioned.

PROTOCOL 3.4 Lactoperoxidase Labeling Cell Surface Proteins

Lactoperoxidase is used to enzymatically introduce radioactive iodine into proteins present on the plasma membrane. The following protocol is a modification of that described by Haustein (1975). This procedure should be performed in a designated fume hood with adequate shielding. Check with your institution's radiation safety officer for specific instructions. All washes containing ^{125}I should be disposed of according to your institution's guidelines for the safe handling of radioactive waste.

Materials

H_2O_2: Freshly prepared, diluted 1:20,000 with PBS from a 30% stock
PBS: Appendix C
Lactoperoxidase (Sigma): 0.2mg/ml in PBS (1 unit/ml)
$Na^{125}I$: Carrier free (Specific activity 17Ci/mg)
Tyrosine: 10mg/ml in PBS (store on ice)
Plastic conical centrifuge tube
Refrigerated bench-top centrifuge

1. Collect the cells in a conical plastic centrifuge tube by centrifugation at 200× *g*.
2. Wash the cells twice by resuspending the cell pellet in 10ml PBS.
3. Resuspend the cells in PBS at a concentration of 5×10^7 cells/ml. Remove a 200μl aliquot and place it in a tube on ice.
4. While on ice, add 500μCi of $Na^{125}I$ to the 200μl aliquot of cells, then 50μl of the lactoperoxidase solution.
5. To initiate the reaction, add 1μl of the freshly prepared H_2O_2 solution. Mix gently. During the next 4min add four 1μl pulses of diluted H_2O_2 at 1min intervals.
6. Terminate the reaction by adding 100μl of the tyrosine solution.
7. Add 10ml of PBS and centrifuge the cell suspension for 5min at 200× *g* at 4°C.
8. Resuspend the cells in 10ml of PBS and centrifuge for 5min at 200× *g*. You are now ready to lyse the cells.

Comments: Do not perform this procedure in the presence of sodium azide as it will inhibit lactoperoxidase. This method will ideally label only proteins that are on the outer surface of the cell. The reaction temperature is kept low to decrease endocytosis.

Dispose of the ^{125}I radioactive liquid waste according to your institutional guidelines.

PROTOCOL 3.5 Labeling Surface Proteins with IODO-GEN®

IODO-GEN® (1,3,4,6-Tetrachloro-3a,6α-diphenylglycouril) (Pierce) is insoluble in aqueous solutions so that the iodination will only proceed at the cell surface. IODO-GEN® was first described as a reagent for the iodination of proteins and cell membranes (Fraker and Speck, 1978). The reagent is very stable; therefore many reaction vessels can be prepared at once and stored indefinitely in a dessicator until use.

This procedure should be performed with adequate shielding. All washes containing ^{125}I should be disposed of according to your institution's guidelines for the safe handling of radioactivity.

Materials

N_2 gas tank and regulator

IODO-GEN® (1,3,4,6-Tetrachloro-3a,6α-diphenylglycouril) (Pierce): 1 mg/ml in chloroform

Glass vials, cover-slips, scintillation vials

PBS: Appendix C

$Na^{125}I$

NaI: 1 M

1. Prepare IODO-GEN® coated surfaces by adding the IODO-GEN® solution to a 2 ml septum-capped glass vial (Wheaton, Vineland, NJ) or other glass surface such as a 12 × 75 mm test tube, 20 ml scintillation vial or glass coverslips. Thoroughly dry the surface under a gentle stream of nitrogen gas in a fume hood.
2. Rinse the reaction vessel with buffer to remove microscopic flakes of IODO-GEN®.
3. Add the cell suspension in the appropriate buffer (see comments below) and then add the desired amount of $Na^{125}I$.
4. Let the reaction proceed 10–30 min with occasional agitation.
5. Transfer the cell suspension from the vessel to a 15 ml plastic conical centrifuge tube. This effectively terminates the iodination reaction.
6. Add cold NaI to a final concentration of 0.25 M for safe handling. Add 10 ml of ice cold PBS and centrifuge the cells at 200× *g* at 4°C for 5 min to remove the unreacted ^{125}I. You are now ready to lyse the cells.

Comments: A ratio of 10 μg or less of IODO-GEN® should be used per 100 μg protein or 100 μg of IODO-GEN® per 10^7 cells. Typically, 500 μl of $Na^{125}I$ are used per 100 μg protein or per 10 cm tissue culture dish. Many different buffer systems can be used, borate, phosphate, Tris, and HEPES.

Since iodination with IODO-GEN® is an oxidative reaction, reducing agents like mercaptans should not be added until after the iodination is completed. Also avoid medium that contains tyrosine and serum.

PROTOCOL 3.6 Non-radioactive Biotinylation of Cell Surface Proteins

Although cell surface labeling followed by immunoprecipitation has historically been performed using ^{125}I, techniques utilizing cell surface biotinylation have many advantages over ^{125}I. Plasma membrane proteins can be surface labeled with the biotinylating reagent sulfosuccinimidobiotin (sulfo-NHS-biotin). This reagent, shown in Figure 3.3, is a water-soluble *N*-hydroxysuccinimide (NHS) ester of biotin. The water solubility results from the sulfonate (—SO_3) group on the NHS ring. Water solubility is an extremely useful feature, eliminating the need to dissolve the reagent in an organic solvent prior to use. Due to its water solubility this compound does not permeate the plasma membrane, restricting the biotinylation to the cell surface (Cole et al, 1987). This technique can be extended to selectively label proteins that are present on the apical or basolateral cell surface (Lisanti et al, 1988).

Materials

PBS$^+$: 0.1 mM $CaCl_2$, 1 mM $MgCl_2$ in PBS

Bovine serum albumin (BSA) solution: 5% BSA in PBS

Sulfosuccinimidobiotin (sulfo-NHS-biotin) (Pierce): 0.5 mg/ml in PBS$^+$

Dulbecco's modified Eagle medium (DMEM): Serum free

Labeling buffer: PBS containing 0.05% Tween 20, 1 M glucose, 10% glycerol, and 0.3% BSA

Wash buffer: 0.05% Tween 20 in PBS

SDS sample buffer (See Appendix C)

1. Wash adherent cells five times with ice-cold PBS$^+$.
2. Treat the cells with the sulfo-NHS-biotin in PBS$^+$ solution at 4°C for 30 min. Make sure that the cells are completely covered by the sulfo-NHS-biotin solution (at least 3 ml/10 cm dish).
3. Wash the cells with ice-cold serum-free DMEM followed by four washes with PBS$^+$.
4. Lyse the cells as detailed in Protocol 3.10. The proteins can be analyzed by SDS-PAGE (Chapter 4), followed by Western blotting (Chapter 7).

Comments: Sulfonated NHS-esters of biotin are soluble in water at concentrations up to at least 10 mM. Another reagent of this type which

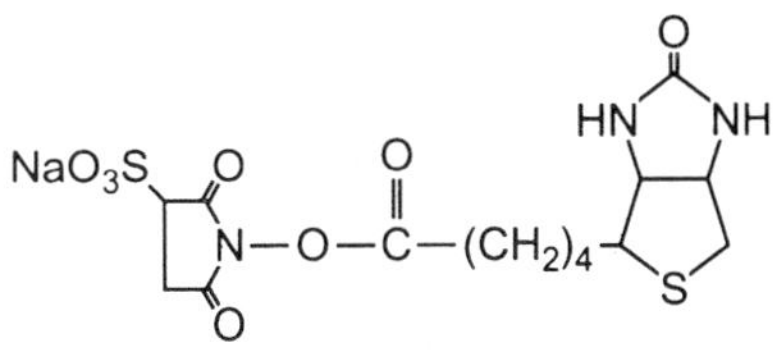

FIGURE 3.3 Structure of sulfo-NHS-Biotin.

can also be used is NHS-LC-Biotin which has a spacer arm of 22.4Å. The spacer serves to reduce steric hindrances associated with binding for biotinylated molecules on one avidin and results in enhanced detection sensitivity. This reagent has become the first choice for surface biotinylations and its use is described in Protocol 3.7. Compatible buffers for these reagents are phosphate, bicarbonate, and borate. High salt buffers may cause the reagent to precipitate. However, the reagent is capable of biotinylating target molecules even when it is not completely solubilized.

The surface biotinylation method has been modified and extended to be used to globally profile the cell surface proteome (Shin et al, 2003). Surface proteins were biotinylated and purified by immobilized avidin columns. The biotinylated proteins were then separated by 2-D electrophoresis and identified by MALDI-TOF. This technique was also used successfully to identify surface proteins of *Helicobacter pylori* (Sabarth et al, 2002).

PROTOCOL 3.7 Domain-Selective Biotinylation and Streptavidin-Agarose Precipitation

Distinct apical and basolateral surfaces are common features of polarized epithelial cells (Hunziker et al, 1991). For selective labeling of the apical or basolateral cell surface, NHS-LC Biotin (Pierce) is added either to the apical or basolateral compartment of a Transwell filter chamber. Some features of the Transwell system are shown in Figure 3.4. The use of permeable supports allows cells to be grown and studied in a polarized state under more natural conditions. Areas of research where Transwells have been used are discussed in Appendix I.

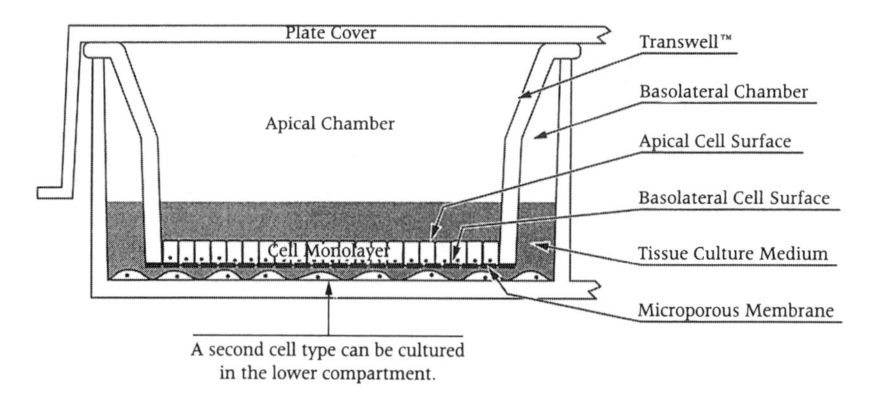

FIGURE 3.4 Polarized cells growing in a Transwell system. The microporous membrane promotes cell attachment and spreading, also allowing cells to be seen during growth.

This method is a double immunoaffinity precipitation. First the target protein is immunoprecipitated with specific antibody; then if it is biotinylated, it is precipitated with streptavidin-agarose.

Materials

Sulfosuccinimidyl-6-(biotinamido) hexanoate (NHS-LC-Biotin) (Pierce): 1.5 mg/ml in PBS^+

PBS^+: 0.1 mM $CaCl_2$, 1 mM $MgCl_2$ in PBS

Transwell inserts for 6 well plates: commercially available from many sources (Costar, Falcon, Nunc).

6 well plates

$Expre^{35}S^{35}S$ labeling mix (Perkin Elmer)

10% SDS (Appendix C)

1% SDS

Glycine: 50 mM in PBS^+

Lysis buffer: 1% Triton X-100 in PBS

Protein A-Sepharose suspension: Appendix C

Streptavidin-agarose beads: (Pierce)

Microcentrifuge and tubes

Scalpel

2× SDS sample buffer: (Appendix C)

1. Choose a cell line that is known to grow on Transwells in a polarized fashion. Culture cells in Transwells until a confluent monolayer is obtained.
2. Metabolically label the cells overnight with 100 μCi/ml of $Expre^{35}S^{35}S$ labeling mix.
3. Wash the monolayers three times with ice-cold PBS^+. Biotinylate for 30 min on ice specifically from the apical or basolateral chamber by adding the NHS-LC-Biotin solution to either chamber.
4. Wash the cells three times with the glycine solution.
5. Cells seeded on Transwells can grow on the sides of the chamber as well as on the filters. For best results, excise the filter from the chamber with a scalpel prior to lysis. Add 1 ml of lysis buffer to the filter and incubate for 30 min at 4°C with constant agitation.
6. Transfer the lysate to a 1.5 ml microfuge tube and centrifuge at maximum speed for 20 min at 4°C.
7. Preclear the supernatant by adding 100 μl of the protein A Sepharose suspension. Perform the immunoprecipitation with a specific antibody as described in Protocol 3.11.
8. Elute the immunoprecipitated proteins from the protein A beads by adding 100 μl of 1% SDS and boiling for 5 min. This step releases all proteins recognized by the specific antibody.
9. Transfer the eluted protein solution to a new microfuge tube and dilute 10-fold with 1% SDS. Capture the biotinylated proteins by incubating the released proteins with streptavidin-agarose beads (20 μl of packed beads) for 1 h at 4°C.
10. Collect the beads by a 1 min centrifugation. Discard the supernatant and add 50 μl of 2× SDS sample buffer and boil the beads for 3 min.

11. Analyze the products by SDS-PAGE followed by enhancing the radioactive signal, drying the gel, and exposing it to X-ray film (Electrophoresis and related techniques are described in Chapter 4). Alternatively, the signal can be detected using phosphorimaging (Chapter 7).

Comments: Labeling surface proteins with iodine forms a covalent bond between tyrosine residues and iodine. Labeling cells with NHS-LC-biotin creates a covalent bond with lysine and arginine residues and biotin. This technique can be modified to detect the cell surface target protein by labeling with NHS-LC-biotin, performing immunoprecipitation, running SDS-PAGE, transferring to a membrane and visualizing the biotinylated protein on the blot with avidin coupled to a detection system.

PROTOCOL 3.8 Labeling Isolated Proteins with Chloramine T

It is often necessary to radiolabel a purified protein for use as a radioactive tracer or for use in binding studies. The Chloramine T method, a chemical oxidation, covalently links radioiodine to tyrosine and to a lesser extent to the histidine residues of protein. The method is rapid and simple, radiolabeling the protein to high specific activity (McConahey and Dixon, 1980). **Perform this procedure in an appropriately designated fume hood. Take precautions to protect personnel from direct radiation.**

Materials

PBS: Appendix C
Sodium phosphate buffer: 0.5M, pH 7.5
$Na^{125}I$: Specific activity 17Ci/mg
Chloramine T: 2mg/ml in water
Sodium metabisulfite ($Na_2S_2O_4$): 2.0mg/ml in water
Tyrosine: 10mg/ml in water (saturated solution)
5ml disposable column
Sephadex G-50
Geiger counter

1. The protein to be labeled should be ~1mg/ml, preferably in 0.5M sodium phosphate buffer, although most other buffers are acceptable. **Avoid Tris buffers and reducing agents.** Add ~25μl of the protein solution to a 1.5ml microfuge.
2. Add 500μCi of $Na^{125}I$ to the protein solution.
3. Initiate the reaction by adding 25μl of freshly prepared chloramine T solution.
4. Incubate at room temperature for 1min.
5. Stop the reaction by adding 25μl of $Na_2S_2O_4$ solution then 25μl of the tyrosine solution.

6. To separate the labeled protein from the free isotope, pass the reaction mixture through a 5–10 ml disposable gel filtration column of Sephadex G-50 equilibrated with PBS. (Sephadex G-50 is adequate for most proteins). Monitor the elution with a Geiger counter. The iodinated protein elutes in the void volume (the first radioactive peak). The free isotope remains on the column which should be disposed of appropriately. After elution of the first labeled peak you may stop eluting.

Comments: Prepare the gel filtration column in advance. Chloramine T labeling should give specific activities between 5–45 μCi/μg. Higher specific activities can be obtained by using more $Na^{125}I$ and less protein. Buffers other than PBS can be used to elute the radiolabeled protein.

Some proteins and peptides may not contain tyrosine residues or the ones that they do have may not be available for iodination. Other proteins may lose their bioactivity when their tyrosine residues are iodinated. A method of iodination that labels lysine residues that can be used as an alternative to radiolabeling tyrosine residues has been used successfully in many projects (Bolton and Hunter, 1973).

B. Lysis: Preparation of the Cell Free Extract

Following the labeling step, the cells are lysed and an extract is prepared. The conditions used for lysis should be gentle enough to retain the antibody binding sites but strong enough to quantitatively solubilize the antigen of interest. Important variables to consider are: salt concentration, type and concentration of detergent, presence of divalent cations, and pH. If a sample is to accurately reflect the state of the cells at the time of lysis, it will be necessary to choose a procedure that will eliminate the actions of proteolytic enzymes. Once the cells are lysed, proteolytic enzymes that were present in a compartmentalized state are released and can come in contact with the proteins in the extract. It is important to inactivate these proteases.

The lysis buffer should contain salt concentrations between 0–1 M, nonionic detergent concentrations between 0.1–2%, divalent cation concentrations between 0–10 mM, EDTA concentrations between 0–5 mM, and pHs between 6–9. In addition, an antiprotease cocktail should always be included. Some frequently used buffers are listed below.

Lysis Buffers

- NP-40 lysis buffer: 50 mM Tris-HCl pH 7.4–8.0, 150 mM NaCl, 1.0% NP-40 (or Triton X-100)
- High salt lysis buffer: 50 mM Tris-HCl pH 7.4–8.0, 500 mM NaCl, 1.0% Triton X-100
- Low salt lysis buffer: 50 mM Tris-HCl pH 7.4–8.0, 10 mM NaCl, 1.0% Triton X-100
- RIPA buffer: 50 mM Tris-HCl pH 7.4–8.0, 150 mM NaCl, 1.0% Triton X-100, 0.5% Sodium deoxycholate (DOC), 0.1% SDS, 2 mM EDTA
- General: 20 mM Tris pH 8.0, 10 mM NaCl, 0.5% Triton X-100, 5 mM EDTA, 3 mM $MgCl_2$

TABLE 3.1 Properties of Selective Protease Inhibitors

Class of protease	Inhibitor	Stock concentration	Effective concentration	Comments[a]
Metallo	EDTA	50 mM	1–5 mM	
	1,10-phenanthroline	100 mM	1 mM	
Serine	Aprotinin	1–10 mg/ml	2–10 μg/ml	
	Phenylmethylsulfonyl fluoride (PMSF)	100 mM	1 mM	Dissolve in isopropanol, store at 4°C,
Cysteine	E-64	1 mM	10 μM	
	Iodoacetamide	100 mM	1 mM	
Serine/cysteine	Antipain	1 mM	1–10 μM	Dissolve in methanol
	Chymostatin	10 mM	10–100 μM	Dissolve in DMSO
	Leupeptin	10 mM	10–100 μM	
Acidic	Pepstatin A	1 mM	1 μM	Dissolve in methanol

[a]Unless otherwise indicated, dissolve the inhibitor in water and store at –20°C.

Add an antiprotease cocktail made up of the reagents listed in Table 3.1 to the lysis buffer immediately before adding it to the cells. A more comprehensive list of protease inhibitors is presented in Appendix Protocol G.2.

For phosphorylation studies, add the phosphatase inhibitors to attain the following final concentrations: NaF 10 mM and Na_3VO_4 1 mM.

PROTOCOL 3.9 Lysis of Cells in Suspension (Continuation of Protocol 3.2)

After the cells have been labeled and following the final wash, the cells are pelleted (see Protocol 3.2). Lysis buffer should be ice-cold and contain antiprotease reagents.

Materials

15 ml plastic conical centrifuge tube

Microcentrifuge and microcentrifuge tubes

1. Wash the cells with PBS or another suitable buffer. Collect the cells by centrifugation, 200× *g* for 10 min.
2. Decant the supernatant and gently disperse the cell pellet before adding the lysis buffer. This step should be performed on ice. The pellet of cells must be dispersed to permit complete lysis. Vortex mix the extract.
3. Transfer the lysate to a 1.5 ml microfuge tube and incubate the lysate on ice for at least 30 min. Remove the insoluble material by centrifugation at maximum speed in a microfuge for 10 min at 4°C.
4. Collect the supernatant. This is your cleared cell lysate.

Comments: Extraneous proteins like serum albumin and immunoglobulins that are present in culture media affect protein concentrations and can distort the migration of cellular proteins during electrophoresis. Therefore, when labeling cells growing in suspension, the cells should be washed prior to lysis with a protein free buffer to remove soluble proteins in the media.

If the washes are radioactive, they should be disposed of appropriately.

PROTOCOL 3.10 Lysis of Adherent Cells (Continuation of Protocol 3.1)

Materials

PBS: Appendix C

Cell scrapers or rubber policemen

Microcentrifuge and microcentrifuge tubes

Lysis buffer: See 3:B

1. Remove the media containing the radiolabel or chase media. If you are following secreted proteins, save the medium for analysis, otherwise discard it according to your institution's guidelines for handling radioactive waste.
2. Wash the cells by adding 10 ml of PBS and gently swirling. Repeat this a total of three times.
3. Keeping the plate on ice after the last wash, add prechilled lysis buffer (1 ml/10 cm dish) and scrape off the cells with either a plastic or a rubber policeman (cell scraper).
4. Transfer the total cell lysate to a microfuge tube and incubate it on ice for at least 30 min. Remove the insoluble material by centrifugation at maximum speed for 10 min at 4°C. Save the supernatant. This is your total lysate.

C. Principles of Immunoprecipitation

Proteins are molecules that are responsible for carrying out most biological functions. To gain insight into how cells work one must study what proteins are present, how they interact with each other and ultimately what they do.

Antibodies as Detection Tools

In the hands of the protein chemist, antibodies are exquisitely specific reagents that are used for identifying target proteins. Antibodies (Ab) consist of two types of polypeptide chains held together by disulfide bonds. The heavy chain has a molecular weight of roughly 55 kDa and the light chain, a molecular weight of approximately 25 kDa. Both heavy and light chains have a constant region and a variable region. Both variable regions are located at the N terminus and which consist of about 100 amino acids. It is these regions that make contact with the antigen. The C-terminal constant region of the molecule is made up of a limited number of sequences and serves to define the antibody subtype. The five human isotypes of antibodies, IgM, IgD, IgG, IgA, and IgE, are defined by the five different heavy-chain types, mu (μ), delta (δ), gamma (γ), alpha (α), and epsilon (ε) respectively. There are two types of light chain, kappa (κ) and lambda (λ) for all the immunoglobulin classes. IgG, the major immunoglobulin in serum, IgD and IgE are all composed of two heavy chains and two light chains. Structurally, IgM is a pentamer of IgG-like molecules while IgA is a dimer. (For an in depth description, consult Harlow and Lane, 1988.)

Polyclonal Antibodies

There are two types of antibodies used as reagents: monoclonal (MAb) and polyclonal. A polyclonal antibody solution is a heterogeneous mixture of antibodies that recognize numerous epitopes on a single antigen. The production of polyclonal antibodies, with the goal of producing a useful laboratory reagent, requires immunizing an animal with an extremely pure antigen. Specific antibodies are produced by many different clones of B-lymphocytes, which react with different determinants on the same antigen molecule. This is the normal heterogeneous immunological response to an antigen resulting in the production of polyclonal antibodies.

Monoclonal Antibodies

Köhler and Milstein (1975) developed methods that allowed for the growth of clonal cell populations that secrete antibodies of defined specificities. These cells are called hybridomas. The antibodies from a single isolated clone are identical (monoclonal) and will react with a specific epitope on the antigen against which it was raised. They can be propagated in vitro, will multiply indefinitely in culture, and will continuously secrete monoclonal antibodies with a defined specificity. A panel of monoclonal antibodies can be produced against a crude population of molecules and then undergo selection/screening on the basis of antigenic properties of the target molecule.

Despite many similarities, each epitope specific monoclonal antibody is a unique protein and presents its own purification challenges. MAbs differ in stability and affinity. These characteristics affect the behavior of individual MAbs in purification and analytical applications. The diversity of amino acids found in the variable region results in a wide range of isoelectric points for these molecules.

Antibody Based Analytical Techniques: Western Blotting and Immunoprecipitation

Antibodies are commonly used to detect antigens in complex mixtures. Some well known immunodetection methods include ELISA (enzyme linked immunosorbent assay), double immunodiffusion, immunoprecipitation, and immunoblotting. If the investigator is in possession of a specific antibody that recognizes the target protein then many analytical methods are possible. Western blotting and immunoprecipitation are two fundamental techniques that are used in association with polyacrylamide gel electrophoresis (PAGE) to identify and enrich proteins in a complex mixture so that they may be further analyzed. Some important characteristics of both techniques are shown in Table 3.2.

Western blots, dot blots, and colony/plaque lifts all require immobilization of the target protein on the membrane. With Western blots, proteins are electrotransferred to a membrane after SDS-PAGE. For dot blots, nondenatured proteins are spotted directly onto a membrane. In colony/plaque lifts, intact bacterial colonies or phage plaques are transferred to a membrane and then lysed to expose the target antigen. After the transfer step, all three methods follow the same basic methodology for detecting the target antigen.

In the technique referred to as immunoprecipitation-Western (IP-Western), immunoprecipitation is followed by Western blotting to increase the sensitivity of detection. Throughout this manual, examples are presented of analytical reactions that are performed prior to or following the separation of proteins by SDS-PAGE.

An antibody directed against one protein antigen in a complex mixture can be used to identify and isolate the specific antigen. This over-simplification understates the power of immunoaffinity recognition. Within the context of this manual, immunoprecipitation is an antibody based method of identifying and purifying a target molecule as first described by Kessler (1975). (The term immunoprecipitation as used in the context of this manual should not be confused with the classical definition of immunoprecipitation, in which an insoluble precipitate is formed by mixing an appropriate ratio of divalent antibody with multivalent antigen.) It is possible to isolate by immunoprecipitation or immunoabsorbent procedures almost any cellular protein to which specific antibodies can be raised, provided that the molecule retains its antigenicity in the extraction system employed. In combination with newer and more sensitive analytical methods, the structure and functional associations with other molecules can be defined for a given protein or class of proteins.

Immunoprecipitation is based on the ability of antibodies that specifically recognize the target antigen to form antigen-antibody complexes that can then be easily collected by capturing the complex onto a solid-phase matrix (Kessler, 1981). Protein A, which has a high affinity for the F_c portion of most Ig molecules, coupled to Sepharose or agarose

TABLE 3.2 Characteristics of Western Blotting and Immunoprecipitation

Western Blotting	Immunoprecipitation
Target protein (usually) denatured	Target protein in its native state
Epitope recognition-potential cross-reactivity	Coprecipitation of associated proteins
Colorimetric and chemiluminescent detection	Radioactive detection (might have to confirm that the antibody recognized a specific protein by Western blotting)

beads, is usually used for this purpose. Specific binding of immune complexes to protein A is extremely rapid, occurring within seconds, and has the added feature of being extremely stable in a variety of solvent systems, whereas nonspecific binding of other proteins is low. (Consult Table 3.3). Due to the macroscopic size of the bead, the complexes are easily collected by centrifugation. Unbound proteins are removed by choosing a wash buffer that will keep the immune-complex intact while removing all macromolecules that are not specifically recognized by the antibody. When used in conjunction with radiolabeling and SDS-PAGE, immunoprecipitation can reveal important characteristics of the immuno-affinity purified target protein. Assays can be performed to determine the presence and quantity of the target protein, its apparent molecular weight, its rate of synthesis and degradation, the presence of posttranslational modifications, and if it is interacting with other proteins, nucleic acids or other ligands. These characteristics are difficult to determine using other techniques. Suitable controls are indispensable for interpreting the results. Preimmune serum or a nonrelevant antibody from the same species is frequently used to demonstrate that the Ab is specifically forming a complex with the target protein.

The protein A molecule, a 42 kDa cell wall component produced by several strains of *Staphylococcus aureus*, contains four high affinity binding sites capable of interacting with the F_c region of IgG from many species. Typical binding capacity of protein A immobilized to Sepharose is 10–15 mg of human IgG per ml of beads.

An alternative to protein A is protein G, a cell wall protein isolated from group G streptococci. Protein G is reported to bind with greater affinity to most mammalian immunoglobulins than does protein A, although there are several species to which protein A has a greater affinity. Because the antibodies bind to proteins A and G through their F_c region, the antigen binding sites of the anibodies remain available for antigen binding.

Many analytical reactions are conveniently performed on the target protein as it exists in the immune complex, prior to SDS-PAGE.

PROTOCOL 3.11 Immunoprecipitation

This protocol is often used following metabolic labeling and the preparation of a total cell lysate.

Materials

Protein A-Sepharose suspension: Appendix C
Microcentrifuge and microcentrifuge tubes
Wash buffer
Tube rocker: Rockers, shakers and gyrotators are fine
2× SDS-PAGE sample buffer: Appendix C
Vortex mixer
Normal Rabbit Ig

1. Preclear cell lysates to remove sticky proteins that could be carried through the procedure by adding 1–5 µl of a nonspecific antibody of the type to be used to immunoprecipitate the antigen of interest (normal rabbit Ig if your antibody is made in rabbit). Add 100 µl of the protein A Sepharose suspension to the lysate and incubate at 4°C on a rocker for at least 30 min. For a minimum preclear, use 100 µl of protein A-Sepharose suspension. This step can also extend overnight. A 10% suspension of fixed protein A-bearing *S. aureus* Cowan I is a less expensive alternative than protein A-Sepharose for the preclearing step.
2. Centrifuge the suspension in a microfuge for 2 min at maximum speed and transfer the supernatant to a new tube. Discard pelleted material as radioactive waste if using radiolabeled lysates.
3. Add an excess of the specific antibody, (usually 1–5 µl) which can be monoclonal or polyclonal, in the form of serum, ascites, culture media, or purified antibody. Add 100 µl of the protein A-Sepharose suspension and incubate at 4°C with constant mixing for at least 1 h. If necessary, add a second bridging antibody. If the specific antibody is a mouse monoclonal or rat monoclonal which has a very low affinity for protein A, a second antibody, which will serve as a "bridge" should be added. The bridging antibody recognizes the first antibody and has a high affinity for protein A. For example, if the specific antibody is a rat monoclonal IgG_{2a} or IgG_{2b}, add 2 µl of a rabbit anti-rat antibody which will act as a bridging reagent between the specific antibody-antigen complex and the protein A Sepharose. See Table 3.3.
4. Centrifuge at maximum speed for 2 min in a microfuge and carefully aspirate off the supernatant and if radioactive, dispose of it according to your institution's guidelines. Alternatively, this material can be used to react with another antibody, (see Protocol 3.12).
5. Resuspend the beads in 1 ml of wash buffer and repeat this step a total of five times. Choosing a wash buffer is similar to choosing a lysis buffer. The complex can be washed with the lysis buffer. Alternatively, the following stringent wash buffer (Rosenberg et al, 1991) can be used: 50 mM Tris-HCl pH 8.0, 0.5 M NaCl, 5 mM EDTA, 0.02% NaN_3, 0.5% Triton X-100, 0.5% DOC, 0.1% SDS.
6. After the final wash, add 50 µl of 2× SDS-PAGE sample buffer directly to the beads, vortex mix, and boil for 5 min. The immunoprecipitated material can be stored at –20°C or loaded directly onto an SDS gel. Prior to loading, briefly centrifuge the tube for 20 sec and load all of the supernatant above the pelleted beads.
7. Following electrophoresis, if phosphorimaging is not an option, if the isotope is ^{35}S, ^{14}C, or ^{3}H, treat the gel with a fluorographic enhancing agent (Protocol 7.20). Dry the gel, expose it to X-ray film, and analyze the data.

Comments: PANSORBIN® (Calbiochem) cells are *S. aureus* that have a coat of protein A.

In parallel, include a control in which the lysate is mock immunoprecipitated with an irrelevant antibody or preimmune antisera. When electrophoresed in parallel with the immunoprecipitated material, nonspecific bands are readily identified, greatly simplifying the analysis.

Washing is important for obtaining a clean result. If you have high background, try using a higher stringency wash buffer. If you do not detect the target protein, try decreasing the stringency of the wash

TABLE 3.3 Binding of Protein G and Protein A to Immunoglobulins from Various Species

	Protein A*	Protein G*
Human IgG_1	++	++
Human IgG_2	++	++
Human IgG_3	00	++
Human IgG_4	++	++
Human IgM	+	00
Human IgD	00	00
Human IgA	+	00
Mouse IgG_1	+	++
Mouse IgG_{2A}	++	++
Mouse IgG_{2B}	++	++
Mouse IgG_3	++	++
Rat IgG_1	+	+
Rat IgG_{2A}	00	++
Rat IgG_{2B}	00	+
Rat IgG_{2C}	++	++
Guinea Pig IgG	++	+
Rabbit IgG	++	++
Pig IgG	++	+
Dog IgG	++	+
Cat IgG	++	+
Chicken IgG	00	00
Bovine IgG_1	00	++
Bovine IgG_2	++	++
Sheep IgG_1	00	++
Sheep IgG_2	++	++
Goat IgG_1	+	++
Goat IgG_2	++	++
Horse IgG_{AB}	+	++
Horse IgG_C	+	++
Horse IgG_R	00	+

*No Binding 00 Weak + Strong ++.

buffer. You may also find that 5 washes are too many and that an acceptable background signal can be achieved with only 3 or 4 washes.

Protein Interaction Analysis

Identifying interacting partners of proteins, proteins that interact with one another, is a way of inferring protein function. As a rule, proteins that interact with one another or are part of the same complex are generally involved in the same cellular processes. Many approaches have been used and are described below.

PROTOCOL 3.12 Sequential Immunoprecipitation: Dissociation and Reimmunoprecipitation of Immune Complexes

Coimmunoprecipitation can provide important insights into the interactions that exist between the target protein and other molecules. Cell surface receptors are often intimately associated with additional proteins which comprise a biologically active complex. It is possible to identify proteins that specifically interact with the target Ag and bind to it and are recognized as part of the immune complex although not through a direct interaction with the antibody. These proteins coprecipitate with the immune complex by way of a noncovalent interaction with the target protein. An example is presented in Figure 3.5, in which the coprecipitating protein is bound to the complex through a lectin-oligosaccharide interaction. If you suspect that associated proteins are coprecipitating with the target protein, the complex can be disrupted and reimmunoprecipitated with a second antibody, if available, directed against the suspected associated protein (Argetsinger et al, 1993).

Materials

Solution A: 50 mM Tris-HCl, pH 7.5, 137 mM NaCl

Solution B: Solution A plus 0.75% SDS, 2% 2-ME, 100 mM DTT, 10 μg/ml aprotinin, and 10 μg/ml leupeptin

Solution C: 50 mM Tris-HCl, pH 7.5, 0.1% Triton X-100, 137 mM NaCl

4× and 2× SDS-PAGE sample buffer (Appendix C)

Protein A Sepharose suspension: Appendix C

Dissociation

1. Perform the final immunoprecipitation wash with 1 ml of Solution A.
2. Resuspend the beads in 100 μl of solution B and boil for 5 min.
3. Dilute the eluted proteins tenfold with solution C.
4. Remove 75 μl and mix with 25 μl of 4× SDS-PAGE sample buffer and boil for 5 min. Analyze by SDS-PAGE followed by phosphorimaging or fluo-

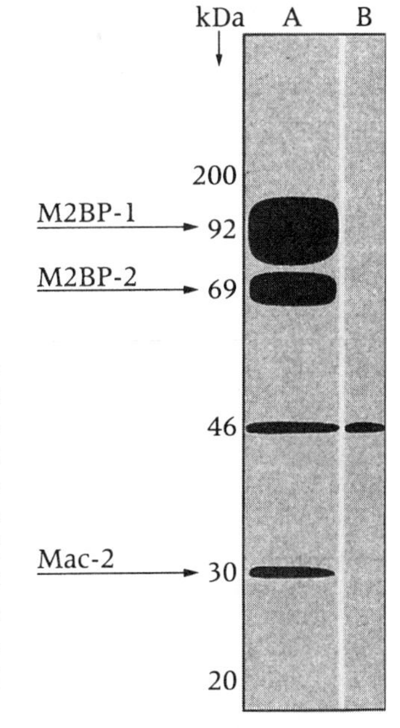

FIGURE 3.5 Coprecipitating proteins. Immunoprecipitates of solubilized extracts from [^{35}S]methionine labeled HT-29 cells were analyzed by SDS-PAGE (10%). (A) immunoprecipitation with M3/38 (anti-Mac-2), (B) immunoprecipitation with isotype-matched control. Migration positions of molecular mass markers (in kDa) are shown at the left. The proteins that coprecipitated with Mac-2 (30 kDa) are shown with arrows. (Adapted with permission from Rosenberg et al, 1991).

rography. This procedure should reveal the target protein and any other proteins coprecipitating with the immune complex.

Reimmunoprecipitation

5. Incubate the remaining sample with a second antibody on ice for 60–90 min and then add 100 µl of the protein A suspension and incubate for an additional 1 h at 4°C.
6. Wash the immune complexes three times with solution C. Add 50 µl of 2× SDS-PAGE sample buffer and boil for 5 min.
7. Analyze the precipitated products by SDS-PAGE followed by fluorographically enhancing the signal in the gel (Chapter 7), drying and exposing to X-ray film. Alternatively, the gel can be exposed to a phosphorimaging screen, significantly reducing the development time.

PROTOCOL 3.13 Eliminating Interfering Immunoglobulin Bands During IP-Western Detection: Analysis of the Immunoprecipitate under Non-Reducing Conditions

Immunoprecipitated proteins are routinely solubilized by addition of a sample buffer containing SDS and a strong reducing agent like β-mercaptoethanol or dithiothreitol followed by boiling (Protocol 3.11 step 6). This procedure causes irreversible denaturation of proteins, including target antigen and the antibody molecules involved in the IP. The immunoglobulin molecules separate into the 50–55 kDa heavy chains and the 25–28 kDa light chains. Upon Western blot analysis, the immunoglobulin chains, especially heavy chain, can react strongly with the secondary antibody which can produce a strong signal that will obscure target bands that migrate in the 50–55 kDa range. The heavy chain signal is often seen as an artifact of the procedure. If the target protein runs as a band of 50–55 kDa it can potentially be obscured by the heavy chain (Wiese and Galande, 2001). If reducing agent is omitted from the sample buffer and the immunoprecipitated product is warmed to 40°C instead of boiling, the Ig molecule migrates as a native tetramer of two heavy chains and two light chains with an apparent molecular weight of 160 kDa.

Materials

2× non-reducing sample buffer: Appendix C

1. Follow Protocol 3.11 (Immunoprecipitation) through step 5.
2. Add 40 µl of 2× SDS-PAGE non-reducing sample buffer.
3. Elute the immune complexes from the Protein A beads by incubation at 40°C for 10 min.
4. Run the gel under non-reducing conditions, transfer, and probe according to the protocol.

Comments: This method can be used to analyze proteins in the 50–55 kDa range provided they are not disulfide linked.

PROTOCOL 3.14 Nondenaturing Immunoprecipitation

Another approach to test for the presence of associated proteins is to perform the cell lysis and immunoprecipitation protocol using a nondenaturing buffer system. The mild, nondenaturing conditions keep the protein complexes intact (Ou et al, 1993). Perform the preclearing, immunoprecipitation and washes in the lysis buffer. All manipulations should be performed at 4°C.

Materials

Iodoacetamide solution: 0.5 M iodoacetamide in PBS

Nondenaturing lysis buffer: 50 mM HEPES, pH 7.5, 200 mM NaCl containing 2% sodium cholate, 1 mM PMSF, 5 μg/ml each of aprotinin and leupeptin

Lysis buffer: 50 mM HEPES, pH 7.5, 200 mM NaCl, the antiprotease cocktail and 0.1% SDS and 1% Triton X-100.

Protein A-Sepharose suspension: Appendix C

1. After labeling, incubate the cells in iodoacetamide solution for 10 min on ice.
2. Lyse the cells in nondenaturing lysis buffer.
3. Perform immunoprecipitation as described in Protocol 3.11 using nondenaturing lysis buffer for antibody incubations and washes.

Comments: Nondenaturing immunoprecipitation is usually followed by a denaturing immunoprecipitation. The beads are incubated with nondenaturing lysis buffer containing in addition 1% SDS and heated at 90°C for 3 min followed by the addition of 2 ml of nondenaturing lysis buffer containing 1% Triton X-100. This immunoprecipitated material is now used in the second, denaturing immunoprecipitation. Do not preclear. Add the desired specific antibody and proceed with the immunoprecipitation as described in Protocol 3.11. This technique is also referred to as sequential immunoprecipitation.

D. Additional Methods to Identify Associated Proteins

It is often important to determine whether the target protein is part of a larger functional complex or network of associated proteins. Sucrose density-gradient centrifugation (rate sedimentation analysis) and chemical cross-linking are two powerful techniques used either alone or together to assess noncovalent protein–protein interactions. Although certain noncovalent protein complexes are known to dissociate during ultracentrifugation in sucrose gradients (Degen and Williams, 1991) such assemblies are often detectable by cross-linking.

Conversely, chemical cross-linking is dependent on the availability of suitable reactive groups on each subunit of a protein complex, a requirement not shared by rate sedimentation analysis.

Sucrose Gradients

Protein–protein interactions can be analyzed following experimental treatment of cells. Total cell lysates or homogenates can be fractionated and analyzed using sucrose gradients. In this way, the sedimentation properties of the target proteins can be assessed. For example, cells treated with tunicamycin, an inhibitor of glycosylation, can be compared with untreated cells to assess the role that glycosylation is playing in protein interactions. In most cases cells are radiolabeled then lysed. The lysate is cleared by a low speed centrifugation to remove undissolved material. The supernatant is then loaded on a sucrose gradient which is typically a linear gradient in the range of 5–30% (w/v). The protein containing supernatant is layered on top and the centrifugation run is for 14–18h at 4°C at 180,000–225,000× *g*. In parallel, protein standards are run on a separate gradient or included on the same gradient as the sample. Fractions are collected and analyzed as described below.

Practical information pertaining to centrifugation is presented in Appendix F. It is not expected that every lab be supplied with specific rotors and equipment. For these reasons this supplementary information should be helpful when adapting a published protocol to the available equipment.

PROTOCOL 3.15 Preparation of Sucrose Gradients

Sucrose gradients are used for analytical studies and also for preparative projects. Linear gradients are prepared with the aid of a gradient maker attached to a pump. Measure the volume capacity of the tube that you choose. If gradient makers are not available you can prepare step gradients which can be extremely useful for preparative studies. Step gradients are formed from a stock solution of 50% sucrose in the desired buffer. Concentrations of 40, 35, 30, 25, and 20% (w/v) are diluted from the stock. These solutions are then layered manually in the centrifuge tube starting with the heaviest solution at the bottom (see Figure 3.6A). It is often helpful to start with a cushion of 2M sucrose at the bottom of the tube so that the heaviest material can be more easily collected. The continuous linear sucrose gradient technique is illustrated below with the Beckman SW50.1 rotor and a partially purified protein, zyxin, an adherens junction protein (Crawford and Beckerle, 1991).

Materials

Ultrapure sucrose: 2M in 20mM Tris HCl, pH 7.5, 100mM NaCl, 5mM EDTA

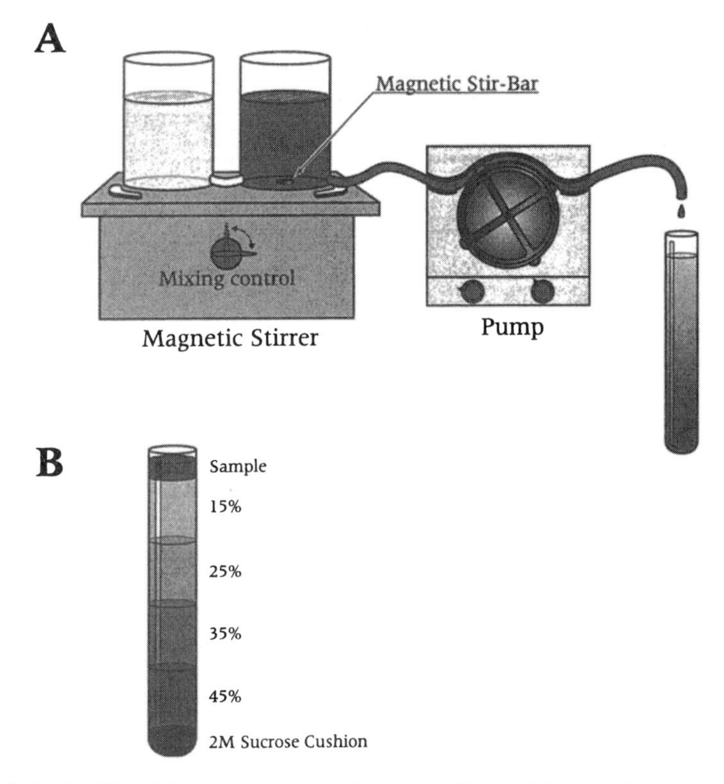

FIGURE 3.6 Panel A is a continuous linear gradient while panel B is a typical step gradient.

Ultracentrifuge, rotor (Beckman SW50.1), and tubes
Refractometer
Gradient solution A: 20 mM Tris HCl, pH 7.5, 100 mM NaCl, 5 mM EDTA, 20% sucrose
Gradient solution B: 20 mM Tris HCl, pH 7.5, 100 mM NaCl, 5 mM EDTA, 5% sucrose

1. Prepare 5 ml gradients of 5–20% sucrose (w/v) using gradient solutions A and B.
2. Layer 100 µl of sample onto the top of the gradient.
3. Centrifuge at 225,000× *g* for 14 h in a Beckman SW50.1 rotor at 4°C.
4. Collect 250 µl fractions for analysis. One drop from each fraction (~20 µl) can be used for measuring the refractive index. Consult Table 3.4 for converting refractive index to molarity and percent sucrose. Fractions can be analyzed by SDS-PAGE and Western blotting (chapters 4 and 7) or for the presence of specific activities and marker enzyme activities.

Comments: Parameters that should be used in choosing the number of fractions to collect are: obtaining an adequate separation; obtaining sufficient material to perform all the planned assays; and not being overwhelmed by too many fractions. You can always go back and collect smaller fractions on your next run.

TABLE 3.4 Density, Refractive Index, and Concentration Data—Sucrose at 20°C, Molecular Weight = 342.3 (Reprinted with permission from "Techniques of Preparative, Zoned and Continuous Flow Ultracentrifugation," Beckman Instruments, Spinco Division).

Density (g/cm³)	Refractive Index, η_D	% by Weight	mg/ml of Solution*	Molarity
0.9982	1.3330	0		
1.0021	1.3344	1	10.0	0.029
1.0060	1.3359	2	20.1	0.059
1.0099	1.3374	3	30.3	0.089
1.0139	1.3388	4	40.6	0.119
1.0179	1.3403	5	50.9	0.149
1.0219	1.3418	6	61.3	0.179
1.0259	1.3433	7	71.8	0.210
1.0299	1.3448	8	82.4	0.211
1.0340	1.3464	9	93.1	0.272
1.0381	1.3479	10	103.8	0.303
1.0423	1.3494	11	114.7	0.335
1.0465	1.3510	12	125.6	0.367
1.0507	1.3526	13	136.6	0.399
1.0549	1.3541	14	147.7	0.431
1.0592	1.3557	15	158.9	0.464
1.0635	1.3573	16	170.2	0.497
1.0678	1.3590	17	181.5	0.530
1.0721	1.3606	18	193.0	0.564
1.0765	1.3622	19	204.5	0.597
1.0810	1.3639	20	216.2	0.632
1.0854	1.3655	21	227.9	0.666
1.0899	1.3672	22	239.8	0.701
1.0944	1.3689	23	251.7	0.735
1.0990	1.3706	24	263.8	0.771
1.1036	1.3723	25	275.9	0.806
1.1082	1.3740	26	288.1	0.842
1.1128	1.3758	27	300.5	0.878
1.1175	1.3775	28	312.9	0.914
1.1222	1.3793	29	325.4	0.951
1.1270	1.3811	30	338.1	0.988
1.1318	1.3829	31	350.9	1.025
1.1366	1.3847	32	363.7	1.063
1.1415	1.3865	33	376.7	1.100
1.1463	1.3883	34	389.7	1.138
1.1513	1.3902	35	403.0	1.177
1.1562	1.3920	36	416.2	1.216
1.1612	1.3939	37	429.6	1.255
1.1663	1.3958	38	443.2	1.295
1.1713	1.3978	39	456.8	1.334
1.1764	1.3997	40	470.6	1.375
1.1816	1.4016	41	484.5	1.415
1.1868	1.4036	42	498.5	1.456
1.1920	1.4056	43	512.6	1.498
1.1972	1.4076	44	526.8	1.539
1.2025	1.4096	45	541.1	1.581
1.2079	1.4117	46	555.6	1.623
1.2132	1.4137	47	570.2	1.666
1.2186	1.4158	48	584.9	1.709

TABLE 3.4 *Continued*

Density (g/cm^3)	Refractive Index, η_D	% by Weight	mg/ml of Solution*	Molarity
1.2241	1.4179	49	599.8	1.752
1.2296	1.4200	50	614.8	1.796
1.2351	1.4221	51	629.9	1.840
1.2406	1.4242	52	645.1	1.885
1.2462	1.4264	53	660.5	1.930
1.2519	1.4285	54	676.0	1.975
1.2575	1.5307	55	691.6	2.020
1.2632	1.4329	56	707.4	2.067
1.2690	1.4351	57	723.3	2.113
1.2748	1.4373	58	739.4	2.160
1.2806	1.4396	59	755.6	2.207
1.2865	1.4418	60	771.9	2.255
1.2924	1.4441	61	788.3	2.303
1.2983	1.4464	62	804.9	2.351
1.3043	1.4486	63	821.7	2.401
1.3103	1.4509	64	838.6	2.450
1.3163	1.4532	65	855.6	2.500
1.3224	1.4558	66	872.8	2.550
1.3286	1.4581	67	890.2	2.864

Fractionating a Sucrose Gradient

The easiest way to fractionate a sucrose gradient is to manually remove equal-volume fractions from the top with a micropipette. Although this method is neither high-tech nor elegant, it almost always works without sticky spills or run-away pumps attached to out-of-control fraction collectors. Alternatively, fractions can be collected from the bottom by carefully making a small hole in the bottom of the centrifuge tube and collecting fractions of equal drop number. Some labs may be equipped with either a commercially available (Buckler Instruments Auto Densi-Flow IIC) or an in-house constructed gradient fractionator. Before using an automatic fractionator on a precious sample or radioactive material, make sure you know how to use it.

The percent sucrose concentrations are determined using a refractometer, and may be converted to density or vice versa (illustrated in Table 3.4). Inclusion of other components in the buffer will cause the values of percent sucrose to deviate from the value of a pure sucrose solution. In most cases this fluctuation can be ignored as the system is more relative than absolute.

The mobility of the target protein is not affected by the presence of protein standards in the gradients.

Sedimentation rate is measured in Svedbergs, denoted as S. The greater the mass, the faster the rate of sedimentation, and the higher the S value. The sedimentation rate is also influenced by the shape of the molecule, since more compact bodies descend faster. Svedberg units are not additive. Some commonly used proteins and their S values are listed in Table 3.5. A much more extensive list appears in the *CRC Handbook of Biochemistry* (Sober, 1963).

TABLE 3.5 Commonly Used Proteins and Their S* Values

Protein	S value
cytochrome c	1.7–2.1
carbonic anhydrase	2.8
ovalbumin	3.55–3.66
bovine serum albumin	4.3
malate dehydrogenase	4.35
horse radish peroxidase	~5
myosin	6.4
murine IgG2b	6.6
lactate dehydrogenase	7.35
catalase	11.4
thyroglobulin	16.5

*Svedbergs.

In addition to examining associations between different proteins in a complex, sucrose gradients can yield valuable biophysical information about the target protein. The relative sedimentation coefficient of the target protein can be estimated graphically by constructing a standard curve and plotting the distance that the target protein has migrated from the top of the gradient. This value is then fitted to the standard curve and compared to the mobilities of known protein standards run under identical conditions. This is illustrated in Figure 3.7.

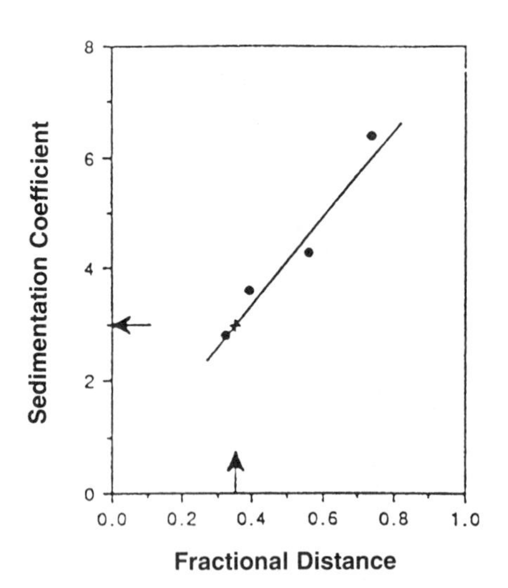

FIGURE 3.7 Estimation of the relative sedimentation coefficient of zyxin. Sample containing zyxin plus standard proteins was centrifuged through a 5–20% sucrose gradient. Sedimentation coefficient values for the standard proteins are as follows: myosin (6.4S), bovine serum albumin (4.3S), ovalbumin (3.55–3.66), carbonic anhydrase (2.8S). The arrow along the horizontal axis marks the fractional distance (0.35) the protein migrated through the gradient. The relative sedimentation coefficient of zyxin is 3.0S ± 0.2S. (Reprinted with permission from Crawford AW and Beckerle MC, 1991).

Glycerol gradients, 5–40% (v/v) can be run in place of the sucrose gradients. Analytical procedures are identical to those used with sucrose gradients.

Chemical Cross-Linking

Biological structures are often composed of complexes or assemblies of proteins which may be relatively stable, such as multi-enzyme complexes, or dynamic, and transitory, like receptor-ligand complexes. Noncovalent protein complexes can often be detected by chemical cross-linking which is dependent on the availability of suitable reactive groups on each unit of the multi-protein-complex. Cross-linking represents the introduction of covalent bridges between neighboring subunits in the native complex. It provides a simple, rapid experimental approach for obtaining information on quaternary structure. Cross-linking generally involves the use of protein-modifying reagents to form a collection of intermolecular dimers. The methodology described below uses intermolecular cross-linking to identify proteins that are in close proximity to each other. Cross-links consisting of more than two components are difficult to analyze. For this reason cross-linking is often followed by immunoprecipitation to capture the target protein and close-proximity molecules that are cross-linked to it.

Cross-linkers are either homobifunctional or heterobifunctional. Homobifunctional cross-linkers have at least two identical reactive groups while the heterobifunctional cross-linkers have two or more different reactive groups. Cross-linkers can be cleavable or noncleavable. They can couple through amines, sulfhydryls, or react nonspecifically. Most of the cross-linkers discussed below are of the homobifunctional primary amine reactive type. The Pierce catalog has a more comprehensive list of cross-linkers. The chemistry of cross-linking is described by Wong (1991).

If the cross-linking reagent is noncleavable and the product is analyzed by immunoprecipitation followed by SDS-PAGE, the cross-linked complexes will appear as new, slower migrating bands compared to the target protein. Analysis of cross-linking experiments is facilitated if the cross-linking bridge has a cleavable bond that permits regeneration of the original monomeric protein subunits from the separated cross-linked complexes. This approach is described in Protocols 3.18 and 3.19.

The existence of cleavable bonds in the cross-bridge makes possible the use of a two-dimensional gel electrophoresis system in which dimers are separated in the first dimension under nonreducing conditions. The bridges are then cleaved and the monomeric subunits are separated in the second dimension. The technique is one form of diagonal SDS-PAGE (discussed in Chapter 4). Subunits that never form cross-links remain on the diagonal, while those that do form cross-links fall below the diagonal in the second dimension. The molecular weights of the individual subunits are compared with molecular weight standards. As will be seen below, disulfide bonds are readily cleavable by mild reduction, and cross-linkers containing cleavable S–S bonds are frequently used.

General Considerations for Cross-Linking

Select cross-linkers on the basis of their chemical reactivities and the compatibility of the reaction with the application. N-hydroxysuccinimide (NHS) homobifunctional cross-linkers, as well as the water soluble analogs *N*-hydroxysulfosuccinimide (sulfo-NHS) esters, react with primary amino groups such as the ε-amine group of lysine or N-terminus amines. At neutral to alkaline pH, the amino group on a particular ligand will react with the NHS ester to form a stable amide bond releasing *N*-hydroxysuccinimide as a byproduct as shown in Figure 3.8. The NHS-ester homobifunctional cross-linkers have steadily replaced the imidoester cross-linkers. Homobifunctional reagents sometimes result in unacceptable levels of polymerization.

NHS-ester cross-linking reactions are commonly performed in phosphate, bicarbonate/carbonate, HEPES, and borate buffers at concentrations between 50–200 mM. The reaction buffer should be free of extraneous amines such as Tris and glycine. The pH should be alkaline. NHS-esters are usually used in a twofold to 50-fold molar excess to protein at concentrations between 0.1–10 mM.

When in doubt about the solvent to use with a particular cross-linker, keep in mind that NHS esters are not very soluble in aqueous solutions. Therefore, dissolve them in a minimal amount of organic solvent such as dimethyl sulfoxide (DMSO) or dimethylformamide (DMF) prior to introducing the cross-linker into the reaction mixture. Sulfo-NHS cross-linkers are water soluble and can be added directly to the reaction buffer. They should be prepared immediately before use to minimize loss of reactivity due to hydrolysis.

Cross-linking reactions are more efficient when performed on concentrated protein solutions. The reaction rate is faster at alkaline pH. There are many more cross-linking options than the ones discussed below.

Typically, cross-linking studies are performed in three ways: (1) adding the cross-linker to live, whole cells; (2) adding the cross-linker after adding extraneous ligand or antibody; (3) adding cross-linker to cell-free homogenates or to defined protein solutions. A cross-linking study should always be performed with appropriate controls. For example, cells or protein solutions should be processed in the presence and absence of the cross-linking agent with all other conditions equal.

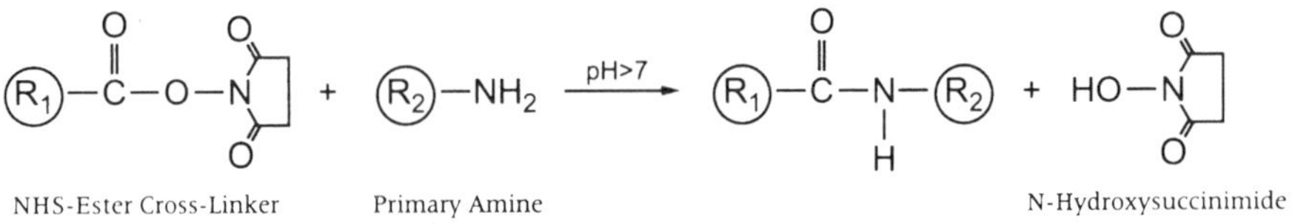

$$R_1-C(=O)-O-N(\text{succinimide}) + R_2-NH_2 \xrightarrow{pH>7} R_1-C(=O)-N(H)-R_2 + HO-N(\text{succinimide})$$

NHS-Ester Cross-Linker | Primary Amine | N-Hydroxysuccinimide

FIGURE 3.8 NHS-ester cross-linking reaction scheme.

PROTOCOL 3.16 Cross-Linking Proteins Added to Cells: Analysis of Receptor-Ligand Interaction

Many cross-linking agents have been used successfully in a wide range of experimental systems. Two will be mentioned: 1-ethyl-3-(3-dimethylaminopropyl)carbodiimide hydrochloride (EDC), and Bis(sulfosuccinimidyl)suberate (BS^3) (Pierce) used at 2mM for 20min at 4°C with conditioned media (Flaumenhaft et al, 1993).

Materials

EDC solution: 15mM EDC freshly prepared in 100mM 4-morpholinoethanesulfonic acid (MES) pH 7.0

PBS^+: See appendix C

Glycine solution: 150mM glycine HCl, pH 7.5

RIPA lysis buffer: 20mM HEPES, pH 7.4, 1.0% Triton X-100, 1% DOC, 0.1% SDS, 0.5M NaCl, 5mM EDTA, 10mM EGTA, 1mM PMSF, 10μg/ml leupeptin

Cell scrapers

1. Incubate adherent cells with the protein of interest for ~90min at 4°C.
2. Transfer the cells to 22°C. Add the EDC solution to a final concentration of 2mM and incubate for 1h. As an alternative, the cross-linker BS^3 can be used at a final concentration of 0.5mM.
3. Terminate the cross-linking reaction by washing the cells three times with cold PBS^+.
4. Incubate the cells for 5min with 10ml of glycine solution.
5. Remove the glycine solution and lyse the cells by adding 1ml of RIPA buffer per 10cm plate. Harvest the lysate with a cell scraper and incubate the lysate for 20min at 0°C.
6. Process the lysate for analysis by immunoprecipitation (Protocol 3.11).

Comments: This protocol can be used for cross-linking exogenously added ligand to cells which are known to produce the specific receptor. EDC, a carbodiimide, couples carboxyls to primary amines. Carbodiimides are unlike the NHS-esters in that no cross-bridge is formed between the molecules being coupled. In the presence of excess cross-linker, polymerization is likely to occur because proteins contain carboxyls and amines.

Control reactions are carried out in parallel by adding cross-linking buffer alone to an aliquot of cells. These cells are processed identically to the cells that were treated with the cross-linker. EDC is the same as EDAC.

PROTOCOL 3.17 Cross-Linking Proteins in Solution

This protocol, using 1-Ethyl-3-(3-dimethylaminopropyl)-carbodiimide (EDC) and *N*-hydroxyl-sulfosuccinimide (NHS), was designed for crosslinking proteins in a cell-free system. The method is illustrated with the specific example of purified actin monomers with actophorin (Maciver et al, 1991).

Materials

EDC solution: Prepare immediately before use, 100 mM EDC (Pierce) in 2 mM 4-morpholinoethanesulfonic acid (MES) buffer, pH 7.5

NHS solution: Immediately before use, 100 mM NHS (Pierce) in potassium phosphate buffer, pH 7.5 (adjust to the desired pH with 1 M KOH)

Glycine solution: 200 mM glycine stock, pH 7.5

Target protein solution

1. With all solutions at 4°C, add the EDC and NHS solutions to the protein solution to yield a final concentration of 1 mM for each cross-linker.
2. After 10 min add more EDC and NHS to give a new concentration of 2 mM and incubate for an additional 10 min at 4°C.
3. Terminate the reaction by adding glycine solution to a final concentration of 9 mM. Analyze by SDS-PAGE, transfer to nitrocellulose membrane, probe with a specific antibody, and detect with a suitable system (see Chapters 4 and 7).

Comment: NHS stabilizes reactive intermediates formed by the reaction of carboxyls with EDC.

PROTOCOL 3.18 Cross-Linking Extraneously Added Ligand to Cells

This technique is illustrated by adding ^{125}I-Interferon-γ to cells demonstrating that IFN-γ effects dimerization of its receptor under physiological conditions (Greenlund et al, 1993).

Materials

PBS /BSA: 0.1% BSA in PBS

Radiolabeled ligand: 1 μM solution (2×10^6 cpm/ml) in PBS

Ligand solution

EDC: 100 mM in 2 mM potassium phosphate buffer, pH 7.5

NHS: 100 mM in 28 mM potassium phosphate buffer, pH 7.5

Glycine solution: 1 M in PBS

2× SDS-PAGE sample buffer: (Appendix C)

Lysis buffer: 1% Triton X-100 with antiprotease reagents in PBS

Microcentrifuge and microcentrifuge tubes

1. Harvest the cells (see Protocols 3.1 and 3.2) and resuspend them in PBS/BSA at a concentration of 10^6 cells/0.2 ml in a microfuge tube. Add 10 μl of the radioactive ligand solution. Divide the cells into two equal fractions and incubate at room temperature for 1 h in the presence or absence of a 100-fold to 1000-fold excess of unlabeled ligand.

2. Add the cross-linking agents (EDC to 10 mM final concentration, and NHS 5 mM final concentration) and incubate the cells for an additional 30 min at room temperature.
3. Terminate the reaction by adding 20 µl of 1 M glycine solution (0.1 M final concentration).
4. Wash the cells three times with cold PBS.
5. Lyse the final cell pellet by adding 20 µl of lysis buffer per 10^6 cells and incubate at 4°C for 1 h.
6. Remove the cell debris by centrifuging in a microfuge at maximum speed for 5 min at 4°C.
7. Add 20 µl of 2× SDS sample buffer to the supernatant, boil for 3 min and analyze by SDS-PAGE.

Comments: This cross-linking system is known to bridge lysine side chains and acidic residues that are within the distance of a peptide bond (Cochet et al, 1988).

PROTOCOL 3.19 Cross-Linking Proteins in Solution Using the Homobifunctional Reagent Dithiobis (succinimidyl propionate) (DSP)

DSP (Pierce) is a thiol-cleavable homobifunctional reagent. This reagent has been successfully used for characterizing cell surface receptors and producing interactions between protein components that make up a specific functional unit. The method described below (Musil and Goodenough, 1993) uses metabolic labeling (Protocols 3.1 and 3.2), cell lysis (Protocols 3.9 and 3.10), and immunoprecipitation (Protocol 3.11) in addition to cross-linking.

Materials

DSP solution: 7.5 mg/ml in DMSO

Glycine solutions: 1 M, pH 9.2 and 1 M, pH 7.2

SDS: 10% (see Appendix C)

2× SDS sample buffer (see Appendix C)

2× nonreducing SDS sample buffer: (see Appendix C)

Incubation buffer: 20 mM HEPES, pH 7.5, 150 mM NaCl, 0.8 mM magnesium sulfate, 2.7 mM $CaCl_2$, 10 mM *N*-ethylmaleimide,10 µM leupeptin, 2 mM PMSF

Ultracentrifuge, rotor and tubes

3 ml syringe

25 gauge needles

Immunoprecipitation buffer: 0.1 M NaCl, 0.5% Triton X-100, 0.02 M Na borate, 15 mM EDTA, 0.02% NaN_3, 10 mM *N*-ethylmaleimide, 2 mM PMSF, pH 8.5.

1. Prepare a lysate from metabolically labeled cells. Pass the lysate through a 25 gauge needle five times. Incubate the lysate on ice for 30 min then

centrifuge at 100,000× *g* for 50 min at 4°C. Collect the supernatant, dilute it with two volumes of incubation buffer, and divide it into two equal fractions.

2. To one fraction add DSP to a final concentration of 500 μg/ml and incubate for 30 min at 4°C. In parallel add an equal volume of DMSO to the second fraction.
3. After 30 min at 4°C terminate the cross-linking reaction by adding 20 μl/ml of the glycine pH 9.2 solution to the fractions. Incubate the samples on ice for an additional 30 min.
4. Add 20 μl/ml of glycine, pH 7.2 solution to each fraction. Heat the samples in SDS (0.6% final concentration) at 95–100°C for 3 min.
5. Dilute with 4 vol of immunoprecipitation buffer and perform immunoprecipitation (Protocol 3.11).
6. Add either 50 μl of nonreducing 2× SDS sample buffer or 50 μl of 2× reducing sample buffer and boil for 3 min.
7. Separate the proteins by SDS-PAGE and analyze the results.

Comments: When analyzed by non-reducing SDS-PAGE, crosslinking should produce a high molecular weight band on the gel which is absent when run under reducing conditions.

Ethylene glycolbis(succinimidyl succinate) (EGS) (Pierce) is another homobifunctional cross-linker that can be used in place of DSP yielding similar results. Cross-links formed with EGS are cleavable at pH 8.5 using hydroxylamine for 3–6 h at 37°C.

For best results it is important to titrate the amount of cross-linker as the extent of cross-linking is dependent on the final concentration of cross-linker.

Analysis of Protein–Protein Interactions

Protein interactions are studied to gain insight into the function of uncharacterized proteins by identifying the proteins with which they form complexes. Protein–protein interactions are identified by studying proteins in their natural context, rather than individually, through systematic identification of physical interactions between proteins. Proteins function through modular entities, also referred to as domains. One protein, with many domains of its own, can be capable of interacting with many different domains on other proteins. Interaction pathways are made up of a network of protein complexes. A protein network can be seen as a graph with proteins at the vertices and interactions as lines connecting the interacting proteins (Schächter, 2002).

Mapping Protein–Protein Contact Sites

The most comprehensive information about both stable and transient protein complexes exists for the yeast proteome of ~6,200 proteins. However, there are many uncertainties even for a small genome. The number, types and sizes of the complexes are often complicated due to the difficulty in unraveling physical interactions from functional links, binary from multiple physical interactions, transient from stable interactions, and direct interactions from indirect physical interactions through intermediates. In addition, the localization of the proteins in the cellular environment may influence the interactions (Sali et al, 2003). Despite these difficulties, techniques have been developed to study protein–protein interactions with exquisite sensitivity. A central question in studying molecular recognition events is determining the functional sequences ("epitopes") recognized by the interacting proteins. Two different types of epitopes have to be considered.

In linear (continuous) binding sites the key amino acids mediating the contacts with the interacting protein are located within one part of the primary structure, usually not greater than 15 amino acids in length. Peptides covering these sequences have affinities for the target proteins that are within the range shown by the intact protein ligand.

In conformational (discontinuous) binding sites, the key residues are distributed over two or more binding regions, which are separated in the primary structure. Upon folding, these binding regions are brought together on the protein surface to form a composite epitope. Even if the complete epitope mediates a high affinity interaction, peptides covering only one binding region, as synthesized in a linear scan of overlapping peptides, generally have very low affinities that often cannot be measured in normal ELISA or Biacore experiments. (From Jerini Array technologies home page.)

Yeast Two-Hybrid Systems

While low-throughput technologies like co-immunoprecipitation, far-Western blots and "pull-downs" are used to study individual protein interactions, the study at the proteome level requires high-throughput technology. Today, protein interaction networks are mostly derived from the yeast two-hybrid system (Fields and Song, 1989). The yeast two-hybrid assay provides a genetic approach to the identification and analysis of protein–protein interactions. This screening system was designed to take advantage of the bi-partite nature of certain transcriptional activators which have both DNA-binding domains and activation domains. These two functional units can also work well apart, as long as they are near one another. The technology involves screening libraries of proteins ("prey") for interactions with a particular protein ("bait"). The bait is fused to a DNA-binding domain of a reporter gene, while the prey is fused to a transcription activational domain. The reporter gene is only activated if the bait and the prey interact, resulting in the increased transcription of a targeted reporter gene due to the proximity of the DNA binding and activation domains.

One major consideration with the yeast 2-hybrid system is the generation of false positives and false negatives. Real biological interactions that are missed (false negatives) arise due to misfolding, inadequate subcellular localization and lack of specific posttranslational modifications. False positives could occur due to the bait protein being an auto-activator, capable of turning on transcription by itself or interacting nonspecifically with a large, random population of prey proteins. Conversely, some prey proteins, which are inherently sticky, may non-specifically be selected by bait proteins. Discarding auto-activator bait proteins and sticky prey proteins significantly reduce the rate of false positives (Schächter, 2002). After identifying a protein: protein interaction the results must be confirmed by independent methods to ensure that the interaction is genuine.

Analyzing Protein Interactions by Fluorescence Resonance Energy Transfer (FRET)

FRET is a general method used to detect protein–protein interactions, based on fluorescence resonance energy transfer between fluorescent tags on interacting proteins. FRET is a non-radioactive process whereby energy from an excited donor fluorophore is transferred to an acceptor fluorophore that is within ~60Å of the excited fluorophore (Wouters et al, 2001). Two fluorophores that are commonly used are variants of green fluorescent protein: cyan fluorescent protein (CFP) and yellow fluorescent protein. Excitation of the donor fluorophore in a FRET pair leads to the quenching of donor emission and an increase in acceptor emission. After excitation of the first fluorophore, FRET is detected either by emission from the second fluorophore or by following the decrease of the fluorescence lifetime of the donor. FRET can be used to make measurements in living cells, allowing the detection of protein interactions at the location in the cell where they normally occur. Transient interactions can be followed with high resolution (Phizicky et al, 2003).

References

Argetsinger LS, Campbell GS, Yang X, Witthuhn BA, Silvennoinen O, Ihle JN, Carter-Su C (1993): Identification of JAK2 as a growth hormone receptor-associated tyrosine kinase. *Cell* 74:237–244

Bolton AE, Hunter WM (1973): The labeling of proteins to high specific radioactivities by conjugation to a ^{125}I-containing acylating agent. *Biochem J* 153:272–278

Cochet C, Kashles O, Chambaz EM, Borrello I, King CR, Schlessinger J (1988): Demonstration of epidermal growth factor-induced receptor dimerization in living cells using a chemical covalent cross-linking agent. *J Biol Chem* 263:3290–3295

Cole SR, Ashman LK, Ev PL (1987): Biotinylation: an alternative to radioiodination for the identification of cell surface antigens in immunoprecipitates. *Mol Immunol* 24:699–705

Crawford AW, Beckerle MC (1991): Purification and characterization of zyxin, an 82,000-dalton component of adherens junctions. *J Biol Chem* 266:5847–5853

Degen E, Williams DB (1991): Participation of a novel 88-kD protein in the biogenesis of murine class I histocompatibility molecules. *J Cell Biol* 112: 1099–1115

Fields S, Song OK (1989): A novel genetic system to detect protein–protein interactions. *Nature* 340:245–246

Flaumenhaft R, Abe M, Sato Y, Miyazono K, Harpel J, Heldin C-H, Rifkin DB (1993): Role of the latent TGF-β binding protein in the activation of latent TGF-β by co-cultures of endothelial smooth muscle cells. *J Cell Biol* 120:995–1002

Fraker PJ, Speck JC (1978): Protein and cell membrane iodinations with a sparingly soluble chloramide,1,3,4,6-Tetrachloro-3α,6α-diphem®lglycouril. *Biochem Biophys Res Comm* 80:849–857

Greenlund AC, Schreiber RD, Goeddel DV, Pennica D (1993): Interferon-gamma induces receptor dimerization in solution and on cells. *J Biol Chem* 268:18103–18110

Harlow E, Lane D (1988): *Antibodies, A Laboratory Manual.* Cold Spring Harbor, NY: Cold Spring Harbor Laboratory

Haustein D (1975): Effective radioiodination by lactoperoxidase and solubilization of cell-surface proteins of cultured murine T-lymphoma cells. *J Immunol Methods* 7:25–38

Hunziker W, Harter C, Matter K, Mellman I (1991): Basolateral sorting in MDCK cells requires a distinct cytoplasmic domain determinant. *Cell* 66:907–920

Kessler SW (1975): Rapid isolation of antigens from cells with a Staphlococcal protein A-antibody absorbent: parameters of the interaction of antibody-antigen complexes with protein A. *J Immunol* 115:1617–1624

Kessler SW (1981): Use of protein A-bearing Staphylococci for the immunoprecipitation and isolation of antigens from cells. *Methods Enzymol* 73: 442–459

Köhler G, Milstein C (1975): Continuous cultures of fused cells secreting antibody of predefined specificity. *Nature* 256:495–497

Lisanti MP, Sargiacomo M, Graeve L, Saltiel AR, Rodriguez-Boulan E (1988): Polarized apical distribution of glycosyl-phosphatidylinositol anchored proteins in a renal epithelial cell line. *Proc Natl Acad Sci* USA 85:9557–9561

Maciver SK, Zot HG, Pollard TD (1991): Characterization of actin filament severing actophorin from *Acanthamoeba Castellanii. J Cell Biol* 115:1611–1620

McConahey PJ, Dixon FJ (1980): Radioiodination of proteins by the use of the chloramine-T method. *Methods Enzymol* 70:210–213

Musil LS, Goodenough DA (1993): Multisubunit assembly of an integral plasma membrane channel protein, gap junction connexin43, occurs after exit from the ER. *Cell* 74:1065–1077

On W-J, Cameron PH, Thomas DY, Bergeron JJM (1993): Association of folding intermediates of glycoproteins with calnexin during protein maturation. *Nature* 364:771–776

Phizicky E, Bastiaens PIH, Zhu H, Snyder M, Fields S (2003): Protein analysis on a proteomic scale. *Nature* 422:208–215

Rosenberg I, Cherayil B, Isselbacher KJ, Pillai S (1991): Mac-2 binding glycoproteins. *J Biol Chem* 266:18731–18736

Sabarth N, Lamer S, Zimny-Arndt U, Jungblut PR, Meyer TF, Bumann D (2002): Identification of surface proteins of Helicobacter pylori by selective biotinylation, affinity purification, and two-dimensional gel elwectrophoresis. *J Biol Chem* 277:27896–27902

Sali A, Glaeser R, Earnest T, Baumeister W (2003): From words to literature in structural proteomics. *Nature* 422:2126–225

Schächter V (2002): Bioinformatics of large scale protein interaction networks. *BioTechniques Suppl* 32:S16–S27

Shin BK, Wang H, Yim AM, Le Naour F, Brichory F, Jang JH, Zhao R, Puravs E, Tra J, Michael CW, Misek DE, Hanash SM (2003): Global profiling of the cell surface proteome of cancer cells uncovers an abundance of proteins with chaperone function. *J Biol Chem* 278:7607–7616

Sober HA (1968): *CRC Handbook of Biochemistry.* Cleveland, OH: CRC Press
Wiese C, Galande S (2001): Elimination of reducing agent facilitates quantitative detection of p53 antigen. *BioTechniques* 30:960–963
Wong SS (1991): *Chemistry of Protein Conjugation and Cross-Linking.* Boca Raton, FL: CRC Press
Wouters FS, Verveer PJ, Bastiaens PIH (2001): Imaging biochemistry inside cells. *Trends Cell Biol* 11:203–211
Wu R, Plopper CG, Cheng P-W (1991): Mucin-like glycoprotein secreted by cultured hamster tracheal epithelial cells. *Biochem J* 277:713–718

4

Electrophoretic Techniques

Introduction to Polyacrylamide Gel Electrophoresis (PAGE)

Since the pioneering separation of human serum proteins, albumin, α-, β- and γ- globulin carried out by Tiselius in 1937, electrophoretic separation of biologically active molecules has been an essential technique in biomedical research. As an analytical tool, electrophoretic characterization of biomolecules has progressed tremendously since the time of Tiselius, becoming more sophisticated, specialized and useful as new types of electrophoretic systems have developed. Today, gel electrophoresis is arguably the single most powerful analytical technique in use. Electrophoresis provides a method for simultaneously analyzing multiple samples or for analyzing multiple components in a single sample. As will become apparent, polyacrylamide gel electrophoresis (PAGE) is a pivotal procedure in protein characterization in the sense that analytical reactions probing the structure and composition of the target protein are carried out before and after the separation of a complex mixture of macromolecules by PAGE. A wealth of information can be obtained using PAGE. Molecular weight determination, purity of proteins, posttranslational modifications, subunit structure, enzyme activity, protein processing, and amino acid sequence are just a few areas that can be investigated using PAGE technology.

Electrophoretic methods have become important tools for analyzing and characterizing macromolecules. Acrylamide is the material of choice for the preparation of electrophoretic gels that are able to separate proteins by size. Acrylamide mixed with bisacrylamide forms a cross-linked network when the polymerizing agent ammonium persulfate (APS) is added. The addition of N,N,N′,N′-tetramethylenediamine (TEMED) accelerates the polymerization and cross-linking of the gel. The pore sizes created in the gel are inversely proportional to the concentration of acrylamide. A 10% polyacrylamide gel will have a larger pore size than a 15% gel. Proteins migrate faster through a gel with high pore sizes. Ideally, the percentage gel selected will place the target protein at the center of the gel.

The most widely used electrophoretic system is called discontinuous or disc electrophoresis. In a discontinuous system, a method first described by Ornstein (1964) and Davis (1964), a nonrestrictive large pore gel, referred to as a stacking gel, is layered on top of a separating gel. Each gel layer is made with a different buffer, and the running buffer is different from the gel buffers. The lower gel, also referred to as the separating gel or resolving gel, usually consists of from 5% to 20% acrylamide and is prepared with pH 8.8 buffer. It is in this gel that the proteins will separate. The separating gel is cast first and allowed to polymerize. Subsequently, a second layer of acrylamide solution, the stacking gel, which has a lower concentration of acrylamide (3–5%), is layered over the separating gel to a height of ~1 cm. The pH of the stacking gel is usually about 2 units lower than the pH of the separating gel. These differences allow the proteins in the sample to be concentrated into stacked bands before they enter the lower separating gel. A sample well comb is inserted into the stacking gel and allowed to polymerize. When the comb is removed, wells for loading samples are formed.

Gels were commonly cast in either tubes or rectangular slabs. The advantage of the slab gel system is that from ten to twenty different samples can be run simultaneously under identical conditions and can be compared. Additional advantages of the slab gel system are that: (1) heat produced during electrophoresis, which can lead to band distortion, is more easily dissipated by the slab gel; (2) the rectangular cross-section is more conducive to densitometric analysis; and (3) the gel can be easily dried for storage and autoradiography.

Single percentage gels consist of one acrylamide concentration throughout the running length. Samples containing a narrow range of protein sizes are well resolved on single percentage gels. For general purposes, when the protein mixture is complex, single percentage gels will not be as useful as the gradient gel format. A gradient gel consists of a gradient of acrylamide concentrations from top (low % acrylamide) to bottom (high % acrylamide). The broad range of the gradient i.e. 4–20%, makes it a good choice for the separation of proteins in a complex mixture containing a wide range of molecular weight polypeptides.

Proteins to be electrophoresed are heated for 2–5 min in sample buffer containing sodium dodecylsulfate (SDS), an anionic detergent (see chapter 6, section B), which denatures the proteins and binds to the uncoiled molecules. SDS binding also confers a negative charge on all of the proteins so that in an electric field they will migrate solely as a function of their molecular weights. With few exceptions, the charge conferred by the SDS will mask any charge that is normally present. Therefore, when treated with SDS, polypeptides become rods of negative charges with equal charge densities or charge per unit length, and their separation is a function of size differences only.

Most proteins contain a number of Cys residues, often linked in pairs to form disulfide bridges. Reducing protocols first used β-mercaptoethanol (2-ME) as the thiol reducer (Laemmli, 1970). In modern times 2-ME was replaced with the more powerful reagent dithiothreitol. Recently, tributylphosphine (TBP) is replacing DTT because it is much more powerful than the other two reagents and is required at low concentrations (3–5 mM; Herbert et al, 1998). When a reducing agent is included in the sample buffer, the sample is run under reducing conditions. Conversely, if the sample does not contain a reducing agent, it is run under nonreducing conditions.

Reduction of the disulfide bonds is only one part of the solution to the problem of the reactivity of free —SH groups in proteins. During electrophoresis, regeneration of disulfide bonds can occur among like and unlike peptide chains, further complicating the situation. A two-step protocol: reduction of disulfide bridges, followed by alkylation of the free —SH groups to prevent the back reaction has been adopted. The alkylating agent that is used for this purpose is iodoacetamide (Herbert et al, 2001). Reduction with DTT followed by carboxymethylation with iodoacetamide is shown schematically in Figure 4.1 (Rademaker and Thomas-Oates, 1996).

Bromophenol blue, a low molecular weight pH sensitive dye, is included in the sample buffer as a tracking dye. Bromophenol blue migrates at the front or leading edge of the electrophoretic run. The progress of the electrophoretic run can be visually monitored by following the movement of the blue dye through the gel. Typically the gel is electrophoresed until the bromophenol blue is within 1 cm of the bottom of the gel.

Glycerol or sucrose is included in the sample buffer, increasing the density of the sample so that it will displace the running buffer when loaded into the sample wells.

Precast gels are increasingly popular alternatives to in-house, hand-cast gels because of their numerous advantages. Precast gels do not require any preparation time, simply unwrap the gel, place it into the electrophoresis tank, load the samples and go. Exposure to toxic substances, especially powdered acrylamide is eliminated. Precast gels have a relatively long shelf life and can therefore be purchased in bulk

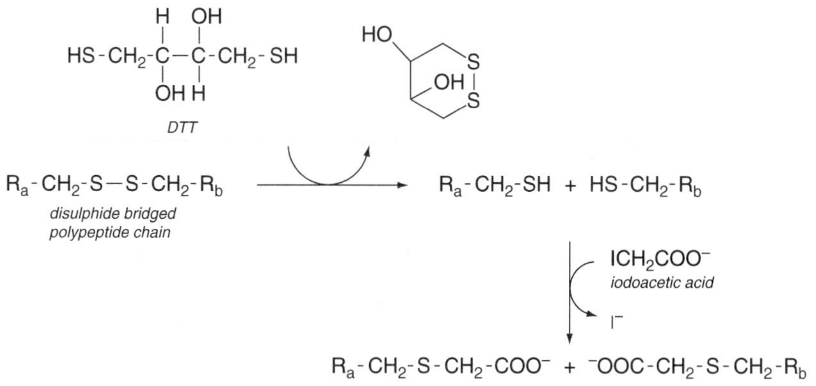

$HS-CH_2-CH(OH)-CH(OH)-CH_2-SH$ *DTT*

$R_a-CH_2-S-S-CH_2-R_b$ (*disulphide bridged polypeptide chain*) $\longrightarrow$ $R_a-CH_2-SH + HS-CH_2-R_b$

ICH_2COO^- (*iodoacetic acid*), I^-

$R_a-CH_2-S-CH_2-COO^- + {}^-OOC-CH_2-S-CH_2-R_b$

FIGURE 4.1 Reduction and carboxymethylation of proteins containing disulfide bridges. (Reprinted with permission of Humana Press, from Rademaker and Thomas-Oates.)

and stored until needed. In the minigel format, the precast gels are economical, reducing sample size requirements. Precast gels give a high level of reproducibility, enabling a degree of standardization within the lab.

The following vendors sell precast SDS-PAGE products: InVitrogen, FMC, BioWhittaker, BioRad, iGels, and Zaxis.

A. Preparation of SDS-Polyacrylamide Gels

PROTOCOL 4.1 Assembling the Plates

Wear gloves to prevent contamination of the plates with skin proteins when setting up the slab gel apparatus.

Materials

Clean and dry plates
Matched spacer set
Clips
1% agarose/water solution

1. Thoroughly clean and dry the plates.
2. Assemble the plate sandwich by inserting matched spacers of uniform thickness made from Perspex or Teflon (shown in Figure 4.2A and B). For SDS-PAGE, useful thicknesses are 0.5, 0.75, 1.0, and 1.5 mm.
3. Seal the plates.

Several methods are used for sealing the plates. The spacers can be lightly and uniformly greased with vacuum grease or vaseline prior to insertion between the plates. The side spacers are assembled to rest flush on the bottom spacer. The plates are now clamped together with metal or plastic clips with pressure being applied directly on the spacers. Examine the plates. If there are any gaps where the

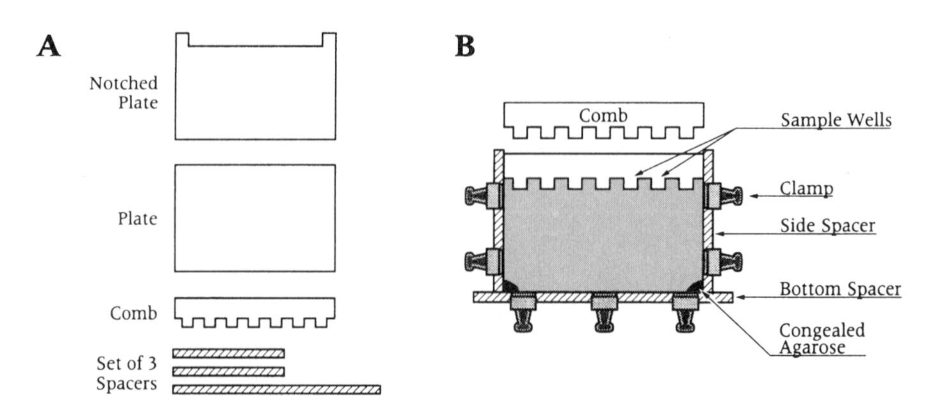

FIGURE 4.2 (A) Components of the slab gel apparatus. (B) The assembled device sealed with agarose.

spacers meet, adjust the spacers. A gap, however small, will lead to a leak.

As an alternative to using grease, a 1% solution of agarose is melted, and a drop is added to the plate assembly so that it is allowed to run down the side spacer. The plate assembly is tilted, the agarose drop moves across the bottom spacer and up the opposite side spacer and the excess is poured off. This procedure can be simplified by gently heating the plate assembly in an oven which will prevent the agarose from congealing between the plates before it has come in contact with all of the spacers. Small triangles of agarose should remain in the two bottom corners. Newer equipment (BioRad and InVitrogen) does not need to be sealed with grease or agarose. If possible, try to avoid greasing the spacers. Grease inevitably gets onto the plates making them difficult to clean.

Choosing the Acrylamide Concentration

To obtain maximum resolution of the target protein on SDS gels, the concentration of acrylamide chosen should ideally place the target protein at the center of the gel. Figure 4.3 plots molecular weight versus migration distance of known proteins in SDS gels at three different acrylamide concentrations. As can be seen from the proteins run on the 5% gel, all proteins with molecular weights of 60 kDa and less will migrate together at the dye front. The linear separation range for a 5% gel run under these conditions is from 60 kDa to at least 200 kDa. For example, if the protein of interest has a molecular weight of 16 kDa, a 15% gel will yield adequate separation and resolution between 10–60 kDa. Consult Figure 4.3 when it is questionable what percentage of acrylamide to use for the visualization of a specific polypeptide on an SDS gel.

PROTOCOL 4.2 Casting the Separating Gel

Materials

Gel plate assembly (See Protocol 4.1)

Acrylamide stock solution 30%: 29.2 g acrylamide, 0.8 g methyl acrylamide per 100 ml

All solutions should be prepared with the highest quality of water available.

The acrylamide stock solution is made up in a total volume of 100 ml (a weight to volume solution, w/v). Filter the solution through Whatman #1 filter paper and store refrigerated.

Preweighed acrylamide and ready to use acrylamide, bisacrylamide solutions are commercially available. If economically possible, premixed acrylamide solutions should be purchased. They are convenient, easy to use, eliminate a handling step, and standardize a common reagent within the lab.

Acrylamide is a known neurotoxin. Gloves and a mask should be worn when handling the powder.

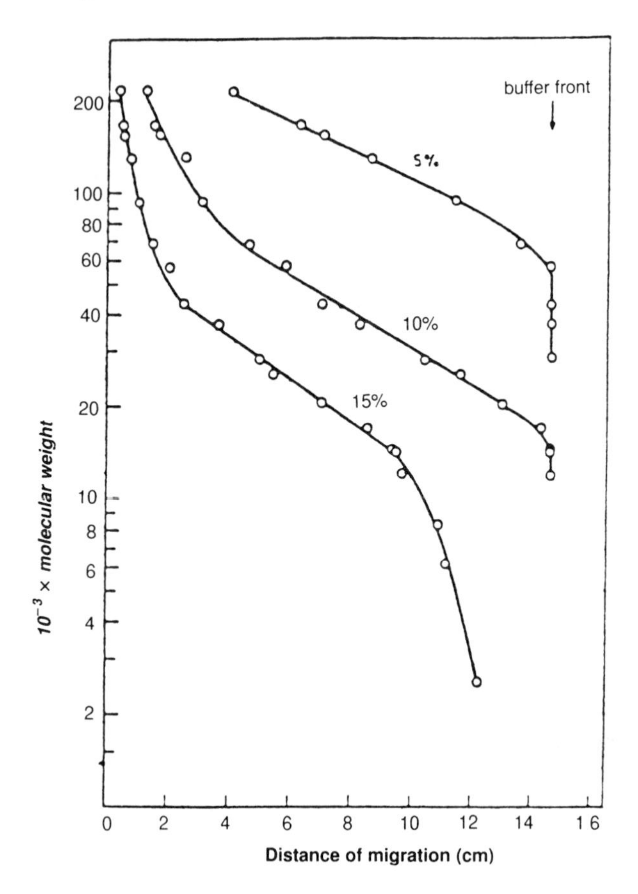

FIGURE 4.3 Calibration curves of $\log_{10}$ polypeptide molecular weight versus migration distance during SDS-PAGE in slab gels using the Laemmli buffer system. The polyacrylamide gels used are uniform concentration 5%, 10%, or 15%. The polypeptide markers in order of decreasing molecular weight are myosin (212KDa), RNA polymerase b′ (165kDa) and b (155KDa) subunits, b-galactosidase (130KDa), phosphorylase a (92.5KDa), bovine serum albumin (68KDa), catalase (57.5KDa), ovalbumin (43KDa), glyceraldehyde-3-phosphate dehydrogenase (36KDa), carbonic anhydrase (29KDa), chymotrypsinogen A (25.7KDa), soybean trypsin inhibitor (20.1KDa), horse heart myoglobin (17KDa), lysozyme (14.3KDa), cytochrome c (11.7KDa), horse heart myoglobin cyanogen bromide cleavage fragments I (8.1KDa), II (6.2KDa), and III (2.5KDa). (Reprinted by permission of Oxford University Press, from Hames and Rickwood, 1981).

The relationship between acrylamide and bisacrylamide is expressed as % Total acrylamide: % crosslinker (%T: %C).

%T is the total acrylamide concentration (acryl + bis) relative to volume (w/v).

%C is the % of T that the crosslinker comprises, i.e., %C is % bisacrylamide relative to the total amount of acrylamide. Therefore, 29.2g acrylamide + 0.8g of bis is 30% T and 2.6% C. A 1:3 dilution of 30:2.6 results in 10% T: 2.6%C.

Separating gel buffer: 1M Tris-HCl, pH 8.8. Store at room temperature

Ammonium persulfate (APS): 10% (w/v). Store the APS solution at 4°C and prepare fresh every two weeks or store in small aliquots at –20°C.

Sodium dodecylsulfate (SDS): 10% (w/v) in water (see Appendix C)

SDS is a bronchial irritant. Wear a mask when weighing it out. It can be very unpleasant if the powder is inhaled. Store the SDS solution at room temperature as it will crystallize in the cold.

N,N,N′,N′-tetramethylene-diamine (TEMED)

Storage buffer: 10 ml Tris-HCl, pH 8.8 (separating gel buffer), 20 ml water

n-butanol (water saturated): 50 ml n-butanol, 10 ml water, shake, then let the phases separate and use the upper phase for the overlay

TABLE 4.1 Separating Gel

Ingredients	5.0%	7.5%	10%	12.5%	15%
Acrylamide solution (ml)	**5.0**	**7.5**	**10.0**	**12.5**	**15.0**
1 M Tris-HCl pH 8.8 (ml)	**11.2**	**11.2**	**11.2**	**11.2**	**11.2**
H_2O (ml)	**13.7**	**11.2**	**8.7**	**6.2**	**3.7**

1. To determine the volumes necessary for the apparatus that you will routinely use, fill the assembled glass plate cassette to the desired height with water, measure the volume, and then for convenience, update the volumes in Table 4.1.
2. Place the plate assembly in a vertical position on a level bench.
3. Select the desired separating gel concentration with the aid of Figure 4.3 and prepare the gel mixture in a 100 ml beaker or flask according to Table 4.1.
4. Add 0.3 ml of SDS solution, 20 µl TEMED, and 100 µl APS. (These volumes will change if the volumes presented in Table 4.1 are scaled up or down). Mix the solution thoroughly by swirling.
5. Pour the complete gel mix between the plates. Leave sufficient space above the gel to accommodate a 1.0 cm high stacking gel and room for the comb.
6. To ensure an even surface, overlay the gel with 1 ml of water saturated n-butanol. The organic solvent n-butanol is less dense than water, does not dilute the top of the gel, and is great for getting rid of bubbles on the gel surface.
7. After polymerization has occurred (30–45 min), pour off the overlay.
8. To store the gel, overlay it with storage buffer to prevent it from drying out. Alternatively, the gel can be used immediately.

Comments: Numerous gels can be prepared at once. Wrap the unused gels in a moistened paper towel and plastic wrap. Unused gels without the stack can be stored at 4°C for up to 1 week.

PROTOCOL 4.3 Casting the Stacking Gel

Materials

Separating gel (from previous protocol)

Stacking gel buffer: 1 M Tris-HCl, pH 6.8

SDS: 10% w/v

TEMED

TABLE 4.2 Stacking Gel

Ingredients	3.0%	5.0%
Acrylamide solution (ml)	1.0	1.67
1M Tris-HCl pH 6.8 (ml)	1.25	1.25
H_2O (ml)	7.7	7.03

Ammonium persulfate (APS): 10% (see appendix C)
Acrylamide: 30% stock solution (see Protocol 4.2)

1. Prepare the Stacking gel solution according to Table 4.2. Use a 3% stacking gel for separating gels less than 10%.
2. Add 0.1 ml SDS, 10 μl TEMED, and 50 μl APS (see Protocol 4.2).
3. Rinse the gel surface with Stacking gel solution and quickly pour off.
4. Add fresh Stacking gel solution, nearly filling the space above the Separating gel.
5. Insert a comb into the Stacking gel solution, making sure that air bubbles are not trapped at the bottom of the sample wells. If there are air bubbles, remove the comb and quickly reinsert it.
6. Let the Stacking gel polymerize for 15–30 min.
7. Carefully remove the comb that forms the sample wells. Rinse with water. Divisions between the wells that have become distorted during the removal of the comb can be straightened by inserting a clean needle or microspatula. Once the stacking gel has polymerized, the gel should be used within 60 min.

PROTOCOL 4.4 Gradient Gels

Gradient gels allow the resolution of a broad range of protein sizes within one slab gel. Polyacrylamide gels that are prepared with a gradient of increasing acrylamide concentration (decreasing pore size) are used both for analysis of complex protein mixtures and for molecular weight estimations using SDS as the dissociating agent. Five to 15% gradient gels offer a wide separation range and provide a good starting point. A linear gradient gel is prepared using a gradient maker which mixes solutions of two different acrylamide concentrations. Gradient gels are cast with acrylamide concentrations that increase from top to bottom. The primary advantage of gradient gels is that proteins are continually entering areas of the gel with decreasing pore size such that the advancing edge of the migrating protein zone is retarded more than the trailing edge, resulting in a marked sharpening of the protein bands. In addition, a gradient gel increases the range of molecular weights that can be fractionated simultaneously on one gel.

Materials

Glass plate assembly (Protocol 4.1)
Solutions: see Protocol 4.2 and Table 4.1
Peristaltic pump

Gradient maker: purchased or constructed in a laboratory workshop from plastic perspex
Magnetic stirrer

1. Assemble the gradient maker and pump. A typical setup for producing linear gradient slab gels is shown in Figure 4.4. Connect the outlet from the gradient maker to the peristaltic pump which uniformly introduces the acrylamide solution between the glass plate assembly through a piece of thin tygon or teflon tubing.
2. Choose and prepare two acrylamide solutions with the desired concentrations of high and low percentage acrylamide (see Protocol 4.2 and Table 4.1).
3. Add the low concentration of acrylamide solution to chamber A. Momentarily open the valve connecting chambers A and B, filling the connection between the two chambers with the lower percentage acrylamide solution. Return the excess solution which had flowed into chamber B to chamber A.
4. Add an equal volume of the high percentage acrylamide solution to chamber B, which also contains a small magnet. Turn on the magnetic stirrer.
5. Open the connection between chamber A and B. Turn on the pump and introduce the mixed acrylamide solution between the glass plate sandwich. The level of acrylamide mixture in the glass plate assembly rises with the higher concentration on the bottom.
6. Once all of the acrylamide mixture has been introduced, overlay the gel with water-saturated n-butanol and leave undisturbed to polymerize for 1 h. After the gel has polymerized, it can be treated like a uniform concentration slab gel as described above.
7. Flush out the gradient maker and tubing with water to prevent the acrylamide solution from polymerizing.

Comments: When the aim is to isolate protein for amino acid sequence analysis, let the separating gel polymerize for at least 4 h prior to loading the sample. This allows the TEMED and ammonium persulfate to react with the gel components, decreasing their chances of reacting with the amino-terminal end of the polypeptide. SDS-PAGE for microsequencing will be described in Chapter 8.

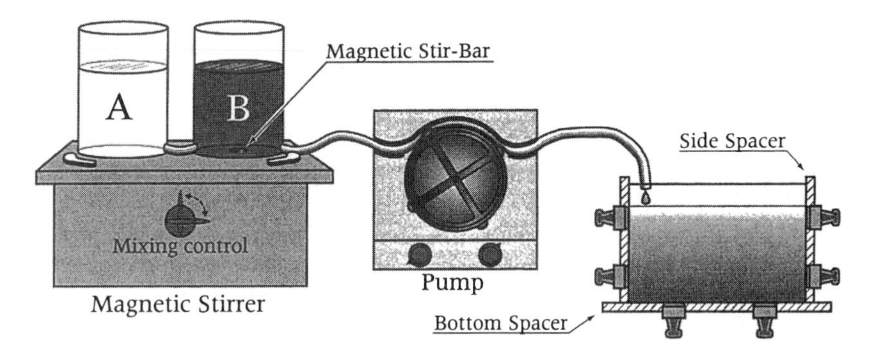

FIGURE 4.4 Preparing a gradient gel.

Numerous gradient gels can be prepared at once. After the separating gels have polymerized, overlay them with storage buffer, then wrap the unused gels in moistened paper towels and plastic wrap. Like single concentration separating gels, unused gradient gels can be stored at 4°C for up to 1 week.

It is simplest and most often cost and labor effective to buy commercially available, precast gradient gels. Many companies offer full lines of gels.

PROTOCOL 4.5 Sample Preparation

Materials

15 ml conical centrifuge tube
Hot plate
Microfuge

Reducing Sample buffer (4×)		**Final Concentrations (1×)**
Sucrose	4.0 g	10% (w/v)
DTT	308 mg	50 mM
SDS	0.8 g	2% (w/v)
1 M Tris-HCl pH 8.8	2.5 ml	0.0625 M
Bromophenol blue	0.001% w/v	0.0004% (w/v)

Adjust the 4× sample buffer to a final volume of 10 ml with water. Warming at 37°C will speed up the solubility process. Once dissolved completely, aliquot the sample buffer and store at –20°C. Glycerol (40%) can be substituted for the sucrose.

Non-reducing sample buffer is prepared exactly like the reducing sample buffer except that the reducing agent, DTT, is ommitted.

1. In most cases add the appropriate amount of 4× reducing sample buffer to your protein sample to obtain a 1× final concentration. Specific protocols may call for the addition of 2× or 1× sample buffer. These solutions can be found in Appendix C.
2. Heat the sample for 2–5 min at 90–95°C. This ensures denaturation of the protein. After heating, allow the sample to cool to room temperature.
3. Remove any insoluble material by centrifugation at 12,000 × *g* for 5 min in a microfuge. Undissolved proteins cause streaking when the gel is stained. The sample may be used immediately or stored frozen at –20°C. After freezing and thawing, vortex mix and recentrifuge the samples before loading them on the gel.

Comments: For optimal visualization of the protein bands, one should have an idea of the protein concentration. When loading the protein samples on the gel it is important not to overload or underload. Overloading will produce a smear and underloading will result in an inability to detect bands of interest.

The behavior of a protein in a gel is a function of the quantity of protein per unit area into which the protein migrates. The protein loaded on a gel can be thought of in μg/cm^2 or μg/well. Therefore, 1–5μg in a band is 9.5 to 48μg/cm^2 in a 1.5mm × 7mm lane and 67 to 333μg/cm^2 in a 0.5 × 3mm lane. Generally, 10–20μg/cm^2/band provides good detection by Coomassie blue staining while providing good resolution. When a complex mixture is to be electrophoresed, about 50–100μg of protein per lane is acceptable. When estimating purity, overloading the gel allows you to visualize the degree to which the target protein is contaminated with other proteins.

In addition to the amount of protein to be loaded, the volume is also important. For best results when using 1.5mm thick slab gels and 7mm wide sample wells use 10–50μl of sample solution. However, in discontinuous systems the method is independent of the sample volume. What actually limits the volume that can be loaded is the capacity of the sample well.

Since molecular weight estimations by SDS-PAGE depend on all polypeptides having the same charge density, which means having the same amount of SDS bound per unit weight of polypeptide, SDS should be present in excess.

PROTOCOL 4.6 Running the Gel: Attaching the Gel Cassette to the Apparatus and Loading the Samples

Materials

Running buffer 1×: 1 liter: 3.03g Tris (0.025M), 14.42g Glycine (0.192M), 1.0g SDS

Running buffer can be prepared as a 10× concentrate and diluted to working strength as needed. The pH does not require adjustment.

Power supply

1× sample buffer (See Appendix C)

Slab gel apparatus

1. First remove the comb and then remove the bottom spacer from the slab gel cassette (see Figure 4.5).
2. Attach the slab gel assembly to the apparatus. When using a notched glass or metal plate, mount the gel such that the notched plate is aligned closest to the upper buffer reservoir.
3. Add running buffer to the upper chamber first. Check for any leaks. If there is a leak, adjust the clamp placements around the glass plates or apply a small amount of grease to the problem area. Do not proceed before stopping the leak. Usually there is a seal between the plate and the apparatus made from a sponge-like material that is cut to the dimensions of the plate. The newer, commercially available equipment has a rubber gasket fitted into the apparatus so that a seal is created when the gel assembly is clamped to it. The glass plates must make a hermetic seal so that buffer in the upper chamber does not leak.

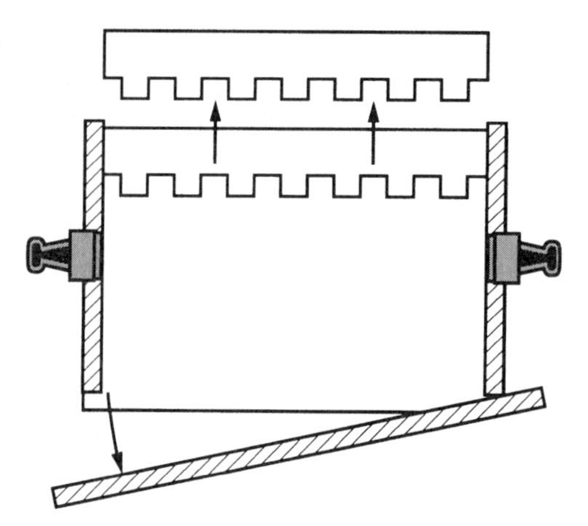

FIGURE 4.5 Removal of the comb and bottom spacer from the cassette.

4. Add running buffer to the lower reservoir and flush out any air bubbles that have been trapped between the plates (see Figure 4.6) with the aid of a long hypodermic needle that is bent to fit under the plates.
5. Using a loading tip, wash the sample wells with a stream of running buffer.
6. Carefully load the samples into the wells with a microsyringe or micropipette. The dense sample will flow into the well displacing the running buffer and form a sharply defined layer.
7. Fill unused wells with 1× sample buffer to prevent distortion of the lanes during the run.
8. Connect the apparatus to the power source so that the anode (+) (usually the red wire and lead) is attached to the bottom reservoir. See Figure 4.6. All of the proteins are negatively charged polyanions due to the SDS and will migrate in the downward direction toward the anode. Therefore, SDS-PAGE is an anionic system.
9. Adjust the power supply to deliver the desired voltage and run the gel in the constant voltage mode.
10. When the bromophenol blue enters the separating gel, increase the current.

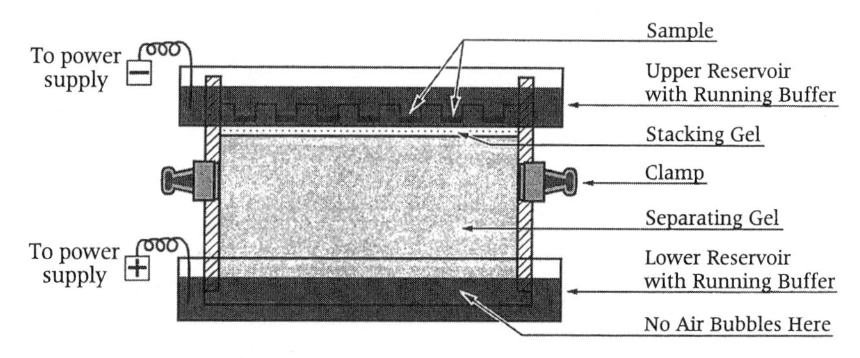

FIGURE 4.6 Mounted gel cassette onto slab gel apparatus highlighting correct connections.

Comments: Electrophoresis adheres to the principle of Ohm's law: voltage = current × resistance or V = I R where the gel is the resistor and the power supply is responsible for both the current in mA and the voltage. SDS-PAGE can be run under constant current or constant voltage. If the current is held constant, then the voltage will increase during the run as the resistance goes up. Most power supplies deliver constant current or constant voltage and have two sets of outlets connected parallel to each other internally. If two gels of identical thickness are to be run simultaneously from the same power supply, they will be connected in parallel. Two examples of this are shown in Figure 4.7. In both cases, the voltage will be the same across each gel. The same voltage setting (constant voltage mode) can be used regardless of the number of gels being run. If the power supply reads 100V then each gel has 100V across its electrodes. The current is the sum of the individual currents going through each gel. Therefore, when running under constant current, it is necessary to increase the current for each gel connected to the power supply. This explains why two identical gels require double the current to achieve the same voltage.

Depending on the thickness and length of the resolving gel, electrophoretic conditions can be chosen such that the run will be completed within a few hours or, for convenience, overnight. A 12cm long 0.75mm thick gel will run ~15h at 4mA constant current. If the system is not well cooled, it should be run slower to avoid distortions and what is called smiling. A 12cm long 1.5mm thick gel will run 15h at 45V constant voltage.

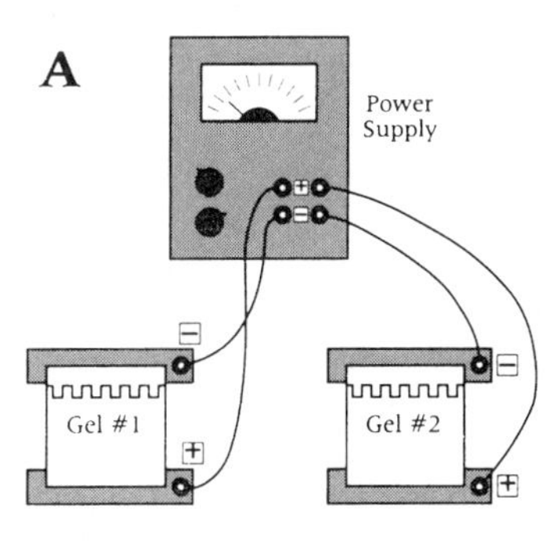

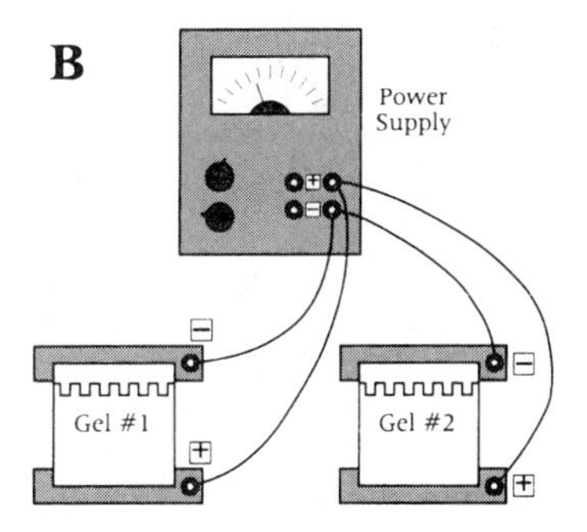

FIGURE 4.7 Two gels connected in parallel to the same power supply. (A) Two gels are connected in parallel to adjacent outlets. (B) Two gels are connected in parallel to one outlet. In both cases the gel currents are additive.

Whenever possible use a constant voltage setting. By using constant voltage both current and wattage will decrease as resistance increases. The wattage decrease (heat) will provide a degree of safety as the run progresses.

Care should be taken to avoid excessive heating, which may occur when the voltage approaches 200V for a 1.5mm thick gel. The use of thin minigels is increasing due to the savings in time and materials without sacrificing resolution. Their running time is proportionally shorter, from 30min to 4h. Mini systems like the BioRad Mini Protean II are recommended to run at a constant voltage of 200V which provides a 45min run. Better results can be obtained by running the gel at 100V, taking ~1h 15min to complete the run.

If there is a slow leak from the upper chamber during an overnight run, you will be unpleasantly surprised to arrive in the lab the next morning to find that the bromophenol blue has only migrated one third to one half of the way down the gel. The buffer level in the upper chamber has fallen below the level of the gel. Current will no longer flow through the gel, stopping the run. Don't despair. Top up the upper chamber and complete the run.

PROTOCOL 4.7 Drying the Gel

Polyacramide gels can be dried to expedite the achievement of certain protocols. Radiolabeled proteins are easily detected in dried gels that are exposed to X-ray film, or phosphorimaging.

Gels can be dried either by vacuum methods or air-dried using cellulose films.

Materials

Gel drier

Whatman #1 filter paper

Plastic wrap

1. To dry a gel using a commercial gel drier equipped with a heating element attached to a vacuum pump and trap, put the gel on a piece of Whatman #1 filter paper, cover it with plastic wrap and place it on the gel drier.
2. Place an extra sheet of filter paper under the sheet with the gel. Most gels that are 1.5mm thick and made from high percentage acrylamide will dry within ~2h.
3. Turn on the power and make sure that the gel is under vacuum. To prevent severe cracking of the gel do not remove it from the gel drier until dry.

Comments: If you are not familiar with the gel drying apparatus ask a colleague for instructions. This will prevent pump oil from being sucked into the trap and a costly repair bill.

Little hands-on time is required when using an air-drying system with cellulose film and frames. However, the gel will take much longer to dry, probably overnight.

After drying, the gel can be exposed to film or analyzed by phosphorimaging.

PROTOCOL 4.8 Separation of Low Molecular Weight Proteins by Tricine-SDS-PAGE (TSDS-PAGE)

The method described in this protocol is a modification of the tricine-based system of Schagger and von Jagow (1987) described by Dayhuff et al (1992) which was designed to improve the resolution of proteins in the low molecular weight range (<10 kDa). The main advantage of this technique is that it can be used prior to electroblotting peptides onto polyvinylidene difluoride (PVDF) membranes, greatly facilitating the sequencing of small polypeptides and proteins described in Chapter 8.

Materials

TABLE 4.3 Acrylamide Solutions*

Acrylamide-bisacrylamide mixture	Percentage acrylamide (w/v)	Percentage bisacrylamide (w/v)
49.5% T, 3% C	48	1.5
49.5% T, 6% C	46.5	3.0

*%T of both monomers (acrylamide and bisacrylamide). %C denotes the percentage concentration of the crosslinker relative to the total concentration T. Store the acrylamide:bisacrylamide solution at 4°C (see Table 4.3).

Gel Buffer: 3.0 M Tris, 0.3% SDS adjust the pH to 8.45 with HCl

Ammonium persulfate (APS): 10% (w/v)

TEMED

For reducing conditions make the 1× sample buffer 50 mM in dithiothreitol

Running Buffer (10×): Tris base 1.0 M, Tricine 1.0 M, SDS 1.0%

Dilute the 10× running buffer to 1× with deionized or distilled H_2O.

1. Choose the desired percentage of separating gel and measure out the required volumes of separating gel and gel buffer solutions according to Table 4.5.

TABLE 4.4 Tris-Tricine-SDS-PAGE Sample buffer

	1 × (Final concentration)	For 10 ml of 2×
Tris HCl, pH 8.45	450 mM	3 ml of 3.0 M solution
Glycerol	12%	2.4 ml
SDS	4%	0.8 g
Coomassie blue G	0.0025%	0.5 ml of a 1% solution
		H_2O to 10 ml

TABLE 4.5 Composition of Separating Gels*

Ingredients	10% T, 3% C	16.5% T, 3% C	16.5% T, 6% C
49.5% T, 3% C	6.1ml	10ml	—
49.5% T, 6% C	—	—	10ml
Gel buffer	10ml	10ml	10ml

*% T denotes the total percentage concentration of both monomers (acrylamide and bisacrylamide). %C denotes the percentage concentration of the crosslinker relative to the total concentration T. Store the acrylamide: bisacrylamide solution at 4°C.

2. Add water to reach a final volume of 30ml.
3. To polymerize the separating gel, add 150µl of a 10% APS solution and 15µl of TEMED. Overlay the gel with 1–2ml of gel buffer diluted 1:3 with water and allow 30–60min for polymerization.
4. Prepare the stacking gel (4% T, 3% C). Measure out 1ml of 49.5% T, 3% C acrylamide solution. Add 3.1ml of gel buffer solution and 8.4ml of water. To polymerize, add 100µl of 10% APS and 10µl TEMED.
5. Rinse the top of the separating gel with complete stacking gel solution. Fill the space above the separating gel with stacking gel solution and insert the comb.
6. Incubate the protein samples for 30min at 40°C in a final concentration of 1× TSDS-PAGE sample buffer.
7. When using the standard gel size of 10 × 14 × 0.15cm, perform electrophoresis at room temperature. Start the run at 30V Let the sample completely enter the stacking gel, then adjust the voltage according to Table 4.6.

Comments: The 10% T, 3% C gel with a 4% T, 3% C stack gives linear separation from ~5–100kDa. This gel type is preferred for preparative purposes because the heat production is low, and proteins elute better than with the other gel types. The resolution below 5kDa is poor.

The 16.5% T, 3% C gel with a 4% T stacking gel is useful in the range from 1–70kDa. The 16.5% T, 6% C gels have their best resolution between 5 and 20kDa. Resolution below 5kDa can be improved by adding urea (Swank and Munkres, 1971). These gradient concentrations and uniform concentrations are commercially available.

Fixing, staining and destaining polypeptides below 2kDa is problematic. Due to their low molecular weight, these molecules may elute out of the gel during fixation, staining and destaining. The original method fixes the gel in 50% methanol, 10% acetic acid for 1h for a 1.5mm thick gel, then staining with 0.025% Serva Blue G in 10% acetic

TABLE 4.6 Electrophoresis Conditions for Different Gel Types

Gel Type	Voltage	Run Time (h)
10% T, 3% C	110V const.	5
16.5% T, 3% C	85V const.	16
16.5% T, 6% C	95V const.	16

acid for 1–2h depending on the thickness. Dayhuff et al (1992) substitute Coomassie Blue R for the Blue G. These authors report that below 2kDa some peptides are not stained. The $ZnSO_4$ staining protocol described in Protocol 4.19 may prove useful for low molecular weight polypeptides.

Tricine buffers and gels are commercially available in single concentration and gradient formats.

When using a pre-cast mini-gel, consult the product insert for recommended running conditions.

This electrophoresis system is reproducible even at some deviation in ionic strength and pH; proteins can be loaded in high ionic strength solutions, e.g., 2M NaCl. In addition, it can tolerate the application of large amounts of protein without adverse overloading effects.

Safety Considerations

Voltages used during electrophoresis are dangerous and potentially lethal.

Safety should be the prime concern. The following short section will reemphasize some safety measures to take when working with high voltages.

- Never remove or insert high voltage leads unless the power supply voltmeter is turned down to zero and power supply is turned off.
- Try to grasp the high voltage leads one at a time and with only one hand. This will decrease the risk of accidentally closing a circuit.
- Inspect all cables for frayed and exposed wires and replace them.
- Have the power supply controls turned down to zero and start with the power supply off. Attach the gel apparatus, turn on the power supply and turn up the voltage, current, or power to desired level. When the run is completed, turn the power supply down to zero then disconnect the gel apparatus one lead at a time.

B. 2-Dimensional (2-D) Gel Systems

The impact that co- and posttranslational modifications have on final gene products can be visualized on a 2-D gel, where a diversity not encoded at the gene level is seen (Gooley and Packer, 1997). Ideally, a technique that can correlate mRNA levels with protein expression, modification and activity should permit the: (1) extraction and high-resolution separation of all proteins, including membrane, extreme-pI and low-copy-number proteins; (2) identification and quantification of each component; and (3) visualization, comparison and analysis of changes in protein expression patterns. Two-dimensional electrophoresis comes close to satisfying many of these criteria.

The reach of 2-D gel separations can be improved by reducing the sample complexity prior to analysis. For example, by analyzing protein subsets and subcellular organelles separately, the analysis and quantitation of low-abundance proteins can be improved.

Two-dimensional (2-D) gel electrophoresis is a two-step process used to separate complex mixtures of proteins into many more components than is possible by standard one-dimensional electrophoresis. Each dimension separates proteins according to different properties. One advantage of 2-D gel electrophoresis is the possibility of studying quantitative variations in cellular protein patterns that may lead to the identification of groups of proteins that are expressed coordinately during a given biological process.

To optimize separation, each dimension must separate proteins according to independent parameters; otherwise, proteins will be distributed across a diagonal rather than across the entire rectangular surface of the gel. The first dimension, which is usually isoelectric focusing, was originally run in the tube gel format. New developments have made this procedure much easier and it is now run mostly using pre-made strips with an immobilized pH gradient. The state of 2-D electrophoresis using immobilized pH gradients (IPG) has been extensively reviewed (Görg et al, 2000). 2-D electrophoresis is illustrated schematically in Figure 4.8. In the first dimensional separation, proteins are focused electrophoretically in a pH gradient so that they move to a position in the gradient where they have no net charge.

For the second dimension, the focused gel is layered horizontally onto the top of a polymerized slab gel containing SDS. A small amount of agarose is used in the grafting process to bond the gels together. The proteins that were separated according to charge in the first dimension migrate out of the focusing gel and into the SDS-PAGE gel that sepa-

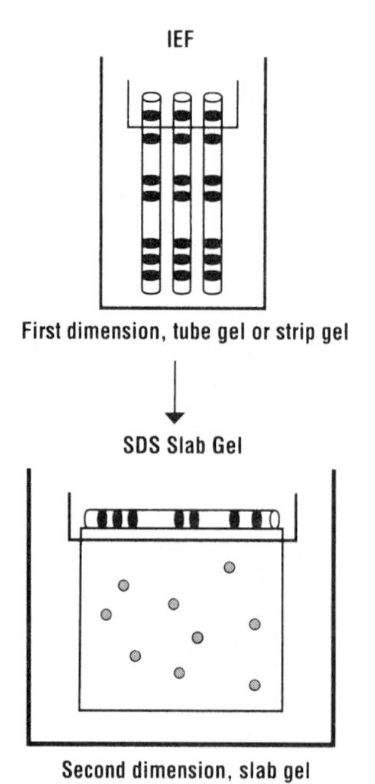

FIGURE 4.8 Schematic representation of 2-D electrophoresis.

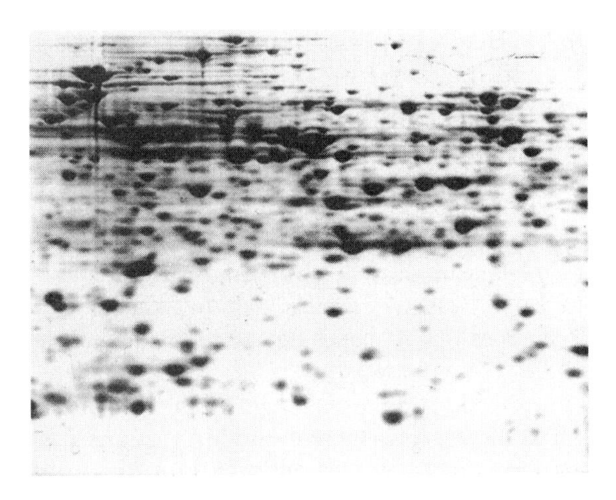

FIGURE 4.9 Two-dimensional electrophoretic patterns of Coomassie blue stained *E.coli* proteins. Proteins were separated using an 11 cm pH 4–7 IPG strip and an 8–16% separating gel. (Figure kindly provided by Bio-Rad Laboratories, Life Science Group, Hercules, CA).

rates proteins according to molecular mass. After the second dimension, when visualized, the proteins appear as spots. The analysis of 2-D gels is more complex than that of standard one-dimensional gels because of the abundancy of spots. A typical result is shown in Figure 4.9.

What does the identification of a proteinaceous spot from a 2-D gel mean? Identification assigns a name or database accession code to a protein spot on a gel. The amino acid sequence of the identified protein is linked to the DNA sequence of its gene, linking proteome to genome. Identification is the first step towards studies on protein posttranslational modification and ultimately to its function.

Software advances have made it possible to directly compare 2-D maps produced from different labs. The reproducibility of reference maps using immobilized pH gradient (IPG) gels has prompted several groups to make their maps available over the World-Wide Web making the SWISS-2DPAGE database a reality.

Reference 2-D maps of different tissues of an organism can reveal proteins unique to a specific tissue and by comparison, determine which ones are common and which ones unique.

Isoelectric Focusing

Isoelectric focusing (IEF) is an extremely sensitive method that separates proteins on the basis of charge (O'Farrell, 1975) and identifies an intrinsic property of a protein, its isoelectric point, the pH at which the net charge on the protein is zero. The isoelectric point (pI) of a denatured protein is determined by its amino acid composition and by any posttranslational modifications. When a protein is placed in a pH gradient and subjected to an electric field, it will initially move toward the electrode with the charge opposite from itself. Proteins that carry a net positive charge will migrate toward the cathode (negative terminal). Negatively charged proteins will move toward the anode (positive terminal). During migration through the pH gradient, the protein will either pick up or lose protons. As it migrates, its net charge and mobility will decrease and the protein will slow down. Eventually, the protein will arrive at the point in the gradient equal to its pI. There, being uncharged, it will stop migrating. At this point it is considered to be "focused", and a narrow, sharply defined band is formed. It is possible to separate proteins that have very close isoelectric points.

IEF can be performed in tubes, slab gels or strips. In most cases, IEF is used as the first dimension of a two dimensional analysis of proteins. However, it can also be used without a second dimension, in which case the slab gel format is preferable.

Originally, pH gradients were established using mixtures of soluble, charged compounds called carrier ampholytes, amino acid polymers that have surface charges corresponding to different pH ranges, which migrated electrophoretically to their isoelectric points. Once reaching their pIs, the ampholytes maintain a constant pH by exerting a strong buffering capacity and provide good conductivity. Since proteins themselves are charged molecules, they can also affect the pH gradient, especially if overloaded. This potential problem has been overcome by using immobilized pH gradients (IPGs). With IPGs, the gradient is generated prior to electrophoresis where the compounds used to construct the pH gradient are chemically immobilized providing an indefinitely stable pH gradient (Righetti, 1990).

The introduction of IPG gels eliminated the problems of gradient instability and low sample loading capacity. The availability of reasonably priced commercially produced IPGs has made 2-D high resolution protein separation available to casual users. Using IPGs as the first dimension of 2-D PAGE it is possible to produce highly reproducible reference maps as well as separate milligram quantities of protein for micropreparative purposes (Herbert et al, 1997).

Much larger amounts of protein can be loaded enabling detection of more of the less abundant protein components of the proteome.

The pI of a protein is valuable information. For example, if the pI of the target protein is known, a buffer system can be rationally designed for the purification of this protein. Assume a protein has a pI of 6.2, it will have a net charge of zero in a solution with a pH of 6.2. As the pH of the buffer rises, the protein will become negatively charged because the carboxyl groups, which had been protonated at the lower pH, become deprotonated. This forms the basis of a separation as the protein is now susceptible to purification by DEAE ion exchange chromatography using a buffer with a pH in the range of pH 7–8. The negatively charged protein will be retained on the positively-charged column. Ion exchange chromatography is described in Chapter 10.

PROTOCOL 4.9 Preparation of the Sample for Isoelectric Focusing

Avoid conditions that can cause chemical modifications to the proteins. Single charge changes and modifications can cause a protein to migrate as more than one isoform, producing two or more bands.

The material to be analyzed can be a whole cell lysate, semi-purified or immunoprecipitated proteins, usually in washed, pelleted form. Dilute samples can be concentrated and then dissolved in sample buffer (see Protocol 4.4).

Materials

Sample buffer: 7M urea, 2M thiourea 2% CHAPS, 0.4% Ampholine pH 6–8, 6.5–9, 8–10.5, 9–11, 5mM tributylphosphine (TBP). Ampholines are supplied as 40% w/v solutions and are available from many commercial sources including Pierce, Pharmacia, and Fluka. Sample buffer can be stored frozen in aliquots.

Sonication buffer: 0.01M Tris-HCl, pH 7.4, 5mM $MgCl_2$, 50μg/ml pancreatic RNase

DNase: 1mg/ml in 0.01M Tris-HCl, pH 7.4, 1mM $MgCl_2$ (This solution can be stored in aliquots at –20°C)

Urea (ultra pure)

Microtip sonicator

1. Suspend the washed, pelleted material in sample buffer. Once sample buffer has been added, the samples should either be run immediately or frozen. For best results the protein content should be ~40μg containing more than 100,000cpm in a volume of 25μl or less.
2. An alternative to the sample buffer is to resuspend the cells in sonication buffer. Sonicate the cell suspension by two 15sec bursts at maximum output with a microtip on ice. After sonication, add DNase to a final concentration of 50μg/ml and incubate the sonicate on ice for at least 5min. Add solid urea to make a final concentration of 9M then add 1 volume of sample buffer.
3. For applying sample to IPG strips use the appropriate size tray channel (Bio-Rad catalog). Pipet the sample evenly along the back edge of the tray channel, except for ~1cm at either end.
4. Place the IPG strip gel side down onto the sample. Make sure that sample does not get on the plastic backing of the strip as it will not be absorbed by the gel material.
5. Take care not to trap air bubbles beneath the strip. If this happens, lift the strip up and down from end to end until the air bubble has moved out from under the gel. Allow the strips to hydrate overnight.

Comments: Nucleic acids in the sample interfere with the isoelectric focusing. To alleviate this problem RNase and DNase are included in the sample preparation. If the sample is relatively pure, use step 1.

Single-Step Extraction/Solubilization Buffer

This extraction buffer solubilizes proteins in a single step and the presence of the isopropanol-isobutanol mixture results in the precipitation of nucleic acids. As a result, no mechanical shearing or enzymatic digestion of the nucleic acids is required (Leimgruber et al, 2002). This eliminates the potential protein-nucleic acid interactions that can lead to vertical streaking in the first dimension.

Solubilization buffer: 2M thiourea, 5M urea, 0.25% CHAPS (Sigma), 0.25% Tween 20 (Bio-Rad), 0.25% SB 3–10 (Sigma), 100mM DTT, 0.25% carrier ampholytes (Bio-Lyte 3–10 (Bio-Rad), Servalyte 3–10 (Serva), Ampholine 3.5–9.5 Amersham, Resolyte 4–8 (BDH) 1:1:1:1, 10% isopropanol, 12.5% water saturated isobutanol, 5% glycerol, 1 mM sodium vanadate, and 1× Complete protease inhibiton cocktail (Roche, Indianopolis, IN).

Comments: Progress has been made on the problem of isolating highly hydrophobic membrane proteins through the continued development of sample preparation protocols based on zwitterionic detergents (Santoni et al, 1999) and organic solvents (Molloy et al, 1999). A single-step extraction solution containing a zwitterionic surfactant like CHAPS or amidosulfobetaine (ASB-14), solubilizing cytosolic and membrane proteins, was used successfully for preparing 2-D maps of rat brain proteins (Carboni et al, 2002).

Replacing DTT with tributylphosphine as the reducing agent was reported to enhance protein solubility during IEF, thereby increasing resolution and recovery (Herbert et al, 1998).

It has been reported that spurious spots can be reduced by performing alkylation of free thiol groups prior to the IEF run (Herbert et al, 2001). The proteins in the focusing gel or strip can be carboxymethylated following reduction by adding a 2-fold molar excess of iodoacetamide (IAA) in relation to the reducer. Failure to reduce and alkylate proteins prior to the second dimension will result in a large number of spurious spots in the alkaline pH region due to "scrambled" disulfide bridges among like and unlike chains (Galvani et al, 2001).

Extraction kits are commercially available to differentially extract proteins based on differential solubility properties of the individual molecules. (BioRad Catalog and Molloy et al, 1998).

PROTOCOL 4.10 Preparation and Running of Isoelectric Focusing Tube Gels

Materials

Urea (ultra pure)

Acrylamide stock solution (30%): 28.4% (w/v) acrylamide, 1.6% (w/v) bisacrylamide CHAPS: 10% (w/v) stock

Ampholytes: pH range 5 to 7 and 3 to 10.

Ammonium Persulfate (APS): 10% (w/v): see Appendix C

TEMED: see Appendix C
Overlay solution: 8 M urea
Cathode buffer: 0.02 M NaOH
Anode buffer: 0.01 M H_3PO_4
Run overlay solution: 9 M urea, 1% Ampholines (0.8% pH range 5 to 7 and 0.2% pH range 3 to 10)
1× SDS-PAGE sample buffer: see Appendix C
Tube gel electrophoresis chamber
Power supply capable of producing 800 V
Long, thin, blunt needle (e.g., 2.5 in., 29 gauge) fitted to a syringe
Shaker

1. Clean and dry the tubes and seal the bottoms with parafilm.
2. Prepare the gel mixture in a 125 ml side-arm flask; dissolve 5.5 g of urea in 1.97 ml of deionized water, 1.33 ml of acrylamide stock solution, 2 ml of CHAPS, and 0.4 ml of Ampholytes, pH range 5 to 7, and 0.1 ml of pH 3 to 10 (to make 2% Ampholytes). These quantities can be scaled up as necessary. Other pH gradients can be produced by blending the ampholytes.
3. Swirl the flask until the urea is completely dissolved then add 10 µl of 10% ammonium persulfate and degas for ~1 min (see Protocol 10.1). Immediately add 7 µl of TEMED and swirl the flask. Fill the tubes to ~5 mm from the top.
4. Add overlay solution and let the gels polymerize for 1–2 h.
5. Remove the overlay solution and replace it with 20 µl of sample buffer and overlay with a small amount of water. Allow the gels to set an additional 1–2 h.
6. Remove the parafilm from the bottom of the tubes. Remove the sample buffer and water from the gels and mount them in the tube gel electrophoresis chamber. Add 20 µl of fresh sample buffer to the gels and fill the tubes with cathode buffer, 0.02 M NaOH.
7. Fill the lower reservoir with anode buffer 0.01 M H_3PO_4. Fill the upper reservoir with cathode buffer.
8. Before loading the sample, pre-run the gels for 15 min at 200 V, 30 min at 300 V, and 30 min at 400 V.
9. Turn off the power. Empty the upper reservoir. Remove the solutions from the surface of the gels and load the sample. Overlay the sample with 10 µl of run overlay solution then cathode buffer. Refill the upper chamber with cathode buffer and run at 400 volts for 12 h then at 800 volts for 1 h.
10. To remove the gel from the tube, insert a long, thin blunt needle (e.g., 2.5 in, 29 gauge) fitted to a syringe filled with water into the bottom of the tube between the glass and gel. While slowly rotating the tube, inject water between the gel and the tube wall. With siliconized glass tubes, the gel will usually slide out. If the gel does not detach itself, perform this procedure at the other end of the tube. If required, attach a 5 ml syringe to the tube via a short piece of tubing. Putting pressure on the syringe, slowly force the gel out of the tube. Perform this over a tray, not over the sink.

PROTOCOL 4.11 Equilibration of the First-Dimension Gel or Strip

Prior to the second dimension, the gel slice, tube gel or IPG strip is equilibrated in a solution to make it compatible with the second dimension separation (SDS-PAGE).

Materials

Equilibration buffer I: Tris-HCl 50mM pH 8.8, 6M urea, 30% glycerol (w/v), 2% SDS (w/v), a trace of bromophenol blue, 1% DTT (w/v)

Equilibration buffer II: Tris-HCl 50mM pH 8.8, 6M urea, 30% glycerol (w/v), 2% SDS (w/v), a trace of bromophenol blue, 4% iodoacetamide (w/v)

1. Incubate the first dimension gel or strip in 10mL of equilibration buffer I for 15min with gentle shaking.
2. Incubate the first dimension gel or IPG strip in 10mL of equilibration buffer II buffer I for 15min with gentle shaking.
3. After equilibration remove the excess liquid and load the strip or gel onto the SDS-PAGE gel (Protocol 4.14).

Comments: Either prior to equilibration or following equilibration, the gel may be stored frozen at –70°C, preferably in a horizontal position. IPG strips not used immediately for the second dimension can be stored between two sheets of plastic film at –70°C for several months. Whenever possible, use commercially available IPG strips or precast gels for ease, convenience and reproducibility.

Flaws with 2-D Analysis

2-D electrophoresis has certain weaknesses that should be kept in mind. General 2-D electrophoresis does not provide an accurate representation of the proteome. Certain classes of "extreme" proteins, those that are very acidic or basic, or very large or small, and poorly soluble membrane proteins may be absent or under-represented (Molloy, 2000). Proteins present in low abundance can be obscured by uninteresting, high-abundance proteins.

Significant protein loss can occur at the focusing step. Hydrophobic proteins are prone to precipitation at their isoelectric point in the first dimension, thereby preventing any further migration in the second dimension. A certain amount of protein is lost during the equilibration step and subsequent transfer from the first to the second dimension. Protein remains in the IPG strip due to insufficient equilibration times and due to wash-off effects. The proteins lost due to wash-off must be primarily located on the surface of the IPG strip, which might have caused a background smear in the second dimension anyway.

2-D Troubleshooting (Görg et al, 2000; www.weihenstephan.de/blm/deg)

Problem: No spots or fewer than expected spots

1. Not enough sample loaded
2. Insufficient sample entered the IPG strip
3. Detection method not sensitive enough
4. Failure of detection reagents

Problem: Streaking and smearing on 2-D gel: Horizontal streaks

1. Sample preparation problems
2. Overloaded sample
3. Nucleic acids bound to the protein
4. Focusing time not optimized

Problem: Vertical streaks

Vertical streaks are related to the second dimension and may be due to various factors, including solubility of the protein at its pI, dust, improper reducing agent, or incorrect placement of the IPG strips.

1. Pinpoint vertical streaks in the background due to dust and DTT are remedied by the addition of iodoacetamide.
2. Broad streaks usually connected to a spot

Problem: Blank stripes in the vertical dimension

1. An air bubble may have been trapped in the agarose that joins the strip to the top of the second dimension gel
2. Blank stripes near pH 7 are often caused by excessive DTT (above 50 mM) in the IPG sample buffer
3. Blank stripes at the cathode can be caused by salt buildup.

Conductivity affects the focusing time with high conductivity, due to buffer salts in the sample or an inherently high conductivity from the sample itself, leading to long focusing times. The optimum salt concentration should be ~10 mM, although 40 mM can be tolerated. If the sample has a high ionic strength, salts can be removed using one of the micro spin columns that are commercially available like Micro Bio-Spin™6 or Micro Bio-Spin 30 for samples ≤75 µl. The simplest approach to focusing highly conducive samples is to be patient, giving the system sufficient time to reach high voltage (BioRadiations, 2002).

2-D gel imaging software is commercially available. For example: PDQuest™2-D Analysis software is available from Bio-Rad Laboratories, and Progenesis™ (www.nonlinear.com/2D/progenesis) is marketed by Nonlinear Dynamics (www.nonlinear.com). A more comprehensive list is presented by Chakravarti et al (2002).

PROTOCOL 4.12 Measuring the pH of the Gel Slices

pI values can be estimated by comparing the migration of target proteins to the migration of standard "landmark" proteins (Bjellqvist et al,

1994). A pI grid can be assembled using the pI values of the landmark proteins which will allow the assignment of pI values to unknown proteins. The pI value of the target protein is a valuable parameter, which can be used to exclude many proteins during database identification analysis.

Materials

Single edge razor blade

Test tubes: 13 × 100 mm

pH meter

1. Run an extra isoelectric focusing tube gel or a lane with no sample (see Protocol 4.20).
2. Following the run, remove the gel from the tube (Protocol 4.20 step 10) and cut the gel into 5 mm sections.
3. Place the pieces into numbered tubes containing 2 ml of water.
4. Shake the tubes 5–10 min and measure the pH.
5. Construct a standard curve of migration distance versus pH. See Figure 4.11. Measure the migration distance of the target protein and read off the pI from the standard curve.

TABLE 4.7 Commercially Available p*I* and Mol wt Standards for 2-D Gels

Protein	p*I*	Mol wt, kDa
Hen egg white conalbumin	6.0, 6.3, 6.6	76.0
Bovine serum albumin	5.4, 5.5, 5.6	66.2
Lentil lectin	8.15, 8.45, 8.65	50
Bovine muscle actin	5.0, 5.1	43.0
Rabbit muscle GAPDH	8.3, 8.5	36.0
Human carbonic anhydrase	6.55	35
Bovine carbonic anhydrase	5.85	30.5
Bovine trypsinogen	9.3	24
Soybean trypsin inhibitor	4.5	21.5
Equine myoglobin	7.0	17.5
Equine cytochrme c	10.25	12.4
Pepsinogen	2.8	

Comments: Alternatively, run standard proteins with known pIs (see Table 4.7). From the positions of these proteins in the gel, construct a pH gradient curve plotting migration distance versus pH. It is then possible to assign a pI value to the target protein by comparing the position of the target protein with the positions of the standard proteins on the curve. pI estimates for denatured, unmodified cytoplasmic proteins separated by 2-D PAGE can usually be estimated to ±0.25 pH units. Eventually, the predicted isoelectric points of identified proteins will function as internal calibration markers.

PROTOCOL 4.13 Nonequilibrium pH Gradient Electrophoresis (NEPHGE)

Nonequilibrium pH gradient electrophoresis (NEPHGE) can be used as an alternative to isoelectric focusing (IEF) in the first dimension to give a high-resolution 2-D procedure, NEPHGE-SDS. Most of the solutions for NEPHGE are the same as those used for IEF (O'Farrell et al, 1977). Advantages of NEPHGE over IEF are that basic as well as acidic proteins are resolved in a shorter time since equilibrium is not achieved. A major difference between IEF and NEPHGE is that in NEPHGE, the proteins migrate toward the cathode with the acidic reservoir on top and the basic reservoir on the bottom. Due to this alignment, the basic proteins lead in the separation.

If not mentioned to the contrary, all solutions and equipment are the same as those described above in Protocol 4.10.

Materials

Tube gel electrophoresis chamber

Power supply

Urea (ultra pure)

Acrylamide stock solution (30%): 28.4% (w/v) acrylamide, 1.6% (w/v) bisacrylamide

NP-40: 10% (w/v) stock

Ampholytes: pH range 5 to 7 and 3 to 10. Ampholytes are supplied as 40% w/v solutions and available from many commercial sources including Pierce, Pharmacia and Fluka.

APS: 10% (w/v): Appendix C

TEMED: Appendix C

Parafilm

Run overlay solution: 8M urea and a mixture of 0.8% pH 5–7 and 0.2% pH 3.5–10

0.02M NaOH

0.01M H_3PO_4

SDS-PAGE sample buffer (1×): see Appendix C

Deionized water

125ml side arm flask

1. Prepare the gel mixture in a 125ml side arm-flask, dissolve 5.5g of urea in 1.97ml of deionized water, 1.33ml of a stock solution, 2ml of NP-40 and 0.4ml of Ampholytes, pH range 5 to 7 and 0.1ml of pH 3 to 10 (to make 2% Ampholytes). These quantities can be scaled up as necessary. Pour the tube gels to a height of 12cm. Use Ampholytes pH 7–10 or 3.5–10 in the gel mixture depending upon the desired pH range.
2. Swirl the flask until the urea is completely dissolved then add 10µl of 10% ammonium persulfate (APS) and degas for ~1min. Add 7µl of TEMED and swirl the flask. If you are using tube gels, fill the tubes that have been sealed at the bottom with parafilm to ~5mm from the top. Gels containing basic Ampholytes tend to polymerize

poorly. Therefore, for gels containing pH 3.5–10 or pH 7–10 Ampholytes use 20 µl of ammonium persulfate and 14 µl of TEMED per 10 ml of gel mixture.

3. Overlay the gels with water and allow 1–2 h for polymerization.
4. Fill the lower reservoir with 0.02 M NaOH and place the tubes in the chamber.
5. Load the samples and add 20 µl of run overlay solution. Fill the tubes with 0.01 M H_3PO_4. Fill the upper reservoir with 0.01 M H_3PO_4. This is the reverse of that used for IEF.
6. Electrophorese for 4–5 h at 400 V without a prerun.
7. At the end of the run, remove the gels from the tubes (Protocol 4.20 step 12) and soak for 2 h in 1× SDS sample buffer and either store frozen or run in the second dimension (see below).

Comments: The proteins in the focusing gel or strip can be carboxymethylated following reduction by incubating the focused gel in equilibration buffers as described above in Protocol 4.11.

PROTOCOL 4.14 2-D Gels—The Second Dimension: SDS-PAGE

This procedure can be used for the second dimension when the first dimension has been IEF (see Protocol 4.10 or 4.12) or NEPHGE.

Materials

Agarose solution: 0.5% agarose in running buffer (25 mM Tris, 192 mM glycine, 0.1% SDS and a trace amount of Bromophenol blue). Heat in a microwave oven to dissolve.

Parafilm

n-butanol

1× SDS-running buffer: see Appendix C

1. Prepare a separating gel (Protocol 4.2). Modify the stacking gel procedure to accommodate a tube gel, strip of gel or IPG. In place of a multi-well comb, pour the stacking gel mixture to nearly fill the space between the plates leaving a 1–2 mm space at the top.
2. Overlay the stacking gel with water-saturated n-butanol and allow it to polymerize for 30–60 min. Remove all unpolymerized stacking gel solution from the surface and rinse with water.
3. Working quickly, layer 1–2 mm of the agarose solution (~75°C) onto the stacking gel. Place the tube gel on a piece of parafilm. Insert the gel between the glass plates with a spatula and bring the strip into contact with the upper edge of the SDS gel. Use the warmed solution of agarose to make sure that there are no air bubbles trapped between the tube gel and the top of the stack. This is a tricky step. Your technique will improve with practice.
4. Allow five minutes for the agarose to solidify. Remove the bottom spacer, and mount the gel cassette onto the electrophoresis apparatus. Add SDS running buffer. Remove any bubbles from the bottom of the plates and

run at 20mA constant current until the dye front reaches the bottom of the gel.

5. You are now ready to analyze the separated proteins according to a procedure of your choice (e.g., autoradiography (Protocol 7.20), microsequencing (chapter 8).

Comments: The slab gel can be either uniform concentration or a gradient. Make sure that the diameter of the tube gel is small enough to allow for easy insertion between the glass plates in the second dimension.

Second dimension gels for 2-D electrophoresis are commercially available.

Fluorescence Two-Dimensional Difference Gel Electrophoresis (2-D DIGE)

A recent innovation in 2-D gel technology is the use of differential in-gel electrophoresis (DIGE), in which two pools of proteins are labeled with different fluorescent dyes (Patton, 2002). Proteins are labeled with spectrally distinct fluorescent dyes like Cyanine-2, (Cy™2), Cy3 or Cy5 prior to electrophoresis. The labeled samples are then mixed and separated within the same 2-D gel and the differently colored labeled proteins are viewed individually by scanning the gel at different wavelengths.

The isoelectric points of the labeled proteins do not vary significantly from their unlabeled counterparts. The minimally labeled proteins can be in-gel digested and analyzed by MS (Lilley et al, 2002). 2-D DIGE was applied to quantify the differences in protein expression between esophageal carcinoma cells and normal epithelial cells (Zhou et al, 2001). The technique is commercially available as the Ettan™ DIGE system from Amersham Biosciences.

PROTOCOL 4.15 Labeling Proteins with Cyanine Dyes (Cy3 and Cy5)

Fluorescent dyes are conjugated to proteins through N-hydroxysuccinimidyl linkages such that ~4–12% of each protein is labeled (Leimgruber et al, 2002).

Materials

Cy3
Cy5
DMSO
Termination buffer: 20mM Tris pH 7.5, 500mm NaCl
Centricon 3 filter assembly (Millipore)

1. Dissolve 1mg of Cy3 or Cy5 in 100µl of DMSO and bring to 1ml with PBS.

2. Prepare the protein solution using ~1 mg of total soluble protein in a solution containing 100 μl of 1 M sodium carbonate pH 9.3, 100 μl of PBS-diluted labeling stock and water to a final volume of 900 μl.
3. Add the protein solution to the tube containing the dye. Seal the reaction vial, protect from light, and agitate at room temperature for 1 h.
4. Terminate the labeling reaction by adding 200 μl of termination buffer and incubate at room temp for 1 h.
5. To separate the labeled protein from the excess unconjugated dye, dilute with equal volume deionized water and concentrate to a final volume of 35 μl using Centricon 3 filter assemblies.
6. To recover the sample, wash the Centricon with 65 μl of sample solubilizing solution.

Comments: Buffers containing primary amines such as Tris and glycine will inhibit the conjugation reaction.

Higher protein concentrations (up to 10 mg/ml) usually increase labeling efficiency.

For a control, reverse the Cy3 Cy5 labeling of the protein extracts. This will rule out differences in the binding of the Cy dyes to specific proteins.

2-D PAGE Databases

2-D PAGE databases contain two components, image data and textual information on spots that have been identified. Image data consists of reference maps which are built from a single master 2-D gel and those which are assembled from different sections of the pH/mol wt range. In both cases the image used as a reference map is a scanned representation of the gel. Textual information generated from a spot that has been identified includes data on the apparent molecular weight, pI, the name of the protein, bibliographic references and cross-references to SWISS-PROT and other databases. The information can be accessed by clicking on a spot on a gel image or by selecting from a list of identified proteins or by keyword searches (Appel et al, 1996). http://www.expasy.ch/ch2d/

The format of SWISS-2DPAGE is similar to SWISS-PROT. Both databases share entry names (on the ID line) and primary accession numbers (on the AC line). Most proteins in SWISS-2DPAGE have been identified by microsequencing. If no entry exists for the target protein a virtual entry can be created. The theoretical pI and M_r for the target protein are computed and its hypothetical position is shown on a map.

PROTOCOL 4.16 Nonreducing-Reducing 2-D Gels

Disulfide bonds between cysteine residues occur infrequently among proteins that reside in a reductive environment such as the cell cytosol (Branden and Tooze, 1991) but are common in all major classes of extracellular proteins (Freedman and Hillson, 1980). Disulfide bond formation in secretory proteins occurs during or shortly after their translocation into the lumen of the endoplasmic reticulum (Freedman, 1984). Disulfide bonds can be interchain, between two separate polypeptide chains, or intrachain, between two cysteine residues on the same chain. Many proteins are composed of subunits associated through interchain disulfide bonds between cysteine residues. When this bond is intact, the protein migrates in an electric field as one band. When the bond is broken by a reducing agent such as 2-mercaptoethanol (2-ME) or dithiothreitol (DTT), proteins that are composed of subunits are dissociated, and each subunit migrates as a unique band in SDS-PAGE.

As a protective agent for keeping molecules in the reduced (—SH) state, between 5 and 20 mM 2-ME is used. Lower concentrations quickly oxidize to the disulfide. The protective ability is lost within 24 h unless the solution is kept in anaerobic conditions. Disulfide formation is accelerated by divalent cations. Therefore a chelating agent such as EDTA at low concentrations 0.1–0.2 mM should be routinely used along with 2-ME or DTT except in situations when the presence of the chelating agent would be detrimental.

2-ME has been associated with artifactual bands that are especially noticeable following silver staining. These potential artifacts can be avoided by using DTT at 25–50 mM, which is also less odorous than 2-ME.

Materials

Slab gel electrophoresis system (see Protocols 4.1–4.5)

Equilibration buffer: 2% SDS, 125 mM Tris-HCl, pH 6.8

2-mercaptoethanol (2-ME): see Appendix C

Equilibration buffer + 2% 2-ME

Overlay solution: 1% agarose in 1× sample buffer (Both a reducing and nonreducing overlay are required)

15 ml plastic conical centrifuge tubes

1. Run two identical samples under nonreducing conditions in a slab gel format and cut out the individual lanes.
2. For the nonreduced control, place one lane in a 15 ml plastic conical centrifuge tube containing nonreducing equilibration buffer and soak for 2 h at room temperature.
3. Place the second lane in a tube containing equilibration buffer plus 2% 2-ME and soak for 2 h at room temperature. This treatment effectively reduces the S-S bonds.
4. For the second dimension, you will need one slab gel for each treated lane. Prior to loading the second dimension gels, boil the first dimension treated lanes for 10 min.
5. Cast a separating gel as described in Protocol 4.2, leaving space, ~8 mm for the equilibrated lane from the first dimension. A stacking gel is not necessary.
6. Add a 1–2 mm layer of hot overlay solution onto the top of the separating gel and quickly insert the lane from the first dimension into the glass plate assembly.

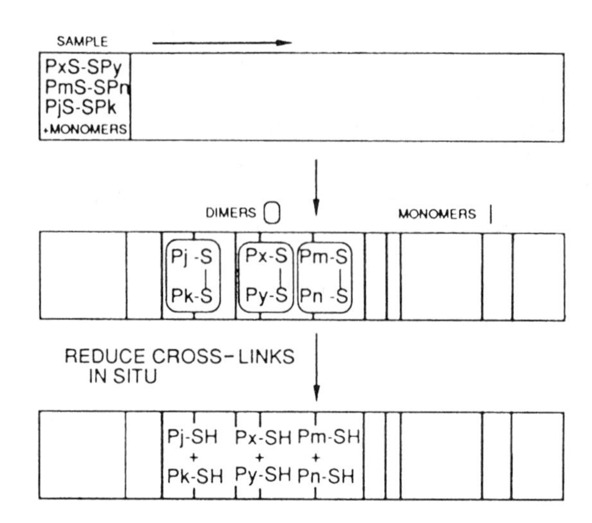

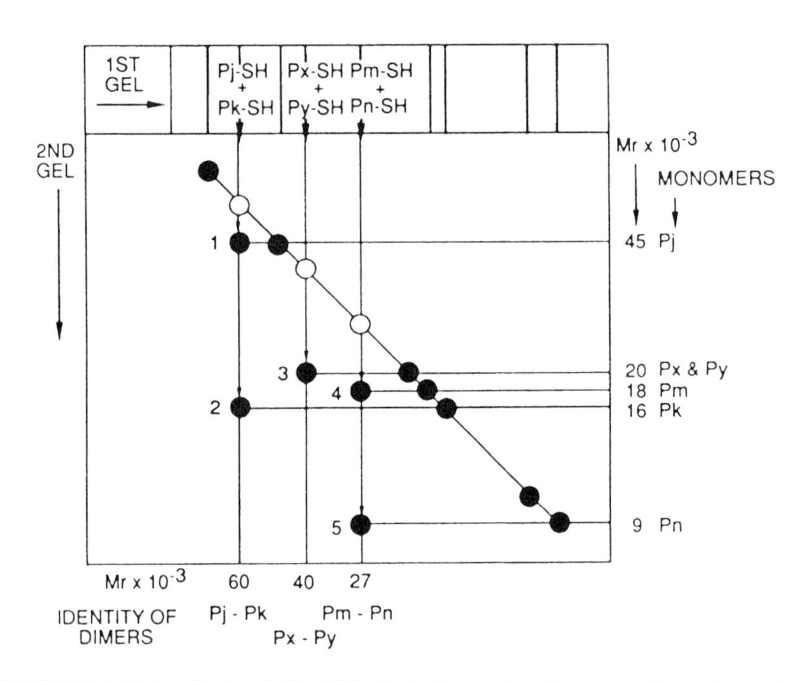

FIGURE 4.10 Analysis of disulfide-linked proteins in a complex mixture by diagonal SDS-gel electrophoresis. The first dimension is run under non-reducing conditions. Upon completion, the first dimension gel is subjected to reduction and run in the second dimension under reducing conditions. The first gel becomes the origin of the second slab gel. Protein subunits that had been disulfide-linked in the first dimension migrate more rapidly as monomers in the second electrophoresis. The shaded circles on the diagonal, representing proteins without disulfide bonds, retain the same relative mobility in both electrophoresis steps and make up the diagonal. The open circles on the diagonal represent the position of disulfide-containing proteins prior to reduction. The black circles below the diagonal represent components of the open circle, which were held together by a disulfide bond. Reprinted by permission of Oxford University Press, from Trant et al, 1989.

7. Perform electrophoresis on the reduced lane in the presence of 0.02% 2-ME in the upper buffer reservoir.

Comments: After electrophoresis, analyze the gels according to individual experimental specifications. Gels may be stained with Coomassie Blue (see Protocol 4.15), transferred to an immobilizing matrix (see Protocol 7.1), or dried and autoradiography performed (see Protocol 7.20). For comparison purposes, the control sample, run under non-reducing conditions in both dimensions, should appear as a diagonal array of spots or a solid diagonal line of varying intensity. A representative result is shown in Figure 4.10. If the sample contained subunits that were held together by disulfide bonds, analysis should reveal spots that have left the diagonal, usually appearing below the diagonal. Diagonal gels are also useful when analyzing proteins cross-linked with an agent sensitive to DTT and 2-ME, described in Protocols 3.18 and 3.19.

Another example of a 2-D format is the 2-D blue-native PAGE (2-D BN-PAGE) system. In this technique, membrane and other functionally intact protein complexes are separated as intact enzymatically active complexes in a first dimension gel. The protein complexes are then subjected to a second dimension denaturing SDS-PAGE gel, separating the complexes into their component subunits. Originally described by Schagger and von Jagow (1991), the method has been refined and is applicable to functional proteomic analysis of protein complexes such as mitochondria or complexes involved in signal transduction (Brookes et al, 2002). 2-D BN-PAGE may prove useful in studying complex cell signaling pathways that involve multiple protein: protein interactions.

C. Detection of Protein Bands in Polyacrylamide Gels

Once the protein bands have been separated on the gel, they can be visualized using various convenient and sensitive methods of detection. If the protein has been radiolabeled prior to the electrophoretic separation, the radiolabeled protein bands can be visualized by pressing the gel against a sheet of X-ray film. As the radioisotope degrades, it produces silver atoms within the silver halide crystals in the film. The bands will appear once the film is developed. This is known as autoradiography. Alternatively, radiolabeled proteins can be visualized by phosphorimaging.

Visualization of the proteins that have not been radiolabeled involves the use of one or more of the many staining procedures that have been developed. There is no single detection method that will identify all proteins quantitatively, independently of protein detection limits of the methods. The choice of staining method is determined by several factors including, desired sensitivity, linear range, ease of use, and the type of imaging equipment available. At present there is no universal detection system. The choice of stain depends on the particular proteins in the gel, their concentrations, and the subsequent uses intended for that protein. When choosing a staining method, an important consideration should be that the staining method does not compromise the chemical integrity and the biological properties of the target proteins if they are to be used following staining.

The staining procedure should combine most, if not all of the following desirable features: (i) simplicity (2 or 3 steps), (ii) speed, (iii) high sensitivity (1–10ng/band), (iv) reproducibility, (v) low cost, (vi) it is non-toxic, (vii) compatibility with subsequent analyses (e.g., sequencing, mass spectrometry, bioassays, immunoblotting). In some cases it may be desirable that the chosen method be reversible.

A variety of staining methods are available for the non-radioactive localization of the separated bands. They may be based on the selective binding of organic dyes (e.g., Coomassie blue dye), non-covalent binding of fluorescent groups (e.g., Sypro dyes, followed by irradiation of the gel with UV light), reduction of silver salts to metallic silver, and selective precipitation of metal cations on the gel matrix. The two most commonly used protein visualization methods are Coomassie blue for protein amounts greater than 10ng per spot and silver staining, with a detection threshold of 0.05ng of protein (Dainese and James, 2001).

Coomassie Brilliant Blue R-250 (CBB) is the most widely used stain for detecting proteins in acrylamide gels. CBB staining is routinely used for staining protein bands following PAGE, and isoelectric focusing. Coomassie blue R-250 is used when proteins are relatively abundant, and the detection of trace/minor proteins is not required. Detection levels are fairly sensitive as a single band of 10ng can be detected. Silver staining is sensitive and can detect sub-nanogram quantities of protein. Heavy metal staining is quick and although not as sensitive as silver, allows quantitative recovery of proteins from the gel for further analysis.

PROTOCOL 4.17 Staining and Destaining the Gel with Coomassie Blue

Two methods are presented. The first method is much more labor intensive than the second, newer, kit-based method.

Materials

Containers: either glass or plastic

Razor blade

Shaker or Rocker

Coomassie Brilliant blue R-250 (CBB) (0.25% w/v) contents per liter: 2.50g Coomassie Brilliant blue R-250, 100ml concentrated acetic acid, 400ml methanol (top off to 1L with deionized water). The stain solution should be filtered through Whatman #1 filter paper before use.

Destain solution: 400 ml methanol, 100 ml concentrated acetic acid (top off to 1 L with deionized water). There is also a Coomassie Blue G. Do not use this compound in this stain recipe.

1. When the electrophoretic run is completed, turn off the power and remove the glass plate sandwich from the gel apparatus. Carefully pry open the glass plates. With a razor blade, cut off the stacking gel and discard.
2. Place the separating gel in a container and add enough stain solution to cover the gel.
3. Gently rock the container for 20–30 min. Gels < 1 mm thick and low percentage acrylamide gels are very delicate. They can be moved using a glass plate or wide spatula for support.
4. Pour off the stain solution and save. The stain solution can also be aspirated off delicate gels.
5. Immerse the gel in destain solution at room temperature with gentle shaking. Destaining usually requires several changes of destaining solution and can take several hours to overnight.

Comments: SDS-PAGE gels are routinely stained for 30 min in a solution of 0.25% CBB in 40% methanol, 10% acetic acid. Fixation and staining occur simultaneously because the acetate-methanol solution is adequate to fix most proteins. Gels are destained in a large excess of 40% methanol, 10% acetic acid until the background is satisfactorily clear, at which time they can be photographed.

To speed up the destaining process, submerge a small piece of sponge that will partially absorb the stain. From time to time the destaining solution can be changed. If your lab is on a very tight budget, the destain solution can be recycled by filtering it through activated charcoal in Whatman #1 paper.

Although the stain can be reused, SDS interferes with the staining process, and, if reused too often, the staining will lose intensity. If you are not satisfied with the staining, the gel may be restained with fresh CBB.

A large quantity of one protein or multiple proteins migrating together may result in a distorted band. To increase the resolution in this area, decrease the amount of sample being loaded and run the gel at a lower voltage.

Sensitivities can approach 40 ng/band in a 10-well mini-gel although 0.5 μg in one band is a more typical protein load for easy visualization (Merril, 1990).

For black and white photography of Coomassie stained gels, contrast can be enhanced with a yellow filter. An image of the destained gel can be obtained using a standard office copy machine. Alternatively, the gel can also be scanned.

Staining at the interface between the stacking and separating gels may occur due to high molecular weight aggregates. To dissolve these complexes and resolve this material, try to: (1) Incubate the protein solution in sample buffer at 55–60°C for 10 min, or 37°C for 1 h (instead of boiling); (2) or, prior to boiling the sample, add solid, ultra-pure urea to make the sample ~6 M in urea.

Coomassie blue may precipitate on the surface of the gel, obscuring faint bands and making the gel esthetically unappealing. Submerge a wad of cotton in destaining solution and carefully wipe the gel surface. Alternatively, Lewis (1996) suggested a 1 minute rinse in 100% methanol with gentle shaking followed by a return to destain solution or water.

PROTOCOL 4.18 Coomassie Staining Using GelCode® Blue

GelCode® blue reagent (Pierce) utilizes the colloidal properties of Coomassie G-250 for staining proteins in polyacrylamide gels. This reagent allows bands to be viewed during the staining process. Use gels that are ≤1 mm thick. GelCode blue stained gels can be used for mass spectrometric analysis (Lim et al, 2002).

Materials

GelCode blue reagent (Pierce)

1. Following electrophoresis, submerge the gel in either a plastic or glass container containing deionized water.
2. Wash the gel 3 × 5 min with 100–200 ml of deionized water. Pour off the last water wash. Mix the GelCode blue reagent prior to use by gently inverting the bottle and pour it into the container making sure that the gel is completely covered. About 20 ml will cover an 8 × 10 cm mini-gel.
3. Stain the gel for 1 h with gentle agitation.
4. Pour off the stain and add ultrapure water to destain the gel. Several water changes may be necessary for optimal results.

Comments: Gels may be left in GelCode blue overnight without increasing the background.

Gels electrophoresed with MOPS or MES running buffers should be prefixed with a solution of 50% methanol, 7% acetic acid for 15 min then washed with water to remove the fixing solution.

Membranes can also be stained with GelCode blue. Wash the blot with water for 1–2 min. Mix the GelCode blue reagent, add it to the blot and incubate 2–5 min on an orbital shaker. Destain the blot with a solution of 50% methanol, 1% acetic acid 2 × 5 min.

Following either brief or extensive destaining, Coomassie stained gels can be restained by the zinc-imidazole method (see Protocol 4.21), often exposing previously undetected protein bands.

PROTOCOL 4.19 Staining Gels with SYPRO® Ruby

SYPRO Ruby is a ready-to-use fluorescent stain for detecting proteins separated in SDS-PAGE gels. It can stain proteins in a variety of gel formats as well as proteins on nitrocellulose and PVDF membranes (Bio-Rad SYPRO Ruby Protein Stains Instruction Manual).

SYPRO Ruby dye is an optimized ruthenium-based metal chelate stain that allows one step, low background staining of proteins in gels without the need for lengthly destaining. The linear range of this dye extends over three orders of magnitude, surpassing silver and Coomassie blue stains in performance. Some proteins that stain poorly with silver stain techniques can often be detected with SYPRO Ruby dye (Steinberg et al, 2000).

SYPRO Ruby is compatible with commonly used microchemical characterization protocols and is ideally suited for use in the identification of proteins by peptide mass profiling using MALDI-TOF mass spectrometry. This stain can be easily incorporated into integrated proteomics platforms that combine automated gel stainers, image analysis work stations, robotic spot excision instruments, protein digestion workstations, and mass spectrometers (Patton, 2000).

Materials

Polypropylene containers for staining (Rubbermaid)
SYPRO Ruby protein gel stain (Bio-Rad)

1. Remove the gel from the cassette or plate ensemble.
2. Place the gel in a clean plastic container. For 1-D gels no fixation step is required.
3. Cover the gel with SYPRO Ruby protein stain. In general, use ~10 times the volume of the gel. Using too little stain reduces sensitivity. Stain the gel for at least 3 hrs for maximum sensitivity. For convenience, gels may be left in stain solution overnight (16–18 hrs) without overstaining.
4. To decrease background fluorescence, rinse the gel in 10% methanol (or ethanol), 7% acetic acid for 30–60 min.
5. Before imaging, wash the gel in deionized water.

Comments: Do not use glass containers as they tend to absorb dye from the staining solution.

Perform all staining and washing steps with continuous gentle agitation.

For 2-D gels: incubate the gel in 10% methanol, 7% acetic acid for 30–60 min prior to staining.

For IEF gels: fix the gel in 40% methanol, 10% trichloroacetic acid for 3 hrs, then wash the gel in water ×3 for 10 min each.

SYPRO Ruby can stain electroblotted proteins on nitrocellulose and PVDF membranes with a sensitivity of 0.25–1.0 ng protein/mm^2 in a slot blot. This translates to between 2 and 8 ng protein per band in routine electroblotting applications.

Viewing and Imaging a SYPRO Ruby-Stained 1-D or 2-D Gel

SYPRO Ruby stained proteins in gels are visualized using a UV or blue light source. The use of a photographic camera or a CCD camera with the appropriate filters (yellow Wratten 9) is essential to obtain high sensitivity. The Bio-Rad Gel Doc™ instruments with CCD cameras and a UV transilluminator work well. For laser based imaging systems check the manual and consult the manufacturer for specific settings. A more comprehensive discussion is given by Kemper et al (2001).

PROTOCOL 4.20 Silver Staining

Silver staining was thought to be incompatible with further analysis due to the use of Ag as an oxidizing agent and glutaraldehyde, a cross-linking agent that blocks free amino groups, as a contrast enhancer (Shevchenko et al, 1996). By altering the procedure and eliminating glutaraldehyde, silver stained proteins could be analyzed by MS, and little chemical modification takes place (Dainese and James, 2001).

The silver staining method presented below utilizes two properties of thiosulfate: image enhancement by pretreatment of fixed gels, and prevention of nonspecific background during development (Blum et al, 1987).

Steps 2–10 should be carried out on a shaker at room temperature (20–25°C). Use a suction system to rapidly remove solutions from the gel.

Materials

Plastic containers

Razor blade

Gel fix solution: 50% methanol, 12% acetic acid

Sodium thiosulfate ($Na_2S_2O_3{\cdot}5H_2O$): 0.2 g/L

Silver nitrate ($AgNO_3$): 2 g/L, 0.75 ml/L of 37% formaldehyde stock solution

Developing solution: Na_2CO_3, 60 g/L, 0.5 ml/L of 37% formaldehyde stock, 4 mg/L $Na_2S_2O_3{\cdot}5H_2O$

Ethanol solutions: 50% and 30%

Methanol: 50%

Glycerol: 3% solution

Shaker or Rocker

1. When the electrophoresis run is complete, turn off the power, remove the glass plate sandwich from the gel apparatus, pry open the glass plates and with a razor blade, cut off the stacking gel and discard it.
2. Submerge the separating gel in a large volume of gel fix solution. The gel can be stored in this solution for several days with no effect on the quality of the staining.
3. Wash the gel in 50% ethanol three times for 20 min each. Gels that are <1 mm thick should be kept in 30% ethanol for the final 20 min wash to prevent gel deformation.
4. Submerge the gel in the solution of sodium thiosulfate for exactly 1 min. Longer submersion times will result in high background staining.
5. Rinse the gel with water three times for 20 sec each.

6. Submerge the gel for 20min in the silver nitrate solution. After this step the gel may appear yellowish. This will not have an effect on sensitivity or background staining.
7. Rinse the gel with water two times for 20sec each. This step removes excess $AgNO_3$ that would otherwise react with thiosulfate added in the next step.
8. Visualize the bands by incubating the gel at room temperature for 10min in developing solution.

The reaction temperature should be between 18° and 25°C. A faint brown precipitate may appear when the developing solution is added. Vigorous shaking will dissolve it completely. After a few minutes, as the free thiosulfate is consumed, the solution may turn slightly brown. In this case replace it immediately. The intensity of the bands will increase continuously. The process can be interrupted any time as soon as the desired intensity and contrast is achieved. A band of 500ng protein will appear after ~30sec, a band of 50ng after ~2min. If the development is allowed to extend longer than 10min, a yellowish background may appear.

9. When you are ready to terminate the developing process, rinse the gel with water twice for 2min each. This removes excess thiosulfate that may decompose in the gel fix solution and cause background staining.
10. Soak the gel for 10min in gel fix solution.
11. Wash the gel in 50% methanol for at least 20min. After this step store the gel at 4°C in 50% methanol.

Comments: The indicated times should be observed exactly in order to ensure reproducible image development.

Before drying (see Protocol 4.7) the silver-stained gels, place the gels in a shaker at 4°C for 30min in 30% methanol followed by 3% glycerol for another 30min.

PROTOCOL 4.21 Reversible Negative Staining of Proteins in Gels with Imidazole and Zinc Salts

A procedure based on the selective precipitation of certain Cu(II) compounds in the gel matrix was introduced by Lee et al (1987). This was extended one year later when it was demonstrated that reversible negative staining of SDS-PAGE separated proteins could be performed with other water soluble heavy metals salts, including zinc (Dzandu et al, 1988). After treatment with heavy metal salts the gels are negatively stained. Treating gels with $ZnSO_4$ results in a semi-opaque background without any coloration. Protein bands appear transparent against a deep white stained background. The limits of detection are in the femtomole range.

Negative staining has several advantages over Coomassie Blue and silver staining; The $ZnSO_4$ method allows quantitative elution and recovery of resolved proteins from the polyacrylamide gel matrix. In addition, gels are stained completely within 30 min, do not require destaining, and can be stored indefinitely without loss of image (Castellanos-Serra and Hardy, 2001).

Heavy metal staining is completely reversible, and the gels can be restained by other methods. A major advantage of the heavy metal staining method is that proteins can be eluted from the gel for further characterization and use. Proteins from reverse-stained gels can be manipulated for analysis as they are in unstained gels.

One disadvantage to heavy metal staining is that it is not easily amenable to quantification using conventional methods such as densitometry. Therefore, quantitative comparison of staining is difficult, and another method should be used to obtain this information.

Materials

Containers: plastic or glass

Imidazole solution: 0.2 M imidazole, 0.1% SDS solution

$ZnSO_4$: 0.2 M solution

Shaker or rocker

1. After electrophoresis, pry open the glass plates, and with a razor blade, cut off the stacking gel and discard it. Rinse the gel for 30–60 sec in distilled water.
2. Immerse the gel in a 0.2 M imidazole, 0.1% SDS solution and shake for 15 min at room temperature.
3. Pour off the imidazole solution and replace it with 0.2 M $ZnSO_4$ for 30 sec. During incubation, monitor the gel for the appearance of negatively stained protein bands against a semi-opaque to white background. Protein patterns seen with $ZnSO_4$ appear within 30 sec. The protein bands will be transparent.
4. Quickly remove the residual salt solution and wash the gel three times for 2–3 min each in water to remove excess reagent. Store the gel in distilled water.

Comments: The zinc staining protocol can also be applied to nondenaturing gels containing no SDS. In these gels the staining is reversed. Protein bands are now stained white against a clear background. The sensitivity of the stain on nondenaturing gels is similar to that of CBB.

Negative staining produces an intermediate sensitivity between that of Coomassie and silver stain.

Following zinc-imidazole staining, the proteins are reversibly fixed in the gel. Fixation should be fully reversed before starting a procedure on the proteins that requires them to be removed from the gel or accessed by enzymes or chemical reagents while in the gel. This can be readily accomplished by incubating the entire gel or the excised band in 0.2 M glycine, 25 mM Tris pH 8.3 for 10 min. The proteins will then be ready for transfer to membranes, in-gel treatments and mass spectrometry (Fernandez-Patron et al, 1995).

PROTOCOL 4.22 Molecular Weight Determination by SDS-PAGE

The mass of a protein is one of the most commonly used attributes in protein identification procedures. Protein mass is determined by the total mass of all the amino acids which constitute the protein, plus the mass of any posttranslational modifications.

The molecular mass (M_r) estimates of proteins separated by SDS-PAGE have been derived by comparing the electrophoretic mobility of the target protein to the mobilities of protein standards of known masses. This method is generally considered accurate within about ±10%. The variability could be explained by: (i) variabilities in binding of SDS to different proteins due to clusters of acidic or basic amino acids, (ii) incomplete unfolding of the protein in the presence of the denaturant. Partial retention of secondary structure in the presence of SDS has been shown to alter electrophoretic mobility (Dianoux, 1992). Molecular weights are best determined when both the band of interest and the standards appear as sharp, narrow bands so that there is no mistake as to where to measure the migration distance. Maximal resolution is obtained with low protein loads. Wide bands present many positions from which to measure (i.e., the top, middle, or bottom of the band). On minigels, 1–2mm may represent several thousand daltons.

Materials

Slab gel, either uniform concentration or linear gradient (see Protocols 4.5 and 4.6)

Molecular weight protein standards (Commercially available, see Table 4.8)

Ruler

Semi-log graph paper

1. In one lane of the gel, run protein standards of known molecular weights and the polypeptide to be characterized in an adjacent lane.
2. Pry open the glass plates and with a razor blade, cut off the stacking gel and discard.
3. Stain and destain the gel. If necessary, perform autoradiography to identify the protein band of interest (see Protocol 7.20).
4. Measure the migration distances of the molecular weight standards, bromophenol blue, and the target protein from the top of the separating gel.
5. Calculate the R_f values. (R_f is the distance that a band has traveled divided by the distance the dye front has traveled.) The R_f value of the tracking dye will always be 1.0. A linear relationship exists between the log of the molecular weight of a polypeptide and its R_f value.
6. To generate a standard curve, plot the R_f values of the protein standards versus the $\log_{10}$ of their molecular weights.
7. Calculate the R_f of the target protein. The $\log_{10}$ of its molecular weight is obtained directly from the standard curve; the antilog is the molecular weight. A standard curve to illustrate these points is presented in Figure 4.11.

TABLE 4.8 Protein Molecular Mass Standards on SDS-PAGE

Polypeptide	Molecular Mass (kDa)
Myosin heavy chain	212
Myosin, rabbit muscle	205
α 2-Macroglobulin	180
β-galactosidase (E. coli)	116
Phosphorylase b	97.4
Phosphorylase a (rabbit muscle)	92.5
Fructose-6-phosphate kinase (rabbit muscle)	84
Transferrin	76
Bovine serum albumin	66–68
Catalase	57.5
Glutamic dehydrogenase	53
Fumarase (porcine heart)	48.5
Ovalbumin	43
GAPDH (rabbit muscle)	36
Aldolase (rabbit muscle)	36.5
Carbonic anhydrase	29–31
Trypsinogen (bovine pancreas)	24
Soybean trypsin inhibitor	20.1
Myoglobin	17
α-lactalbumin (bovine milk)	14.4
Lysozyme (egg white)	14.3
Cytochrome c	11.7
Aprotinin	6.5
Insulin β-chain	3.5
Bacitracin	1.45

Comments: In some cases, mass estimations for posttranslationally modified proteins from gels can give an apparent mass 50% greater than that predicted from the database (Gooley et al, 1997).

A standard curve can be extrapolated to the y-intercept, the point that represents the molecular weight exclusion limit of that particular gel.

Some polypeptides commonly used as protein molecular weight markers are listed in Table 4.8. Many of them are commercially avail-

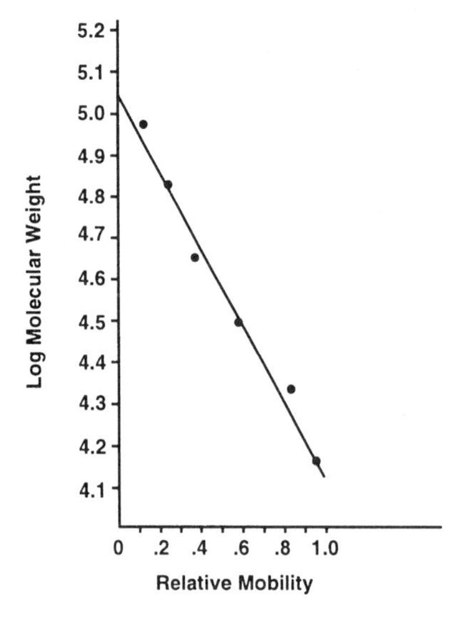

FIGURE 4.11 Relative electrophoretic mobility vs log molecular weight. Protein standards were run on a 12% polyacrylamide gel (12% T 2.6% C). Relative mobilities of proteins were determined relative to the mobility of the tracking dye, bromophenol blue (relative mobility, 1.0). The y-intercept was 5.06, indicating that the theoretical gel exclusion limit is 115KDa. (Figure kindly provided by Bio-Rad Laboratories, Life Science Group, Hercules, CA).

able separately or as components of kits. Some suppliers (Amersham, Rainbow markers and BioRad, Kaleidoscope standards) market prestained, colored molecular weight markers that allow you to monitor the progress of electrophoretic separations during the run. These markers also transfer to nitrocellulose and PVDF membranes simplifying alignment. The molecular weights of certain standards may vary slightly from supplier to supplier.

D. Recovery of Proteins from the Gel

After proteins have been separated by SDS-PAGE, it is frequently necessary to extract, or elute them from the gel for further study which may include amino acid composition analysis, sequence determination, and partial enzymatic or chemical cleavage.

For optimal efficiency of elution, it is desirable to have a simple technique that quantitatively extracts the protein from the gel and avoids any chemical modification to it. Many techniques have been described for eluting proteins from polyacrylamide gels (Hager and Burgess, 1980). The simplest method is elution by diffusion, described below. A second, slightly more efficient method, which does not require special equipment, is to place the mashed gel pieces in a dialysis bag with a molecular weight cut-off suitable for retaining the protein of interest and applying an electric current.

PROTOCOL 4.23 Excising the Protein Band from the Dried Gel

Radiolabeled proteins present in a dried gel can be identified and further analyzed by first excising the protein band and then extracting the protein. When performing further analysis, do not treat the gel with fluors to enhance the fluorographic image as this may inhibit subsequent reactions.

Materials

New single edged razor blade
Light box
Dried gel (see Protocol 4.8)
X-ray film
X-ray film processor

1. Make alignment marks on the paper backing and expose the dried gel to X-ray film.
2. Develop the film and align it over the dried gel. If the band is easily identifiable, cut it out of the gel with a new single edge razor blade.

Alternatively, staple the film to the paper backing. Place this sandwich film-side down on a light box and trace the outline of the target band on the back of the paper. If the film is needed, remove the staples and save. Wearing gloves and using a new single-edge razor blade, put the dried gel on a hard, flat surface and excise the band.

PROTOCOL 4.24 Extracting the Target Protein from the Dried Gel

Materials

Vortex mixer (shaker)
Elution buffer: 50 mM ammonium bicarbonate, pH 7.3–7.6
Disposable tissue grinder (Kontes)
Screw cap 1.5 ml plastic microfuge tubes and a microfuge
2-mercaptoethanol (2-ME)
SDS: 10% (w/v): see Appendix C

1. Place the gel slice in a small amount of freshly prepared elution buffer for 3–5 min.
2. Peel off the paper backing and transfer the pieces of gel to a Kontes disposable tissue grinder tube and add 500 µl of elution buffer per band.
3. Incubate the gel slice for 5–10 min in elution buffer then grind it until the pieces are small enough to pass through a 200 µl pipette tip.
4. Transfer the gel suspension to a 1.5 ml screw-cap microfuge tube.
5. Rinse the tissue grinder and transfer the residual gel bits to the microfuge tube (total volume should be 1 ml).
6. Add 50 µl of 2-ME and 10 µl of 10% (w/v) SDS, vortex mix, and then boil for 3—5 min.
7. Incubate the tube at 37°C for at least 90 min with shaking. This step can extend overnight, in which case incubate at room temperature.
8. Centrifuge the tube in a microfuge at 12,000 × *g* for 5 min. Carefully, transfer the supernatant to a new tube and measure the recovered volume.
9. Add a volume of ammonium bicarbonate, 2-ME, and SDS so that the final combined volume of supernatant is 1.2 ml. Reextract the gel by boiling and reincubating at 37°C (see steps 6 and 7 above).
10. Clarify the total eluate by microfuge centrifugation for 10 min at maximum speed to pellet any debris and small pieces of gel. At this stage, the extracted protein can be stored frozen at –70°C.

E. Identification of Enzyme Activity in Polyacrylamide Gels

Identification of a bioactive protein following PAGE or IEF is often an effective method of identifying enzymes in crude preparations. Proteins have varied sensitivities to SDS treatment. For proteins that do not retain their enzymatic activity after treatment with SDS, it is sometimes possible to recover enzymatic activity by renaturing the protein. Renaturation can be accomplished by removing the SDS and denaturing agents, like urea and guanidine-HCl, thereby enabling the reformation of the native protein structure. However, there appears to be no universally effective procedure for enzyme renaturation.

Overlays that contain the necessary auxillary molecules for bioactivity can be brought into direct contact with slab gels. Alternatively, the proteins can be blotted to membranes which are then probed for a specific enzyme activity. McLellan and Ramshaw (1981) detected 17 different enzymes with their specific stains following blotting from a single polyacrylamide gel to DEAE-cellulose paper. Both general and specific approaches are presented in this section so that they may be successfully adapted to specific enzyme activities that may be encountered during a project.

General Considerations

Before separating the proteins and performing the assay to localize the activity, consider the properties of the enzyme, pH optimum, cofactor requirements, and inclusion of cations and cosubstrates. Determine the protein concentration of the extract (see chapter 5 section B). Run ~100 μg of a crude mixture or 1–10 μg of a purified protein per lane. Usually a compromise will have to be made between optimal protein resolution and enzyme activity. Remember, too little protein will make it difficult to detect the enzyme and overloading will result in smearing, tailing and band distortion.

One important consideration when assaying for a particular enzyme is to be sure that the product represents the desired enzymatic activity. Therefore, controls should be included to answer the following questions: (1) Is the reaction product dependent on the substrate? (2) Do specific enzyme inhibitors abolish the reaction? (3) Does the denatured, inactivated enzyme react in the reaction process? (4) Does omission of an essential cofactor abolish enzyme activity?

A comprehensive review of staining for enzymatic activity after gel electrophoresis was reported by Gabriel and Gersten (1992) and is extensively referenced. These authors describe many different techniques used for a wide variety of enzymes.

PROTOCOL 4.25 Localization of Proteases: Copolymerization of Substrate in the Separating Gel

In this protocol the substrate is included as a component of the separating gel. The localization of proteolytic enzymes in gels is illustrated using the specific example of gelatin-degrading proteases (Fisher et al, 1989). Standard 10% T: 2.6% C acrylamide concentration is used that is copolymerized with gelatin.

Materials

Gelatin: Type A gelatin from porcine skin (Sigma)

SDS-Polyacrylamide gel (see Protocol 4.2) co-polymerized with 0.33 mg/ml gelatin in the separating gel

PBS (see Appendix C)

Sample buffer: 0.062 M Tris-HCl, pH 6.8, 2% SDS, 10% glycerol

Rinse buffer: 2.5% Triton X-100

Reaction buffer: 0.05 M Tris-HCl, pH 7.5, 0.2 M NaCl, 0.01 M $CaCl_2$, 1 μM. $ZnCl_2$, 1 mM Phenylmethyl sulfonyl fluoride (PMSF), 0.02% NaN_3 and 0.005% Brij 35

Stain solution: 0.25% Coomassie Blue R in 40% isopropanol

Destain solution: 7% acetic acid

1. Wash adherent cells three times with PBS and prepare whole cell extracts by scraping the cells in sample buffer (which does not contain reducing agents or bromophenol blue), 1 ml per 10 cm dish.
2. Centrifuge the extract at 1000 × *g* to remove insoluble material and store on ice for not longer than 1 h until loading on the substrate gels. Do not heat.
3. Assay the protein concentration of the cell extract using a method described in Chapter 5. Adjust the protein concentration to 1 μg/μl. Routinely, 10 μl is loaded per lane.
4. After electrophoresis, rinse the gel twice with at least 50 ml of rinse buffer for 20 min at room temperature, then incubate for 18 h at 37°C in reaction buffer.
5. Stain the gel for 1 h then destain. The gelatinases appear as clear bands on a blue background

Comments: This method is also referred to as zymography. It can be used to determine the effects of protease inhibitors on a target protease. Major classes of proteases and working concentrations of their inhibitors are listed below.

Protease	*Inhibitor*
Serine proteases	PMSF (PMSF is also added to the sample buffer for solubilizing the cells)
Metalloproteases	EDTA (10 mM) and 1,10-phenanthroline (0.3 mM)
Aspartic proteases	pepstatin A (1 mM)
Cysteine proteases	leupeptin (1 mM)

Incubate samples for 30 min at 4°C prior to electrophoresis with specific protease inhibitors.

PROTOCOL 4.26 Identification of Protease Inhibitors: Reverse Zymography

This protocol (Lokeshwar et al, 1993) is a modification of the method described above and is used for the detection of naturally occurring specific protease inhibitors found in cells or conditioned media.

Materials

Separating gel (from Protocol 4.2) with 0.75 mg/ml gelatin copolymerized

Gel rinse buffer: 50 mM Tris-HCl, pH 7.5, 0.1% Triton X-100

Substrate buffer: 50 mM Tris-HCl, pH 7.5, 0.2 M NaCl, 0.01 M $CaCl_2$, 1 μM $ZnCl_2$, 1 mM PMSF, 0.02% NaN_3, and 0.005% Brij 35

Proteolytic enzyme: 0.1 mg/ml trypsin in 50 mM Tris-HCl, pH 7.5

Staining solutions: 0.1% CBB, 0.25% CBB

Destain: 7.0% acetic acid

1. Separate the proteins by SDS-PAGE (Protocol 4.5) using gels that have been copolymerized with gelatin (0.75 mg/ml).
2. Rinse the gel twice in gel rinse buffer, 20 min each and incubate the gel with a mixture containing the substrate buffer and an active proteolytic enzyme (this could be a metalloproteinase, gelatinase, collagenase or any enzyme at ~1–2 units/ml) for 1–3 h at 37°C. Use enough solution to completely cover the gel.
3. Pour off the enzyme-containing solution and incubate the gel for 16–18 h at 21°C with only substrate buffer.
4. Stain the gel with 0.1% Coomassie Blue R for 30 min, followed by 0.25% Coomassie Blue R for 30 min. Destain with 7% acetic acid. Areas of the gel containing protease inhibitors (in this case gelatinase inhibitors) appear as dark blue bands against a paler blue background.

Comments: Staining and destaining times can be modified to maximize the signal-to-noise ratio.

PROTOCOL 4.27 Locating the Enzyme Activity: Reacting the Gel with Substrate Solution after Electrophoresis

This method is an alternative to using the copolymerized substrate method described in the previous Protocol. The enzyme activity is assayed in the presence of 0.1% SDS which does not appear to alter the activity of the proteases (Garcia-Carreno et al, 1993).

Materials

Sample buffer: 0.0625 M Tris-HCl, pH 6.8, 2% SDS

Substrate solution: 2% casein in 50 mM Tris-HCl, pH 7.5

Fix/stain solution: 40% ethanol, 10% acetic acid, 0.1% Coomassie Blue R

Destain: 40% ethanol, 10% acetic acid

Molecular weight markers

1. Prepare the samples in sample buffer. Do not boil the samples. Include a lane with molecular weight markers for easy comparison of the molecular weight of the activity zone.
2. Perform electrophoresis using a 12.5% acrylamide gel in a narrow width gel cassette, 0.7 mm, at 15 mA constant current for 90 min at 10°C.

3. After electrophoresis, immerse the gel in 50 ml of substrate solution for 30 min at 4°C. This step allows the substrate to diffuse into the gel at reduced enzyme activity.
4. Incubate the gel for 90 min at 25°C to allow digestion of the protein substrate by the active fractions.
5. Rinse the gel with water 3 times for 5 min each at room temperature. Immerse the gel for 2 h in fix/stain solution. Longer periods of incubation will result in reduced resolution of the activity zone.
6. Destain the gel.

Comments: Proteinase activities are identified as clear zones on a blue background. Molecular weight marker bands and proteins other than proteinases have a higher intensity of blue color than the background staining which is due to the undigested casein.

This method can be modified to search for proteinaceous protease inhibitors. After electrophoresis and incubating the gel with substrate (step 3), immerse the gel in a solution of 0.1 mg/ml of trypsin in 50 mM Tris-HCl, pH 7.5, for 30 min at 4°C. (Any other protease can be used, changing the conditions as necessary.) Wash the gel with distilled water and proceed as described in the protocol above, from step 4 to completion. The presence of proteinaceous protease inhibitors is demonstrated by dark zones produced by the staining of the undigested substrate. This is due to the inhibition of the exogenously added protease by the protease inhibitor in the gel.

Slightly thicker gels, 0.8 mm, may also be used, but more time will be needed for all steps.

If the presence of SDS inhibits enzyme activity, it should be removed following electrophoresis. Wash the gel for 30 min at 4°C in 50 mM Tris-HCl, pH 7.5, 2.5% Triton X-100, without substrate.

Some enzymes need activators such as Ca^{2+} or EDTA. Therefore, these factors must be included in the substrate solution.

Other substrates such as hemoglobin and BSA can also be used.

PROTOCOL 4.28 Detection of β–glucuronidase Activity in Polyacrylamide Gels

The hydrolytic enzyme β–glucuronidase is robust and active over a wide pH range and in the presence of many detergents and ionic strength conditions. These properties enable it to be detected in-gel following denaturing SDS-PAGE. The fluorogenic substrate ELF® 97-β-D-glucuronide (Molecular Probes) is hydrolyzed to the highly fluorescent product, ELF 97 alcohol (Kemper et al, 2001).

Materials

Sample buffer: 0.0625 M Tris-HCl, pH 6.8, 2% SDS

Renaturation buffer: 0.1% Triton X-100 in 50 mM Na_2HPO_4, pH 7.0

Reagent solution: 25 μM ELF 97-β-D-glucuronide (Molecular Probes) in 50 mM Na_2HPO_4, pH 7.0

1. Prepare the samples in sample buffer. Do not boil the samples. Include a lane with molecular weight markers for easy comparison of the molecular weight of the activity zone.
2. After electrophoresis, incubate the gel in renaturation buffer for two 10 min washes.
3. Incubate the gel in reagent solution at room temperature for 60 min to overnight.
4. To visualize the bands illuminate the gel with a 302 nm UV-B transilluminator. β–glucuronidase activity is visualized as green-fluorescent bands.

Comments: After staining for β–glucuronidase the gel can be counterstained with SYPRO Ruby dye allowing the simultaneous detection of total proteins as red bands in the same gel.

The enzyme activity could not be visualized after electroblotting to PVDF membranes.

Identification of DNA Binding Proteins—Gel Shift Assay

Nondenaturing, native PAGE is similar to SDS-PAGE systems. The nondenaturing system is used when the activity or structure of the protein of interest is to be maintained in a native, active state. **Do not boil the proteins.** In native PAGE, resolution will be sacrificed, and proteins will migrate according to their net charge and molecular weight.

The gel mobility shift assay, also referred to as electrophoretic mobility shift assay (EMSA) is widely used in studies of DNA binding proteins, such as transcription factors, that are components of cell and nuclear extracts (Revzin, 1989). This method allows highly sensitive detection of sequence-specific DNA-binding proteins by electrophoretic separation of free and protein-bound probes. In a typical experiment, a radioactively labeled DNA fragment containing the sequence of interest is incubated with a whole cell or nuclear extract. The mixture is separated electrophoretically and the results are analyzed after the gel is dried and film is exposed. A specific DNA complex may be seen which migrates slower than the free DNA fragment.

In a crude extract many proteins may interact nonspecifically with DNA. In fact, if only the labeled probe is added to the extract, all of the DNA will be found at the top of the gel, bound in large complexes with many protein molecules. To overcome this problem, a nonspecific, unlabeled carrier DNA, calf thymus DNA, salmon sperm DNA, or poly (dI-dC) is added along with the labeled, specific probe. The nonradioactive, nonspecific carrier DNA binds the nonspecific proteins. The optimum amount of competitor is determined empirically. Using a series of reactions with increasing amounts of nonspecific DNA, the level of radioactivity at the top of the lane diminishes while the intensities of the specific complex and of the free DNA probe rise. However, if too much nonspecific DNA is added, it will compete for the specific protein of interest, and the intensity of the specific complex will decrease.

Interactions between a small protein and a nucleic acid are best studied using polyacrylamide gels. For proteins with molecular weights ranging from 15 to 500 KDa and DNA fragments from 10 to 300 base pairs a 4–6% polyacrylamide gel works well.

PROTOCOL 4.29 Gel Shifts

Gel shifts are also known as electromobility shift assays (EMSA). Expertise in advanced techniques of molecular biology is recommended.

Materials

DNA probe: Double stranded oligonucleotides or restriction fragments are usually end labeled (Maniatis et al, 1982), radioactively tagging the oligonucleotide with ^{32}P. Ideally, the probe should be of the highest possible specific activity ~1×10^6 cpm/pmol and gel purified.

Poly (dI-dC): 1 mg/ml in water

EMSA buffer (10 ×): 0.1 M Tris-HCl, pH 7.5, 0.5 M NaCl, 50 mM $MgCl_2$, 10 mM EDTA, 50% glycerol, 10 mM DTT (add last)

TEMED

APS: 10% (w/v): see Appendix C

Acrylamide stock 30%: see Table 4.1

Running buffer (10 ×): 108 g Tris, 55 g boric acid, 40 ml 0.5 M EDTA, made up to 1 liter

Filter paper: Whatman 3MM

Microcentrifuge and tubes

Bromophenol blue (0.0004% w/v)

1. To a 1.5 ml microcentrifuge tube, add 2 µl 10 × EMSA buffer, 2 µg poly (dI-dC), 15 µg protein extract, and ^{32}P labeled probe. Bring the reaction mixture to a final volume of 20 µl with water. If you are using a nuclear extract use 1–5 µg protein.
2. Incubate the reaction at room temperature for 20 min.
3. Prepare a nondenaturing 4% acrylamide gel by mixing 10 ml 30% acrylamide stock, 1.9 ml 10× running buffer, 300 µl APS, 45 µl TEMED, and 62.9 ml water. Fill the plates with the acrylamide solution and insert the comb. A stacking gel is not necessary. For optimal results, use a comb with slots that are >7 mm wide.
4. Prepare 1× running buffer from the 10× stock.
5. Prior to loading the samples, pre-run the gel for 30–60 min at 100 V.
6. Load the samples. In one lane include EMSA buffer plus the bromophenol blue solution to follow the progress of the run. Electrophorese at 30–35 mA until the Bromophenol blue reaches the bottom of the gel (~1.5 h). On a 4% gel, the bromophenol blue runs comparable to a 40mer oligo.
7. Disassemble the plates by first removing the side spacers then inserting a spatula to pry apart the plates. The gel should adhere to one of the

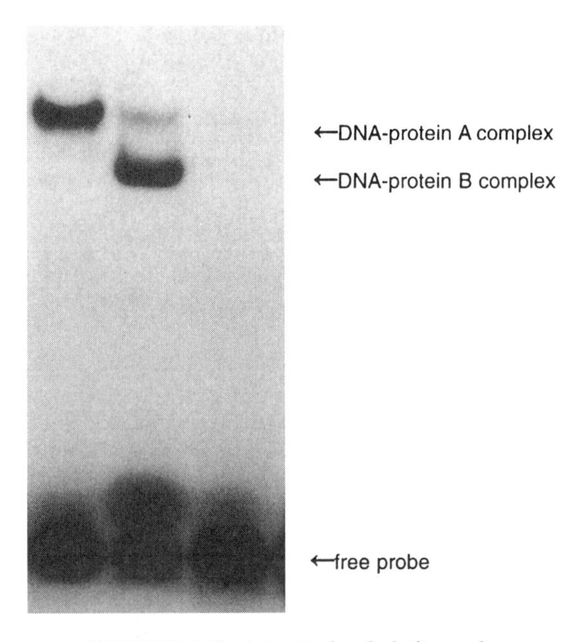

FIGURE 4.12 A typical gel shift result.

plates. Place a dry piece of Whatman 3MM on the gel. Turn the plate, gel, and paper up-side-down so that the plate is on top. At the corner of the bench, move the gel slowly toward you and let the paper and gel come off of the plate. Due to the low acrylamide concentration of the gel, great care should be taken during the disassembly of the plates and while putting the gel on Whatman 3MM paper.

8. Cover the gel with plastic wrap. Dry the gel and expose to film at –70°C overnight. Alternatively, the dried gel can be analyzed by phosphorimaging.

Comments: The oligonucleotides or restriction fragments that are to be used as probes must be pure. Probes should be gel purified. They can be used for up to one month. The probe should be between 10–300 base pairs. The longer the probe, the greater the chance of nonspecific binding of proteins in the extract with sequences that flank the specific binding site. A typical gel shift result is shown in Figure 4.12.

Verify sequence specificity with competitor oligos and mutant probes.

Test a variety of binding conditions with particular emphasis on the concentrations of the nonspecific competitor.

Some variables that have an effect on the DNA-protein complex are the composition of the running buffer, particularly the salt concentration, the amount of protein and its purity, and the specific activity of the labeled probe. All of these factors can be tinkered with to obtain best results in your system.

Preparative gel retardation could be an important complement to DNA affinity columns. This technique could be used as the final step in a purification protocol. Electrophoresis may allow separation of complexes from hard-to-remove contaminants.

References

Appel RD, Sanchez JC, Bairoch A, Golaz O, Ravier F, Pasquali C, Hughes GJ, Hochstrasser DF (1996): The SWISS-2DPAGE database of two-dimensional polyacrylamide gel electrophoresis, its status in 1995. *Nucleic Acids Res* 24:180–181

BioRadiations (2002): 2-D Electrophoresis: Troubleshooting isoelectric focusing. 108:42–43

Bio-Rad SYPRO Ruby Protein Stains Instruction Manual, 4006173 Rev A

Bjellqvist B, Basse B, Olsen E, Celis JE (1994): Reference points for comparisons of two dimensional maps of proteins from different human cell types defined in a pH scale where isoelectric points correlate with polypeptide compositions. *Electrophoresis* 15:529–539

Blum H, Beier H, Gross HJ (1987): Improved silver staining of plant proteins, RNA and DNA in polyacrylamide gels. *Electrophoresis* 8:93–99

Branden C, Tooze J (1991): *Introduction to Protein Structure.* New York: Garland

Brookes PS, Pinner A, Ramachandran A, Coward L, Barnes S, Kim H, Darley-Usmar VM (2002): High throughput two-dimensional blue-native electrophoresis: a tool for functional proteomics of mitochondria and signaling complexes. *Proteomics* 2:969–977

Carboni L, Piubelli C, Righetti PG, Jansson, B, Domenici E (2002): Proteomic analysis of rat brain tissue: comparison of protocols for two-dimensional gel electrophoresis analysis based on different solubilizing agents. *Electrophoresis* 23:4132–4141

Castellanos-Serra L, Hardy E (2001): Detection of biomolecules in electrophoresis gels with salts of imidazole and zinc II: a decade of research. *Electrophoresis* 22:864–873

Chakravarti DN, Chakravarti B, Moutsatos I (2002): Informatic tools for protein profiling. *Biotechniques Computational Proteomics Supplement* 32: S4–S15

Davis BJ (1964): Disc electrophoresis. *Ann NY Acad Sci* 121:321–349

Dayhuff TJ, Gesteland RF, Atkins JF (1992): Electrophoresis, autoradiography and electroblotting of peptides: T4 gene 60 hopping. *BioTechniques* 13:500–503

Dainese P, James P (2001): Protein identification by peptide-mass fingerprinting. In: *Proteome Research: Mass Spectrometry.* Berlin: Springer-Verlag, p. 111

Dianoux AC, Stasia MJ, Garin J, Gagnon J, Vignais PV (1992): The 23-kilodalton protein, a substrate of protein kinase C, in bovine neutrophil cytosol is a member of the S100 family. *Biochemistry* 25:5898–5905

Dzandu JK, Johnson JF, Wise GE (1988): Sodium dodecyl sulfate-gel electrophoresis: staining of polypeptides using heavy metal salts. *Anal Biochem* 174:157–167

Fernandez-Patron C, Calero M, Collazo PR, Garcia JR, Madrazo J, Musacchio A, Soriano F, Estrada R, Frank R, Castellanos-Serra L, Mendez E (1995): Protein reverse staining: high-efficiency microanalysis of unmodified proteins detected on electrophoresis gels. *Anal Biochem* 224:203–211

Fisher SJ, Cui T, Zhang L, Hartman L, Grahl K, Guo-Yang Z, Tarpey J, Damsky CH (1989): Adhesive and degradative properties of human placental cytotrophoblast cells in vitro. *J Cell Biol* 109:891–902

Freedman RB (1984): Native disulfide bond formation in protein biosynthesis: evidence for the role of protein disulfide isomerase. *Trends Biochem Sci* 9:438–441

Freedman RB, Hillson DA (1980): Formation of disulfide bonds. In: *The Enzymology of Post-translational Modification of Proteins,* Freedman RB, Hawkins HC, eds. London: Academic Press

Gabriel O, Gersten DM (1992): Staining for enzymatic activity after gel electrophoresis. *Anal Biochem* 203:1–21

Galvani M, Hamdan M, Hertbert B, Righetti PG (2001): Alkylation kinetics of proteins in preparation for two-dimensional maps: a matrix assisted laser desorption/ionization-mass spectrometry investigation. *Electrophoesis* 22:2058–2065

Garcia-Carreno FL, Dimes LE, Haard NF (1993): Substrate-gel electrophoresis for composition and molecular weight of proteinases or proteinaceous proteinase inhibitors. *Anal Biochem* 214:65–69

Gooley AA, Ou K, Russell J, Wilkins MR, Sanchez JC, Hochstrasser DF, Williams KL (1997): A role for Edman degradation in proteome studies. *Electrophoresis* 18:1068–1072

Gooley AA, Packer NH (1997): The importance of protein co- and posttranslational modifications in proteome projects. In: Wilkins MR, Williams KL, Appel RD, Hochstrasser DF (eds) *Proteome Research: New Frontiers in Functional Genomics.* Berlin: Springer-Verlag, p. 86

Görg A, Obermaier C, Boguth G, Harder A, Scheibe B, Wildgruber R, Weiss W (2000): The current state of two-dimensional electrophoresis with immobilized pH gradients. *Electrophoresis* 21:1037–1053

Hager DA, Burgess RR (1980): Elution of proteins from sodium dodecyl sulfate-polyacrylamide gels, removal of sodium dodecyl sulfate, and renaturation of enzyme activity: results with sigma subunits of E. coli RNA polymerase, wheat germ DNA polymerase, and other enzymes. *Anal Biochem* 109:76–86

Hames BD (1981): An introduction to polyacrylamide gel electrophoresis. In: *Gel Electrophoresis of Proteins: A Practical Approach,* Hames BD, Rickwood D, eds. Oxford: IRL Press

Herbert BR, Sanchez J-C, Bini L (1997): Two-dimensional electrophoresis: the state of the art and future directions. In: *Proteome Research: New Frontiers in Functional Genomics,* Wilkins MR, Williams KL, Appel RD, Hochstrasser DF, eds. Berlin: Springer-Verlag

Herbert BR, Molloy MP, Gooley AA, Walsh BJ, Bryson WG, Williams KL (1998): Improved protein solubility in two-dimensional electrophoresis using tributylphosphine as reducing agent. *Electrophoresis* 19:845–851

Herbert B, Galvani M, Hamdan M, Olivieri E, McCarthy J, Pedersen S, Righetti PG (2001): Reduction and alkylation of proteins in preparation of two-dimensional map analysis: why, when, and how? *Electrophoresis* 22:2046–2057

Kemper C, Steinberg TH, Jones L, Patton WF (2001): Simultaneous, two-color fluorescence detection of total protein profiles and β–glucuronidase activity in polyacrylamide gel. *Electrophoresis* 22:970–976

Laemmli UK (1970): Cleavage of structural proteins during the assembly of the head of bacteriophage T4. *Nature* 227:680–685

Lee C, Levin A, Branton D (1987): Copper staining: a five minute protein stain for sodium dodecyl sulfate-polyacrylamide gels. *Anal Biochem* 166:308–312

Leimgruber RM, Malone JP, Radabaugh MR, LaPorte ML, Violand BN, Monahan JB (2002): Developing of improved cell lysis, solubilization and imaging approaches for proteomic analyses. *Proteomics* 2:135–144

Lewis SA (1996): Removal of Coomassie blue precipitates from polyacrylamide gels. *BioTechniques* 21:820

Lilley KS, Razzaq A, Dupree P (2002): Two-dimensional electrophoresis:recent advances in sample preparation, detection and quantitation. *Curr Opin Chem Biol* 6:46–50

Lim JH, Bustin M, ogryzko VW, Postnikov YV (2002): Metastable macromolecular complexes containing high mobility group nucleosome-binding chromosomal proteins in HeLa nuclei. *J Biol Chem* 277:20774–20782

Lokeshwar BL, Selzer MG, Block NL, Gunja-Smith Z (1993): Secretion of matrix metalloproteinases and their inhibitors (tissue inhibitor of metalloproteinases) by human prostate in explant cultures: reduced tissue inhibitor of metalloproteinase secretion by malignant tissues. *Cancer Res* 53:4493–4498

Maniatis T, Fritsh EF, Sambrook J, eds. (1982): *Molecular Cloning.* Cold Spring Harbor, New York: Cold Spring Harbor Laboratory

McLellan T, Ramshaw JAM (1981): Serial electrophoretic transfers: a technique for the identification of numerous enzymes from single polyacrylamide gels. *Biochem Genet* 19:648–654

Merril CR (1990): Gel-Staining Techniques. *Meth Enzymol* 182:477–488

Miele L, Cordella-Miele E, Mukherjee AB (1990): High level bacterial expression of uteroglobin, a dimeric eukaryotic protein with two interchain disulfide bridges in its natural quaternary structure. *J Biol Chem* 265:6427–6435

Molloy MP, Herbert BR, Walsh BJ, Tyler MI, Traini M, Sanchez JC, Hochstrasser DF, Williams KL, Gooley A (1998): Extraction of membrane proteins by differential solubilization for separation using two-dimensional gel electrophoresis. *Electrophoresis* 19:837–844

Molloy MP, Herbert BR, Williams KL, Gooley AA (1999): Extraction of *E.coli* proteins with organic solvents prior to two-dimensional electrophoresis. *Electrophoresis* 20:701–704

Molloy NP (2000): Two-dimensional electrophoresis of membrane proteins using immobilized pH gradients. *Anal Biochem* 280:1–10

O'Farrell PH (1975): High resolution two-dimensional electrophoresis of proteins. *J Biol Chem* 250:4007–4021

O'Farrell PZ, Goodman HM, O'Farrell PH (1977): High resolution two-dimensional electrophoresis of basic as well as acidic proteins. *Cell* 12:1133–1142

Ornstein L (1964): Disc electrophoresis-I Background and theory. *Ann NY Acad Sci* 121:321–349

Patton WF (2000): Making blind robots see: the synergy between fluorescent dyes and imaging devices in automated proteomics. *BioTechniques* 28: 944–957

Patton WF (2002): Detection technologies in proteome analysis. *J Chromatogr* B771:3–31

Rademaker GJ, Thomas-Oates J (1996): Analysis of glycoproteins and glycopeptides using fast-atom bombardment. *Methods Mol Biol* 61:231–241

Revzin A (1989): Gel electrophoresis assays for DNA-protein interactions. *BioTechniques* 7:346–355

Righetti PG (1990): *Immobilized pH Gradients: Theory and Methodology.* Amsterdam: Elsevier

Santoni V, Rabilloud T, Doumas P, Rouquie D, Mansion M, Kieffer S, Garin J, Rossignol M (1999): Towards the recovery of hydrophobic proteins on two-dimensional electrophoresis gels. *Electrophoresis* 20:705–711

Schagger H, von Jagow G (1987): Tricine-sodium dodecyl sulfate-polyacrylamide gel electrophoresis for the separation of proteins in the range from 1 to 100KDa. *Anal Biochem* 166:368–379

Schagger H, von Jagow G (1991): Blue native electrophoresis for isolation of membrane protein complexes in enzymatically active form. *Anal Biochem* 199:223–231

Shevchenko A, Wilm M, Vorm O, Mann M (1996): Mass spectrometric sequencing of proteins silver-stained polyacrylamide gels. *Anal Chem* 68:850–858

Steinberg TH, Chernokalskaya E, Berggren K, Lopez MF, Diwu Z, Haugland RP, Patton WF (2000): Ultrasensitive fluorescence protein detection in isoelectric focusing gels using a ruthenium red metal chelate stain. *Electrophoresis* 21:486–496

Swank RT, Munkres KD (1971): Molecular weight analysis of oligopeptides by electrophoresis in polyacrylamide gels with sodium dodecyl sulfate. *Anal Biochem* 39:462–477

Tiselius A (1937): A new apparatus for electrophoretic analysis of colloidal mixtures. *Trans Faraday* Soc 33:524–531

Trant RR, Casiano C, Zecherle (1989): Cross-linking protein subunits and ligands by the introduction of disulfide bonds. In: *Protein Function,* Creighton TE, ed. Oxford: IRL Press

Weber K, Osborn M (1985): Proteins and sodium dodecyl sulfate: molecular weight determination on polyacrylamide gels and related procedures. In: *The Proteins, Vol. 1.* Neurath H, Hill RL, eds. New York: Academic Press, pp. 179–223

Wirth PJ (1989): Specific polypeptide differences in normal versus malignant breast tissue by two-dimensional electrophoresis. *Electrophoresis* 10:543–554

General References

Ausubel FM, Brent R, Kingston RE, Moore DD, Seidman JG, Smith JA, Struhl K, eds. (1993): *Current Protocols in Molecular Biology.* New York: John Wiley and Sons, Inc.

Hames BD, Rickwood D, eds. (1981): *Gel Electrophoresis of Proteins: A Practical Approach.* Oxford: IRL Press

5

Getting Started with Protein Purification

Introduction

Protein purification is generally a multi-step process exploiting a wide range of biochemical and biophysical characteristics of the target protein, such as its source, relative concentration, solubility, charge, and hydrophobicity. The ideal purification strives to obtain the maximum recovery of the desired protein, with minimal loss of activity, combined with the maximum removal of other contaminating proteins.

Proteins are fragile molecules that denature readily at extremes of temperature and pH. Each protein offers its own unique set of physicochemical characteristics. The methods used for protein purification should be mild, to preserve the native conformation of the molecule and its bioactivity. In most cases, having a reliable assay to be used as the means of following the target protein is essential.

There is no set procedure for isolating proteins. Purification schemes should be tailored to take advantage of the biochemical properties of the target protein, as well as the cellular properties of the tissue that provides the most abundant source of material. Whenever possible try to reduce the complexity of the sample. For example, isolate specific protein subsets or subcellular organelles thereby enriching for the low abundance target molecule.

When designing a purification protocol one should aim for the following: (1) high recovery; (2) highly purified end product; (3) reproducibility, within the lab, in other labs and also when either scaled up or down; (4) economical use of reagents; and (5) convenience with regard to time.

The chemical structure and physical properties of the protein are the two key parameters used to develop most purification protocols. Isoelectric point (pI), pH stability, and charge density are some of the properties of proteins that can be exploited during purification. A number of separation techniques are capable of resolving proteins on the basis of differences in net charge. These include isoelectric focusing (Chapter 4), ion exchange chromatography (Chapter 10), and chromatofocusing.

In general, the most successful isolation procedures involve only a few steps, chosen to give the highest yields. Rarely will a single technique fulfill the requirements of any specific separation. More frequently, two, three, or more steps are needed for the purification of the desired protein.

Before beginning to purify a protein it is important to have an idea of the degree of purity that is necessary for the intended use of the target protein. Often, a 90–95% enrichment will suffice because additional steps decrease the yield and may not provide the desired purity of the final product. Concern for purity will vary greatly depending on whether the protein is to be used for kinetic, sequence or crystallographic studies, or is to be injected into humans.

Protein purification is routinely assessed by summarizing the results of each purification step as specific activities, total units, total protein, and yields. Activity units are calculated based on the assay system that is used to track the target protein. An example of expressing the data is shown in Table 5.1 using the purification of 3-hydroxy-3-methylglutaryl-CoA (HMG-CoA) synthase from ox liver. This bookkeeping is a means of critically evaluating each step for yield and purification. If a step results in a major loss of activity, it should be changed.

When undertaking any purification, always begin with enough starting material so that a workable amount of final product can be isolated. Delays during a multistep protocol should be avoided. The stability of the target protein will determine the time necessary for the protocol. For example, in cases where stability is limited due to the inherent lability of the protein or copurifying proteolytic activity, the speed of operation can be much more important than the protein purity. Therefore, there will be instances when it will be beneficial to rush one step after another, sacrificing purification, rather than carrying out each step to perfection and taking a long time in the process.

TABLE 5.1 Purification of Mitochondrial HMG-CoA Synthase from Ox Liver*

	Volume (ml)	Total activity (units)	Protein concn. (mg/ml)	Specific activity (units/mg)	Purification (fold)	Yield activity (%)
1 Crude extract (from 300 g of liver)	1000	183.0	42.6	0.0043	(1)	(100)
2 $(NH_4)_2SO_4$ fractionation and desalting	650	171.6	12.7	0.021	4.9	93.8
3 DEAE-cellulose chromotography	575	168.5	6.9	0.042	9.8	92.1
4 Blue Sepharose chromotography	155	158.3	2.4	0.43	100	86.5
5 Substrate elution from cellulose phosphate	41	118.1	3.2	0.90	209	64.5
6 Chromatofocusing	61	71.8	1.3	0.91	212	39.2

*Reprinted with permission, from Lowe and Tubbs, 1985.

Whenever possible, fractionation steps should be arranged to follow one from the other without extensive manipulation between steps. Some typical protocols that do not involve intermediate treatments are listed below:

Salt precipitation → Gel filtration → Ion exchange chromatography

Salt precipitation → Hydrophobic interaction chromatography → Ion exchange chromatography → Reversed phase HPLC

Ion exchange chromatography → Hydrophobic interaction chromatography → Affinity chromatography → Gel filtration chromatography

Organic solvent extraction → Affinity chromatography

Affinity chromatography → Salt precipitation → Gel filtration chromatography

The positions of the steps in the above strategies are predicated on the volume and ionic strength of the sample. In a typical purification protocol, the first step after extracting the target protein in soluble form from the starting material could be ion exchange chromatography, which has excellent resolving power that can concentrate the target protein from a dilute starting solution. The target protein is usually eluted with high concentrations of salt. An interim desalting step can be avoided if the next step is hydrophobic interaction chromatography in which the protein is loaded onto the column in a high salt containing solution. The eluate from this step, now free of lipids and other potential problem-causing contaminants, can be put onto an affinity column, if one is available. Using an immobilized ligand, the target protein is specifically bound, and the contaminants are washed away. The protein of interest is then eluted. Often, this step is the most powerful in the scheme, being good enough to stand alone when a large quantity of semi-purified protein is desirable.

The final step in a purification protocol is frequently size exclusion chromatography, which, in addition to contributing to the purity of the final product, also accomplishes two other important goals. Size exclusion chromatography will yield the Stokes radius of the specific protein which can be used as a close approximation of the protein molecular weight. Secondly, the final step can be used to adjust or change the buffer, transferring the protein into a solution that is compatible with the intended usage.

A. Making a Cell Free Extract

Whatever the starting material may be, sample preparation is the most important step for successful analysis and purification. Starting material for a protein purification project may come from a variety of sources: animal tissue like liver or skeletal muscle, cells in culture, plants, yeast and bacterial to name a few. Bacteria and yeast are relatively easy to culture and can be grown to high densities. However, bacteria, and to a certain degree yeast, lack the ability to carry out post-translational modifications that are required to produce a mature, functional recombinant protein. Mammalian cell culture, despite its inherent difficulties may be the only alternative for producing large quantities of the desired recombinant protein.

Typically, most mammalian cells are anchorage dependent by nature although some cell lines like Chinese hamster ovary (CHO) and baby hamster kidney (BHK) have been adapted for suspension culture for growth in fermentors. Anchorage dependent cells are cultured in glass or plastic bottles, flasks and plates. Scaling up is straightforward but not efficient. Over the past several years various strategies for increasing the growing area have been put into practice. Some of these methods include microcarrier beads and different plate designs. The next step would be a scale-up to some type of bioreactor that can support anchorage dependent cell culture.

The choice of raw material from which the protein is to be isolated may not be a relevant decision because the lab may be working exclusively on a particular organism or tissue. When appropriate, choose a raw material that is available in sufficient quantities. Prior to starting a protein purification project, try to optimize the amount and relative purity of the protein of interest. Many proteins are routinely isolated from cultured cell lines, which are often transformed and tumorigenic. Larger amounts of starting material can be obtained by injecting the cell line into a susceptible mouse strain and then using the resulting solid tumor or ascites fluid as the raw material. When purifying proteins from bacteria, it may be possible to induce the bacteria to overproduce the protein of interest by either genetic manipulation or by controlling the quantity of specific nutrients or factors available. Knowledge of the function of a protein and when and where it is expressed in development can be used to synchronize a population of cells that should be harvested at the specific time when the production of the protein of interest is enhanced.

Cellular Disruption

The first step in making a cell free extract is to choose a method that will disrupt the cells to efficiently release the target protein into the aqueous supernatant in a form conducive to subsequent purification steps. Ideally, the extraction method should release the target protein and leave as many contaminants behind as possible. Extraction should minimize degradation of the target by enhancing its stability. Initial preparation steps should be performed as fast as possible and at low temperatures.

Homogenization of cultured cells is frequently performed in a hypotonic buffer to facilitate cell breakage and to keep the target protein in an active form. If the cells are not broken properly, much of the protein of interest will not be released and will be removed with the residue, making an otherwise excellent source appear impractical. The disruption treatment should be no more vigorous than necessary, decreasing the possibility of denaturing labile components.

The stabilization of enzymes and proteins against irreversible inactivation is a major concern. Minimizing protein inactivation is a crucial part of the purification and storage of the target protein. Irreversible inactivation is not only a function of external agents such as heat, detergents, or pH, it also depends on the nature of the protein. In general, biological material should be processed for extraction quickly, using procedures that will minimize endogenous degradation. Fresh tissue should be chilled on ice, cleaned, and minced quickly. Extraction efficiency is often improved when smaller tissue pieces are used to increase the surface area exposed to the solvent. Whenever possible, the fresh tissue should be extracted immediately. If necessary, the tissue can be quick frozen and stored for later use. In this case the tissue should be thawed directly in the extracting solvent. This helps to avoid degradation from proteolytic enzymes released by the freeze-thaw cycle. Once the small tissue pieces are in the extracting solvent, homogenation at low temperature (0–4°C) with a Polytron homogenizer operated (see below) at low speeds for short times can be used to further disperse the tissue.

Extraction Buffer Composition

The composition of the extraction buffer is an important consideration. For soluble proteins, a first approximation would be a buffer consisting of 20–50 mM phosphate, pH 7–7.5, or 0.1 M Tris-HCl pH 7.5, 0.1 M NaCl. Also included would be stabilization additives, EDTA (1–5 mM), β-mercaptoethanol (2-ME) (5–20 mM), sucrose and protease inhibitors.

Protease Inhibitors

Intact cells contain the elements for their own destruction, which are compartmentalized and released in response to specific physiologic signals. When an extract is prepared, this delicate balance is destroyed, and proteolytic enzymes can now mix with cell contents. In the context of protein purification, proteases, enzymes that cleave peptide bonds (also referred to as proteinases), are unwanted, potentially destructive contaminants that need to be inactivated and removed.

There are many different classes of proteolytic enzymes, not all of which can be completely inhibited. Over the last several years, protease inhibitors, compounds that specifically inhibit proteolytic enzymes, have been developed that act on the various classes of proteases. From a practical standpoint, it is best to use a mixture of inhibitors affecting the different classes of proteases. As a starting point, the following cocktail of leupeptin, EDTA, pepstatin, and PMSF has been suggested (Keesey, 1987). Selected protease inhibitors and a range of their working concentrations are also listed (see Table 5.2).

Phenylmethylsulfonyl fluoride (PMSF) (Gold, 1967) was introduced to replace the highly dangerous first generation serine protease inhibitor diisopropyl fluorophosphate (DFP). Add PMSF directly to the solution containing the serine proteases. Once inhibited, the enzymes are irreversibly inactivated and no more PMSF is needed.

TABLE 5.2 Protease Inhibitors

Inhibitor	Specificity	Final concentration
Leupeptin	broad spectrum	0.5–10 μg/ml
EDTA-Na2	metalloproteases	5–10 mM
Pepstatin A	acidic proteases	0.7–10 μg/ml
Aprotinin	serine proteases	50 μg/ml
PMSF	serine proteases	0.2–2 mM
Benzamidine HCl	serine proteases	100 μg/ml
Soybean	Trypsin-like trypsin inhibitor (SBTI) enzymes	100 μg/ml

Methods of Cell Disruption

Cells in culture may require harsh means to be disrupted. Two of these methods are sonication and the use of a Polytron disrupter, which lyses most cells and causes extensive release of mitochondrial and lysosomal enzymes. The various options for cellular disruption should permit a selection of a technique appropriate for the starting material. Some frequently used methods are described below.

Bead Mill Homogenizers

Bead mill homogenizing is a form of grinding that works well with cells that are hard to disrupt like yeast, spores, bacteria and plant tissue. In bead milling, a large number of small glass beads are vigorously agitated with the starting material at 3000–6000 oscillations/min. Disruption occurs by a shearing and crushing action as the beads collide with the cells. After treatment, the homogenate is separated from the beads by filtration. Beads can be reused after washing. Most bead mills are restricted to sample sizes of 3.0 ml or less. Three companies that sell these devices are: Mini-Bead Beater, BioSpec Products, Bartlesville, OK; H. Mickle, Middlesex, England; B Braun Biotech, Bethlehem, PA.

High Pressure Homogenizers

High pressure homogenizers have been used for disrupting microbial cells. This type of homogenizer induces cell lysis by forcing cell suspensions through a narrow orifice under high pressure. The cells are sheared as they pass through the orifice. The efficiency of disruption is directly proportional to the pressure. This method is both gentle and thorough. However, its use is restricted by its limited sample size range. A popular high pressure homogenizer is the French press (American Instrument Co., Silver Spring, MD).

Nitrogen Decompression

This method is best suited for mammalian cells and some plant cells. The animal tissue requires some pretreatment to form a homogeneous suspension capable of passing through the discharge valve of the vessel. The sample is placed in a stainless steel vessel, sealed and pressurized to around 2000 psi using a tank of compressed nitrogen gas. The vessel can be immersed in an ice bath. The gas is allowed to dissolve into the aqueous media and the intracellular volume of the cells.

The cell suspension is then allowed to flow out of the pressure vessel. As the pressure in the sample goes from high pressure to atmospheric pressure, the dissolved gas in the cells comes out of solution, and the cells explosively decompress. Disruption also occurs by shearing as the sample solution passes through the outlet valve. Two manufacturers are: Parr Instruments Moline, IL; and Kontes, Vineland, NJ.

Ultrasonic Disintegrators
Ultrasonic disruption, commonly referred to as sonication, breaks apart cells by generating intense sonic pressure waves in a liquid suspension. The pressure waves cause streaming in the liquid, and, under the right conditions, microbubbles form, grow and coalesce, vibrate violently and eventually collapse. The implosions of the bubbles generate shock waves sufficient to disrupt cells. Sonication is routinely performed in an ice bath as the ultrasonic probe generates considerable heat. The tip of the probe should be placed under the surface of the liquid to prevent foaming which occurs when the tip is not placed deep enough in the liquid. Using various sized probes, this instrument adequately accommodates samples from as little as < 1 ml up to a liter. Manufacturers of ultrasonic disruptors are: Branson Sonic Power Co., Danbury, CT; Braun Biotech Allentown, PA; and Heat Systems-Ultrasonics, Plainview, NY.

Pestle and Tube Homogenizers
As a group, these types of instruments consist of test tubes, usually made of glass, into which is inserted a tight-fitting pestle made of a similar material. The walls of the test tube and pestle can be smooth or have a ground finish. Most tissues must first be chopped or cut into small pieces before being suspended in a four- to tenfold excess volume in the test tube. The pestle is manually worked to the bottom of the tube, tearing and fragmenting the tissue as it is forced to pass between the sides of the pestle and the wall of the tube. The shearing action is repeated as the pestle is withdrawn. This type of cellular disruption is popular because the equipment is inexpensive and small enough to be sterilized if necessary, and the method is gentle. Some names of homogenizers are Potter, Potter-Elvehjem and Dounce. Pestle homogenizers are available from many manufacturers including: Bellco Glass, Vineland, NJ; Kimble/Kontes Vineland, NJ; Wheaton Industries, Milville, NJ.

Freezing and Thawing
Cells that have rigid cell walls can sometimes be ruptured by several cycles of freezing to –20° to –30°C and thawing. The formation of large ice crystals serves to rupture intracellular membrane structure. Freezing should be performed slowly to produce large ice crystals. Advantages to this method are that it can be scaled up or down, is inexpensive, and does not require any special equipment.

Dehydration
Dehydration with acetone or ethanol has been the starting point for protein. Dried powders of many tissues are commercially available. The solvent dehydration method is relatively rapid and prepares the

tissues in a powder form. In this method, a large volume of cold acetone or 100% ethanol is slowly added to a tissue or a microbial paste while agitating the mixture in an explosion-proof blade homogenizer or blender. The solvent is replaced with fresh solvent and blending is continued. The suspension is recovered by filtration and the powdered, dehydrated tissue is dried and placed in a tightly sealed container.

Clarification of the Extract

In general, protein samples derived from tissue homogenates or cell lysates require the removal of particulate contamination. Simple filtration or centrifugation effectively removes insoluble components. However, creative use of these techniques frequently improves the overall purification protocol. The use of ultrafiltration, precipitation and extraction techniques can help produce a simplified scheme and a more concentrated sample.

After cell disruption, the extract is clarified by removing insoluble material by centrifugation. Liquid is trapped within the precipitated residue, and there will be a small loss related to the proportion of residue volume to total volume. As a rule, half the volume of the residue is trapped liquid. To realize an 80% recovery, the volume of the supernatant should be about twice the volume of the residue. Therefore, when making an extract, use at least two volumes of a suitable extraction buffer. Keep in mind that although the amount of material is slightly increased by using more extraction buffer, the extract will be in a larger volume and more dilute. The volume of extraction buffer will be a compromise between maximum extraction and minimal volume of the extract.

The protein of interest can be enriched by differential centrifugation. If the protein is soluble in the crude extract, then insoluble material can be removed by centrifugation. Conversely, if the protein is insoluble, membrane bound or is found exclusively in an organelle, the soluble proteins in the supernatant are removed, and the target protein (in the pellet!) can then be extracted in soluble form from the sedimented fraction. This alone may give a significant enrichment and often will remove contaminating molecules that may prove difficult to remove during later steps. This enrichment is often accomplished by differential centrifugation (Appendix F) which can be fine-tuned to obtain a relatively rich starting material.

Following centrifugation, the extract may be turbid or a floating layer may have formed at the air-buffer interface. At this point, a coarse filtration step is recommended. The extract can be passed through a loosely packed plug of glass wool or cheesecloth. Alternatively, the extract can be mixed with Celite, which consists mainly of SiO_2 and introduces an inert material with a large surface area, efficiently trapping fine particulate material. The suspension is then cleared by filtering it through a Büchner funnel under slight suction.

Generating Extracts from Whole Cells and Nuclei

The following two protocols illustrate the preparation of crude extracts from lymphocytes and nuclei.

PROTOCOL 5.1 Nuclear Extracts

A widely used method for preparing nuclear extracts involves hypotonic swelling of cells, mechanical shearing of cytoplasmic membranes, and isolation of the nuclear pellet followed by high salt extraction (Dignam et al, 1983).

Materials

Cell scraper

PBS (see Appendix C)

Hypotonic buffer A: 10 mM HEPES, pH 7.9, 10 mM KCl, 1.5 mM MgCl2, 0.5 mM PMSF, 0.5 μg/ml of leupeptin/Pepstatin A/Antipain/Aprotinin, and 1 mM DTT

Dounce hand-held homogenizer: (Wheaton)

Low salt buffer: 20 mM HEPES, pH 7.9, 20 mM KCl, 1.5 mM MgCl2, 0.5 mM PMSF, 0.5 μg/ml protease inhibitors, 1 mM DTT, 20% glycerol

High salt buffer: 20 mM HEPES, pH 7.9, 20 mM KCl, 1.5 mM MgCl2, 0.5 mM PMSF, 0.5 μg/ml protease inhibitors, 1 mM DTT, 20% glycerol (The high salt buffer is the same as the low salt buffer except KCl = 0.8–1.4 M)

Buffer D: 20 mM HEPES, pH 7.9, 100 mM KCl, 0.2 mM EDTA, 0.5 mM PMSF, 0.5 mM DTT, 20% glycerol

1. Begin the procedure with 10^8 or more cells. If the cells are adherent, rinse the cells with PBS then use a cell scraper to remove them from the dish.
2. Collect the cells by centrifugation (1500 rpm × 10 min) and resuspend and wash the pellet twice with cold PBS. Determine the packed cell volume (PCV) by visual estimation.
3. Resuspend the cell pellet in five PCV of hypotonic buffer A and pellet the cells.
4. Swell the cells by resuspending them in 2 PCV of hypotonic buffer A and incubate on ice for 10–15 min.
5. Homogenize the cells using a Dounce glass homogenizer fitted with the Type B pestle (the loose fitting one) until the cells are disrupted. This can be monitored microscopically. Usually 10–20 strokes are sufficient depending on the fragility of the cells.
6. Centrifuge at 3,300 × g for 15 min at 4°C to collect the nuclear pellet.
7. Resuspend the crude nuclear pellet in 1/2 nuclear volume of low salt buffer.
8. While the nuclei are on ice, slowly add high salt buffer to 1/2 the packed nuclear volume. The final salt concentration should be derived empirically. Stir the suspension for 30 min at 4°C, then centrifuge as in step 6.
9. Dialyze the supernatant against buffer D.
10. Centrifuge the nuclear extract at 25,000 × g for 30 min to pellet nuclear debris. Save the supernatant and store at –70°C in small fractions.

Comments: Keep solutions and extracts ice cold.

- Use solutions that are not more than one week old.
- Judiciously include protease inhibitors.

- Test various high salt extraction conditions for optimal recovery of desired factors.
- Avoid multiple freeze-thaw cycles of nuclear extracts.
- Avoid over homogenizing the hypotonic cell suspension.

PROTOCOL 5.2 Total Lymphocyte Extract

Under carefully controlled conditions, some nonionic detergents solubilize the plasma membrane and most cellular organelles, releasing the cytoplasm but leaving the nuclei and most of the cytoskeleton intact.

Materials

Lymphocytes: Isolated from peripheral blood or tissue culture cell lines
PBS: see Appendix C
Lysis buffer: 1% w/v NP-40 in PBS containing 1 mM EGTA
PMSF: 200 mM stock solution in isopropanol

1. Harvest lymphocytes by centrifugation at 250 × *g* for 5 min at room temperature.
2. Resuspend the pellet ~ 10^7 cells in 0.5 ml PBS. Cool the cell suspension.
3. Add 10 μl of PMSF and 0.5 ml of cooled lysis buffer to the lymphocytes.
4. Mix very gently by slowly inverting the tube and incubate on ice for 30 min.
5. Centrifuge the mixture at 200 × g for 10 min at 4°C. Transfer the supernatant to an ultracentrifuge tube and centrifuge at 100,000 × *g* for 30 min at 4°C.
6. If not to be used immediately, store the supernatant at –20°C or lower.

Comments: Up to 2×10^8 lymphocytes/ml can be solubilized in the presence of 2% w/v detergent.

PROTOCOL 5.3 Subcellular Fractionation

One way to enrich for the target protein is by subcellular fractionation. Once the starting material has been homogenized it is fractionated by differential centrifugation. The scheme described in the following protocol was used for rat kidney and modified to be used for several other tissues (Pattel et al, 1993).

Materials

Homogenization buffer: 50 mM HEPES/NaOH pH 7.5, 0.33 M sucrose
Low speed and ultra-centrifuge with the appropriate rotors

1. Homogenize the tissue or cells in homogenization buffer.
2. Centrifuge the homogenate at 600 × *g* for 3 min at 4°C.
3. Remove the supernatant. The pellet consists of nuclei and cell debris.
4. Centrifuge the supernatant at 6,000 × *g* for 10 min.

5. Remove the supernatant. The pellet consists of lysosomes and mitochondria.
6. Centrifuge the supernatant at 40,000 × *g* for 30 min.
7. Remove the supernatant. The pellet consists of microsomes.
8. Centrifuge the supernatant at 100,000 × *g* for 90 min.
9. Remove the cytosolic fraction from above the light microsomal and ribosomal pellet.

Comments: Perform all steps at 4°C.

This method can easily be extended to enrich for proteins that reside within a subcellular organelle by adding additional steps.

Subcellular Markers

Table 5.3 can be used as an aid for following the subcellular distribution of your target protein. Experimental fractions are assayed for the presence of specific marker proteins. The distribution of the target protein can be compared to the distribution of the standard enzyme proteins listed in the table. In most instances, the target protein will not be found exclusively in one fraction due to leakiness introduced during the manipulations carried out in previous steps.

TABLE 5.3 Marker Enzyme Subcellular Distribution

Enzyme	Location	Reference
Lactate dehydrogenase	Cytosol	Neilands, 1955
Glutathione-S-transferase	Cytosol	
Carbonic anhydrase	Cytosol	
5′-nucleotidase	Plasma membrane	Ray, 1970
b-Galactosidase	Lysosome	
Acid phosphatase	Lysosome/golgi	
Glucose-6-phosphatase	Endoplasmic reticulum	Morre, 1971
Cytochrome p450 reductase	Endoplasmic reticulum	
NADPH cytochrome c reductase	Microsome	Sottocasa et al, 1967
Galactosyltransferase	Golgi	Morre et al, 1969
DNA	nucleus	
Actin	Cytoskeleton	
Tubulins	Cytoskeleton	
Vinculin	cytoskeleton	
P38	Nuclear membrane	Ramsby et al, 1994
α-connexin	Gap junction	

B. Protein Quantitation

Accurate determination of protein concentrations is fundamental for all quantitative measurements of biochemical interactions. If it is critical for the next step in a purification protocol to know the protein concentration, you should usually not delay to measure protein concentration. In most cases, small samples are put aside at every step and protein quantitation measurements performed later. The protein quantitation assay should be rapid, reliable and resistant to potentially interfering substances. The times when accurate protein estimation is needed are: (1) when a fractionation step is critically dependent on the protein concentration; (2) when it is necessary to know whether a particular step has really removed much unwanted protein; (3) to assess the specific activity and follow the progress of the purification. From experience, you will develop intuitive skills to know how the purifications are progressing, i.e., how much protein is contained in a particular precipitate, and the approximate protein concentration corresponding to a peak on a chart recorder.

A quick, useful estimation of protein concentration is to simply read the OD (optical density) at 280nm of the protein solution using ultraviolet spectrophotometry with quartz cuvettes. Proteins absorb at 280nm due to the presence of tyrosine and tryptophan residues. Since the content of these two amino acids varies markedly from protein to protein, the extinction coefficient, usually expressed either as $E^{1\%}{}_{280}$ or $E^{1mg/ml}{}_{280}$ varies considerably (Gill and von Hippel, 1989). Most proteins fall in the range of OD_{280}-0.4-1.5 for a 1mg/ml solution. Therefore, although highly inaccurate, as a first approximation, a protein solution that has an OD_{280} of 1 translates roughly to 1mg/ml of protein. Crude preparations represent the average of a large number of different proteins so generalizations can be made more easily than for pure solutions. Be aware of interfering substances; e.g., nucleic acids that absorb in the same range and some commonly used detergents, i.e., Triton X-100 absorbs at 280nm. If the extinction coefficient of the pure protein is known, then the reading at 280nm can provide an accurate measure of protein concentration (Scopes, 1987). See Appendix C for a more detailed explanation of the use of extinction coefficients.

Due to the importance of knowing the protein concentration during analysis and purification, four methods are presented for determining protein concentration. When choosing a method, give consideration to potential interfering substances, availability of instrumentation, and desired sensitivity.

The Bradford method (Bradford, 1976) is widely used, replacing the classical Lowry method (Lowry et al, 1951). The bicinchoninic acid (BCA) method is also described. One major advantage of the BCA method is that it can be performed in the presence of detergents. In addition, a more sensitive method based on the generation of a fluorescent product (Jones et al, 2003) is presented for determining protein concentrations below the detection limits of the Bradford and BCA assays. Other methods of protein quantitation in the nanogram range have been reported. For example, a method using o-phthalaldehyde has been reported to detect 10ng of protein (Weidekamm et al, 1973).

Accurately determining protein concentrations at the nanogram level is difficult. A semi-quantitative method (Crawford and Beckerle, 1991) compares the intensity of a band in PAGE with bands of known concentration. Protein concentrations can be estimated from a standard curve of the densitometric intensities of known amounts of a protein that has been resolved on SDS-PAGE and stained with Coomassie blue (see chapter 4).

The Bradford Method

The Bradford protein assay (Bradford, 1976) involves the addition of an acidic dye, Coomassie Brilliant Blue G-250, to the protein solution. The Coomassie blue dye binds to basic and aromatic amino acids resulting in a shift of the absorbance maximum from 465nm (brownish) to 595nm (blue). The method is based on the production of a standard curve. OD_{595} values of experimentally generated fractions are fitted to the standard curve, and protein content is determined. A list of commonly used biochemicals and the maximum concentrations if appropriate that are allowed in the Bradford assay is presented below.

- Acetate, 0.6M
- Acetone
- Adenosine, 1mM
- Amino Acids
- Ammonium sulfate, 1.0M
- Ampholytes, 0.5%
- Acid pH
- ATP, 1mM
- Barbital
- BES, 2.5M
- Boric Acid
- Cacodylate-Tris, 0.1M
- CDTA, 0.05M
- Citrate, 0.05M
- Deoxycholate, 0.1%
- Dithiothreitol, 1M
- DNA, 1mg/ml
- EDTA, 0.1M
- EGTA, 0.05M
- Ethanol
- Eagle's MEM
- Earle's salt solution
- Formic acid, 1.0M
- Fructose
- Glucose
- Glutathione
- Glycerol, 99%
- Glycine, 0.1M
- Guanidine-HCl
- Hank's salt solutiont
- HEPES buffer, 0.1M
- KCl, 1.0M
- Malic acid, 0.2M
- $MgCl_2$, 1.0M
- Mercaptoethanol, 1.0M
- MES, 0.7M
- Methanol
- MOPS, 0.2M
- NaCl, 5 M
- NAD, 1mM
- NaSCN, 3M
- Peptones
- Phenol, 5%
- Phosphate, 1.0M
- PIPES, 0.5M
- Polyadenylic acid, 1mM
- Polypeptides (MW $<$ 3000)
- Pyrophosphate, 0.2M
- rRNA, 0.25mg/ml
- RNA, 0.4mg/ml
- total RNA, 0.30mg/ml
- SDS, 0.1%
- Sodium phosphate
- Streptomycin sulfate, 20%
- Triton X-100, 0.1%
- Tricine
- Tyrosine, 1mM
- Thymidine, 1mM
- Tris, 2.0M
- Urea, 6M
- Vitamins

PROTOCOL 5.4 Bradford Standard Assay

The Bradford assay detects proteins with molecular weights greater than 3–5kDa. The amount of protein present in an experimental sample is determined by performing a simple colorimetric reaction and comparing the results with those obtained from standard amounts of protein. This particular protocol is recommended for 20–140µg protein (200–1400µg/ml).

Materials

Plastic disposable cuvettes

Bradford dye reagent concentrate: (Bio-Rad)

Diluted dye reagent: 1 part Dye Reagent Concentrate plus 4 parts distilled water. Filter through Whatman #1 or equivalent filter. The diluted dye reagent may be used for approximately 2 weeks when kept at room temperature.

Test tubes: glass 13 × 100 mm

Spectrophotometer

Protein standard: 1 mg/ml BSA or IgG in water

1. Construct a standard curve by preparing four dilutions from a 1mg/ml protein stock solution of BSA or IgG. This is accomplished by adding 0 (the blank), 10, 20, 40, and 80 µl of protein stock solution to individual tubes and adjusting the volume to 100 µl with water. The tubes will contain 0–80 µg of total protein in 100 µl. It is recommended to prepare a standard curve each time that the assay is performed.
2. Pipet 0.1 ml of the experimental samples into clean, dry test tubes. Also prepare two blanks.
3. Add 5 ml of the diluted dye reagent to all test tubes.
4. Mix the tubes by gentle vortexing. Avoid foaming.
5. Let the tubes stand at room temperature for at least 5 min but not longer than 1 h.
6. Auto zero the spectrophotometer by reading the reagent blanks (prepared in step 2) against each other at 595 nm in the reference and sample cells of your double beam (if you have one) spectrophotometer.
7. With a blank in the reference cell of the spectrophotometer, measure the OD_{595}. The color of the blank should be rusty-brown but will contribute to the absorbance at OD_{595}. Alternatively, all tubes can be read versus water (in the reference cell) and the readings corrected for the blank. Preferably, use plastic disposable cuvettes. The dye will bind to quartz and glass but can easily be removed by washing with methanol or ethanol.
8. Construct the standard curve by plotting OD_{595} of the standards versus µg protein of the unknown. From the graph, determine the concentration of unknown protein by reading the values from the standard curve. Calculate the protein concentration of the experimental samples. Be sure to include the dilution factor. For example, if 100 µl of a 1:10 dilution of the unknown protein solution gives a reading comparable to 60 µg from the standard curve, the concentration of the unknown is 60 µg/100 µl = 600 µg /ml × 10 (the dilution factor) = 6 mg/ml. A representative standard curve is shown in Figure 5.1.

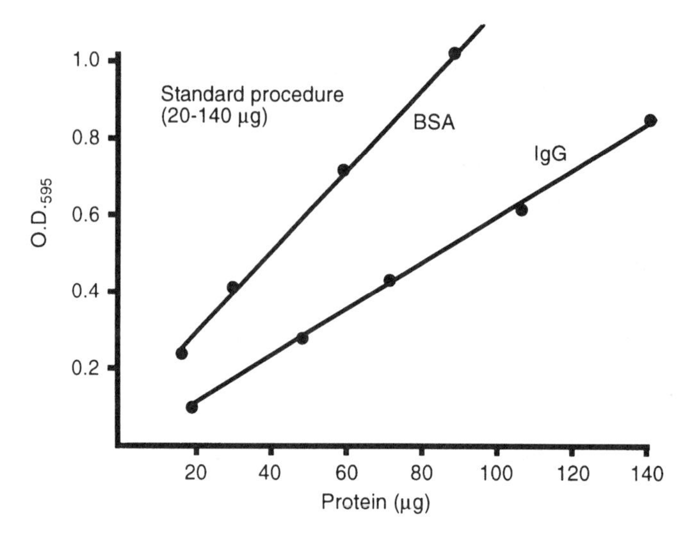

FIGURE 5.1 Standard curve for the Bradford standard protein assay. Two proteins, BSA and bovine IgG, are shown to illustrate the variation between different proteins in the assay. (Figure kindly provided by Bio-Rad Laboratories, Life Science Group, Hercules, CA).

PROTOCOL 5.5 Bradford Microassay

The Bradford microassay is generally used for dilute solutions in the 1–20 µg of protein range.

Materials

Glass test tubes

Disposable 1 ml plastic cuvettes

Bradford dye reagent concentrate: (Bio-Rad)

Protein standard: 1 mg/ml BSA or IgG in water

Spectrophotometer

1. Prepare two blanks and protein standards of 1, 3, 5, 8, and 15 µg each in 0.8 ml of water.
2. Pipet 0.8 ml of the unknowns into clean, dry test tubes. The 0.8 ml sample can consist totally of the unknown protein solution if you believe it to possess a low protein concentration, or it may contain a dilution if the protein is concentrated.
3. Add 0.2 ml of the concentrated dye reagent.
4. Mix the tubes, being careful to avoid foaming. Incubate at room temperature for 5 min.
5. Zero the spectrophotometer at OD_{595} using the two blanks. Read the OD values versus the blank in the reference cuvette.
6. Plot a curve using the values obtained from the standards OD_{595} versus µg protein as illustrated in Figure 5.2. Calculate the protein concentration of the unknowns from the standard curve and the dilution factor if necessary.

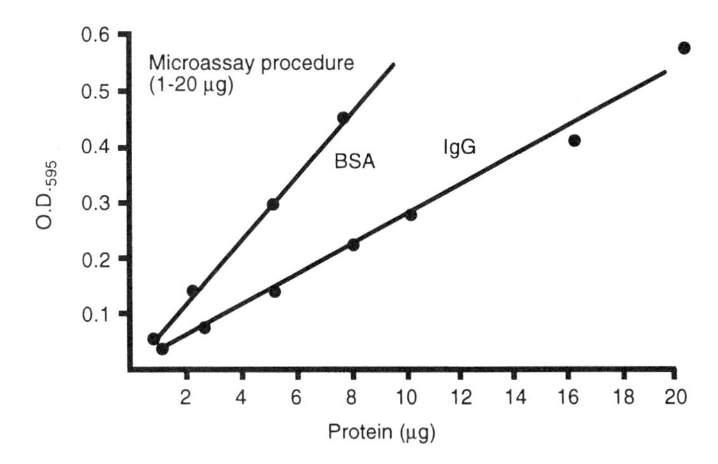

FIGURE 5.2 Standard curve for the Bradford microassay. Two protein standards are used, BSA and bovine IgG. (Figure kindly provided by Bio-Rad Laboratories. Life Science Group, Hercules, CA).

Comments: A standard curve for the Bradford microassay is shown in Figure 5.2. Both Bradford assays will only detect proteins with molecular weights greater than 3–5kDa.

If the OD_{595} value of an experimental sample is greater than the highest standard, try adding 0.8ml of H_2O and 0.2ml of concentrated dye solution to the sample. If this places the OD value of the unknown on the standard curve, then there is no need to prepare a new, more diluted sample of the same fraction. Consider this a twofold dilution.

The Bradford method can also be performed using a plate reader with a final volume of 200µl. To use this tecnique, scale down the volumes given for the micro method.

PROTOCOL 5.6 Protein Determination Using Bicinchoninic Acid (BCA)

When protein is placed in an alkaline system containing Cu^{+2}, a colored complex forms between the peptide bonds of the protein and the copper atoms. Bicinchoninic acid (BCA) is a compound that forms a complex with cuprous ion (Cu+1) in an alkaline environment resulting in a stable, highly colored chromophore with an absorbance maximum at 562nm (Smith et al, 1985). This reaction forms the basis of the BCA protein quantitation system. The BCA reagent shows greater tolerance than both the Bradford and the Lowry protein assays toward commonly encountered interfering compounds in buffer systems. The BCA protein assay is favored for determining protein concentrations in the presence of detergents. The list below shows substances and concentrations that are compatible with the BCA protein assay. However, it is sensitive to interference from reducing sugars.

Compatible Substances for the BCA Protein Assay

0.5% NP-40	6.0M urea
0.5M HEPES	40% Sucrose
1.0M NaCl	1.0M Glycine, pH 11.0
0.1M NaOH	50mM glucose
3% Ammonium sulfate	50mM EDTA
1% Triton X-100	0.2mM DTT
1% Octyl-β-Glucoside	4.0M Guanidine HCl
2.0M Sodium acetate, pH 5.5	1% Lubrol®
1% SDS	1% CHAPS
1% Brij®-35	

Incompatible Substances

100mM EDTA	1mM DTT	20% ammonium sulfate

Materials

Reagent A: 1% BCA-Na_2, 2% $Na_2CO_3 \cdot H_2O$, 0.16% Na_2 tartrate, 0.4% NaOH, and 0.95% $NaHCO_3$. If needed, make the appropriate addition of NaOH (50%) or solid $NaHCO_3$ to adjust the pH to 11.25.

Reagent B: 4% $CuSO_4$ $5H_2O$

Reagents A and B are stable indefinitely at room temperature.

Standard Working Solution (S-WR): Mix 100vol of Reagent A with 2vol of Reagent B. The S-WR should be green.

Spectrophotometer

Cuvettes: 1ml plastic disposable

Protein Standard: BSA 1mg/ml in H_2O. Store aliquotted at –20°C.

1. Follow the instructions in step 1 of the Bradford assay (Protocol 5.5).
2. Add 2ml of S-WR to all test tubes.
3. Incubate at 37°C for 30min. Color development begins immediately. Cool the samples to room temperature and then measure absorbances at 562nm versus the reaction blank.
4. Construct a standard curve by plotting the absorbance at 562nm of the standard versus μg protein. Determine amounts of protein in the experimental samples from the standard curve. Knowing the volume of experimental sample used and the dilution used (if any) calculate the protein concentration.

Comments: The BCA reagents are also marketed in kit form by Pierce. This method can be downscaled to be performed using a plate reader. The method can also be used in a modified form with increased sensitivity (Smith et al, 1985) to assay the protein concentrations of dilute solutions (0.5–10μg/ml).

When constructing the standard curve, dilute the standards with the same diluent used for the unknowns.

PROTOCOL 5.7 NanoOrange® Protein Quantitation Assay: A Fluorescence-Based Assay of Proteins in Solution

Fluorescence based techniques offer higher sensitivity, lower background signals and a wider dynamic range than absorbance based techniques. There are two general methods for fluorescence based protein quantitation. One method involves non-fluorescent reactive dyes that couple with protein amines to form fluorescent, covalent adducts, exemplified by fluorescamine (Bohlen et al, 1973). The second method uses dyes that exhibit fluorescence enhancement upon non-covalent interaction with hydrophobic regions of proteins or detergent coated proteins (Haugland, 1996).

The protein quantitation assay described below is based on the dye, NanoOrange®, which is essentially nonfluorescent in aqueous solution but becomes intensely fluorescent upon binding to detergent-coated proteins or hydrophobic regions of proteins (Jones et al, 2003). As little as 100 ng/ml protein could be detected in a 200 μl volume, yielding a sensitivity of 20 ng protein per sample using a fluorescence microplate reader. The upper limit of the assay range is approximately 10 μg/ml using either a microplate reader or a fluorometer.

Materials

NanoOrange protein quantitation reagent (Molecular Probes Eugene, OR)

Assay diluent: 10 mM Tris-HCl, pH 7.5, with a proprietary mixture of anionic detergents (Molecular Probes)

BSA standard: 2 mg/ml in water

Glass test tubes 13 × 100 disposable or Microfuge tubes

Acrylic cuvettes or acrylic microplates

Fluorometer

1. Prepare the working NanoOrange reagent solution by diluting the dye 500-fold into assay diluent as described in the manufacturer protocol.
2. Dilute BSA standards into the NanoOrange reagent in assay diluent using 2.5 ml for 13-mm disposable glass test tubes or 250 μl for microfuge tubes. Include a tube with no protein as a background fluorescence control.
3. Protect all tubes from light throughout the procedure to reduce photobleaching.
4. Heat the samples for 10 min at 96°C and cool for 20 min at room temperature, protected from light.
5. Briefly mix the samples and transfer them to disposable acrylic cuvettes (2 ml volume) or acrylic microplate wells (200 μl volumes).
6. Read fluorescence intensities at 485 nm excitation and 590 nm emission settings or suitable filters.
7. Determine sample protein concentrations from the standard curve.

Comments: A list of tolerance levels for potential contaminants is presented in Jones et al (2003). However, the high sensitivity of the assay makes it possible to dilute most potential contaminants to acceptable levels.

C. Manipulating Proteins in Solution

The extract containing the protein of interest should be treated with care as it represents an investment of many hours. Therefore, techniques that are designed to manipulate the extract into a form conducive for assaying or for the next step in the purification protocol or for storage with maximum retention of activity are described.

Stabilization and Storage of Proteins

Stabilization of proteins to withstand conditions that lead to denaturation has been found necessary in many cases. A widely used method is the inclusion of glycerol in buffer solutions. Levels of glycerol from 10–50% w/v have been used with the higher concentrations reserved for storage. Protocols run in the presence of glycerol as a stabilizer are more time consuming due to the high viscosity of glycerol.

High salt concentrations stabilize enzymes and generally inhibit proteases. Many enzymes are typically supplied in 50% ammonium sulfate. If an ammonium sulfate fractionation is being carried out (Protocol 5.11), leave the protein in as high a concentration of ammonium sulfate as possible, that is to say, store the material as a "wet pellet".

Dilute enzyme solutions lose activity quickly. Therefore, try to store the protein at a concentration of at least 1 mg/ml. A small amount of purified protein can adsorb to the walls of the container. Extreme dilution of protein may lead to instability. The addition of bovine serum albumin (BSA) can act as an enzyme stabilizer. Used at concentrations up to 1 mg/ml, BSA will prevent the adsorption of the protein of interest onto container walls and increase the stability of proteins.

Frozen Storage

During the freezing process many events occur. First, free water freezes and ice crystals grow which can be destructive to membranes and organelles. The least soluble solute will then precipitate. If the solute is one component of the buffer, the pH will markedly change before complete solidification takes place. The higher the protein concentration relative to buffer salts, the more it will be capable of acting as a buffer itself and counteracting drastic pH shifts that may occur during freezing. If the temperature is low enough, all degradative processes stop, and the sample can theoretically be kept indefinitely. A temperature below –50°C reached as quickly as possible is recommended (snap freezing). Normal deep freeze temperatures are usually suitable for overnight storage. Care should be taken when using a frost-free freezer as the defrost cycles could damage the protein. Freezing at –10° to –15°C is probably no better than not freezing at all. When thawing, the rule is the faster the better. Immerse the container in warm 40–50°C water shaking frequently. Remove the container when there is still a small piece of ice remaining.

Try to avoid repeated cycles of freezing and thawing. Store purified proteins in small portions and thaw individual samples once, as needed. Alternatively, the protein may be stored under conditions in which it does not freeze, such as high glycerol concentration. Many restriction endonucleases used in molecular biology are supplied as glycerol solutions.

Protecting Cysteine Residues

Cysteine residues are susceptible to modification, especially oxidation, during purification and storage. Normally, within the living cell in a reducing atmosphere, the presence of other sulfhydryl-containing molecules like glutathione, protect these groups. However, when exposed to high oxygen tensions, several reactions are possible including disulfide bond formation, partial oxidation to a sulfinic acid, and irreversible oxidation to a sulfonic acid.

Formation of a disulfide bond requires another sulfhydryl group to be in the vicinity. Disulfide bond formation is accelerated in the presence of divalent cations which activate oxygen molecules and complex with sulfhydryls. For these reasons two protective actions are routinely taken. Metal ions are removed from the solution with the inclusion of the chelating agent EDTA. A sulfhydryl-containing reagent such as β-mercaptoethanol (2-ME) or dithiothreitol (DTT, Cleland's reagent) (Cleland, 1964) is added to the buffer. Routinely, 2-ME at 5–10 mM or DTT at 1–5 mM is added to the buffer along with 0.1–0.2 mM EDTA.

Concentrating Proteins from Dilute Solutions

Frequently, a protein solution is not in the optimal state for the next purification step. The target protein might be in the wrong buffer, have too much salt, or be too dilute. Concentrated solutions can be diluted easily. However, the converse is often not as simple or as rapid. Biologically active material present in a large volume, like conditioned media, must be concentrated prior to proceeding. Concentration of dilute protein solutions involves the removal of water which can be accomplished in a number of ways. The chosen method will depend on a particular application and specific requirements. Precipitation methods are best suited for protein concentrations above 1 mg/ml. The lower the protein concentration in solution, the more difficult it will be to quantitatively recover the target protein. It has been reported that trichloroacetic acid (TCA) does not precipitate protein in the range of 1–25 μg/ml (Bensadoun and Weinstein, 1976). Methods that do not involve precipitation are ultrafiltration, dialysis, evaporation, adsorption, and lyophilization. The method you choose for protein concentration should be based on speed, protein stability, familiarity, and convenience. Once the cell extract has been reduced to a manageable volume, the real purification can begin.

Adsorption

Concentration of a dilute protein solution can be achieved by using an ion exchanger. Be sure that the protein of interest is totally adsorbed. Although it does not matter if other proteins are adsorbed, it will be advantageous if unwanted proteins do not adhere. For example, a dilute protein solution eluted from a DEAE matrix at pH 6 at low ionic strength will probably adsorb to a small column of DEAE at pH 8 and can then be eluted with a small volume of high salt containing buffer.

PROTOCOL 5.8 Recovery of Protein by Ammonium Sulfate Precipitation

Solubilities of proteins are decreased by high concentrations of neutral salts (salting out). The protein of interest can often be salted out of the crude extract with enrichments of from two- to tenfold. In addition to purification, the target protein is now present in a reduced volume.

Salting out by adding sufficient amounts of ammonium sulfate is a convenient, nondenaturing way to concentrate a protein, provided that the solution is not too dilute to start with. If the starting protein concentration is less than 1 mg/ml, you may try ultrafiltration prior to salting out. Selective precipitation with ammonium sulfate is also used as an early purification step described in Protocol 5.11.

Materials

Ammonium Sulfate: Solid ultrapure
Magnetic stirrer

1. Dissolve 60 g solid ammonium sulfate for every 100 ml of solution. This produces an 85% saturated solution which should quantitatively precipitate most proteins.
2. Keeping the suspension at 4°C, stir the solution for 30 min then centrifuge the slurry for 15 min at 10000 × *g*.
3. Decant the supernatant and save.
4. Solubilize the pellet in the smallest allowable volume of buffer. You may need to dialyze this solution before proceeding further.

Comments: Assay the supernatant for the presence of target protein before discarding.

Ultrafiltration

Ultrafiltration, also referred to as reverse osmosis, is a process in which solvents and solutes, up to a certain critical size, are forced through a barrier membrane by a higher pressure on one side of the membrane than the other. There is always a flow of solvent moving through the barrier in the same direction as the smaller solutes that are also able to pass through the membrane. Ultrafiltration uses pressure to force liquid through a membrane. This technique is primarily used for concentrating proteins in dilute solutions. It can be used as a purification step, but it only provides two fractions: bigger molecules found in the retentate, and smaller molecules passing through the membrane and found in the ultrafiltrate.

Ultrafiltration membranes come in a wide range of pore sizes; however, the cutoff values are not absolute. Rather than refer to the pore size of an ultrafiltration membrane, it is more common to use a nominal molecular weight cutoff (NMWC) for the membrane. The NMWC is defined as the minimum molecular weight globular protein which will not pass through the membrane. The pore sizes are not uniform and will show a normal distribution around the mean pore size. A proportion of molecules with molecular weights close to the stated cutoff size will pass through the membrane; the remainder will be retained and stay behind. The NMWC of the membrane used should be significantly less than the molecular weight of the target protein, (usually ≥20% less). If the protein of interest is either much smaller or larger, by at least 30–50%, than the cutoff size, then the majority of it will appear in either the ultrafiltrate or the retentate.

Ultrafiltration slows as the protein concentration builds up at the surface of the membrane, forming a gel-like layer. This layer can be minimized by stirring or by applying a cross-flow. Fouling of the membrane occurs when particles or macromolecules become adsorbed to the membrane or become embedded in the pores.

Several companies market two-compartment centrifuge tubes that are separated by a membrane allowing small molecules to pass while retaining the high molecular weight combining centrifugation with an ultrafiltration semi-permeable membrane (Pall Filtron Corporation). An example of this system is shown in Figure 5.3. The devices are available in a wide range of molecular weight cut-offs from 3–100 kDa. Most of these units are suited for small volumes (<5 ml) and are often used for concentration or removal of a low molecular weight contaminant rather than for purification.

Ultrafiltration can be used to remove salts or change the buffer composition of the protein solution, a process referred to as diafiltration. A concentrated protein solution is diluted with water or a specific buffer and ultrafiltration is continued until the retentate reaches the desired ionic strength or pH.

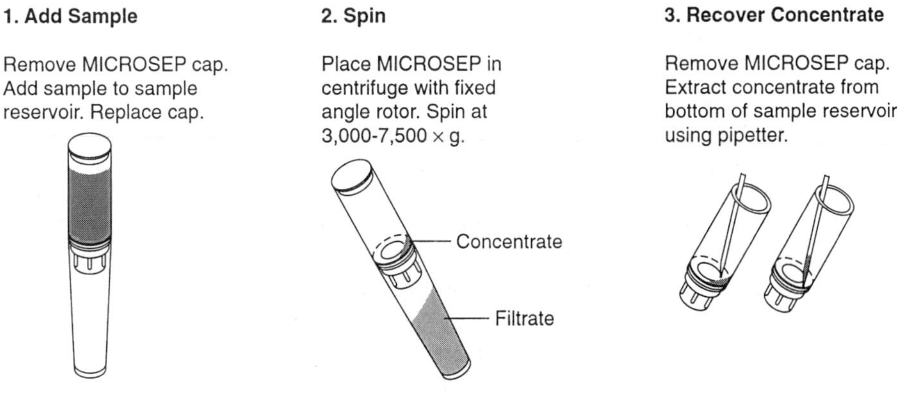

1. Add Sample

Remove MICROSEP cap. Add sample to sample reservoir. Replace cap.

2. Spin

Place MICROSEP in centrifuge with fixed angle rotor. Spin at 3,000-7,500 × g.

Concentrate

Filtrate

3. Recover Concentrate

Remove MICROSEP cap. Extract concentrate from bottom of sample reservoir using pipetter.

FIGURE 5.3 Microsep™ centrifugal concentrator. (Figure kindly provided by Pall Filtron corporation.)

Lyophilization

Lyophilization, also known as freeze drying, is a sublimation process where water and other frozen solvents are removed under vacuum in a progressive and careful dehydration. This results in the drying of the solute components of the solution, proteins and any buffer components that were not previously removed by dialysis. Routinely, a protein solution to be lyophilized is first dialyzed against water to remove salts and buffer components. Using a bath consisting of dry-ice and acetone, the protein solution is placed in a special flask and shell frozen" by slowly rotating the flask in the ice bath to maximize the surface area. Once frozen, the flask is mounted on the lyophilizer (VirTis, or Labconco) and the valve is opened putting the contents of the flask under vacuum. Depending on the volume of the solution, allow from 2h to overnight. After a short time, check the contents of the flask to make sure it has not thawed. If there is any liquid in the flask, take the flask off line by closing the valve then slowly release the vacuum to the flask. Refreeze the flask and remount it onto the lyophilizer. When the outside of the flask reaches room temperature the freeze drying process is complete.

Phosphate buffers are not ideally suited for lyophilization since the pH will drop on freezing which may result in the denaturation of the target protein. Solutions prepared with volatile buffers such as ammonium bicarbonate are preferable, minimizing interference in subsequent purification steps.

Dialysis

Dialysis is a separation process that takes advantage of osmotic forces between two liquids or a liquid and a solid. Dialysis is used for removing excess low molecular weight solutes and simultaneously equilibrating the sample in a new buffer, and as a means of concentrating a dilute solution. Dialysis tubing is a semi-permeable membrane, usually made from cellulose acetate, available in a wide range of dimensions and nominal molecular weight cut-offs (NMWC) allowing molecules below a certain molecular weight to freely equilibrate on both sides of the membrane. Practical removal of a dialyzable component from within the bag cannot be accomplished without changing the dialysate solution at least once.

Solutions may be concentrated by "dialysis" against a high molecular weight solid hydroscopic substance. The solution is placed in dialysis tubing which is then coated with an inert, high molecular weight hydroscopic substance that "pulls" water out of the bag. Commonly used materials are polyethylene glycol (MW 20,000+) and aquacide (Calbiochem). Dry Sephadex can also be used. It is not as messy as PEG but more expensive. A 10ml solution can be concentrated to less than 1ml within an hour by coating the dialysis bag with aquacide and occasionally stripping off the hydrated gel from the outside of the bag. Care should be taken not to bring the contents of the tubing to dryness, which would result in loss of the protein by irreversible absorption to the dialysis membrane.

PROTOCOL 5.9 Preparation of Dialysis Tubing

Materials

Dialysis tubing: Spectrum Medical Industries
Equilibration buffer: 2% sodium bicarbonate, 1 mM EDTA
1 mM EDTA
Storage solution: 50% ethanol

1. Select tubing with a diameter suitable for the job and cut the required length from the role.
2. Cut the tubing into convenient lengths.
3. Boil the tubing in a large beaker for at least 10 min in equilibration buffer. To keep the tubing submerged during boiling, partially fill an Erhlenmeyer flask with water and place it in the beaker.
4. Rinse the tubing thoroughly with distilled water.
5. Boil for 10 min in 1 mM EDTA and rinse in distilled water.
6. Store the tubing at 4°C in the storage solution making sure that the tubing is always submerged to prevent drying.

Comments: Alternatively, the dialysis tubing can be used directly from the dry roll. After wetting the tubing with water, tie a knot in one end. If available, dialysis clips can be used. Sample is loaded into the tubing. The tubing is closed either by tying a knot or using a clip. High concentrations of salt in the sample will cause water to enter the tubing before the salt can exit, resulting in a volume increase during the early stages of dialysis. If there is a high solute concentration in the tubing, space should be left for expansion. High pressure in the tubing might result in leakage through the knots. Place the loaded dialysis tubing in a beaker containing the dialysis buffer and a magnetic stirrer.

A solution containing $(NH_4)_2SO_4$ can be dialyzed without agitation. Float the dialysis tubing in a tall, graduated cylinder. The relatively dense $(NH_4)_2SO_4$ exits the tubing and sinks to the bottom of the cylinder, leaving the solute concentration very low in the vicinity of the tubing. When removing the sample from the tubing, care should be taken to avoid spillage and sample loss.

Slide-A-Lyzer® Dialysis Cassettes (Pierce) are ready-to-use disposable devices that feature a silicon gasket with a 10,000 molecular weight cut-off (MWCO) membrane on either side, forming a hermetically sealed sample chamber. The increased surface-to-volume ratio makes dialysis faster than conventional tubing. An accessory buoy floats the cassette during dialysis and also serves as a stand for the cassette during sample loading and recovery.

The cassettes, which come in different sizes, can accommodate sample volumes between 0.5–15 ml. Their use is shown in Figure 5.4.

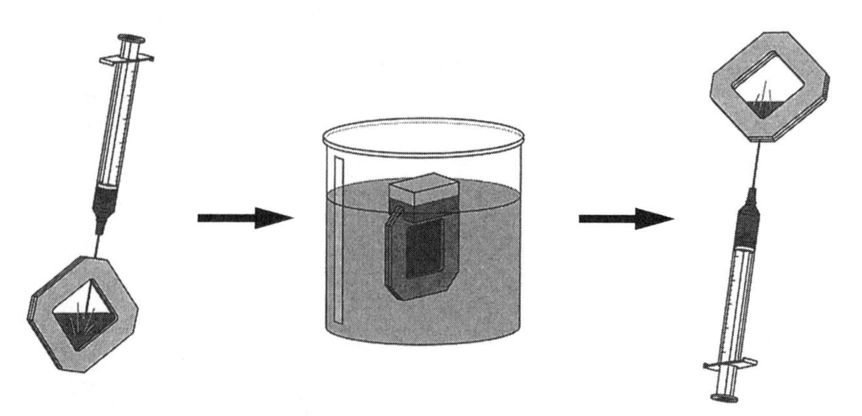

FIGURE 5.4 Depiction of a Slide-A-Lyzer† Dialysis Cassette accommodating sample between 3 and 15 ml. Patent of Pierce, #5503741.

Changing the Buffer by Gel Filtration

Gel filtration can be used for removing salts from the protein solution and for changing the buffer. The final volume following gel filtration will probably be larger than the initial volume. If the protein concentration of the solution is already low and the volume large, it may be advantageous to concentrate the protein by ammonium sulfate precipitation (Protocol 5.8) and then use a small desalting column. The medium chosen should totally exclude the target protein. A suitable quantity of Sephadex G-25 or Biogel P-30 is packed in a column and equilibrated with the desired new buffer (see Chapter 10 for details). The sample volume should not be greater than one-fifth that of the volume of gel in the column so that the change can be accomplished in one pass. Small columns should be packed with finer grade gel beads for optimum resolution (see Table 5.4).

Dissolve an ammonium sulfate precipitate with a minimal amount of buffer, then dilute with an equal volume of buffer prior to desalting on a column. Protein concentrations should not be greater than 30 mg/ml.

Small volumes of <0.5 ml can be desalted rapidly with virtually no dilution with a spin column (Protocol 10.4). The protein emerging from

TABLE 5.4 Dimensions of Columns for Desalting Using Sephadex G-25*

Column dimensions [cross-section] area (cm^2) × height (cm)]	Column vol. (ml)	Sample size (ml)	Flow rate (ml/h)	Time for completion (min)
1 × 8	8	1–1.5	40	7
6 × 10	60	2–10	200	15
8 × 30	240	10–30	250	30
8 × 60	480	30–80	250	45

*Data from Scopes (1987), page 19. Reprinted with permission of Springer-Verlag.

the column can be conveniently monitored by UV absorption. Small, prepacked desalting columns are commercially available.

D. Precipitation Techniques

Protein precipitates are aggregates of protein molecules large enough to be visible and to be collected by centrifugation. The distribution of hydrophilic and hydrophobic residues at the surface of a protein determines its solubility properties. The solubility of a protein is the result of polar interactions with the aqueous solvent, ionic interactions with salts and repulsive electrostatic interactions between like charged molecules. The solvent properties of water can be altered by changing ionic strength and pH. The addition of miscible organic solvents, other inert solutes and polymers with temperature variation can all be manipulated to cause selective precipitation.

Isoelectric precipitation can occur when the overall charge on a protein is near zero, minimizing electrostatic repulsion and causing molecules to attract each other. In a crude mixture, different molecules with similar properties can coprecipitate. Since most isoelectric precipitates of crude extracts are aggregates of many different proteins, they may also include protein-nucleic acid complexes. In addition, if the initial composition of an extract is changed, a desired protein may not exhibit the same solubility behavior if its partners in precipitation are absent. The increase in solubility with increasing salt concentration is known as salting in.

PROTOCOL 5.10 Salting Out with Ammonium Sulfate

Salting out with ammonium sulfate is a technique that is used as an early step in a purification scheme. Even if a significant purification is not achieved, broad ammonium sulfate cuts are at least a volume reducing step. A large volume of extract can be reduced to a volume suitable for subsequent purification steps (described earlier for quantitive recovery of proteins in a complex mixture).

Protein precipitation using ammonium sulfate is achieved by dehydration in the microenvironment of the protein molecule. In solution, a large number of water molecules are bound to the sulfate ion (SO_4^{-2}), reducing the amount of water available to interact with the protein molecules. At a particular concentration of $(NH_4)_2SO_4$, an insufficient quantity of unbound water will remain to keep a given protein species in solution, resulting in the precipitation of that protein.

The solubility of ammonium sulfate is constant between 0–30°C. A saturated solution in water is ~4M with a density of 1.235g/ml compared to a protein aggregate density in this solution of ~1.29g/ml. This difference makes it possible to pellet out the aggregates by centrifugation.

Like most steps in a protein purification protocol, $(NH_4)_2SO_4$ fractionation is a compromise between recovery and purity. Keep the temperature low to increase stability and decrease solubilities. It is best to operate at a neutral pH between 6–7.5.

After deciding to perform an ammonium sulfate fractionation the next decision is what percentage saturation to try. According to the volume of the protein solution, perform a quick calculation and add solid ammonium sulfate to reach the final desired concentration by consulting Table 5.5. Material precipitating prior to 25% $(NH_4)_2SO_4$ saturation is generally particulate, preaggregated or very high molecular weight protein. The best conditions for enriching a specific protein from a complex mixture are reached empirically. As a first approximation the following percentages should be helpful: 0–25%, 25–4.0%, 40–60%, 60–80%, and the 80% supernatant.

Materials

Ammonium sulfate
Magnetic stirrer

1. Weigh out the appropriate amount of solid ammonium sulfate according to the starting volume (see Table 5.5). Break up any lumps and slowly add the solid ammonium sulfate to the protein solution at 4°C while stirring. Rapid addition of the salt may lead to very high local concentrations and excessive precipitation of undesired proteins. Do not stir vigorously as this could cause denaturation. The degree of foaming is a good indication whether stirring is too vigorous.
2. After the salt has dissolved, continue stirring for 30 min to allow equilibrium to be reached between dissolved and aggregated proteins.
3. Centrifuge at 10,000 × *g* for 10 min at 4°C. At higher salt concentrations somewhat more centrifugation may be needed.
4. Decant and save the supernatant, record the volume and calculate how much ammonium sulfate will be needed for the next cut.

TABLE 5.5

Initial concentration of ammonium sulfate, % saturation	Final concentration of ammonium sulfate, % saturation: 10	20	25	30	33	35	40	45	50	55	60	65	70	75	80	90	100
	Grams solid ammonium sulfate to be added to 1 L of solution																
0	56	114	144	176	196	209	243	277	313	351	390	430	472	516	561	662	767
10		57	86	118	137	150	183	216	251	288	326	365	406	449	494	592	694
20			29	59	78	91	123	155	189	225	262	300	340	382	424	520	619
25				30	49	61	93	125	158	193	230	267	307	348	390	485	583
30					19	30	62	94	127	162	198	235	273	314	356	449	546
33						12	43	74	107	142	177	214	252	292	333	426	522
35							31	63	94	129	164	200	238	278	319	411	506
40								31	63	97	132	168	205	245	285	375	469
45									32	65	99	134	171	210	250	339	431
50										33	66	101	137	176	214	302	392
55											33	67	103	141	179	264	353
60												34	69	105	143	227	314
65													34	70	107	190	275
70														35	72	153	237
75															36	115	198
80																77	157
90																	79

*Reprinted with permission, from Kaplan, 1955.

5. Dissolve the precipitate in a suitable buffer. The amount of buffer needed to dissolve the pellet will be roughly one to two times the volume of the pellet itself. If all the precipitate does not dissolve, insoluble material should be removed by centrifugation as this is probably denatured or particulate material.

PROTOCOL 5.11 Precipitation with Acetone

Many proteins can be precipitated by the addition of water-miscible organic solvents such as ethanol and acetone. Addition of ethanol or acetone to an aqueous protein extract will lead to protein precipitation due to a reduction in water activity. Precipitation occurs more readily when the pH of the solution is close to the pI of the target protein. Also, a larger protein will precipitate at lower concentrations of organic solvent than a smaller protein with otherwise similar properties.

As the concentration of the organic solvent increases, the dissolvent power of water for a charged, hydrophilic protein decreases. The net effect on cytoplasmic and other water soluble proteins is a solubility decrease resulting in aggregation and precipitation. Some hydrophobic, membrane bound proteins are not precipitated by organic solvents but are rather solubilized from the membrane by organic solvents as the solvent displaces the water molecules from around the hydrophobic patches of the protein.

In water-solvent mixtures, the interaction of water with proteins is reduced because water becomes ordered around solvent molecules. This promotes interprotein electrostatic and dipolar interactions. Water ordered around hydrophobic patches on proteins is displaced by solvent molecules, increasing entropy. This is energetically favorable, and proteins with large hydrophobic patches tend to stay dissolved in organic solutions.

Organic solvent precipitation is routinely carried out at 0°C and can be incubated at subzero temperatures since the miscible solvents form mixtures that freeze well below 0°C. The addition of the organic solvent causes heat evolution so care should be taken to keep the temperature at or below zero. The two most popular solvents in use are ethanol and acetone. Both are completely water-miscible, do not react with proteins and are relatively safe with regard to flammability and noxious vapors. The first addition should be performed slowly with efficient cooling. The protein concentration can be 5–30 mg/ml but the salt concentration should be around 0.05–0.2 M.

Most proteins will precipitate in the range of 20–50% v/v acetone. The first cut should be between 20–30% acetone or ethanol. There is a problem in defining percentages by volume. If 50 ml of acetone is added to 50 ml of water the final volume will be 95 ml. The volume loss is due to the formation of hydrated solvent complexes that occupy a smaller volume than their constituent components. Therefore, when giving percentages by volume, one should add, "assuming additive volumes."

Materials

Acetone

1. Add acetone to the protein solution to reach the desired percentage and incubate at 0°C for 15min.
2. Collect the precipitated material by centrifugation in a pre-cooled rotor at 3000 × *g* for 10min. Save the supernatant.
3. Redisolve the precipitate in a minimal volume of cold buffer. Do not add excess buffer. If the precipitate does not dissolve, there is probably denatured protein present.
4. To reach the next precipitation value, add more solvent to the supernatant from step 2. A formula for calculating the amount of organic solvent to add is:

$$\text{Volume to add to 1 liter to take \% from x to } y = \frac{1000(y - x)}{100 - y}\ \text{ml}$$

At 50% solvent concentration, only proteins of molecular weight less than 15,000Da are likely to remain in solution.

Comments: Addition of large volumes will result in dilution of the protein. Therefore, the starting protein concentration should be >1mg/ml.

Precipitation with Polyethylene Glycol (PEG)

The use of nonionic polymers for protein purification deserves wider application. The method is simple and results in products which appear to retain their native configuration as evidenced by measurements of biological activity (Fried and Chun, 1971).

Polyethylene glycol (PEG), available in a variety of degrees of polymerization, can be regarded as a polymerized organic solvent. Solutions of this polymer up to 20% w/v are not too viscous to preclude their use. The types most commonly used for protein precipitation are PEG-2000, PEG-4000, and PEG-6000. PEGs of higher molecular weights are not recommended because their high viscosity in water makes pelleting difficult. The useful fractionation range is between 8–20% w/v.

Although PEG is not easy to remove from a protein fraction, low levels of PEG are compatible with many procedures.

PROTOCOL 5.12 PEG Precipitation

The protocol described below is slightly unconventional in that all precipitable proteins are removed from the original extract in the first step. The ionic strength of the buffer may have an effect on the precipitation of proteins. PEG concentrations greater than 14.5% can be used. Once optimal conditions have been determined, intermediate concentrations can be skipped.

Materials

PEG-6000: 50% (w/v) PEG-6000 in distilled water warmed to 37°C to speed up the solubilization
Phosphate buffer: 55 mM potassium phosphate buffer, pH 7.0

1. Carry out all additions at 4°C with stirring. To the protein solution, add PEG solution to a final concentration of 35% (w/v) and stir for 30 min.
2. Let the suspension stand for 5 min then centrifuge for 30 min at 35,000 × *g* at 4°C.
3. Discard the supernatant and dissolve the pellet in 10 ml of phosphate buffer. Centrifuge for 30 min at 35,000 × *g* at 4°C. Discard the pellet. The supernatant should be clear.
4. Now you are ready to fractionate. Add the PEG solution to the supernatant to a concentration of 2.5% PEG. Allow the suspension to stand for 5 min, then centrifuge for 30 min at 35,000 × *g* at 4°C. The precipitate represents Fraction I. Resuspend the precipitate in the desired buffer.
5. In the manner described in step 4, prepare fractions of 4.5%, 6.5%, 8.5%, 10.0%, and 14.5% PEG.

Comments: Care should be taken when using high concentrations of PEG because the high density may cause inadequate pelleting. In addition, when using high concentrations of PEG on crude extracts, a phase separation can occur, with a protein-rich heavy phase separating from a lighter phase above.

PROTOCOL 5.13 Removal of PEG from Precipitated Proteins

This method is an example of the use of ion exchange chromatography used in a batch-wise format. Refer to Chapter 10 for more details.

Materials

15 ml plastic conical centrifuge tubes
Phosphate buffer: 10 mM potassium phosphate buffer, pH 7.0
Ammonium sulfate: see Appendix C
DEAE cellulose
KCl: 0.1 M in phosphate buffer

1. Redissolve the PEG precipitated proteins in phosphate buffer. Reprecipitate the proteins with a 35% (w/v) $(NH_4)_2SO_4$ solution. Repeat the ammonium sulfate precipitation.
2. Redissolve the second $(NH_4)_2SO_4$ precipitate in phosphate buffer and add it to a 1 ml suspension of DEAE-cellulose equilibrated in phosphate buffer. Mix the DEAE cellulose and protein solution. Let the cellulose settle to the bottom of the tube and remove the supernatant. PEG remains unabsorbed to the DEAE cellulose under these conditions.
3. Recover the absorbed proteins by adding 10 ml of the KCl solution to the DEAE cellulose. Mix the tube gently. Let the DEAE cellulose settle to the

bottom and collect the protein containing supernatant. The protein solution can be dialyzed and lyophilized.

Comments: An alternative method for removing PEG is to precipitate the proteins with ice cold 20% ethanol and centrifuge the suspension to collect the protein (Fried and Chun, 1971). PEG will remain in solution. The proteins can be redissolved in buffer, dialyzed and lyophilized as desired.

Precipitation by Selective Denaturation

The object of a selective denaturation step is to create conditions in which the protein of interest remains stable and soluble while the other components of the extract are denatured and precipitate. Ideally, one is looking for the maximum loss of unwanted proteins while recovering >90% of the protein of interest. It is possible to choose a temperature that completely denatures one protein while leaving another protein unaffected.

The behavior of a particular protein can be determined experimentally by using small-scale trials at 5°C intervals between 35–65°C for a specific time period but usually not more than 5 min. Following the heat treatment, the samples are rapidly cooled and centrifuged to remove precipitate material. The supernatant is then assayed. It is advisable to perform a heat denaturation step in the presence of ammonium sulfate which will make the proteins slightly more resistant to proteolysis at the higher temperatures. The trade-off is that higher temperatures may be required because the salt may stabilize the proteins.

The buffer composition and pH of the solution must be carefully noted to insure reproducibility as small differences can have major effects on denaturation.

The following three methods denature proteins. They are often used to concentrate the dilute solution prior to analysis by SDS-PAGE.

PROTOCOL 5.14 Recovery of Protein from Dilute Solutions by Methanol Chloroform Precipitation

This method can be used for both soluble and hydrophobic proteins and is not affected by the presence of detergent, lipid, salt, buffers and 2-ME (Wessel and Flugge, 1984).

Materials

Methanol

Chloroform

Nitrogen tank

1. Add 0.4 ml of methanol to 0.1 ml of protein solution.
2. Vortex mix, then centrifuge for 10 sec at 12,000 × *g*.
3. Add 0.1 ml chloroform (0.2 ml when the sample contains a high concentration of phospholipid), vortex mix and centrifuge at 12,000 × *g* for 10 sec.
4. Add 0.3 ml of water, vortex mix and centrifuge 1 min at 12,000 × *g*.
5. Remove the upper layer and discard.
6. Add 0.3 ml of methanol to the lower phase (chloroform phase). Mix and centrifuge 2 min at 12,000 × *g*.
7. Remove the supernatant and dry the protein pellet under a stream of nitrogen.

Comments: This method can be scaled up following the proportions given in the protocol.

For electrophoresis, dissolve the pellet directly in 50 µl of 1 × SDS sample buffer.

Recoveries are of the order of 90–100% for amounts of protein 40–120 µg in 0.1 ml of the original solution.

PROTOCOL 5.15 Recovery of Protein by Trichloroacetic Acid (TCA) Precipitation

This method removes interfering salts as well as some nondialysable contaminants. In some cases the precipitated proteins may prove difficult to completely redissolve.

Materials

Trichloroacetic acid (TCA): 100% (w/v)

Ethanol: ether: 1:1 (v/v)

1 × SDS sample buffer: see Appendix C

Ammonium hydroxide

Tris: 1 M unbuffered

1. Using the 100% (w/v) TCA stock solution, adjust the volume of the protein solution to a final concentration of 10% (v/v) TCA and incubate on ice for 30 min. Centrifuge at 12,000 × *g* for 5 min.
2. Discard the supernatant. To remove the TCA, rinse the pellet with ethanol: ether (1:1 v/v) then centrifuge at 12,000 × *g* for 5 min.
3. Dissolve the protein precipitate in 1 × SDS sample buffer and boil for 5 min. The sample should be blue in color. If it is yellow, then the pH is too low due to residual TCA which may interfere with the electrophoretic separation. Adjust the pH by adding microliter amounts of either unbuffered Tris or NH_4OH until the sample turns blue.
4. Analyze the sample by SDS-PAGE. Alternatively, the boiled samples can be stored at –20°C.

PROTOCOL 5.16 Concentration of Proteins by Acetone Precipitation

Acetone precipitation of proteins is a technique used to concentrate proteins from dilute solutions that avoids the problem of acid neutralization.

Materials

Acetone: stored at –20°C

1 × SDS sample buffer: see Appendix C

1. Add nine volumes of prechilled –20°C acetone to the protein solution. Mix and incubate at –20°C for 10 min to overnight.
2. Collect the precipitated protein by centrifugation at 4°C for 5 min at 12,000 × *g*.
3. Remove the acetone and air dry the protein pellet.
4. Dissolve the protein in 1 × SDS sample buffer and boil for 3 min.
5. Analyze the samples by SDS-PAGE or store them at –20°C.

Comments: In all of the precipitation protocols it is important to check that the proteins of interest are quantitatively precipitated. Once reconstituted in sample buffer the solution should be centrifuged to remove any undissolved material which could cause streaking during electrophoresis.

What to Do When All Activity Is Lost

Infrequently, we face the nightmare of losing our activity after a purification step-not just recovering only 10%, but not recovering anything! If this happens, a careful analysis is in order. This is where the bookkeeping is very helpful. A proteolytic enzyme may have been carried through the purification protocol. As protein concentrations diminish during the protocol, the unwanted protease will have a greater chance of hydrolyzing the target molecule. Analyze and compare an aliquot of the prestep material with fractions in which you expect the target protein to be present using SDS-PAGE and suitable detection. If the target protein is degraded, you will see new low molecular weight bands. Also assay the prestep aliquot to determine if it has retained activity.

In some cases, proteins bind with great avidity to a particular matrix. Go back to the last column used where the protein might still be bound and change the elution conditions. Include increased ionic strength in the buffer. If this is unsuccessful, include chaotropic agents and detergents. If the protein is radiolabeled or an antibody to it is available, scoop out a little bit of the column gel and boil it in sample buffer, then run it out on SDS-PAGE and see if the presence of the target protein can be detected. In this way you will at least know if the protein is present.

The purer the protein and the smaller the quantity, the greater the chances that it will stick to the sides of the storage vessel, especially if the molecule is hydrophobic. Try to extract the storage vessel with acetonitrile or methanol then look for the presence of the target protein.

It is also possible that more than one component is required for activity and that the last step has stripped away a necessary cofactor. A way to test for this is by performing a mixing assay, combining fractions to see if the activity will reappear.

References

Bensadoun A, Weinstein D (1976): Assay of proteins in the presence of interfering materials. *Anal Biochem* 70:241–250

Bohlen P, Stein S, Dairman W, Udenfriend S (1973): Fluorometric assay of proteins in the nanogram range. *Arch Biochem Biophys* 155:213–220

Bradford MM (1976): A rapid and sensitive method for quantitation of microgram quantities of protein utilizing the principle of protein-dye binding. *Anal Biochem* 72:248–254

Burton K (1956): A study of the conditions and mechanism of the diphenylamine reaction for the colorimetric estimation of deoxyribonucleic acid. *Biochem J* 62:315–323

Cleland WW (1964): Dithiothreitol, a new protective reagent for SH groups. *Biochemistry* 3:480–482

Crawford AW, Beckerle MC (1991): Purification and characterization of zyxin, an 82,000-Dalton component of adherens junctions. *J Biol Chem* 266: 5847–5853

Dignam JD, Lebowitz RM, Roeder RG (1983): Accurate transcription initiation by RNA polymerase II in a soluble extract from isolated mammalian nuclei. *Nucleic Acids Res* 11:1475–1486

Fried M, Chun PW (1971): Water-soluble nonionic polymers in protein purification. *Methods Enzymol* 22:238–248

Gill SC, von Hippel P (1989): Calculation of protein extinction coefficients from amino acis sequence data. *Anal Biochem* 182:319–326

Gold AM (1967): Sulfonylation with sulfonyl halides. *Methods Enzymol* 11:706–711

Haugland RP (1996): Detection and quantitation of proteins in solution. In: *Handbook of fluorescent Probes and Research Chemicals, Sixth Edition*, Spence MTZ ed., Eugene, OR: Molecular Probes.

Jones LJ, Haugland RP, Singer VL (2003): Development and characterization of the NanoOrange protein quantitation assay: a fluorescence-based assay of proteins in solution. *BioTechniques* 34:850–861

Kaplan N (1955): *Methods in Enzymology, Volume I.* New York: Academic Press.

Keesey J, ed. (1987): *Biochemica Information.* Indianapolis, IN: Boehringer Mannheim Biochemicals

Low SH, Tang BL, Wong SH, Hong W (1994): Golgi retardation in Madin-Darby canine kidney and Chinese hamster ovary cells of a transmembrane chimera of two surface proteins. *J Biol Chem* 269:1985–1994

Lowe DM, Tubbs PK (1985): 3-Hydroxy-3-methylglutarl-coenzyme: a synthase from ox liver. *Biochem J* 227:591–599

Lowry OH, Rosebrough NJ, Farr AL, Randall RJ (1951): Protein measurement with the Folin phenol reagent. *J Biol Chem* 193:265–275

Morre DJ (1971): Isolation of Golgi apparatus. *Methods Enzymol* 22:130–148

Morre DJ, Merlin LM, Keenan TW (1969): Localization of glycosyl transferase activities in a Golgi apparatus-rich fraction isolated from rat liver. *Biochem Biophys Res Commun* 37:813–819

Neilands JB (1955): Lactic dehydrogenase of heart muscle. *Methods Enzymol* 1:449–455

Nimmo-Smith RH (1961): p-Nitrophenyl-β-glucuronide as substrate for β-glucuronidase. *Biochem Biophys Acta* 50:166–169

Patel D, Hooper NM, Scott CS (1993): Subcellular fractionation studies indicate an intracellular localization for human monocyte specific esterase (MSE). *Br J Haematol* 84:608–614

Peterson GL (1979): Review of the Folin phenol quantitation method of Lowry, Rosebrough, Farr and Randall. *Anal Biochem* 100:201–220

Ramsby ML, Makowski GS, Khairallah EA (1994): Differential detergent fractionation of isolated hepatocytes: biochemical, immunochemical and two-dimensional gel electrophoresis characterization of cytoskeletal and noncytoskeletal components. *Electrophoresis* 15:265–277

Ray TK (1970): A modified method for the isolation of the plasma membrane from rat liver. *Biochim Biophys Acta* 196:1–9

Rome LH, Crain LIZ (1981): Degradation of mucopolysaccharide in intact isolated lysosomes. *J Biol Chem* 256:10763–10768

Schnaitman C, Greenwalt JW (1968): Enzymatic properties of the inner and outer membranes of rat liver mitochondria. *J Cell Biol* 38:158–175

Scopes RK (1987): *Protein Purification Principles and Practice, Second Edition.* New York: Springer-Verlag

Smith PK, Krohn RI, Hermanson GT, Mallia AK, Gartner FH, Provenzano MD, Fujimoto EK, Goeke NM, Olson BJ, Klenk DC (1985): Measurement of protein using bicinchoninic acid. *Anal Biochem* 150:76–85

Snoke JE (1956): Chicken liver glutamic dehydrogenase. *J Biol Chem* 223:271–276

Sottocasa GL, Kuylenstierna B, Ernster L, Bergstrand A (1967): An electron-transport system associated with the outer membrane of liver mitochondria. *J Cell Biol* 32:415–438

Weidekamm E, Wallach DFH, Fluckiger R (1973): A new sensitive, rapid fluorescence technique for the determination of proteins in gel electrophoresis and in solution. *Anal Biochem* 54:102–114

Wessel D, Flügge UI (1984): A method for the quantitative recovery of proteins in dilute solution in the presence of detergents and lipids. *Anal Biochem* 138:141–142

Worwood M, Dodgson KS, Hook GFR, Rose FA (1973): Problems associated with the assay of arylsulphatases A and B of rat tissues. *Biochem J* 134:183–190

General References

Birnie GD ed. (1972): *Subcellular Components, Preparation and Fractionation.* London: Butterworth

Cooper TG (1977): *The Tools of Biochemistry Wiley.* New York: Joseph P. Wiley

Harris ELV, Angal S eds. (1990): *Protein Purification Methods-A Practical Approach.* New York: Oxford University Press

Jakoby WB ed. (1971): *Enzyme Purification and Related Techniques.* New York: Academic Press

Roe S ed. (2001): *Protein Purification Techniques, Second Edition.* New York: Oxford University Press

6

Membrane Proteins

Introduction

The plasma membrane regulates the exchange of information between the cell and its environment through signaling mechanisms and the regulation of the transport of ions and solutes into and out of the cell. The cell's intracellular membrane network isolates and connects the different subcellular organelles and the cytoplasm. The membrane that surrounds each subcellular organelle has characteristic functional properties so that the total number and types of proteins and lipids in a given membrane are highly variable and exquisitely specific for the organelle.

Integral membrane proteins are amphipathic, composed of regions that are hydrophobic and hydrophilic. The amphiphilic nature of membrane proteins that enables them to localize in the membrane makes them difficult to isolate and study. It has been estimated that 20–30% of the human genome encodes membrane proteins (Wallin et al, 1998). However, less than 1% of the proteins of known structure are membrane proteins (Wu and Yates, 2003).

Analytical methods for the study of membrane proteins have improved over the past few years. A challenging area remains the solubilization of intact membrane proteins for further analysis. Unfortunately, no single strategy provides a global solution for all membrane proteins. Conditions are typically optimized for each membrane enriched sample.

Following cell disruption in a neutral, isotonic buffer that does not contain detergents, many proteins are associated with insoluble components of the cell extract. Integral membrane proteins, material trapped within an organelle, and other proteins that are strongly associated with the cytoskeletal matrix will all be found in the particulate fraction. When one considers that the cytosol is responsible for roughly 75% of cellular protein, a protein exclusively found in the precipitate will be enriched fourfold over the crude extract. This section discusses strategies for dissociating a membrane into its individual components in such a way that the protein of interest retains its bioactivity and becomes amenable to standard purification and analytical procedures. Since there is no single procedure to characterize the different types of membrane proteins, several methods will be presented to begin this process.

A useful preliminary step in isolating membrane proteins is the removal of contaminating cytosolic proteins. The subcellular membranes of eukaryotic and prokaryotic cells constitute the sources of membrane proteins. Therefore, the usual first step in the isolation of a membrane protein is to obtain a highly purified membrane fraction. The starting material can be further enriched if one knows that the target protein is exclusively associated with a specific subcellular membrane fraction, be it plasma membrane, mitochondria or nuclear.

A. Peripheral Membrane Proteins

The strategy for purification of a particular membrane protein will depend initially on whether it is a peripheral or integral membrane protein. Membrane proteins have been classified into integral and peripheral membrane proteins (Singer and Nicolson, 1972). Peripheral or extrinsic membrane proteins are membrane-bound through their noncovalent interactions with the hydrophilic regions of integral membrane proteins or with the headgroup region of the lipid bilayer. Examples of peripheral proteins are illustrated in Figure 6.1. In prac-

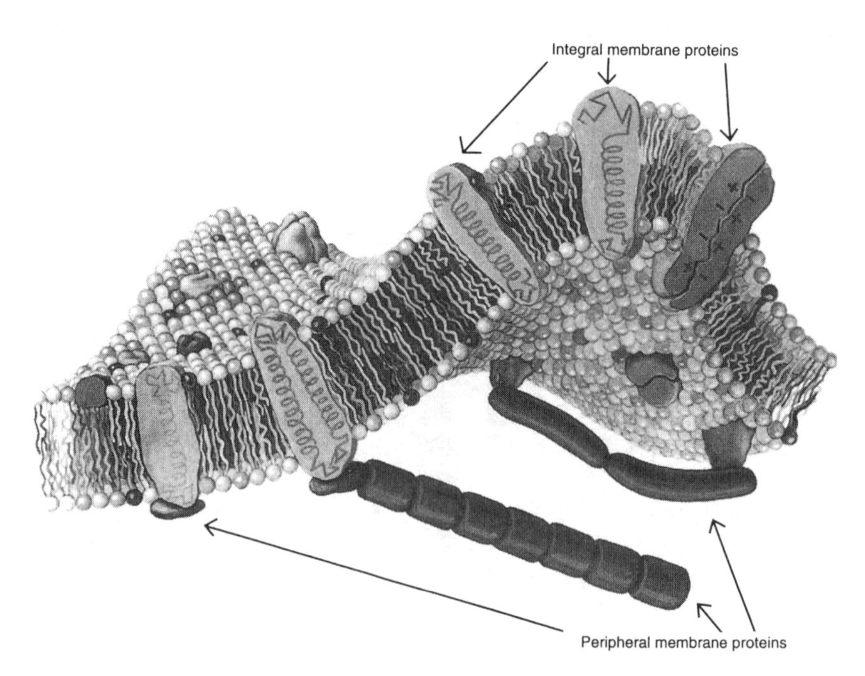

FIGURE 6.1 Integral and peripheral membrane proteins are shown associated with the plasma membrane. The integral membrane proteins span the bilayer. Peripheral membrane proteins are associated noncovalently with the polar head groups of lipids and integral membrane proteins. (Artwork ©1989 Sally J. Bensusen/Visual Science Studio.)

TABLE 6.1. Peripheral Proteins

Treatment	Protein	Reference
1M NaCl	Human erythrocyte glyceraldehyde-3-phosphate dehydrogenase	Tanner and Boxer, 1972
1M NaCl	Human platelet glycoprotein I (thrombin receptor)	Okumura et al, 1978
2M NaI	Pig kidney Na^+-dependent ATPase	Rendi, 1970
0.5M NaSCN	*Spirillum itersonii* ferrochelatase	Dailey and Lascelles, 1974
0.5M $NaClO_4$	Beef heart mitochondria NADH dehydrogenase I	Dooijewaara et al, 1978
0.2M LiSCN or 0.2M Li acetate	*Micrococcus lysodeikiticus* NADH dehydrogenase II	Collins and Salton, 1979
6M guanidine HCl	Human peripheral membrane proteins	Steck, 1972
8M urea	Human erythrocyte peripheral proteins	Juliano and Rothstein, 1971

*Reprinted from Bennett, 1982.

tice, the usual experimental approach is empirical. If the biological activity remains in the sedimentable fraction after the membrane is subjected to a treatment that would solubilize peripheral membrane proteins, then one proceeds on the assumption that the object protein is integral.

Peripheral membrane proteins can be separated from the membrane by procedures used to dissociate soluble protein-protein interactions without total membrane disruption. Once separated, peripheral membrane proteins resemble soluble proteins and can be handled as such. Many membrane-associated proteins can be released/solubilized by physical, chemical, or enzymatic treatments and can then be purified using conventional methods. Methods used successfully include:

- High salt extraction with chaotropic agents
- Ultrasonication
- Alkali extraction with or without metal chelators (EDTA)
- Low ionic strength dialysis
- Phospholipase treatment

Some specific examples of nondetergent extraction conditions that have been used successfully to dissociate peripheral proteins from membranes are presented in Table 6.1 (Bennett, 1982).

High salt treatments result in decreased electrostatic interactions between proteins and charged lipids. The chaotropic ions I^-, ClO_4^-, and SCN^- act by disordering the structure of water. They disrupt hydrophobic bonds near the surface of membrane structures and promote the transfer of hydrophobic groups from an apolar environment to the aqueous phase. At low concentrations, chaotropic agents cause selective solubilization; at higher concentrations they lead to protein inactivation.

Denaturing agents like urea and guanidine hydrochloride diminish the hydrophobic interactions that play a crucial role in holding together the protein tertiary structure (Creighton, 1984), breaking noncovalent interactions if used at high concentrations (6–10 M).

When a reagent is added to the membrane suspension, a large decrease in turbidity, noticeable by eye, can be taken as indicative of extensive lipid bilayer collapse.

A number of anions were ranked in terms of relative effectiveness for extracting proteins from the plasma membrane of *Bacillus subtilis* (Hatefi and Hanstein, 1969). This series is usually expressed as: $SCN^- > ClO_4^- > NO_3^- > I^- > Br^- > SO_4^- > CH_3COO^-$. Most effective extracting agents are in general, the most effective protein denaturants. Nevertheless, it may be assumed that some proteins can refold with subsequent resumption of bioactivity after exposure to conditions that lead to extensive unfolding. A word of caution: high salt concentrations can also lead to artifacts. For example, an endoplasmic reticulum protein implicated in the synthesis of secretory proteins which was originally thought to be dislodged by high salt was subsequently shown to be released by the action of salt-activated protease (Warren and Dobberstein, 1978; Meyer and Dobberstein, 1980). Once removed from their normal membranous environment, peripheral membrane proteins behave like water-soluble proteins.

PROTOCOL 6.1 Alkali Extraction

The influence of pH on the solubilization of several membrane systems has been noted. Figure 6.2 illustrates the differential solubility from

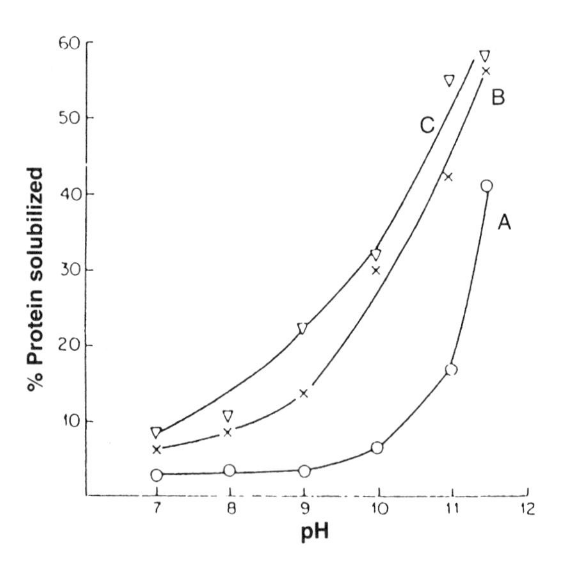

FIGURE 6.2 The effect of pH on the solubility of membrane proteins. The percentage of protein solubilized represents the fraction of the total protein found in the 105,000 × *g* supernatant. Curve A (o): submitochondrial particles from beef heart. Curve B (x): rat liver microsomes. Curve C (∇): Plasma membranes from *Bacillus subtilis*. (Reprinted with permission from Penefsky and Tzagoloff, 1971).

three different membranes. The titration of charged groups in the membrane may play a role in the dissociation of membrane components. Although extensive solubilization occurs at high pH, these extreme conditions may result in a loss of activity of many enzymes (Penefsky and Tzagoloff, 1971). However, there have been proteins, for example, succinic dehydrogenase (King, 1963), that have been successfully solubilized from membranes between pH 8–11.

Materials

Resuspension buffer: 0.1 M Tris acetate, pH 7.5, 0.25 M sucrose
NaOH: 10 M
Ultracentrifuge

1. Incubate the membrane fraction in resuspension buffer at a protein concentration of 10 mg/ml and divide it into 5 equal fractions.
2. Adjust the membrane suspensions to the indicated pH (pH 8, 9, 10, 11) with NaOH and incubate them at 4°C for 30 min. Leave two fractions untreated.
3. Centrifuge the fractions at 105,000 × *g* for 60 min.
4. Remove the supernatant and determine the protein concentration (chapter 5). Compared to the total membrane suspension (the untreated fraction), this represents the percentage of protein solubilized. Using a bioassay or antibody, determine if the target protein is solubilized by any of the treatments, and if so, is it still bioactive.
5. Neutralize the fraction containing the target protein and continue with the purification.

Comments: This experiment can be performed as a pilot study. Once conditions have been determined to solubilize the target protein in bioactive form, this method can be scaled up.

PROTOCOL 6.2 High pH Membrane Fractionation

High pH conditions have been used for biochemical fractionation of peripheral membrane proteins from the integral membrane proteins that are embedded in the lipid bilayer. Electron micrographs revealed that mechanical agitation in a high pH buffer prevented resealing of membrane structures, favoring the presence of membrane sheets with free edges (Howell and Palade, 1982). In contrast to detergents and organic solvents, high pH disrupts sealed membrane structures without denaturing the lipid bilayer or extracting integral membrane proteins.

Materials

Ultracentrifuge
Ti 50 rotor, Beckman Instruments
Teflon pestle with an AA glass homogenizer (A.H. Thomas, Phila, PA)

1. Resuspend the membrane fraction at a protein concentration of <2 mg/ml in a solution of 100 mM Na_2CO_3, pH 11.3.

2. Homogenize the suspension using a Teflon pestle in an AA glass homogenizer with 6–8 strokes.
3. Mix the suspension by vortexing three times during 30 min while keeping the temperature at ~4°C.
4. Pellet the membrane fraction by centrifugation for 60 min at 100,000 g_{av} in a Ti 50 rotor (Beckman).

Comments: The Na_2CO_3 supernatant consists of the content of the membrane organelles, while the pellet consists of the membrane fraction.

The starting material from the original description was rat liver Golgi fraction that was isolated by sucrose gradient centrifugation (Howell and Palade, 1982).

Other investigators have substituted passage of the resuspended material in step 2 above with 5 passes through an insulin syringe (Wu et al, 2003).

B. Integral Membrane Proteins

Integral membrane proteins, also referred to as intrinsic membrane proteins, interact with the hydrophobic moieties of the phospholipid bilayer (see Figure 6.1). Integral membrane proteins from sources as diverse as the bacterial inner membrane, the endoplasmic reticulum, the inner mitochondrial membrane, and the thylakoid membrane of chloroplasts all have one or more characteristic runs of apolar amino acids that, in the final structure, span the lipid bilayer (von Heijne and Gavel, 1988). The classification of integral membrane proteins has been extended. A Type I membrane protein contains the COOH terminus in the cytosol. A Type II membrane protein contains the NH_2-terminus in the cytosol. There are also membrane proteins that contain both COOH and NH_2 termini in the cytoplasmic domain. The complicated structure within the membrane is organized in a manner opposite to a soluble protein in an aqueous milieu; the domain of the protein embedded in the membrane in contact with the hydrophobic lipid tail is composed of mostly hydrophobic amino acids. In order to obtain an integral membrane protein in a monomeric state, it is necessary to disrupt the lipid bilayer, and in order to prevent reaggregation, it will be important to satisfy a protein's requirement for hydrophobic interactions. Such a structure can only remain thermodynamically stable while the correct hydrophobic environment is maintained.

Organic Alcohol Extraction of Peripheral Membrane Proteins

When examining the interaction between alcohols and cell membranes, it was observed that alcohol partitioning is strongly dependent on its chain length. This suggested that shorter-chain alcohols easily disrupt hydrophobic interactions taking place at the membrane-aqueous interface (Rowe et al, 1998). Methanol is short-chained, polar, water miscible and compatible with mass spectrometric analysis. Since the membrane bilayer was suspended in an aqueous buffer, an optimal concentration of methanol should reduce the phospholipids-water interfacial tension to zero, resulting in a single phase solution and miscible extraction of membrane proteins (Blonder et al, 2002). The use of methanol did not interfere or impede tryptic digestion.

PROTOCOL 6.3 Butanol Extraction

Solubilization, by definition, involves the disintegration of the lipid bilayer. Solubilization for the purpose of isolating membrane proteins is usually accomplished by treating the membranes with detergents or extracting them with organic solvents (van Reswoude and Kempf, 1984). Butanol extraction using a biphasic system is a technique for solubilizing proteins from membranes into dilute aqueous buffers (Penefsky and Tzagoloff, 1971). The low solubility of n-butanol in water combined with its lipophilicity minimally denatures proteins.

Materials

n-butanol

1. Add a volume of n-butanol equal to that of the suspension and maintain the temperature at 4°C.
2. Mix and centrifuge at 500 × *g* or 10 min. Two phases should be obtained. The upper phase consists of butanol and membrane lipids. The lower aqueous phase contains solubilized proteins. Lipid-rich material localizes at the interface. Denatured proteins precipitate in the aqueous phase.
3. Collect the butanol and the aqueous phases separately.
4. Dialyze (Chapter 5) the aqueous phase against a large volume of water. The butanol phase may contain extremely hydrophobic proteins.

Comments: Other organic solvent extraction systems have been described. A chloroform/methanol extraction protocol enabled the isolation of hydrophobic membrane proteins from chloroplasts (Ferro et al, 2000). The extracted hydrophobic proteins can be easily analyzed by standard 1-D SDS-PAGE.

PROTOCOL 6.4 Single-Phase Butanol Extraction

The miscibility of n-butanol with small amounts of aqueous solutions, without forming a two-phase system favors an easy separation of the solubilized fraction (Nelson et al, 1977).

Materials

3 ml syringe

n-butanol

Diethyl ether

1. Start with a membrane preparation whose protein concentration is at least 3 mg/ml and can be as much as 20 mg/ml in water, 5 mM $CaCl_2$, or other hypotonic buffer.

2. Load a syringe with the membrane suspension and inject it into 100 ml of n-butanol at 0°C stirring vigorously. The final concentration of the aqueous component is 2%.
3. Mix the suspension for 30 min then centrifuge at 20,000 × *g* for 10 min then recentrifuge the supernantant at 20,000 × *g* for 10 min.
4. Add 500 ml of diethyl ether to the butanol supernatant and incubate for 30 min at 0°C.
5. Centrifuge for 10 min at 10,000 × *g*. The proteolipid will be in the precipitate.
6. Resuspend the pellet in the desired buffer and assess the purity and bioactivity of the recovered material.

Comments: You would be very fortunate indeed if this method purified your target membrane protein. However, the ease and simplicity of this technique make it appealing and worth a try.

Despite their biological importancc and natural abundance, a large-scale proteomic analysis of integral membrane proteins has been difficult. The main obstacle for a large-scale mass spectrometric analysis of integral membrane proteins is the inability to achieve the dissolution of these hydrophobic proteins from the phospholipids bilayer while maintaining their solubility throughout the entire isolation and separation process while avoiding reagents that may interfere with mass spectrometric analysis. To this end, efficient extraction of hydrophobic membrane proteins was achieved using a combination of techniques. High pH fractionation, thermal denaturation and organic solvent-assisted solubilization were all used to enable a large scale LC-MS/MS analysis of a membrane subproteome (Blonder et al, 2002).

C. Detergents

A wide variety of chemicals can be classified as detergents. Detergents are also known as tensides, soluble amphiphiles, soaps, and surfactants. Detergents are a class of polar lipid molecules that are soluble in water. They have a bipartite structure with a hydrophobic portion that is more soluble in oil or other hydrocarbon solvents and a hydrophilic portion which is more soluble in water. For the purposes of this manual the term detergent will be used to describe any amphiphilic compound that has the property of disrupting the structure of a biological membrane by solubilizing the membrane lipids. The characteristic behavior of a detergent in solution is that while it dissolves as the monomeric species at low concentrations, at higher concentrations it is present in solution in the form of aggregates known as micelles. The structure of a micelle is such that the hydrophobic regions of the detergent molecules are associated together at the interior of the micelle, and only the hydrophilic regions of the molecules are exposed to the surrounding aqueous medium.

Detergents have been used to solubilize membrane proteins to enable the investigator to examine the structure of membrane proteins and their interactions with each other. The role of the detergent is twofold: (1) to inhibit intra-or inter-protein hydrophobic interactions and (2) to prevent the loss of hydrophobic integral membrane proteins through aggregation or adsorption (Blonder et al, 2002). Detergent micelles mimic the lipid bilayer and provide an environment which is suitable for the hydrophobic portions of membrane proteins and lipids. This provides a suitable environment for individual membrane proteins, leaving their tertiary and quaternary structure intact. Each protein or protein complex will ideally partition into separate micelles, behaving as a distinct entity and can be studied using the same methods employed for soluble proteins.

Membrane proteins are routinely solubilized by identifying a suitable detergent and the best conditions for its use. Detergents are capable of displacing a tightly bound membrane protein by dissolving the membrane and then replacing the membrane with aliphatic or aromatic chains which form the lipophilic part of the detergent. Once the protein has been solubilized, standard fractionation procedures can be used in the presence of concentrations between 0.05–1.0% of a nonionic detergent. Most proteins can tolerate levels of 1–3% (w/v) of the non-ionic detergent Triton X-100 and still retain complete bioactivity.

There are certain techniques that are not compatible with the presence of detergent. Salting out with ammonium sulfate causes Triton X-100 to separate as a floating layer. When using detergents, the proteins may be present in micellular form leading to artifactual migration on gel filtration (different molecular weights) columns.

Elucidation of protein-protein interactions and their functional role is often difficult in the natural lipid bilayer habitat of membrane proteins. A useful approach is to solubilize membrane proteins or membrane protein complexes in such a way as to keep them in a native-like state and to study functional properties under well defined conditions. Many different detergents have been used successfully for this purpose.

When a membrane is solubilized with a detergent, the amphiphilic membrane components (i.e. the lipids and the integral membrane proteins) become incorporated into the detergent micelles. These components can partition between micelles in a time and detergent concentration-dependent manner. When there are more micelles than protein molecules, no micelle will have more than one protein molecule.

Detergents that have been used for biological studies fall into three groups: (1) ionic, which consist of a hydrocarbon chain with a strongly acidic or basic polar headgroup; (2) nonionic, which have a polyoxyethylene polar region; and (3) bile salts or other steroid-based detergents such as CHAPS and digitonin which are relatively rigid.

The list of detergents is deliberately selective, being restricted to compounds readily available and widely used (Table 6.2).

Recently, a series of novel zwitterionic detergents has been reported that show an improvement in solubilizing integral membrane proteins (Henningsen et al, 2002). Both 4 octylbenzol amidosulfobetaine and myristic amidosulfobetaine were better than CHAPS at solubilizing membrane proteins. Their structures are shown in Figure 6.3.

TABLE 6.2 Detergents

Detergent	CMC (mM)	Monomer MW	Micelle MW	Structure	Comments
Sodium deoxycholate (DOC)	1–4	415	4,200	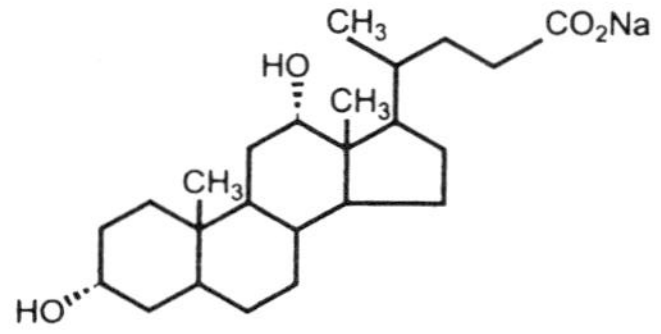 CH_3 CO_2Na HO CH_3 CH_3 HO	Keep pH > 8.0
Sodium cholate	0.57	431	1,800	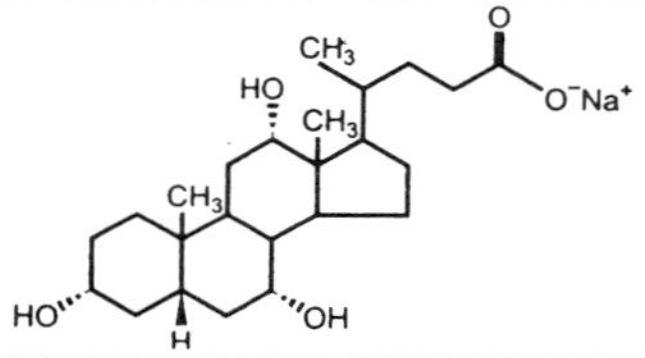 O CH_3 O^-Na^+ HO CH_3 CH_3 HO H OH	Keep pH > 8.0
CHAPS	8–10	615	6,150	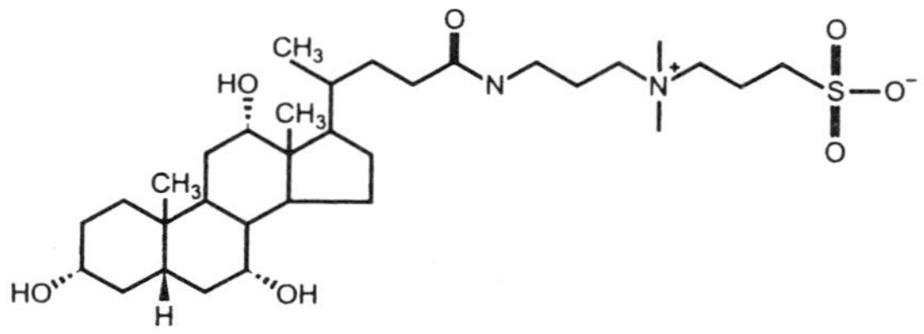 O CH_3 N N^+ O S O^- O HO CH_3 CH_3 HO H OH	Amphoteric and easily dialyzed
CHAPSO	8–10	631	6,940	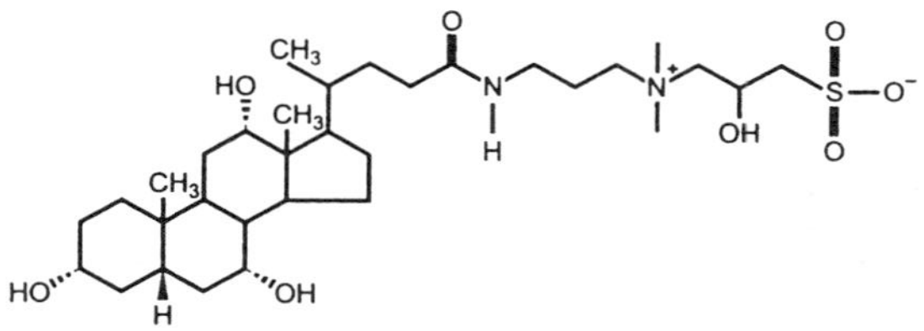 O CH_3 N H N^+ OH O S O^- O HO CH_3 CH_3 HO H OH	Zwitterionic

Digitonin	—	1,229	70,000	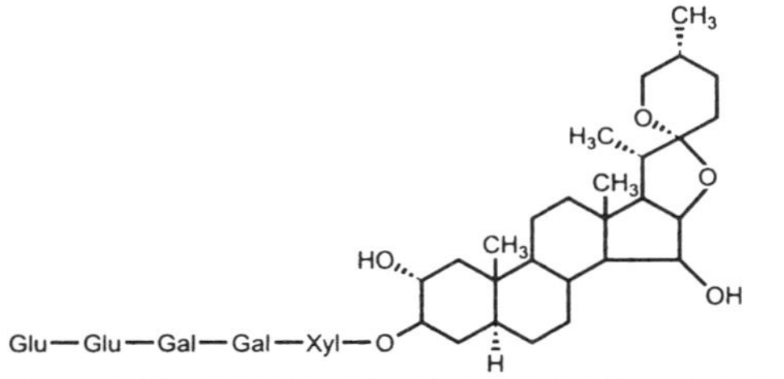 Glu—Glu—Gal—Gal—Xyl—O— (steroid: CH_3, H_3C, CH_3, CH_3, HO, OH, O, O, H)	
NP-40	0.3	603		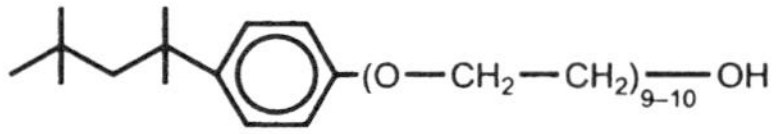 $-(O-CH_2-CH_2)_{9-10}-OH$	Similar to Triton X-100
Octyl glucoside (OG)	25	292	8,000	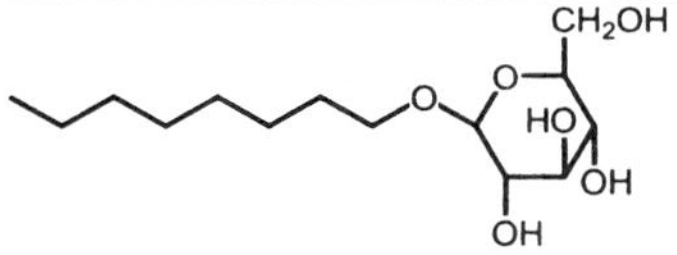 CH_2OH, O, HO, OH, OH	
Zwittergent 3–14	0.3	364	30,000	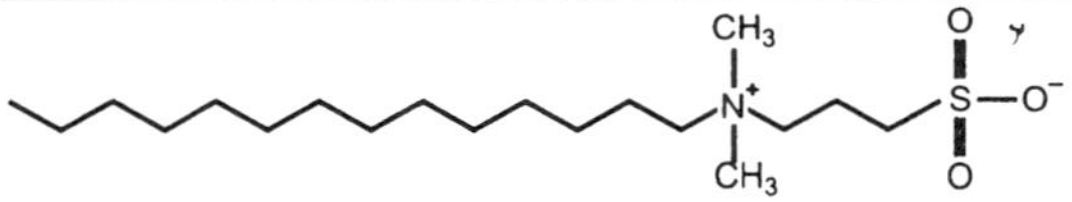 CH_3, N^+, CH_3, $S(=O)_2-O^-$	
Brij 35	0.09	1,168		 $-O-(CH_2-CH_2-O)_{23}-H$	
Triton X-100	0.3	650	90,000	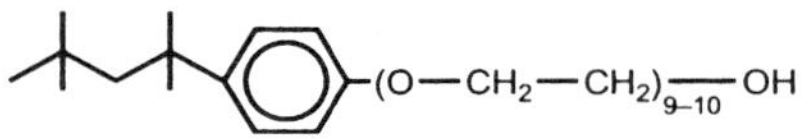 $-(O-CH_2-CH_2)_{9-10}-OH$	Equivalent to NP-40

Continued

TABLE 6.2 *Continued*

Detergent	CMC (mM)	Monomer MW	Micelle MW	Structure	Comments
Triton X-114	0.21	537		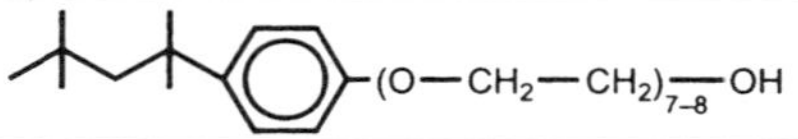 $-(O-CH_2-CH_2)_{7-8}-OH$	Cloud point 23°C
Lubrol PX	0.1	582	64,000	 $(O-CH_2-CH_2)_{9-10}-OH$	
Tween 20		1,228		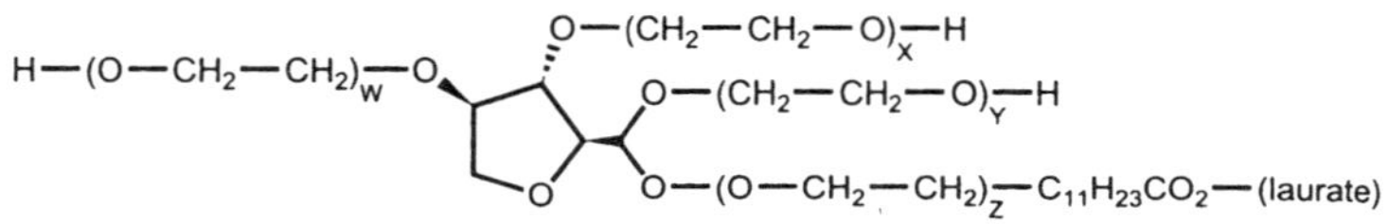 $H-(O-CH_2-CH_2)_W-O$; $O-(CH_2-CH_2-O)_X-H$; $O-(CH_2-CH_2-O)_Y-H$; $O-(O-CH_2-CH_2)_Z-C_{11}H_{23}CO_2-$ (laurate); O	
Tween 80	0.01	1,310	76,000	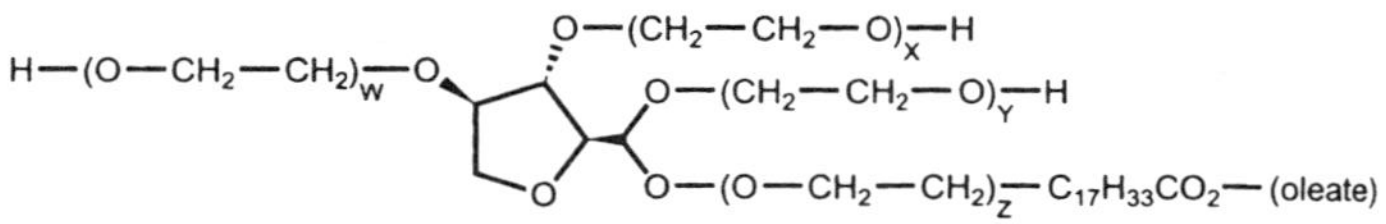 $H-(O-CH_2-CH_2)_W-O$; $O-(CH_2-CH_2-O)_X-H$; $O-(CH_2-CH_2-O)_Y-H$; $O-(O-CH_2-CH_2)_Z-C_{17}H_{33}CO_2-$ (oleate); O	
Sodium dodecylsulfate (SDS)	8*	288	18,000	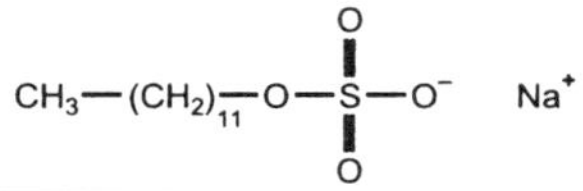 $CH_3-(CH_2)_{11}-O-S(=O)_2-O^-$ Na^+	Anionic and highly denaturing
Cetyltrimethylammonium bromide (CTAB)		365	62,000	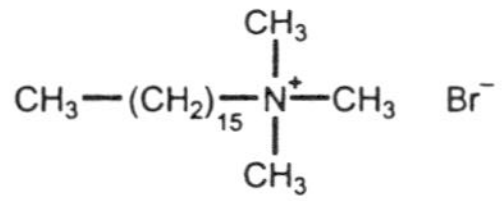 $CH_3-(CH_2)_{15}-N^+(CH_3)_3$ Br^-	Cationic and highly denaturing

*The CMC varies with ionic strength. Where I = 0.15 CMC = 1.1.

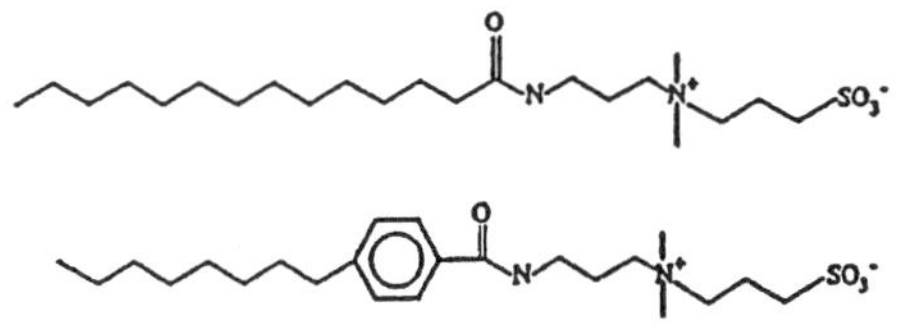

ASB14

C8Ø

FIGURE 6.3 New zwitterionic detergents ASB14 and C8Ø. (Adapted with permission from Rabilloud et al, 1999.)

Properties of Detergents

Many properties of detergents can be described by three values: the critical micelle concentration (CMC), the micelle molecular weight, and the hydrophile-lipophile balance (HLB).

Critical Micelle Concentration (CMC)

At low concentrations in aqueous solution, detergents exist as monomers of single unclustered molecules. Above a characteristic limit, called the critical micelle concentration (CMC), which is unique for each detergent, micelles or clusters of detergent molecules are formed. Micelle structure is a consequence of the dual nature of the detergent molecule. In an aqueous solution, detergent molecules are organized such that the hydrophobic portions are in contact with each other in the micellar core shielding themselves from contact with water molecules, and the hydrophilic portions form a shell in contact with the aqueous environment. The number of monomers that come together to form a micelle is referred to as the aggregation number (N). All detergent molecules added to a solution already at its CMC will be incorporated into micelles. Solutions above the CMC will be composed of micelles in equilibrium with monomer at a concentration equal to the CMC.

Micelle Molecular Weight

Micelles and monomers interact dynamically with constant exchange between monomers in the bulk solution and micelles. When formed by a particular detergent, the micelles will have a characteristic molecular weight. In some cases micelles are penetrated by water. A micellar molecular weight can be experimentally established by techniques used to determine the molecular weight of biological macromolecules. The aggregation number is calculated by dividing the micellar molecular weight by the molecular weight of the detergent monomer. Detergents with large micelle molecular weights will be difficult to remove from a solution by dialysis.

Both the CMC and the micelle molecular weight will vary depending upon the composition of the buffer. Varying the salt concentration, temperature, and pH will all have an effect on the CMC. In particular, SDS (sodium dodecyl sulfate) is dramatically affected by temperatures below 20°C often resulting in the SDS crystallizing out. DOC (deoxycholate) is insoluble below pH 7.5.

Hydrophile-Lipophile Balance (HLB)

The HLB attaches a numerical value to the overall hydrophilic properties of a detergent. Values above 7 indicate that the detergent is more soluble in aqueous solutions than in organic solutions. For biological applications, values of 12.5 and higher are needed. The HLB number can be used to get an idea of how denaturing a particular detergent can be. Values between 12 and 16 are relatively nondenaturing, while values above 20 are more likely to be denaturing.

Certain nonionic detergents are effective at temperatures higher than the critical micelle temperature, above which detergent crystals will form micelles. At a characteristic temperature referred to as the cloud point, detergents will undergo a phase separation. Of particular note is Triton X-114. Its low cloud point has been exploited in a method of protein and lipid solubilization based on phase separation (Bordier, 1981). This method is detailed in Protocol 9.38.

Classification of Detergents

Detergents are commonly classified on the basis of the charge and/or nature of the hydrophilic portion (head group) of the molecule and the flexibility and/or chemical nature of the hydrophobic portion. Head groups may be anionic, zwitterionic, nonionic, or cationic. Zwitterionic head groups contain both negatively charged and positively charged moieties.

Ionic Detergents

The charged detergents have strongly acidic or basic polar head groups, form relatively small micelles, and have high CMC values. Ionic detergents have a tendency to denature proteins at concentrations below their CMC. Cationic detergents cooperatively bind to proteins at close to their CMCs, which is approximately tenfold higher than for the anionic detergent SDS.

Nonionic Detergents

Nonionic detergents are popular choices for membrane work because they are usually mild in their effects on protein structure, enzyme activity and protein-protein interactions. A disadvantage in their use is that they are difficult to remove from the solubilization mixture because of their low CMC.

Bile Salts

Bile salts consist of a rigid steroid ring structure with hydroxyl groups located on the side of the molecule only and have an ionic group at one end of the molecule. A concern when using deoxycholate and cholate is that the pK of their acidic groups is close to the physiological pH range (pKa = 6.2 for DOC, pKa = 5.2 for cholate). The insoluble acid forms of these detergents precipitate at even slightly acid pH. Therefore, it is advisable to use these detergents at a pH of 7.5 or higher.

Detergent Solubilization

Detergent solubilization of proteins is not a routine procedure. Finding the best agent for a particular situation often depends on trial and error. Knowledge of the physicochemical properties of detergents may make the process of choosing a detergent easier and remove some of the guesswork. The primary requirement for the detergent is that it does not affect the biological activity of the target protein.

A wide variety of detergents have been used in the isolation and study of integral membrane proteins. Detergent extraction of a membrane protein occurs in three steps.

1. Detergent binds to the membrane and lysis occurs.
2. The membrane is solubilized in the form of detergent-lipid-protein complexes.
3. The detergent-lipid-protein complexes are further solubilized to give detergent-protein complexes and detergent-lipid complexes.

At extremely low concentrations of detergent, monomers partition into the membrane without gross alteration in membrane structure. As the concentration of detergent is increased, the structure of the membrane is grossly changed, leading to lysis. At detergent-to-protein ratios of ~1:1, slowly sedimenting complexes are produced. At this point the species being generated are usually large heterogeneous complexes of lipids, detergent and protein with molecular weights of 0.5 to 1 million daltons. Increasing the detergent-to-protein ratio from 10:1 to 20:1 leads to the formation of protein-detergent complexes and to mixed micelles of lipid and detergent. Residual interactions between proteins may not be dissociated at this point. Artifactual interactions between proteins may exist.

Irrespective of the detergent used, solubilized membrane proteins are often inactivated at high detergent-to-protein ratios. This is due to disruption of the quaternary structure of solubilized membrane proteins into subunits as a function of detergent concentration or progressive removal of residual lipid attached to the solubilized detergent membrane protein complexes.

Before attempting to solubilize the target protein one must define the criteria that will be used to distinguish between soluble and insoluble membrane proteins after detergent treatment. The most widely accepted criteria is that a protein is considered solubilized if it is recovered in the supernatant after centrifugation for 1h at 105,000 $\times g$ (Hjelmeland and Chrambach, 1984).

Choose a detergent that will be appropriate for the manipulations to be carried out. Different detergents may be preferred at certain stages in a purification scheme. Detergents can be easily changed. Methods include gel filtration into a buffer containing the new detergent or centrifugation of the protein through a sucrose density gradient containing the new detergent (Helenius et al, 1977).

Choosing a Detergent

Before beginning pilot experiments aimed at finding a suitable detergent, conduct a literature search. If the exact conditions have not been previously worked out, try to adapt a method that was used in a closely related system.

Prior to using a detergent in an experimental protocol, be aware of the following questions:

- Are there alternatives to using a detergent for a particular procedure?
- What are the experimental conditions, and how will they affect the detergent? Will the detergent be soluble in the experimental buffer system? Will it be above its CMC? Some important information to keep in mind: SDS will precipitate in the presence of K^+ and Mg^{+2}; sodium cholate forms insoluble complexes with divalent metals; concentrated solutions of SDS will precipitate at low temperatures; ionic detergents like SDS tend to denature proteins by destroying their secondary, tertiary and quaternary structure, although antibody activity and some enzyme activities are retained at low concentrations (less than 0.1% SDS); octyl and nonyl phenol detergents are easily iodinated and should not be present during the radioiodination of proteins.
- Will it be necessary to remove the detergent? If yes, choose a detergent with a high CMC (>1 mM) and low micelle molecular weight that can be dialyzed and/or removed by established methods.
- Will the detergent interfere with UV-VIS absorption, fluorescence or any other detection system? The Triton series of detergents strongly absorb in the UV region.
- Will the detergent interfere with charge related procedures like isoelectric focusing or ion exchange chromatography? If separation techniques for exploiting charge differences are to be employed, e.g., ion exchange chromatography, charged detergents should be avoided. Ionic detergents bound to proteins affect the protein's native charge. Bound nonionic detergents do not produce this effect.
- Is the detergent suitable for gel filtration analysis? When proteins are separated according to size by gel filtration, detergents with smaller micelle size aggregation number (<30) yield better resolution because the size difference between protein-containing and non-protein-containing micelles is greater.
- How will the detergent affect the rest of the system? Conversely, how will the detergent be affected by the system?
- How much detergent should be used? The optimal detergent concentration should be determined experimentally for each system. For an initial solubilization, use this rough guideline: detergent-to-protein (w/w) ratios covering the range 10:1 to 0.1:1.
- What is the cost? Can a less expensive detergent with comparable properties be substituted for an expensive one?

Choice of Initial Conditions

Ionic strength of the solubilization medium is a critical consideration. In the absence of a specific requirement for low ionic strength, a high ionic strength of between 0.1 and 0.5 M KCl are suggested.

The buffer composition is also important. To assure adequate buffering capacity the concentration of the buffer should be at least 25 mM and the pH close to the pK of the buffering compound.

In the absence of reasons to the contrary, solubility trials should be carried out at 4°C.

PROTOCOL 6.5 Differential Detergent Solubilization

Detergent fractionation is useful for confirming the subcellular distribution of proteins, monitoring compartmental redistribution of target proteins, or enhancing the detection of low-abundance species (Ramsby and Makowski, 1999). This method employs sequential extraction of cells with detergent-containing buffers to enrich cellular proteins into distinct fractions. Some advantages of this method are that it is applicable to use with limited amounts of biomaterial, it is simple, highly reproducible, and does not require an ultracentrifuge. Proteins that are intrinsically distributed in more than one subcellular location may be differentially regulated. Compartment-specific changes in protein localization may be of regulatory significance. Differential detergent extraction yields four fractions:

1. Cytosolic proteins and soluble cytoskeletal elements.
2. Membrane and organellar proteins.
3. Nuclear membrane and soluble nuclear proteins.
4. Detergent-resistant proteins.

Record the volumes of all fractions and store them at –70°C.

Materials

Digitonin extraction buffer: 10 mM PIPES, pH 6.8, 0.015% (w/v) digitonin, 300 mM sucrose, 100 mM NaCl, 3 mM $MgCl_2$, 5 mM EDTA, anti-protease pellet (Roche)

Triton X-100 extraction buffer: 10 mM PIPES, pH 7.4, 0.5% (v/v) Triton X-100, 300 mM sucrose, 100 mM NaCl, 3 mM $MgCl_2$, 5 mM EDTA, anti-protease pellet (Roche)

Tween-40/deoxycholate extraction buffer: 10 mM PIPES, pH 7.4, 1% (v/v) Tween-40, 0.5% (v/v) deoxycholate (DOC), 3 mM $MgCl_2$, anti-protease pellet (Roche)

Cytoskeleton solubilization buffers: Non-reducing buffer-5% SDS, 10 mM sodium phosphate, pH 7.4, Reducing buffer-5% SDS, 10 mM sodium phosphate, pH 7.4, 10% (v/v) β-mercaptoethanol

Teflon smooth-walled glass homogenizer

1. Wash the cells twice with PBS.
2. Add ice-cold digitonin extraction buffer to the washed cells and incubate on ice with occasional swirling for 10 min.

3. If the cells are growing in a monolayer, incubate with digitonin extraction buffer without agitation. After 5 min remove the solution. Neither cells nor debris should be evident in the supernatant from this treatment.
4. Remove the contents of the dish or flask with a cell scraper. Centrifuge the suspension at 500 × *g* for 10 min at 4°C.
5. Resuspend the digitonin-insoluble pellet in ice-cold Triton X-100 extraction buffer in the same volume used for the digitonin extraction.
6. Incubate on ice with gentle agitation for 30 min.
7. Centrifuge the suspension for 10 min at 5000 × *g*. Remove and save the supernatant. The supernatant fraction is enriched in membranes and organelles.
8. Resuspend the pellets from step 6 in half the volume used in step 4 with Tween-40/deoxycholate extraction buffer using a Teflon smooth-walled glass homogenizer (five medium speed strokes).
9. Incubate on ice with gentle agitation for 10 min.
10. Centrifuge the suspension at 7,000 × *g* for 10 min at 4°C. Remove and save the supernatant.
11. Resuspend the pellet with ice-cold PBS. Mechanically shear the DNA by using the homogenizer (10 strokes) followed by centrifugation at 12,000 × *g* for 10 min.
12. Resuspend the pellet in either reducing or non-reducing cytoskeleton solubilization buffer. This fraction should be rich in nucleic acid.

Comments: Digitonin is a steroidal compound believed to interact with cholesterol in the plasma membrane and outer mitochondrial membranes, leaving the inner mitochondrial membrane intact and resulting in membrane permeabilization and release of soluble cytoplasmic components. The inner mitochondrial membrane is resistant to digitonin because of its low β–hydroxy sterol content (Mackall et al, 1979).

Digitonin treatment of a cell monolayer (step 3) should leave a perforated layer of cells remaining on the culture dish. The cells should appear intact as judged by phase contrast microscopy and exhibit no apparent change in cell or nuclear morphology. The digitonin/EDTA extraction yields ~35% of total cellular protein.

Triton X-100, is a non-ionic detergent that solubilizes membrane lipids thereby releasing organellar contents. The Triton X-100 extract is enriched for membrane and organelle proteins and extracts ~50% of total cellular protein.

DOC is a weakly ionic detergent that destroys nuclear integrity, and solubilizes actin and other cytoskeletal elements. The Tween/DOC buffer extracts ~5% of total cellular protein and contains, almost exclusively, nuclear proteins.

For studies in which nuclear parameters are of no interest the Tween/DOC step can be omitted. If the starting material is small, it might be warranted to omit this step.

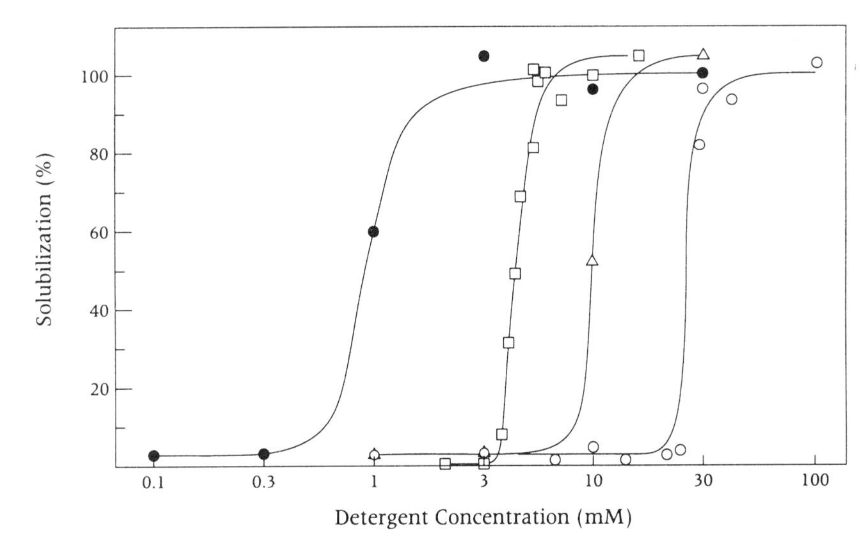

FIGURE 6.4 Solubilization of IgE receptor from rat basophilic leukemia cells as a function of the concentration of several detergents. Triton X-100 (•), CHAPS (□), sodium cholate (△), octyl glucoside (○). (Adapted with permission from Rivnay and Metzger, 1982).

PROTOCOL 6.6 Solubilization Trial

This protocol is designed to determine conditions that lead to the solubilization and the preservation of function of the target protein. Figure 6.4 demonstrates progressive solubilization of the IgE receptor with four different detergents (Rivnay and Metzger, 1982). The detergent that yields the highest activity of the target protein in the supernatant should be chosen for future studies.

Materials

Detergent stock solutions: 10% w/v

Magnetic Stirrer

1. Suspend the particulate protein preparation in 50 mM buffer containing 0.15 M KCl at a protein concentration of 10 mg/ml if possible, and incubate at 4°C.
2. Prepare stock solutions of detergent in the same buffer as that used for the protein suspension at a concentration of 10% (w/v) except for digitonin which must be prepared at 4% (w/v) due to limited solubility.
3. Add detergent stock to the protein preparation to make a final concentration of 5 mg of protein per ml. A series of detergent concentrations should include 0.01%, 0.03%, 0.1%, 0.3%, 1.0%, and 3.0% (w/v) for each detergent tested.
4. Stir individual aliquots gently with a magnetic stirrer for 1 h at 4°C, then centrifuge at 100,000 × *g* for 1 h at 4°C.
5. Remove the supernatant from each sample and resuspend the pellet in a volume of buffer equal to the volume of the supernatant fraction.
6. Assay both supernatant and resuspended pellet for functional activity and protein concentration.

Comments: The detergent that yields the highest activity in the supernatant as well as the highest total activity should be chosen for further work. If, individually, all detergents tested prove unsuccessful, a number of mixtures of detergents can be considered.

If activity is very low following detergent extraction, additives known to stabilize solubilized proteins can be tried. Glycerol at concentrations between 10–50% (v/v); 1 mM DTT or 5 mM 2-ME; 1 mM EDTA; 1 mM PMSF, leupeptin 10 μg/ml and pepstatin 10 μg/ml at acidic pH.

Protein-to-Detergent Ratio

Many proteins that are solubilized by detergents have well-defined detergent-to-protein ratios that result in optimum solubility. This ratio may be experimentally evaluated by solubilization at several different detergent concentrations. Protein solubilization occurs at or near the CMC for most detergents. Usually optimum detergent-to-protein ratios for low concentrations of protein will be slightly higher than those for high concentrations of protein. Results are expressed graphically by plotting solubilized activity on the ordinate versus detergent concentration on the abscissa as demonstrated in Figure 6.5A. Two results are expected, either the percentage of solubilized protein (activity) will increase and plateau at and beyond a given detergent concentration for each protein concentration, or the soluble activity will rise and fall to yield an optimum detergent concentration for each protein concentration. If the protein concentration is known, the data can be expressed as a function of the detergent-to-protein ratio, as demonstrated in

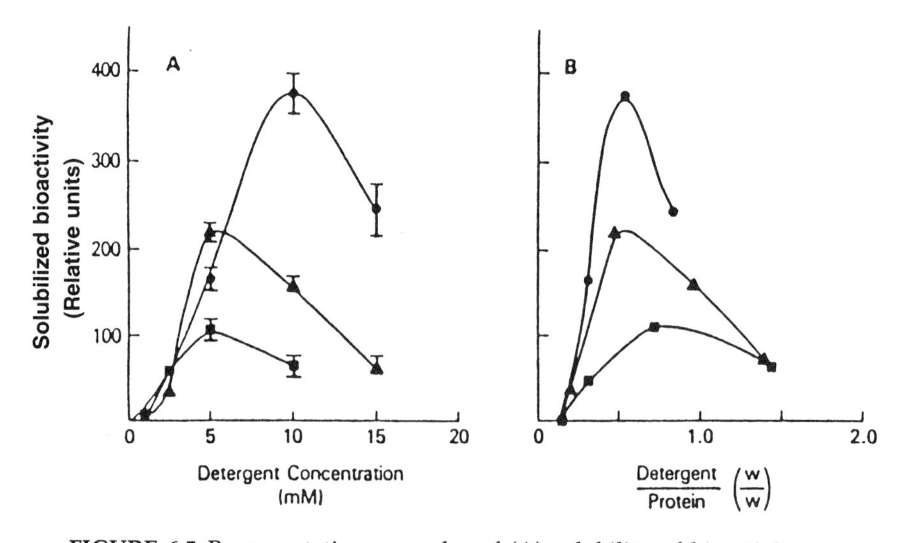

FIGURE 6.5 Representative examples of (A) solubility of bioactivity versus detergent concentration and (B) solubility of bioactivity versus detergent/protein ratio. Protein concentrations for both panel A and B are (mg/ml) 11.1(•), 6.5(▲), and 4.3(■). (Reprinted with permission from Hjelmeland and Chrambach, 1984).

Figure 6.5B. This will allow the choice of a suitable detergent concentration (Hjelmeland and Chrambach, 1984).

Detergent Removal

Removing detergents from protein solutions can be inefficient and time consuming. It can also result in large protein losses. However, separating unbound detergent from hydrophobic protein may become necessary. The ease with which detergents can be removed from protein solutions depends on a number of factors: (1) the properties of the detergent itself; (2) the hydrophobic/hydrophilic characteristics of the protein and (3) the composition of the buffer. Methods that are mentioned below are discussed in greater detail by Furth et al, (1984).

Removal of Ionic Detergents

- Gel filtration chromatography using Sephadex G-25 (or comparable matrix).
- Add urea to a concentration of 8M, then bind the detergent to an ion exchange column. Run the protein solution in the presence of 8M urea. The protein should flow through. Dialyze to remove the urea.
- When using ionic detergents with low micelle size and high CMC, dilute the solution as much as possible and dialyze. To increase the exchange rate, a mixed bed resin can be added to the dialysate.

Removal of Nonionic Detergents

- If possible, dilute the solution, then dialyze against a solution containing DOC. Remove the DOC by dialysis.
- Velocity sedimentation into detergent-free sucrose. Stephens, (1985) applied solubilized membranes in the presence of 2–17 times the CMC of NP-40 to a 20–60% linear sucrose gradient and sedimented to equilibrium.
- Bind the protein to an affinity matrix or ion-exchange resin; wash extensively to remove the detergent, then elute the protein (see Chapter 10).
- Bio-Beads SM-2 (Bio-Rad) are neutral, macroporous polymeric beads with a high surface area for adsorbing organics with MW < 2000 from aqueous solution. Simply add the Bio-Beads to the solution containing the nonionic detergent and stir gently.

Extracti-Gel® D

Extracti-Gel® D is an affinity matrix marketed by Pierce for the removal of unwanted detergents from proteinaceous solutions (Pierce Catalog and Handbook, 2002). Small detergent molecules enter the gel matrix where they interact with a specifically designed ligand which effectively removes them from solution.

Once packed, the column should be thoroughly washed and equilibrated with a suitable buffer. The product will remove a variety of detergents from solutions of different buffers or salt concentrations and can be used between pH 3.5–10 (see Table 6.3).

TABLE 6.3 Use of Extracti-Gel® D*

Detergent	Capacity (mg/ml gel)	Binding Conditions
CHAPS	50	50mM Tris, pH 9.0
SDS	80	50mM Tris, pH 9.0
Triton X-100	57	100mM Phoshate buffer, pH 7.0
Brij-35	80	100mM Phoshate buffer, pH 7.0

*Extracti-Gel® D can be regenerated and reused. (Data from *Pierce Catalog and Handbook: Life Science and Analytical Research Products*, 1994.)

The support matrix has a size exclusion limit of 10kDa. If the target molecule is 10kDa or less it can enter the pores of the gel and interact with the affinity ligand, increasing the chance of loss. Therefore, this matrix is not recommended for molecules less than 10kDa.

When using dilute solutions with respect to the target molecule, 50µg/ml or less, it is recommended to use a carrier molecule, such as BSA at 0.1%.

References

Bennett JP (1982): Solubilisation of membrane-bound enzymes and analysis of membrane protein concentration. In: *Techniques in Lipid and Membrane Biochemistry.* Holland:Elsevier

Blonder J, Goshe MB, Moore RJ, Pasa-Tolic L, Masselon CD, Lipton MS, Smith RD (2002): Enrichment of integral membrane proteins for proteomic analysis using liquid chromatography-tandem mass spectrometry. *J Proteome Res* 1:351–360

Bordier C (1981): Phase separation of integral membrane proteins in Triton X-114. *J Biol Chem* 256:1604–1607

Collins MLP, Salton MRJ (1979): Solubility characteristics of *Micrococcus lysodeikticus* membrane components in detergents and chaotropic salts analyzed by immunoelectrophoresis. *Biochim Biophys Acta* 553:40–53

Creighton TE (1984): *Proteins: Structures and Molecular Properties.* New York: W.H. Freeman

Dailey HA, Lascelles J (1974): Ferrochelatase activity in wild-type and mutant strains of *Spirillum itersonii.* Solubilization with chaotropic agents. *Arch Biochem Biophys* 160:523–529

Dooijewaara G, Slater EC, Van Dijk PJ, De Bruin GJM (1978): Chaotropic resolution of high molecular weight (type I) NADH dehydrogenase, and reassociation of flavin-rich (type II) and flavin-poor subunits. *Biochim Biophys Acta* 503:405–424

Ferro M, Seigneurin-Berny D, Rolland N, Chapel A, Salvi D, Garin J, Joyard J (2000): Organic solvent extraction as a versatile procedure to identify hydrophobic chloroplast membrane proteins. *Electrophoresis* 21:3517–3526

Furth AJ, Bolton H, Potter J, Priddle JD (1984): Separating detergents from proteins. *Methods Enzymol* 104:318–328

Hatefi Y, Hanstein WG (1969): Solubilization of particulate proteins and nonelectrolytes by chaotropic agents. *Proc Natl Acad Sci USA* 62:1129–1136

Helenius A, Fries A, Kartenbeck J (1977): Reconstitution of Semliki Forest virus membranes. *J Cell Biol* 75:866–880

Henningsen R, Gale BL, Straub KM, DeNagel DC (2002): Application of zwitterionic detergents to the solubilization of integral membrane proteins for two-dimensional gel electrophoresis and mass spectrometry. *Proteomics* 2:1479–1488

Hjelmeland LM, Chrambach A (1984): Solubilization of functional membrane proteins. *Methods Enzymol* 104:305–318

Howell KE, Palade GE (1982): Hepatic Golgi fractions resolved into membrane and content subfractions. *J Cell Biol* 92:822–832

Juliano RL, Rothstein A (1971): Properties of an erythrocyte membrane lipoprotein fraction. *Biochim Biophys Acta* 249:227–235

King TE (1963): Reconstitution of respiratory chain enzyme systems. *J Biol Chem* 238:4037–4051

Mackall J, Meredith M, Lane MD (1979): A mild procedure for the rapid release of cytoplasmic enzymes from cultured animal cells. *Anal Biochem* 95: 270–274

Meyer DI, Dobberstein B (1980): A membrane component essential for vectorial translocation of nascent proteins across the endoplasmic reticulum: requirements for its extraction and reassociation with the membrane. *J Cell Biol* 87:498–502

Nelson N, Eytan E, Notsani B-E, Sigrist H, Sigrist-Nelson K, Gitler C (1977): Isolation of a chloroplast *N,N′*-dicyclohexylcarbodiimide-binding proteolipid, active in proton translocation. *Proc Natl Acad Sci USA* 74: 2375–2378

Okumura T, Hasitz M, Jamieson GA (1978): Platelet glycocalicin. Interaction with thrombin and role as thrombin receptor of the platelet surface. *J Biol Chem* 253:3435–3443

Penefsky HS, Tzagoloff A (1971): Extraction of water-soluble enzymes and proteins from membranes. *Methods Enzymol* 22:204–219

Pierce Catalog and Handbook: Life Science and Analytical Research Products (2002): Rockford, IL: Pierce Co.

Rabilloud T, Blisnick T, Heller M, Luche S, Aebersold R, Lunardi J, Braun-Breton C (1999): Analysis of membrane proteins by two-dimensional electrophoresis: comparison of the proteins extracted from normal or Plasmodium falciparum-infected erythrocyte ghosts. *Electrophoresis* 20:3603–3610

Ramsby ML, Makowski GS (1999): Differential detergent fractionation of eukaryotic cells. In: *Methods in Molecular Biology, Vol, 112: 2-D Proteome Analysis Protocols,* Link AJ ed.

Rendi R (1970): Na^+, K^+-requiring ATPase. Preparation and assay of a solubilised Na^+-stimulated ADP-ATP exchange activity. *Biochim Biophys Acta* 198:113–119

Rivnay B, Metzger H (1982): Reconstitution of the receptor for immunoglobulin E into liposomes. *J Biol Chem* 257:12800–12808

Rowe ES, Zhang F, Leung TW, Parr JS, Guy PT (1998): Thermodynamics of membrane partitioning for a series of n-alcohols determined by titration calorimetry: role of hydrophobic effects. *Biochemistry* 37:2430–2440.

Singer SJ, Nicolson GL (1972): The fluid mosaic model of the structure of cell membranes. *Science* 175:720–731

Steck TL (1972): Selective solubilization of red blood cell membrane proteins with guanidine hydrochloride. *Biochim Biophys Acta* 255:553–556

Stephens RE (1985): Evidence for a tubulin-containing lipid-protein structural complex in ciliary membranes. *J Cell Biol* 100:1082–1090

Tanner MJA, Boxer DH (1972): Separation and properties of the major proteins of the human erythrocyte membrane. *Biochem J* 129:333–347

van Renswoude J, Kempf C (1984): Purification of integral membrane proteins. *Methods Enzymol* 104:329–339 von Heijne G, Gavel Y (1988): Topogenic signals in integral membrane proteins. *Eur J Biochem* 174:671–678

Warren G, Dobberstein B (1978): Protein transfer across microsomal membranes reassembled from separated membrane components. *Nature* 273: 569–571

Wu CC, MacCoss MJ, Howell KE, Yates JR (2003): A method for the comprehensive proteomic analysis of membrane proteins. *Nature Biotechnol* 21:532–538

7

Transfer and Detection of Proteins on Membrane Supports

Introduction

Electro-transferring proteins from an SDS-PAGE gel to a membrane, known as Western blotting, combines the resolution capabilities of electrophoretic protein separation with the specificity of immunological identification in a rapid and highly sensitive format. For optimal results, the macromolecules must be transferred efficiently from the gel to the membrane. The proteins that are now immobilized on the membrane can be identified and visualized by using specific and sensitive immunobiochemical detection techniques.

A. Transfer of Proteins to Membrane Supports

Once transferred onto the membrane, a wide variety of reactions may be carried out on the immobilized proteins which would otherwise have proven to be difficult or impossible in the gel. Therefore, Western blotting has become a highly popular and convenient method for analyzing and characterizing proteins. Figure 7.1 demonstrates the steps that are involved and some of the options that are available once the target protein has been immobilized on the membrane.

The specific recognition and strong binding affinity between antigen and antibody is the basis of the technique. The target protein is located by a specific antigen-antibody (polyclonal or monoclonal) reaction. A second antibody, which recognizes the primary antibody, is used to aid in locating the position of the primary antigen-antibody complex. The species-specific second antibody can be labeled in a number of ways to produce the final visual image.

In addition to antigen-antibody based methodology, the target proteins can be detected by utilizing other high affinity biochemical interactions. Lectin-glycoprotein, protein-nucleic acid, biotin-avidin, and ligand blotting have all been used successfully.

Blotting, followed by immunobiochemical detection, can be used to assay fractions collected from chromatography columns or for monitoring any other purification step in which locating the protein of inter-

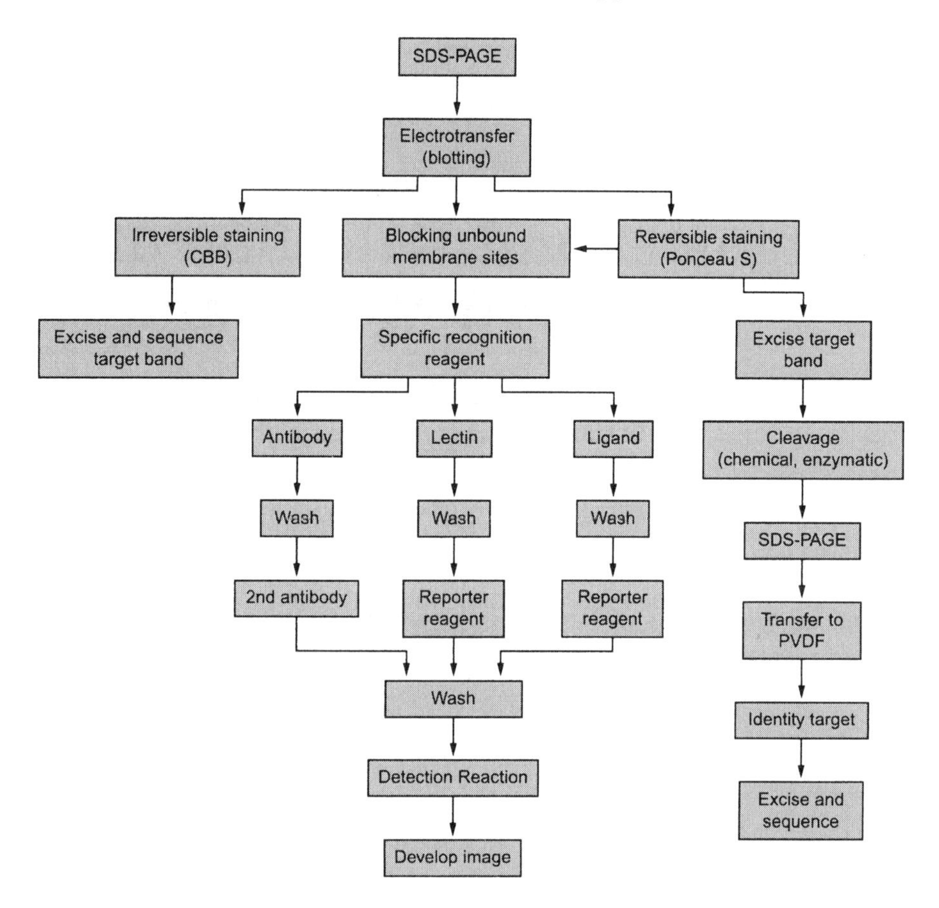

FIGURE 7.1 Blotting flowchart.

est is desired. An in-depth treatment of Western blotting was presented by Gershoni (1988).

PROTOCOL 7.1 Transfer of Proteins to Nitrocellulose or Polyvinylidene Difluoride

Binding of a protein to a blotting membrane depends on the distribution of the charged and hydrophobic amino acid residues on the surface of the molecule. Proteins can be immobilized on a number of different types of membranes. The most widely used are polyvinylidene difluoride (PVDF), nitrocellulose (NC) and nylon membranes which all have capacities of ~150 μg protein/cm^2. Not all membranes are created equal. It may be necessary to identify the optimal membrane suitable for the detection of a specific protein (Balasubramanian and Zingde, 1997). Four different membranes are listed below. Some of them are available in more than one pore size.

Hybond®-C Nitrocellulose Membrane, Amersham Cat # RPN 303C
Hybond-ECL Nitrocellulose Membrane, Amersham Cat # RPN 202D

Hybond-C Extra Supported Nitrocellulose Membrane, Amersham Cat # RPN 303E

Immobilon™-P PVDF Transfer Membrane, Millipore Cat # IPVH00010

When the SDS-PAGE run is nearing completion, the membrane should be prepared for the transfer.

Materials

Transfer buffer: 25 mM Tris (3.03 g/L), 192 mM Glycine (14.42 g/L), Methanol 10% (v/v) should not exceed 20%

Membranes: PVDF or NC

Plastic containers

Electrophoretic transfer chamber

Power supply

Filter paper: Whatman 3MM

Paper tweezers

Sponges

1. Cut a piece of NC or PVDF membrane slightly larger than the dimensions of the gel and carefully hydrate it in transfer buffer (see Figure 7.2). (If transferring to PVDF, the membrane must be soaked in 100% methanol for 10 sec prior to hydrating it in transfer buffer.) Incubate the membrane in transfer buffer for 5 min.
2. Cut two pieces of Whatman 3MM filter paper slightly larger than the membrane and soak them in transfer buffer. Sponges (part of the transfer apparatus) are also saturated with transfer buffer.
3. The orientation of the membrane and the gel in relation to the electrodes is shown in Figure 7.2. Place the membrane between the (+) electrode and the gel. The proteins, which are all negatively charged due to the SDS treatment, will migrate toward the positive electrode and will be immobilized on the membrane. Make sure that there are no air bubbles between the gel and the membrane. Air bubbles will cause bald spots

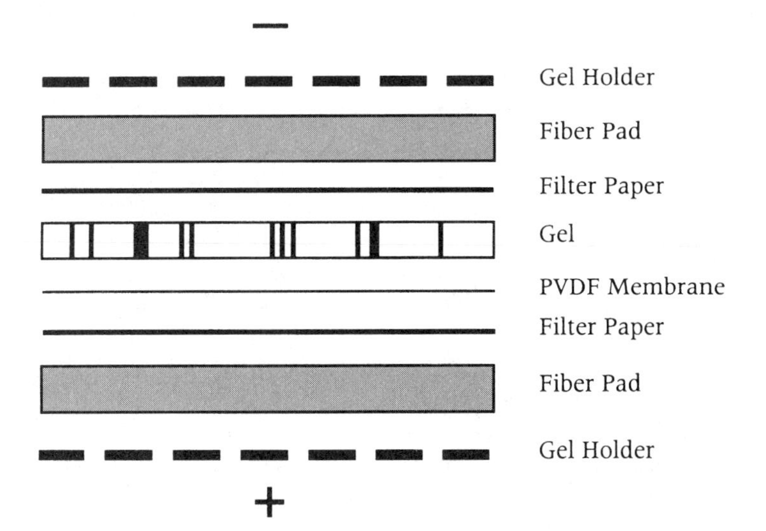

FIGURE 7.2 Gel-membrane orientation.

which are very annoying if they happen to occur directly over the target protein.

4. Mini-gel transfers are routinely run at 25 V for 60–90 min. This is usually sufficient to transfer most proteins up to 100 kDa. The greater the mass of the protein the more time needed for transfer. As a general rule, the temperature of the transfer buffer should not exceed 20°C. To avoid overheating, the transfer can be performed in a coldroom, or with a cooling system (if this capability exists) or the transfer buffer can be pre-cooled before use by storing it in a coldroom.
5. When the transfer is complete, turn off the power and remove the cassette containing the gel and membrane. With a pencil, mark the side of the membrane to which the proteins were transferred. You will need to know this for optimum detection.

Comments: Do not handle the membranes with bare hands. Fingerprints will show up. Wear gloves and handle the membrane with paper tweezers.

There are many recipes for transfer buffers. SDS and methanol are two additives commonly used to enhance transfer. Some protocols call for the addition of between 10–20% methanol, others call for up to 0.1% SDS (Towbin et al, 1979). Methanol is frequently included when transferring low molecular weight proteins. Methanol in the buffer removes SDS from the detergent-protein complexes and increases the affinity between the NC membrane and the SDS-free proteins. Low concentrations of SDS, ~0.1%, aid in the transfer of high molecular weight proteins (>100 kDa) but may interfere with the binding of proteins to the membrane. All proteins do not uniformly transfer and bind to the membrane. For the majority of cases, the standard transfer buffer should be successful.

Highly glycosylated and/or charged proteins often do not electroblot efficiently and the standard general protein stains (e.g. Coomassie Blue) often do not visualize glycoproteins well.

Optimal conditions for transfer of specific proteins should be derived experimentally. Transfer efficiency can be monitored by staining the gel after transfer with Ponceau S as described in Protocol 7.6.

Stained proteins in gels can also be transferred to NC or PVDF membranes. The transferred proteins remain stained during immunodetection, providing a set of background markers for protein location and size determination (Thompson and Larson, 1992). Gels are stained with Coomassie blue (0.25% in 25% isopropanol, 10% acetic acid) and destained in 10% acetic acid. Before electro-transfer, the gel must be soaked for 1 h in 50 mM Tris-HCl, pH 7.5, 1% SDS to recoat the protein/dye complex with SDS, followed by equilibration for 15 min in transfer buffer.

Protein bands can be detected directly, inside the polyacrylamide gel. UnBlot™, a new technology developed by Pierce, is marketed in kit form that enables the immunodetection of specific target proteins inside the gel, skipping the transfer and the blocking steps. This system allows for stripping and reprobing for different targets. The system also permits immunodetection and staining for total proteins in the same gel. Sensitivity down to 1 ng of target protein can be reached.

The recommended gel thickness for use with this kit is 0.75–1.5 mm. A word of caution, be sure to check the product specifications. The UnBlot™ system does not work well with all commercially available precast gels.

Kodak has introduced a kit that allows the simultaneous production of up to 10 Western blots from a single protein gel. Based on proprietary technology, each identical filter can be screened with one or more different antibodies, improving speed and accuracy of multiple antibody screening. Compatible with existing Western blotting transfer systems, the kit eliminates the inconsistency of running multiple gels or same blot stripping and reprobing.

Troubleshooting Western Blots

1. *High background.* The membrane was not suitably blocked prior to probing. Try different blocking agents, higher concentrations and longer incubations.
2. *Too much signal.* Reduce the amount of protein loaded onto the gel. Retiter the primary antibody to reduce the signal. The film may be overexposed. Try a shorter film exposure or let the chemiluminescent signal decay for 15–30 minutes and repeat the exposure.
3. *No signal.* The proteins may not have transferred from the gel. To confirm protein transfer, run a lane with colored markers. If they can be visualized on the membrane following the transfer procedure then the proteins probably transferred successfully. A "no-signal" result may be due to poor quality antibody. Be sure the antibody recognizes its target in the Western blot system. An additional lane containing a positive control could be helpful. Did you make a silly mistake? Was the chemiluminescent reagent prepared properly? Omission of the second antibody or trying to detect a mouse monoclonal first antibody with a goat anti-rabbit HRP will produce a blank film.
4. *White bands "ghostbands".* Blank bands on film are caused by the depletion of chemiluminescence substrate at sites of excess antigen or antibody. Try reducing the amount of antigen loaded and/or the antibody concentrations.

PROTOCOL 7.2 Enhanced Capture of Small Histidine Containing Polypeptides on Membranes in the Presence of $ZnCl_2$

Small hydrophilic polypeptides, less than 10 kDa may be difficult to capture on Western blot membranes using standard protocols. The addition of 2 mM $ZnCl_2$ was shown to enhance the capture of polypeptides by almost 10 fold (Maillet et al, 2001). However, to immobilize the proteins on the membrane a post-transfer fixation step must be added.

Materials

Transfer buffer plus 2mM $ZnCl_2$
2% glutaraldehyde
5mg/ml $NaBH_4$

1. Perform the transfer using transfer buffer, plus 2mM $ZnCl_2$.
2. Following transfer, fix the membrane for 5min in 2% glutaraldehyde.
3. Incubate the membrane for 5min in 5mg/ml $NaBH_4$.

Comments: The presence of $ZnCl_2$ in the transfer buffer does not appear to affect the transfer of other proteins nor does the $ZnCl_2$ interfere with subsequent microsequencing. This protocol could have broader applicability for use with recombinant histidine-tagged polypeptides (see Chapter 11) that do not bind well to membranes using standard transfer protocols.

PROTOCOL 7.3 Dot Blots

Dot blots are produced by spotting solutions of nondenatured proteins directly onto a membrane (Hawkes et al, 1982). Dot blots are useful for screening multiple fractions generated from monoclonal antibody production, column chromatography or sucrose gradients. In addition, since the proteins that are immobilized on the membrane are native, the specificity of the antibody reaction may differ from that obtained when denatured proteins are electrotransferred and probed with the same antibody.

Materials

PVDF or NC membrane
Foreceps for paper
Micropipette
Transfer buffer (see Appendix C)
Soft lead pencil

1. Wearing gloves, cut out a piece of membrane from the roll or sheet and with a soft lead pencil make a grid with labeled columns and rows so that the spots can be identified.
2. To immobilize native proteins on either NC or PVDF membranes moisten the membrane with transfer buffer and spot 1 µl of sample onto the membrane. The membrane can now be processed like a Western blot.

PROTOCOL 7.4 Thin-Layer Chromatography Blotting

Thin-layer chromatography (TLC) is the most widely used method in the detection, separation, and monitoring of phospholipids and glycosphingolipids. Methods have been developed that take advantage of the excellent separation of lipids on high performance TLC (HPTLC) plates and the transfer and immobilization of lipids onto polyvinylidene difluoride (PVDF) membranes (Taki and Ishikawa, 1997). Modifications of the basic method described below can be used to purify lipids on a microscale, perform mass spectrometric analysis on complex lipids, perform binding assays of microorganisms, detect enzymes, and examine lipid-protein interactions.

Materials

HPTLC plates: Silica Gel 60, E. Merck, Darmstadt, Germany
Drying oven or hair drier
Glass microfiber filter sheet (GF/A; Whatman)
PVDF membrane
Blotting solvent: Isopropanol/0.2% aqueous $CaCl_2$/methanol, 40/20/7 by volume
Household clothes iron

1. Separate the glycosphigolipids on a plate precoated with Silica Gel 60.
2. Dry the plate then submerge it in blotting solvent for 20 sec.
3. Place a PVDF membrane and then a glass microfiber filter sheet (GF/A; Whatman) over the plate.
4. Using a household clothes iron at 180°C, press the whole assembly evenly for 30 sec.
5. To block and probe the membrane, proceed to Protocol 7.12.

Comments: Phospholipids are transferred to a PVDF membrane by the same procedure. There is linear dose-dependent transfer of lipids from the HPTLC plate to PVDF membrane for 0.1–3.0 μg of each glycosphingolipid and for 2–10 μg of each phospholipids.

B. Staining the Blot

Protein blots are frequently stained for total protein (Hancock and Tsang, 1983) or for glycoproteins (Stromqvist and Gruffman, 1992) prior to the blocking step.

PROTOCOL 7.5 Total Protein Staining with India Ink

Materials

Wash buffer (PBS-T): 0.15 M NaCl, 0.01 M Na_2HPO_4/NaH_2PO_4 buffer, pH 7.2, 0.3% Tween 20
India ink solution: Pelikan Fount India drawing ink for fountain pens (Pelikan AG, D-3000 Hanover 1, Germany or Pelikan 17 black), 1% solution in wash buffer
Glass container
KOH solution: 1%
PBS (see Appendix C)
Shaker

1. Following electrotransfer (Protocol 7.1), place the blot in a glass container and incubate for 5 min in the KOH solution at 20°C.

2. Rinse the blot twice for 10min each with PBS.
3. Wash the blot four times for 5min each time with wash buffer at 37°C, shaking. Rinse the blot with water between each wash. The quantity of buffer used for the washes will depend on the size of the blot.
4. Stain the blot by submerging it in India ink solution with agitation for 15min to 18h depending on the sensitivity required.
5. Destain by rinsing the blot in multiple changes of PBS over 30min.

Comments: The sensitivity of this staining method varies from protein to protein. For example, 80ng of β-galactosidase will stain with the intensity of 520ng of ovalbumin. Protein bands will appear as black bands on a gray background. The blots can be stored in wash buffer (step 3) for at least one month with no loss of sensitivity or resolution.

The brief KOH treatment in step 1 will enhance subsequent staining with India ink.

PROTOCOL 7.6 Reversible Staining with Ponceau S

Ponceau S staining of proteins on the blot is a reversible procedure as the stain can be readily removed by washing the blot with water. Although it is not a very sensitive method, it is used primarily to locate major bands which will be excised for microsequencing or to confirm that a protein has been successfully transferred to the inert matrix prior to subsequent analysis. Ponceau S staining is performed prior to blocking the blot.

Materials

Shaker or rocker

Plastic containers

Ponceau S solution: 0.1g Ponceau S, 1ml glacial acetic acid, 100ml H_2O

1. Soak the blot in Ponceau S solution for 5min with shaking.
2. Destain the blot by rinsing in water for 2min. The blot can be photographed or photocopied. If desirable, mark the molecular weight standards and the location of any other band of interest either with a pencil mark, pin hole or with indelible ink.
3. Completely destain the membrane by soaking in water for 10min. You may then continue with other detection protocols.

PROTOCOL 7.7 Irreversible, Rapid Staining with Coomassie Brilliant Blue

One important application for the following quick protocol is to use the stained strips as a visual aid for alignment purposes. Do not block the strips.

Materials

Razor blade
Stain solution: 0.1% Coomassie blue R in 50% methanol
Destain solution: 50% methanol, 10% acetic acid
Shaker
TTBS (Appendix C)

1. Following electrotransfer (Protocol 7.1), cut vertical strips from both sides of the blot and wash them in deionized water for 5 min. Store the remainder of the blot in TTBS until ready for further use.
2. Submerge the strips in stain solution for 5 min.
3. Soak the strips in destain solution for 5–10 min with gentle shaking at room temperature. Change the destaining solution three or four times.

PROTOCOL 7.8 Staining Immobilized Glycoproteins by Periodic Acid/Schiff (PAS)

The periodic acid/Schiff method (PAS) of staining glycoproteins in polyacrylamide gels has been a commonly used technique (Zacharius et al, 1969). The main drawbacks of PAS staining are that it is not very sensitive and is time consuming, requiring numerous incubations and washing steps. However PAS staining does offer some advantages, the most important of which is that it is nondiscriminating, i.e., it stains all carbohydrates containing diol groups that can be oxidized by periodic acid. Protein blotting increases the sensitivity of PAS staining, and the method can be used to determine whether a specific band is a glycoprotein. Glycoproteins containing carbohydrate moieties that are easily oxidized by periodate are stained more readily than those that are resistant to oxidation. To characterize the type of carbohydrate other more specific methods must be used.

Materials

Periodic acid solution: 1% periodic acid, 3% acetic acid
Schiff's reagent (Sigma)
Sodium bisulfite: 0.5% w/v
Aluminum foil

1. Following electrotransfer (Protocol 7.1), wash the blot with water for 5 min then incubate in periodic acid solution for 15 min.
2. Wash the blot with water for 15 min changing the water several times.
3. Add Schiff's reagent and stain the membrane for 15 min in an aluminum foil covered container (so no light gets in).
4. After staining, wash the blot for 5 min in the sodium bisulfite solution and then wash with water to visualize the bands. Keep the blot protected from the light.
5. Air dry the membrane and store it wrapped in aluminum foil to prevent bleaching.

Comments: The complete staining procedure from the time the electrotransfer is completed should take less than one hour.

C. Recovery of Proteins from the Blot

The ability to recover proteins offers a simple, high-resolution approach to isolating individual proteins from complex mixtures. Recovery of proteins from blots is desirable for many purposes such as using the protein as an immunogen, or for subsequent purification prior to protein sequencing. Contaminating salts and detergents can be easily removed from the membrane by washing. Many factors may influence the recovery of proteins from a blot. Elution efficiencies have been found to be inversely proportional to the molecular weight of the protein. Heat generated during the transfer may contribute to strong protein-membrane associations making elution more difficult. Exposure to extremes of pH during blot processing also affects protein-membrane interactions. Staining with Coomassie Brilliant Blue or Amido Black prior to elution from NC prevented protein desorption by acetonitrile (Parekh et al, 1985). Acetonitrile might degrade the membrane.

The choice of solvent system depends on the goal of the project. Solvent systems with acetonitrile or n-propanol maintain protein structure and antigenic character necessary for immunogen preparation. If structural characterization is desired, detergents and buffer salts should be avoided.

PROTOCOL 7.9 Recovery of Proteins Using an Organic Solvent System

Two methods for recovery of polypeptides from PVDF and NC membranes are presented. Different concentrations of acetonitrile are used to elute proteins from either PVDF or NC membranes.

Materials

Ponceau S: 0.1% w/v Ponceau S in 1% v/v acetic acid

Razor blade

PVDF elution buffer: 40% acetonitrile in 0.1M ammonium acetate buffer, pH 8.9

Rocker or shaker

Microfuge tubes

Microfuge

1. To locate the band of interest, stain the membrane (see Protocol 7.6) with Ponceau S for 5min then destain with distilled water. The use of Ponceau S has been found to minimize protein alteration so that when eluted, the protein would be a suitable immunogen. Avoid

drying the membrane. Elution is more efficient from NC membranes than PVDF.

2. Cut out the band of interest from the PVDF membrane and incubate it in a microfuge tube with ~500 µl of elution buffer rotating or shaking for 3 h at 37°C. Use 20–30% acetonitrile for eluting proteins from nitrocellulose (Montelaro, 1987).
3. Centrifuge the tube for 10 min at maximum speed in a microfuge. Remove and save the supernatant.
4. Add 250 µl of fresh elution buffer to the tube and centrifuge again for 5 min. Pool the second supernatant with the first.
5. Lyophilize the extract to remove the volatile solvent and buffer. This step concentrates the protein and permits its resolubilization in a buffer of choice.

Comments: Proteins eluted from the membrane can be used as immunogens, antigens in radioimmunoassays, and ELISA (Enzyme Linked Immunosorbent Assay) procedures, as well as substrates for peptide mapping, protein sequencing, and amino acid composition analysis.

Alternatively, a slightly harsher method for eluting proteins from blots has been reported (Yuen et al, 1989). When the optimum priority is obtaining protein for subsequent analytical determinations and not for retaining structure and antigenic character, 70% isopropyl alcohol and 1–5% TFA can be used to elute the protein from the membrane.

When the proteins are concentrated, the NC will form a precipitate that can easily be removed by centrifugation.

PROTOCOL 7.10 Recovery of Proteins from the Blot Using a Detergent-Based Solvent System

When the goal is to obtain pure protein in a solvent system compatible with subsequent proteolytic and analytical manipulations, a detergent based extraction buffer can be used (Simpson et al, 1989).

Materials

Ponceau S: 0.1% w/v Ponceau S in 1% v/v acetic acid

Elution buffer: 50 mM Tris-HCl, pH 9.0, 2% SDS, 1% Triton X-100, and 0.1% DTT

1. To locate the band of interest, stain the proteins that have been transferred to a PVDF membrane (see Protocol 7.6) with Ponceau S for 5 min then destain with distilled water. The use of Ponceau S has been found to minimize protein alteration so that when eluted, the protein would be a suitable immunogen. Avoid drying the membrane.
2. Cut out the band of interest from the PVDF membrane and incubate it in a microfuge tube with 500 µl of elution buffer at room temperature for 1–3 h.

3. Centrifuge the tube for 10 min at maximum speed in a microfuge. Remove and save the supernatant.
4. Rinse the tube with 250 μl of fresh elution buffer and centrifuge again for 5 min.
5. Pool the second supernatant with the first. At this juncture, the protein is ready to be further characterized.

PROTOCOL 7.11 Blocking the Blot

Common to all immuno-detection procedures is a preliminary blocking step to prevent nonspecific binding of the detecting system to the membrane. Many substances have been used for quenching or blocking the blot. Currently, a popular choice is 5% instant nonfat dry milk made up in Tris buffered saline (TBS).

Materials

Tris buffered saline (TBS): see Appendix C

Blocking solution: 5% Instant nonfat dry milk solution in TBS

Shaker

Foreceps

1. Disassemble the transfer apparatus and remove the cassette.
2. With foreceps, transfer the blot to the blocking solution.
3. Incubate the blot for 1 h at room temperature or overnight at 4°C with constant shaking.

Comments: A solution of 5% instant nonfat dry milk in TBS has consistently yielded a cleaner background than other blocking solutions such as 1–5% bovine serum albumin (BSA) or fish gelatin. It is also inexpensive. However, it has been found that certain proteins that bind carbohydrate are not detected when the blot is quenched with the milk solution. Nonfat dry milk contains a variety of proteins and carbohydrates, not all of which are well characterized. Some may interact with the proteins transferred to the membrane or with the particular antibodies and could contribute to increased levels of nonspecific background. If backgrounds are high using 5% dry milk, try reducing the concentration to 1% and including 0.5% BSA in membrane wash steps to keep the membrane blocked throughout the whole immunoblotting procedure (Sheng and Schuster, 1992). Alternative blocking solutions are 3% BSA with 0.05% Tween 20 or a 5% BSA solution in TBS.

PROTOCOL 7.12 Exposing the Blot to Primary Antibody

Incubation of the blot with a solution of primary antibody can be performed in plastic or glass containers, or "Seal-a-Meal" bags. For optimal processing of blots, especially with a rare antibody, the use of a rolling cylinder is recommended (Rollins Scientific Corporation, if available). The cylinder should be of sufficient circumference to accommodate the blot without overlapping. The lid of the cylinder should seal well and be flush with the outside of the container such that the inner walls of the container are uniformly covered with solution. The blot is placed against the inner wall of the cylinder with the side of the blot carrying the transferred proteins facing inward. This system has many advantages over the use of a rectangular container or a rotary shaker. Small volumes of reagent can be used, a big plus when the antibody is scarce or expensive. Since the membrane is constantly in motion, there will be no localized areas of reagent depletion, and the washing steps will be very efficient, resulting in reduced backgrounds and increased detection sensitivity. Rocking shakers work quite well also.

To minimize nonspecific reactions, antibodies should be used at the highest dilution that will still produce a strong positive signal. The antibody can be in the form of purified monoclonal, polyclonal, whole serum, ascites fluid or tissue culture supernatant. Routine working dilutions of antibody range from 1:10–1:10,000. As a first approximation try: for polyclonal antibody (serum), 1:100–1:1000; for hybridoma tissue culture supernatants, 1:10 to 1:100; for monoclonal ascites fluid >1:1000; for purified polyclonal or monoclonal antibody 1–5 μg/ml. The appropriate dilution of Ab is usually determined experimentally employing serial tenfold dilutions to produce the optimal signal to noise ratio.

Materials

Shaker

Containers

Tween Tris buffered saline (TTBS): 10 mM Tris-HCl, pH 8.0, 150 mM NaCl, 0.05% Tween 20

Instant nonfat dried milk

1. Prepare an appropriate dilution of the desired antibody in a solution of TBS containing 1% instant nonfat dried milk, or a 1% BSA solution in TBS when milk is undesirable.
2. Following the blocking step (Protocol 7.11), incubate the blot in the antibody solution for 1 h at room temperature with constant shaking.
3. Wash the blot 3 times with TTBS for 5 min each at room temperature. The blot is now ready to be incubated with the second antibody.

Comments: Tween 20 is included in the wash buffer because it reduces nonspecific hydrophobic interactions between antibody and the immobilized proteins. Monoclonal antibodies may fail to react with the blotted proteins because the antigenic determinants that they recognize have been destroyed by conditions required for sample preparation (detergents, heat, electrophoresis, reducing agents, etc). The affinity of an antibody or ligand for a partially denatured, blotted protein may be considerably less than for its native counterpart.

Provided that the antibody is indeed monoclonal, staining of multiple bands suggests recognition of a common sequence or of repeated determinants that are present in more than one protein. Therefore, when using a monoclonal, do not expect one band exclusively. Monoclonal antibodies can be used to follow the maturation of a particular protein, in which case multiple bands that will vary in intensity will be visualized if the antibody recognizes different forms of the target protein during processing.

Blots can be stripped of the first antibody and then be reprobed. A 30 min incubation in 0.2 M glycine-HCl, pH 2.5, containing 0.05% Tween 20 will erase the blot. Following stripping, the blot should be neutralized by rinsing in two changes of TBS. A more robust stripping protocol is presented in Protocol 7.21.

Ligand Blotting

Ligand blotting is technically similar to immunoblotting. Specific proteins are detected on the membrane by their ability to bind a specific ligand rather than an antibody. Molecules other than antibodies can be used as specific, sensitive probes for proteins that have been immobilized on NC or PVDF. This method and its variations has also been referred to as Far Western blotting. Ligand blotting allows for a diverse array of biochemical analyses including: (1) modification of blotted proteins, (2) competitive binding assays, and (3) *in vitro* reconstitution experiments. The ligand can be another protein or peptide, a nucleic acid (Southwestern blotting), or any molecule that binds with high affinity. Specific examples of ligand blotting are the detection of calcium-binding proteins by the use of $^{45}Ca^{2+}$, the use of ^{125}I-labeled human chorionic gonadotropic hormone (hCG) to probe a blot for hCG receptor complexes (Roche and Ryan, 1989), and bungarotoxin to analyze proteolytic fragments of the acetylcholine receptor (Wilson et al, 1984).

Proteins can be modified chemically or enzymatically prior to electrophoresis or on the blots after transfer. For example, the effects of exposing or removing glycosyl groups from glycoproteins (Gershoni et al, 1983) or phosphate groups from phosphoproteins can be studied (Fernandez-Pol, 1982). Ligand specificity can and should be demonstrated on blots by including an appropriate control based on competitive inhibition. For example, lectin binding can be inhibited by cognate sugars, a suitable control to demonstrate specificity (Gershoni and Palade, 1983). Calmodulin binding can be inhibited by chelation of Ca^{2+} (Flanagan and Yost, 1984).

In addition to probing blots with molecules, intact cells and bacteria have been used as probes to detect cell-surface attachment proteins which have been separated by SDS-PAGE and transferred to NC (Hayman et al, 1983; Boren et al, 1993).

Epitope mapping, the determination of the linear sequence of ligand binding sites on a protein, is a valuable contribution to the analysis of protein structure (see Chapter 8). A protein is digested chemically or enzymatically, producing peptides that are resolved by SDS-PAGE and transferred to NC. The blot is then probed with the desired ligand or monoclonal antibody. In this way, the specific antigenic determinant, ligand binding domain, structural domains such as DNA binding sites (Southwestern blots), receptor binding sites, or toxin binding sites can be defined.

A sufficiently sensitive means of detecting the bound ligand must be available. This may be by direct identification of a radioactive or visible label, or by means of an indirect system. These alternatives are discussed in detail below.

It should be emphasized that the binding of a particular ligand to a protein on a blot does not in itself constitute sufficient evidence that a specific binding protein of physiological significance has been identified. Additional confirmatory observations should be sought.

Enrichment of the Target Protein

If the target protein is present as a minor cellular component, it may be necessary to carry out a brief subcellular fractionation procedure to increase the specific activity of the target protein in the extract. A mitochondrial, nuclear, cytosolic, or crude membrane fraction can be prepared by centrifugation. In this manner it is possible to increase the chances of detecting the rare target protein (Soutar and Wade, 1989).

An alternative approach is to solubilize the whole tissue or subcellular fraction containing the target protein and perform a purification step on a small scale that will both concentrate and enrich the extract for the specific protein. Absorption and elution of the extract from an ion-exchange medium such as DEAE-cellulose, or a specific affinity matrix, are rapid procedures. This methodology is described in Chapter 10.

Southwestern Blotting

Southwestern blotting is a powerful technique for identifying and characterizing DNA-binding proteins by their ability to bind to specific oligonucleotide probes (Bowen et al, 1980). Using a suitably labeled stretch of DNA as a probe, the identity and molecular weight of a specific DNA binding protein can be revealed. Most investigators working in this field use ^{32}P-labeled oligonucleotide probes. However, technological advances in Southwestern analysis and non-radioactive preparation of oligonucleotide probes have made it possible to use oligonucleotides labeled with digoxigenin (DIG)-dUTP to obtain results identical to those obtained using ^{32}P-deoxyribonucleoside triphosphate (dNTP)-labeled probes (Dooley et al, 1992). An improved method was described (Handen and Rosenberg, 1997), focusing on improved stability of nuclear extracts and increasing the sensitivity of the oligonucleotide probe.

Far Western Blotting

The Far Western blot technique is a form of ligand blotting discussed in the following protocol. In Far Western blotting, protein/protein interactions are examined. Proteins are separated on SDS-PAGE and transferred to a membrane as in a normal Western blot. The blot is incubated with a candidate protein and the interaction is detected either by antibody directed against the added protein or by direct detection if the added protein had been prelabeled. Methods have been developed to increase the signal to noise ratio by denaturing/renaturing the immobilized proteins after transfer (Guichet et al, 1997).

PROTOCOL 7.13 Ligand Binding

To illustrate ligand binding, the specific example of laminin is presented. Laminin is a well characterized component of the extracellular matrix. Exciting work has shown that laminin binds to dystrophin associated glycoproteins and may be involved in Duchenne muscular dystrophy (Ibraghimov-Beskrovnaya et al, 1992; Ervasti and Campbell, 1993).

Materials

Radiolabeled ligand
Unlabeled ligand
TTBS: See Appendix C
Razor blade
Shaker

1. Load a slab gel with duplicate samples to create identical panels as shown in Figure 7.3. Separate proteins by SDS-PAGE (Protocol 4.5) and electrotransfer to PVDF (Protocol 7.1). Block the blot (Protocol 7.11) and cut it in half with a razor blade.
2. Radiolabel laminin with [^{125}I]. (See Protocol 3.8 for iodinating proteins with Chloramine T).
3. Incubate panel A with 0.1 μg/ml of [^{125}I] laminin (~1.7 μCi/μg). Incubate panel B with the same amount of labeled laminin, containing in addition, a 1000-fold excess (w/v) of unlabeled laminin.
4. Incubate both blots for 1 h at room temperature on a rocker. Wash extensively with TTBS and perform autoradiography (Protocol 7.22) to visualize the bands.
5. Compare the lanes that were incubated with 1000-fold excess of unlabeled laminin to the lanes that were incubated without the competing unlabeled ligand.

Comments: The possibilities of this technique are limitless. Fibronectin, calmodulin, and any other protein can be used. The method can be extended by using peptides that inhibit the ligand binding interaction, effectively mapping the site on the protein responsible for the interaction.

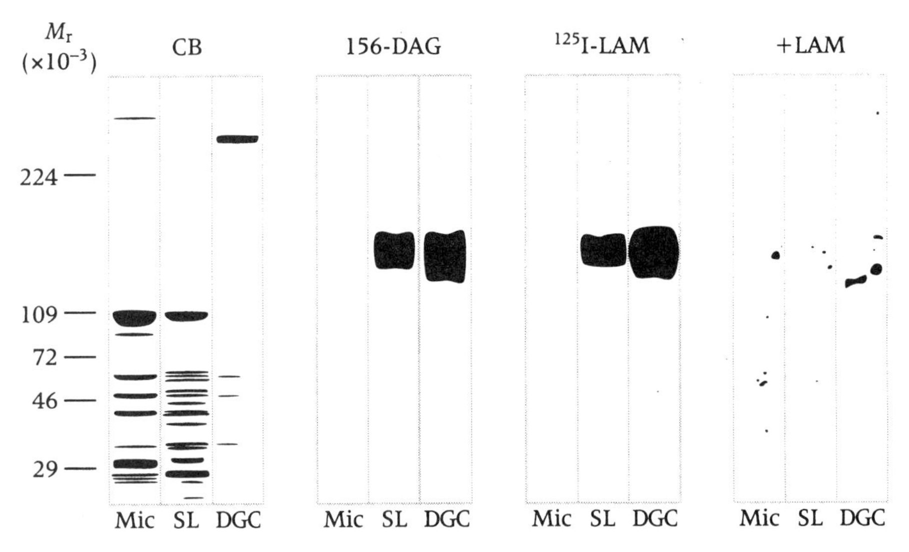

FIGURE 7.3 Laminin binding to dystroglycan (DAG). Four identical panels consisting of proteins derived from crude rabbit skeletal muscle membranes (Mic), sarcolemma membranes (SL), or dystrophin-glycoprotein complex (DGC) were prepared following electrophoresis and transfer to nitrocellulose. (CB) Coomassie blue-stained SDS gel. (156-DAG) Immunoblot stained with monoclonal antibody to 156kDa DAG. (^{125}I-LAM) Panel incubated with 0.1 µg/ml ^{125}I-labeled laminin (~1.7 µCi/mg). (+LAM) Panel incubated with ^{125}I-labeled laminin in the presence of a 1000-fold excess of unlabeled laminin. (Adapted with permission from Ibraghimov-Beskrovanaya et al, 1992.)

PROTOCOL 7.14 Lectin Blotting

Lectins are carbohydrate-binding molecules used to investigate cell-surface sugars and the changes that these sugars undergo during cell growth, differentiation, and malignancy (Lis and Sharon, 1984). They are also useful aids for the structural characterization of carbohydrate moieties on glycoproteins. As will be shown in Chapter 9, lectins are excellent reagents for the separation and purification of glycoproteins and glycopeptides. With few exceptions, most lectins are themselves glycoproteins.

Using a panel of lectins to probe proteins that have been transferred to PVDF or NC membranes is a rapid means for determining if a particular protein is glycosylated and furthermore, to identify the specific oligosaccharide(s) that are present on the glycoprotein. Results from this analytical approach can form the basis of a purification step. For instance, if wheat germ agglutinin (WGA) binds to the protein of interest on the blot, then WGA-agarose can be used as an immobilized affinity ligand in a purification step. This technique is discussed in Chapter 9.

Lectin overlays were originally performed following electrophoresis with radiolabeled lectin being subsequently added directly to SDS gels

(Burridge, 1978). The methodology has improved so that it is no longer necessary to radiolabel the lectin, and blots are used in place of gels. Most well characterized lectins are commercially available, conjugated to reporter enzymes like horseradish peroxidase or alkaline phosphatase, biotinylated or conjugated to digoxygenin. A list of commonly used lectins and their sugar specificities is presented in Table 7.1.

Materials

NC or PVDF membrane

TTBS: 0.05% Tween-20, 150mM NaCl, 10mM Tris-HCl, pH 7.5

Blocking solution: 5% BSA in TTBS

Lectins (conjugated to horseradish peroxidase)

Staining solution: 0.06ml 30% H_2O_2 in 100ml of 0.2M NaCl, 20mM Tris-HCl, pH 7.5 and 60mgs of 4-chloro-l-napthol (CN) in 20ml of ice-cold methanol.

1. Electrotransfer proteins to either NC or PVDF membranes as described in Protocol 7.1.
2. Immerse the blot in blocking solution for 1–2h.
3. Incubate the blot for 1h with a 1:2000 dilution of peroxidase-conjugated lectin (Sigma or any other supplier) in TTBS. If necessary, add the appropriate cation required by the lectin (see Table 7.1).
4. Wash the blot extensively with TTBS and incubate the blot in freshly prepared staining solution.

TABLE 7.1 Lectins

Lectin	Carbohydrate Specificity	Metal Requirement
Concanavalin A (Con A)	α-Man > α-Glc	Mn^{2+}, Ca^{2+}
Lens culinaris Agglutinin (LCA)	α-D-Man > α-D-Glc	Mn^{2+}, Ca^{2+}
Galanthus nivalis Agglutinin (GNA)	Terminal Man	Mn^{2+}
Soybean Agglutinin (SBA)	N-acetylgalactosamine	
Dolichos biflorus Agglutinin (DBA)	α-GalNAc	Ca^{2+}
Helix pomatia Agglutinin (HPA)	α-D-GalNAc	
Peanut Agglutinin (PNA)	Gal(β1-3)GalNAc	
Sambucus nigra Agglutinin (SNA)	Neu5Acα2-6Gal or Neu5Acα2-6GalNAc	
Maakia amurensis Agglutinin (MAA)	sialic acid α2-3Gal (does not bind α2, 6-linked sialic acid)	
Amaranthus caudatus Lectin (ACA)	NeuAc-Gal(β-1-3)-GalNac and Gal(β-1-3)-GalNAc bound to Ser or Thr	
Datura stramonium Agglutinin (DSA)	Gal(β1-4)GlcNAc	
Wheat Germ Agglutinin (WGA)	β-D-GlcNAc, NeuAc	Mn^{2+}, Ca^{2+}
Ulex europaeus Agglutinin I (UEA)	α-L-Fuc	Mn^{2+}
Aleuria aurantia Lectin (AAA)	Fucα1-6GlcNAc	
Erythrina cristagalli Lectin (ECL)	Gal(β1-4)GlcNAc	
Phytohaemagglutinin (PHA)	D-GalNAc	Mn^{2+}, Ca^{2+}
Ricinus communis Agglutinin I (RCA)	β-D-Gal	
Bandeira simplicifolia I (BS-I)	α-D-Gal	Ca^{2+}, Mg^{2+}
Bandeira simplicifolia II (BS-II)	α-D-GlcNAc	Ca^{2+}, Mg^{2+}
Lotus	α-L-Fuc	
Jackalin	α-D-Gal	

5. Terminate the staining by washing the membrane 5 times over 30 min with distilled water.

Comments: Lectins can be purchased that are conjugated to digoxygenin or to biotin. In these cases the appropriate detection systems should be used.

Include the appropriate controls for the specificity of lectin binding. Perform the blotting procedure in duplicate. On one blot, include the specific inhibiting sugar at 0.1–0.5 M in the final wash prior to incubating the blot with the lectin and also when incubating the blot with the lectin. This treatment should inhibit the lectin-glycoprotein interaction, and result in a reduction of the intensity of staining compared to the results when the competing sugar is omitted. For a list of lectins, their carbohydrate specificities and appropriate sugars see Table 7.1.

If you suspect that the target glycoprotein contains sialic acid, treat the glycoprotein with *Vibrio cholera* neuraminidase prior to SDS-PAGE. This treatment, which removes sialic acids should result in a loss of reactivity with *Maackia amurensis* agglutinin (MAA) (Wang and Cummings, 1988) and *Sambucus nigra* agglutinin (SNA) (Shibuya et al, 1987) and a gain of reactivity with PNA.

PROTOCOL 7.15 Bacterial Protein Overlay Analysis

The following method has been used successfully to examine the affinity of *Helicobacter pylori* for gastric epithelium mediated through blood group antigens (Boren et al, 1993). The digoxygenin system specifically employed to analyze bacteria-epithelial cell interactions utilizes bacteria that are labeled with digoxygenin, that are then detected with anti-digoxygenin-alkaline phosphatase antibody conjugate.

Materials

TBS: See Appendix C
Blocking solution
Nitro blue tetrazoleum (NBT)
5-bromo-4-chloro-3-indolyl phosphate (BCIP)
H. Pylori labeled with DIG-NHS

1. Electrotransfer the glycoproteins from SDS polyacrylamide gels to nitrocellulose membranes (Protocol 7.1).
2. Block the membranes (Protocol 7.11) overnight at 4 C°. The authors used a 1% solution of a commercially available blocking reagent (Roche Molecular Biochemicals) in TBS.
3. Wash the filter twice with TBS and incubate 8 h at room temperature with a suspension of *H. pylori* P466 (0.1 OD_{600}) that has been labeled with digoxigenin-3-*O*-succinyl-E-aminocaproic acid N-hydroxysuccinimide ester (DIG-NHS-ester; Roche Molecular Biochemicals) (Falk et al, 1993).
4. Wash the filters six times for 5 min each with TBS.

5. Add alkaline phosphatase conjugated sheep antibody to DIG (Roche Molecular Biochemicals) diluted 1:2000 and incubate for 1h at room temperature.
6. Wash five times with TBS and develop with NBT and BCIP.

Comments: This methodology has been extended to a bacterial *in situ* adherence assay. Labeled bacteria were applied to tissue sections and allowed to adhere. The sections were then washed and developed with a fluorescein isothiocyanate-conjugated sheep antibody to digoxygenin. This new application further emphasizes the great potential of bacterial protein overlay analysis.

D. Detection of the Target Protein

The final stage in the localization of the target protein on the immobilized membrane is the detection step, which can be autoradiographic, chemiluminescent, or the visualization of a colored precipitate. Ideally, the intensity of the resultant band that is produced should correlate to the concentration of the detected protein.

Immunodetection of the target protein complex can take many forms. After reacting a specific antibody, ligand or lectin tagged with a reagent like biotin, fluorescein or DIG with the membrane bound immobilized proteins, the target protein complex is detected by incubation with a molecule (usually an antibody) which itself is modified for detection. The first generation detection systems were radiolabeled antibodies and radiolabeled Protein A. However, the last few years have seen a gradual decrease in the use of radioisotopes as more labs are seeking alternatives without compromising the sensitivity offered by radioactive detection. Non-radioactive detection systems offer safety and environmental benefits, simplify regulatory compliance, and lower the costs for storage and disposal. Today, there is a growing list of nonisotopic alternatives which are finding increasing use.

Nonisotopic systems have been developed which use a second antibody that recognizes the primary antibody and has the added feature of being biotinylated or carrying a reporter enzyme, usually alkaline phosphatase (AP) or horseradish peroxidase (HRP), which catalyze reactions that generate either a colored precipitate or chemiluminescence. Chemiluminescent detection has rapidly become the detection method of choice, replacing radioactive methods without compromising sensitivity.

Protein A

Protein A has been useful as a general detection reagent for IgGs of most mammalian species (see Table 3.3). It binds to the F_c portion of the Ig molecules. Since only one molecule of labeled protein A binds to each primary antibody molecule, the detection sensitivities are much lower than those obtainable with labeled second antibody having many potential binding sites on the target antibody molecule. In addition, protein A does not detect all isotypes or subtypes of the primary antibody (see Table 3.3). However, the lower sensitivity of protein A can be advantageous when second antibody detection generates too many bands. Protein A can also be biotinylated.

Second Antibody Conjugate

The most commonly used secondary antibodies are species-specific, that is to say they recognize and bind to antibodies or fragments of antibodies originating in a specific species of animal. For example, if the primary antibody is a mouse monoclonal, then the second antibody could be rabbit anti-mouse, or goat anti-mouse. To detect the antigen of interest on the blot, the second antibody can be radiolabeled, biotinylated or conjugated to an enzyme that will produce a colored product or chemiluminescence.

Biotin Avidin System

Biotin, also known as Vitamin H has a molecular weight of 274. It has an extraordinarily high affinity for avidin, a 67kDa glycoprotein with an affinity constant $>10^{15} M^{-1}$, which is about one million times more powerful than most antigen-antibody reactions, making the avidin-biotin bond essentially irreversible. Four biotin molecules can bind to each avidin molecule. This high affinity has been used in a large number of systems. Many biotin molecules can be covalently coupled to each protein molecule giving that protein the ability to bind avidin and enabling the biotinylated protein to bind more than one avidin molecule, effectively amplifying the signal. When biotinylation is performed under gentle conditions, the biological activity of the protein can be preserved. Avidin can be covalently linked to different ligands such as fluorochromes and enzymes. The biotin/avidin system has proved useful in the detection of antigens and glycoconjugates using biotinylated antibodies and biotinylated lectins.

Since avidin is a glycoprotein, it contains oligosaccharides, which under certain circumstances, can lead to unwanted binding to other molecules (i.e., lectins). Streptavidin, (MW 60,000) a product of *Streptomyces avidinii,* has properties similar to avidin, and in practice can be used in the same manner as avidin. It has two important advantages; it is not a glycoprotein and therefore will not nonspecifically bind to molecules through carbohydrate interactions, and its isoelectric point is close to neutrality so that it bears few strongly charged groups at neutral or the slightly alkaline pH of most detection systems. Consequently, streptavidin has replaced avidin as the preferred ligand.

Avidin and streptavidin are available commercially, conjugated to either HRP or AP, and the bands of interest are detected colorometrically.

PROTOCOL 7.16 Biotinylation of Proteins

A number of biotin derivatives are available for conjugation to a variety of functional groups. One commonly used derivative is biotinyl-N-hydroxysuccinimide ester (NHS-biotin), which predominantly reacts with amino groups of lysine.

Materials

NHS-biotin
DMSO

1. Dissolve NHS-biotin in DMSO at a concentration of 1 mg/ml.
2. Dissolve the protein in either PBS or 0.1 M $NaHCO_3$ pH 9.0 at a concentration of 2 mg/ml.
3. Mix the NHS-biotin and protein at a ratio of 1:8 (biotin:protein) and incubate at room temperature for 4 h.
4. Dialyze overnight against PBS (see Protocol 5.10).

PROTOCOL 7.17 Purifying and Biotinylating Antibodies from Immunoblots

Antibodies that bind to a given polypeptide band can be purified from immunoblots. These band-specific antibodies, although recovered in small amounts, have high specificity and can be used for many techniques. The antibodies can also be biotinylated while still bound to the antigen, thereby protecting the active antigen-binding sites of the antibodies (Peranen, 1992).

Materials

Na_2HPO_4: 1 M unbuffered
TTBS: 10 mM Tris-HCl, pH 8.0, 150 mM NaCl, 0.05% Tween 20
Elution buffer: 5 mM glycine, pH 2.3, 400 mM NaCl and 0.2% BSA
HEPES buffer: 0.1 M HEPES, pH 8.0
Biotin solution: sulfo-LC-NHS-biotin, 0.5–2 mg/ml, (Pierce) in 0.1 M HEPES, pH 8.0
Glycine solution: 20 mM

1. Prepare the sample for SDS-PAGE. Using a preparative gel comb which has one large well, load the sample. Run the gel (Protocol 4.5) and transfer the proteins to a PVDF membrane as described in Protocol 7.1.
2. Block the blot (Protocol 7.11). Incubate the blot with primary antibody and wash three times with TTBS.
3. To determine the exact location of the antigen of interest, cut longitudinal strips from the ends of the blot. Often, a strip from the middle is also helpful. Store the main body of the blot in PBS, 0.02% sodium azide at 4°C and continue to process the strips for antigen detection.
4. After the band of interest has been visualized, align the strips with the untreated portion of the blot and excise the horizontal area of the blot that contains the antigen and primary antibody.
5. If the antibody is to be biotinylated, wash the strip three times in HEPES buffer.
6. Incubate the strips in 5 ml of freshly prepared biotin solution for 2 h at room temperature. Wash five times with HEPES buffer, followed by two 5 min washes with 20 mM glycine, and two washes with water.

7. To elute the antibody, incubate the filter with 500 μl of elution buffer for 1 min then transfer the eluate to a new tube.
8. Neutralize the solution with 25 μl of 1 M Na_2HPO_4.
9. Repeat the elution using 800 μl of elution buffer.

Comments: Recoveries should be ~30–40% of the absorbed antibody. The cross-linking agent used for the biotinylation is chosen because of its water solubility and because it has an extended spacer arm that limits steric hindrance.

Enzymatic Detection Methods

Enzymatic methods of signal detection offer many advantages over systems based on radionuclides. Conjugated to a second antibody, the enzymes horseradish peroxidase (HRP) and alkaline phosphatase (AP) have been used extensively for immunodetection. The reactions catalyzed by AP and HRP produce a colored precipitate on the membrane at the site of antibody binding. Alternatively, the reporter enzymes can catalyze chemiluminescence reactions, the results of which are captured on X-ray film. This system can detect as little as 10–20 pg of specific antigen. AP and HRP, conjugated to an appropriate second antibody, are commercially available from many suppliers. Some preparations have even been pre-absorbed to reduce unwanted cross-reactions. There are no specific precautions necessary in storage, handling, or disposal of these reagents. The enzyme-antibody conjugates are prepared at a suitable dilution suggested by the manufacturer and then incubated with the blot for 1 h. Washing is performed identically to that between the first and second antibody. The banding pattern is now ready to be visualized.

Horseradish Peroxidase

Historically, horseradish peroxidase (HRP) was the first enzyme conjugate used for immunological detection on blotted membranes. The results obtained when HRP was used to generate a colored product (Geysen et al, 1984) were not as sensitive as with other detection systems, and there were additional limitations. The bands faded when exposed to light, the enzyme is inhibited by azide, and there is nonspecific color precipitation due to endogenous peroxidase activities.

PROTOCOL 7.18 Colorimetric Detection with Diaminobenzidine, 3,3′,4,4′-tetraaminobiphenyl) (DAB)

Diaminobenzidine, 3,3′,4,4′-Tetraaminobiphenyl) (DAB) is a sensitive substrate which produces a brown reaction product. Its sensitivity can be enhanced by adding cobalt or nickel ions to the substrate solution. The drawback in using DAB is that the reaction is rapid and difficult to control, often resulting in high background.

Materials

DAB solution: See step 1 below
PBS (see Appendix C)
Hydrogen peroxide stock solution: See step 2 below

1. After incubating the blot with HRP conjugated to the second antibody, wash the blot × 3 in TTBS.
2. Immediately before developing the blot, dissolve 6mg of DAB (use the tetrahydrochloride form) in 10ml of 50mM Tris, pH 7.6. For metal ion enhancement, add 1ml of 0.3% (w/v) $NiCl_2$ or $CoCl_2$. A small precipitate may form. Quickly filter the suspension through Whatman #1 filter paper.
3. Add 10µl of 30% H_2O_2 (The stock solution is usually supplied as a 30% solution and should be stored at 4°C.)
4. Add the reaction mix to the blot. Develop with gentle mixing until the bands are suitably dark, typically 1–5min.
5. Stop the reaction by rinsing the blot with PBS.

Comments: When using an HRP detection system, do not include sodium azide as a bacteriacide in the buffers. If NaN_3 was mistakenly included in the buffer, try washing the blot with a newly prepared buffer that does not contain NaN_3. This treatment may save the experiment.

DAB has been reported to be carcinogenic. Wear gloves and a mask when handling it and carefully dispose of solutions. DAB is also available in preweighed tablet form (Sigma) making handling easier.

A major disadvantage to using DAB is the necessity of using the substrate immediately. The reaction process begins as soon as the H_2O_2 solution is added.

Another substrate used with HRP is 4-chloro-l-napthol (4-CN), a substrate used for the chromogenic detection of HRP. Although not as sensitive or as stable as DAB, 4-CN does not produce an excessive background, and the blue-purple color that is generated photographs well (Wordinger et al, 1983). When used, the product that is formed tends to solubilize over time and must be photographed for a permanent record. The colored product is soluble in alcohol and other organic solvents.

Dissolve 4-CN in ethanol or methanol prior to adding it to the H_2O_2-containing substrate buffer, then filter.

PROTOCOL 7.19 Colorimetric Detection Using Alkaline Phosphatase

Alkaline phosphatases (AP) are a widely distributed family of enzymes found in many species and tissues. They have optimal enzymatic activity at pH 9.0–9.6 and are activated by divalent cations. AP-conjugated second antibodies catalyze color development using the substrates nitro blue tetrazolium (NBT) and 5-bromo-4-chloro-3-indolyl phosphate (BCIP).

Materials

TTBS (see Appendix C)

AP buffer: 100 mM Tris-HCl, pH 9.5, 100 mM NaCl, 5 mM $MgCl_2$

AP substrate: Add 33 µl of nitro blue tetrazolium (NBT) (stock solution 50 mg/ml in 70% dimethylformamide (DMF)) and 16 µl of 5-bromo-4-chloro-3-indolyl phosphate (BCIP) (50 mg/ml in DMF) to 5 ml of AP buffer.

1. After incubating the blot with AP conjugated to a second antibody, an avidin derivative or protein A or G, wash the blot three times in TTBS.
2. Perform one final wash in AP buffer. This wash serves to raise the pH and provides the divalent cation necessary for the reaction.
3. Pour off the AP buffer of the final wash and add the substrate solution with gentle shaking.
4. When you are satisfied with the intensity of the band(s), stop the reaction by washing the blot with two to three changes of distilled water. This is a good time to photograph the blot as the stain is still intense.

Comments: Bound AP dephosphorylates the substrate BCIP and the reaction product generates an intense purple/black precipitate with NBT at the reaction site. Bands usually begin to appear within 30 seconds. In most cases, the desired intensity is reached within 10 min. The reaction can be allowed to proceed for a longer time depending on the intensity of the background. The blot can be stored wet or dry. An accurate record can be obtained by photocopying the blot.

Unprobed blots can be stored dry for extended periods of time after the blocking and wash steps.

PROTOCOL 7.20 Enhanced Chemiluminescence

Chemiluminescent detection applied to Western blotting is a light emitting nonradioactive method for the detection of immobilized specific antigens. Chemiluminescence occurs when the energy from a chemical reaction is emitted in the form of light. HRP is often used to catalyze the oxidation of luminol in the presence of H_2O_2. Luminol is oxidized in alkaline conditions. Immediately following oxidation the luminol is in an excited state, which then decays to the ground state via a light-emitting pathway. Certain compounds can be added to the reaction mixture to enhance the light emission from the HRP-catalyzed oxidation of luminol. Therefore, by labeling the second antibody with HRP, it is possible to detect immobilized antigens specifically. Other systems use dioxetane substrates and alkaline phosphatase (AP).

Chemiluminescent detection offers a number of advantages over radioactive methods and unamplified chromogenic substrates such as DAB. The methodology is convenient and sensitive, at least fourfold more sensitive than unamplified chromogenic systems. The detection of the signal is on X-ray film providing a permanent, hard-copy record that will not fade or crack over time. The intensity of the reaction can be adjusted by re-exposing the blot for longer or shorter time intervals.

The blot can be easily stripped and reprobed, allowing the membrane to be reused several times. For these reasons chemiluminescent detection is the method of choice.

Materials

HRP conjugated second antibody

Chemiluminescent reagents

TTBS: 10mM Tris-HCl, pH 8.0, 150mM NaCl, 0.05% Tween 20

Plastic wrap

X-ray film

Foreceps

Whatman #1 filter paper

1. Following electrophoretic transfer to PVDF membrane, block nonspecific (see Protocol 7.12) sites with either 5% instant nonfat dry milk in TTBS or 5% BSA in TTBS for at least 1h at room temperature.
2. Incubate the blot in the appropriate dilution of primary antibody for 1h.
3. Wash the blot three times for 5min each with TTBS.
4. Incubate the blot in HRP conjugated second antibody. This antibody-enzyme conjugate is usually obtained commercially. Follow the supplier's instructions for dilution and incubation time.
5. Wash as in step 3.
6. Immediately prior to performing the chemiluminescence reaction, prepare the reagent. (This usually entails mixing two solutions.) For the Renaissance reagent (NEN duPont) two solutions are provided. Mix equal volumes of each.
7. Using foreceps, remove the membrane from the final wash and place it on Whatman # 1 filter paper to remove excess liquid. Do not dry the membrane. Quickly transfer the membrane to a clean glass or plastic container and add the freshly prepared chemiluminescent reagent from step 6.
8. Incubate the membrane in the chemiluminescent reagent with shaking for exactly 60sec making sure that the surface of the membrane is covered.
9. Working quickly, remove excess liquid from the membrane by gently blotting with Whatman #1 filter paper. Do not let the blot dry. Wrap the membrane in plastic wrap and expose to X-ray film with the protein side of the membrane next to the X-ray film.

Comments: The ideal exposure time will be reached empirically. This can range from a few seconds up to an hour depending on the signal-to-noise ratio. For the first exposure, try 30 seconds. Many manufacturers market chemiluminescent detection systems under various trade names: Renaissance™ (DuPont NEN) and ECL™ (Amersham).

If it is desirable to reprobe a blot that was previously developed by chemiluminescence, the antibody that led to the production of the previous signal must be removed. The chemiluminescent reaction itself is short lived and will not pose a problem.

Blots that are stored wet can be reprobed with a different antibody, and bands can be visualized with the same or a different detection system. Chemiluminescence has the advantage of being a temporary signal that can be specifically removed. The alkaline phosphatase colorimetric detection system can also be reprobed, but the original bands will remain, although somewhat faded.

Blots that have been allowed to dry may also be reprobed. Chemicon introduced the Blot Restore membrane Rejuvenation Kit for the reuse of PVDF and NC membranes that have been developed with chemiluminescent or color substrates and let dry. The kit permits reanalysis of a dried membrane with the original or different antibodies, saving time and money.

PROTOCOL 7.21 Stripping and Reprobing the Blot

Stripping of antibodies and chemiluminescence substrates from the blot has been found to be feasible. Stripping the antibodies from the membrane allows the investigator to reprobe the blot with another antibody enabling one to compare the banding patterns of up to four different probes on the same blot.

Materials

Chemiluminescence reagents

TTBS: 10 mM Tris-HCl, pH 8.0, 150 mM NaCl, 0.05% Tween 20

Stripping buffer: 62.5 mM Tris-HCl pH 6.7, 2% SDS and 100 mM 2-mercaptoethanol

X-ray film

Film cassette

X-ray developer

1. Perform the Western Blot as described above. After film exposures have been completed satisfactorily the membrane is stripped taking care not to allow the membrane to dry.
2. Wash the membrane 4 × 5 min with TTBS.
3. Incubate the membrane for 30 min at 50°C in stripping buffer in a sealed container.
4. Wash the membrane 6 × 5 min in TTBS.
5. To be sure that the old signal has been removed, incubate the blot in chemiluminesence reagent for 1 min and expose the membrane to film for 1 min to 1 hr.
6. Wash the membrane again for 4 × 5 min in TTBS. The membrane is now ready for reuse. Begin to reprobe at the blocking step.

Comments: Chemicon markets a kit, *ReBlot™ Plus,* that quickly removes antibodies and their corresponding chemiluminescent signal. The kit contains a mild and a strong stripping solution. The strong solution is recommended for use on membranes with a high intensity signal or when the mild solution is not sufficient.

Detection of Radiolabeled Proteins

Radiolabeled proteins immobilized in SDS gels can be detected photographically by placing the dried gel or blot in contact with X-ray film.

The optimum exposure time is determined empirically, depending on the amount of radioactivity present in each band. You might have to choose between visualizing major or minor radioactive bands. Several exposures of different durations may be required. Long exposures will reveal minor bands, but if major bands are present they may spread, possibly obscuring the portion of the film where the minor bands are located.

PROTOCOL 7.22 Direct Autoradiography

Autoradiography is the direct exposure of film by beta particles or gamma rays.

Materials

Gel drier
X-ray film
Film cassettes
X-ray developer
Radioactive ink

1. Dry the gel onto a sheet of Whatman 3MM filter paper (Protocol 4.12). Use two sheets of filter paper to minimize potential radioactive contamination of the gel drier.
2. With radioactive ink, make a nonsymmetrical array of spots on the paper backing. This will facilitate aligning the exposed film with the dried gel.
3. Place the dried gel in direct contact with X-ray film in a radiographic cassette and expose the film with the gel for the desired length of time.
4. Develop the film according to the manufacturer's instructions. Use an automatic developer if available.

Comments: The gel can be stained and destained prior to drying.

Radioactive protein markers (bought commercially or prepared in-house) can also be used to assist in orienting the gel and film.

If an automatic developer is unavailable, a manual film developing protocol is presented in Appendix H.

Direct autoradiography of dried gels using Kodak X-GMAT X-ray film will give a film image absorbance of 0.02 A_{540} units which is just visible above background in 24 h as follows:

Isotope	dpm/cm^2
^{14}C	6,000
^{35}S	6,000
^{125}I	1,600
^{32}P	500

The film image absorbance is proportional to sample radioactivity. ^{3}H is not detected since the low-energy β-particles do not penetrate the gel matrix. However, fluorography can detect ^{3}H and increase the sensitivity of ^{14}C and ^{35}S over that possible using direct autoradiography.

In all cases where X-ray film is used, intimate, immovable contact between the gel or membrane and the film and intensifying screen is essential during the autoradiographic exposure period. Film should not be drawn rapidly from the box or handled in any way that would cause a static discharge. Kodak X-OMAT AR film can be used for all isotopes and exposure methods and is recommended for general use. X-ray film from other companies; Agfa, Fuji, Amersham (Hyperfilm™), and DuPont (Reflection™), are also excellent, and, if the price is competitive, should be used.

Removing Unwanted Background Signal from X-ray Film

High background, shading, overexposed bands and spotting are common problems encountered with film exposure. High background can be caused by exposing the film for too long, inefficient blocking, inappropriate enzyme labeled probe concentration, or antibody concentration. Spotting and shading occur when enzyme conjugates form complexes and precipitate on the blot. Pierce sells a kit, Erase-It™, that can correct many of these problems without the need to re-expose the blot to film or re-do the entire experiment. The kit can be used for any application using x-ray film including, Western, Northern, Southern blots, gel-shift assays, and ribonuclease protection assays. The Erase-It reagent reduces the signal evenly over the film so that relative densitometric values are consistent. The procedure takes only a few minutes and can be used with any brand of film.

Indirect Autoradiography (for ^{32}P and ^{125}I)

Exposure of the X-ray film with intensifying screens improves autoradiographic detection of ionizing radiation. When using the older model conventional screens, for optimal image sharpness, a single screen (like Dupont Corex Lighting Plus) is placed against the film as shown in Figure 7.4. The entire sandwich is then placed at –70°C for film exposure. Emissions from the sample pass to the film producing the usual direct autoradiographic image, but emissions that pass completely through the film are absorbed by the intensifying screen where they produce multiple photons which, as they return, re-expose the film. In this fashion, the sensitivity is increased, resulting in shorter exposure times and use of smaller amounts of radioactivity (Laskey, 1980).

Intensifying screens save time and improve resolution when used with both strong and weak emitters. Improvements in the phosphors and design of the screens have increased the efficiency of autoradiography while reducing exposure times, and extending their application to weak emitters. With conventional screens, the emitted particles must penetrate the film before coming in contact with the phosphor in the screen, which weak beta emitters are unable to do.

Eastman Kodak Co. has developed a line of intensifying screens that can be used with weak beta emitters. The manufacturer claims that the use of these screens will preclude the need for fluorography. The new generation BioMax Transcreens are designed such that emitted par-

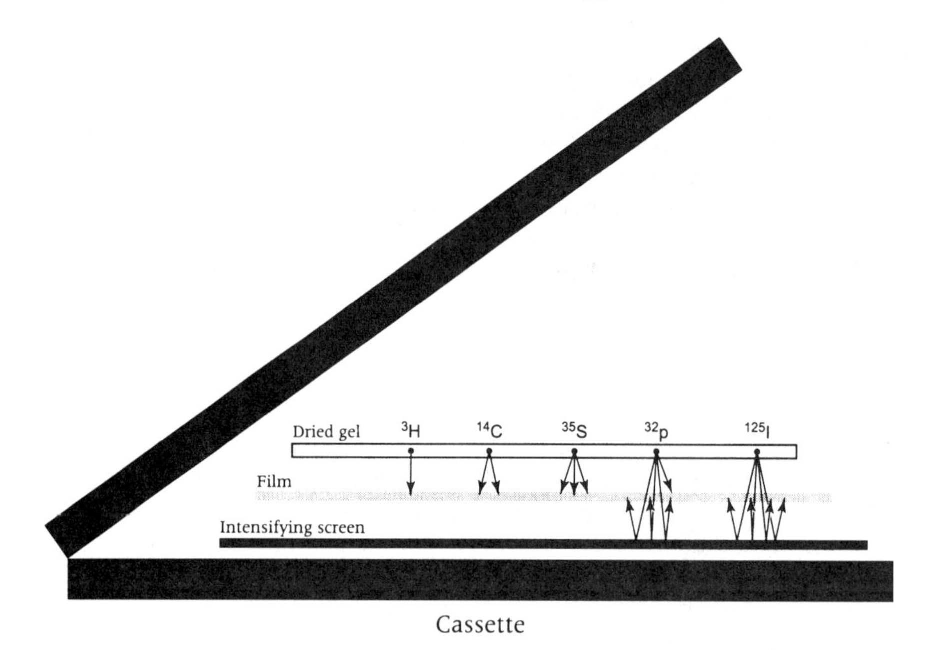

FIGURE 7.4 Orientation of film and intensifier in the cassette.

ticles excite the phosphor first, producing light that penetrates a thin, clear plastic base before hitting the film emulsion. The BioMax TranScreen Low Energy (LE) screen is penetrable by weak beta emitters like ^{3}H, ^{35}S, ^{14}C, ^{33}P, and ^{59}Fe. Use of the LE screen decreases exposure time and eliminates the need for fluorography.

Intensifying screens absorb energy from room light. The subsequent emission of that energy as light by a screen can contribute to the formation of background fog on a film placed against that screen. To avoid this effect, place the screen in the dark several hours before starting an autoradiographic exposure.

Fluorography

Fluorography is the exposure of film by secondary light generated by the excitation of a fluor or a screen by a beta particle or gamma ray. In fluorography, an organic scintillant is infused into the gel before drying (Bonner and Laskey, 1974). Fluorography is almost always used for the detection of tritium and also improves the detection of other low- and medium-energy beta emitters such as ^{14}C and ^{35}S. The β-particles interact with the scintillant in the gel, converting the emitted energy of the isotope into visible light which increases the proportion of energy detected by the X-ray film. Because the low-energy β-particle produced from ^{3}H decay cannot penetrate the gel and expose the film directly, fluorography is essential for detecting proteins labeled with ^{3}H. For other low β–emitters such as ^{14}C and ^{35}S, fluorgraphy enhances the detection by approximately 15-fold compared to direct autoradiography.

Excellent enhancing solutions are commercially available, En^{3}hance™ (Dupont NEN), Amplify™ (Amersham) and Enlightening™ (Dupont NEN). The use of these reagents has all but replaced the original PPO method. Simply soak the gel in the enhancing solution for 30–60 min prior to drying. Figure 7.5 compares the signals obtained when Enlightening™ is compared to the PPO-DMSO method.

Long exposures carried out at temperatures between –70°C and –80°C stabilize latent-image formation. At low temperatures the fluorographic detection of ^{3}H can be increased by a factor of four or more times over room temperature exposure. Similarly, the detection of ^{32}P and ^{125}I using intensifying screens is dramatically improved when the exposures are made at low temperatures. The placement of the screen relative to the film and dried gel or blot is shown in Figure 7.4. However, exposure at low temperatures is only effective in light exposure situations when fluors and intensifying screens are used. Direct exposure to ionizing radiation is not improved by low temperature treatment (Table 7.2).

PROTOCOL 7.23 Phosphorimaging

Storage phosphor imagery uses a screen which contains barium-fluorohalide crystals possessing rare earth compounds (e.g., BaFBr:Eu^{2+}) that store incident β-particle energy in their electronic structures. The crystals are then irreversibly excited by the ionizing radiation into an intermediate energy state, creating a latent image. A laser provides additional electronic energy that puts the crystal into a pho-

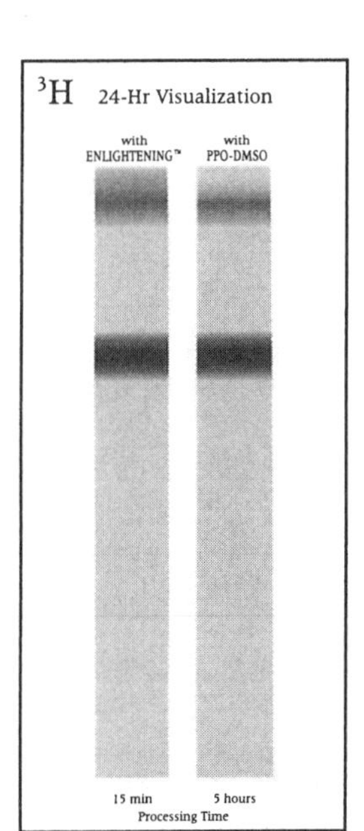

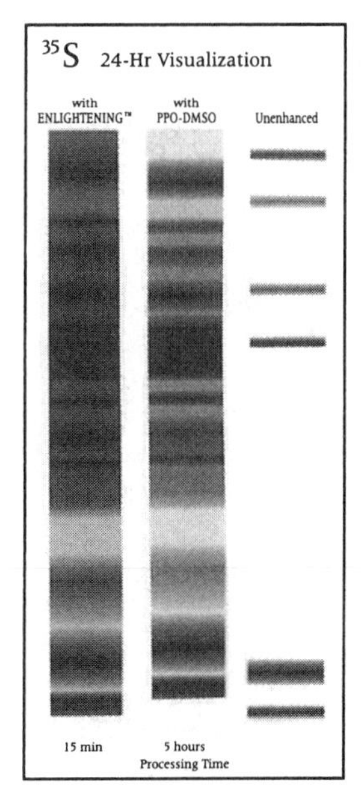

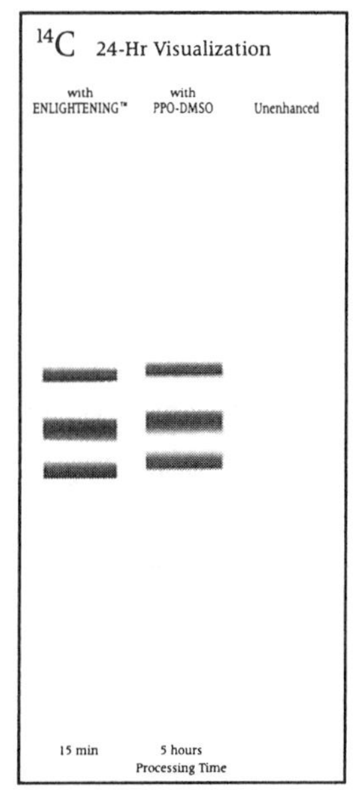

FIGURE 7.5 Autoradiography enhancer. Three gels with proteins labeled with [^{3}H] (Panel A), [^{35}S] (Panel B), and [^{14}C] (Panel C) are presented to illustrate the image enhancement and time savings using a fluorographic enhancer (in this example, Enlightening™ from DuPont). Adapted with permission from DuPont NEN.

TABLE 7.2 Isotopic Detection on X-ray Film

Isotope	Type of method	dpm/cm^2 required for detectable image (A_{540} = 0.02) in 24h	Enhancement over direct autoradiography
^{125}I	Indirect autoradiography	100	16
^{32}P	Indirect autoradiography	50	10.5
^{14}C	Fluorography	400	15
^{35}S	Fluorography	400	15
^{3}H	Fluorography	8000	>1,000

(Reprinted by permission of Oxford University Press, from Hames and Rickwood, 1981.)

tochemically unstable state, releasing blue light. This emission can be captured and quantified with a charge-coupled device (CCD) camera. The screen is then erased by photobleaching and is ready for reuse.

Phosphor screen imaging technology has several advantages over film based detection of radioactive proteins. Phosphor screens are placed on the samples in ambient light. Exposure takes place at room temperature. Results are in digital form and can be analyzed qualitatively and quantitatively using Image Quant software. The nearly linear dynamic range of phosphor imaging covers 5 orders of magnitude. Film covers 2.7 orders of magnitude. Fast imaging times, no requirements for chemicals, darkroom or special treatment make it very appealing. The major drawback is the cost of the apparatus and the screens.

Materials

Phosphorimager

Storage Phosphor screen cassette

1. Make sure the imaging plate has been blanked.
2. Place the dried gel in the cassette.
3. Place the erased, white side of the imaging plate on the dried gel and close the cassette.
4. While exposing, store the cassette at room temperature. Do not place at –70°C.

Comments: the sample must be absolutely dry or you risk damaging the expensive imaging plate. Phosphorimaging is ~10–100 fold more sensitive than x-ray film autoradiography. ^{14}C deposited on filter paper or nitrocellulose will produce good quality images.

Phosphorimaging is not sensitive to ^{3}H.

References

Balasubramanian N, Zingde SM (1997): Even highly homologous proteins vary in their affinity for different blotting membranes. *BioTechniques* 23:390–396

Bonner WM, Laskey RA (1974): A film detection method for tritium labeled proteins and nucleic acids in polyacrylamide gels. *Eur J Biochem* 46:83–88

Boren T, Falk P, Roth KA, Larson G, Normark S (1993): Attachment of *Helicobacter pylori* to human gastric epithelium mediated by blood group antigens. *Science* 262:1892–1895

Bowen B, Steinberg J, Laemmli UK, Weintraub H (1980): The detection of DNA-binding proteins by protein blotting. *Nucleic Acids Res* 3:1–20
Burridge K (1978): Direct identification of specific glycoproteins and antigens in sodium dodecyl sulfate gels. *Methods Enzymol* 50:54–64
Dooley S, Welter C, Blin N (1992): Nonradioactive Southwestern analysis using chemiluminescent detection. *BioTechniques* 13:540–543
Ervasti JM, Campbell KP (1993): A role for the dystrophin-glycoprotein complex as a transmembrane linker between laminin and actin. *J Cell Biol* 122:809–823
Falk P, Roth KA, Boren T, Westblom TU, Gordon JI, Normark S (1993): An *in vitro* adherence assay reveals that *Helicobacter pylori* exhibits cell lineage-specific tropism in the gastric epithelium. *Proc Natl Acad Sci USA* 90:2035–2039
Fernandez-Pol JA (1982): Immunoautoradiographic detection of epidermal growth factor receptors after electrophoretic transfer from gels to diazo-paper. *FEBS Lett* 143:86–92
Flanagan SD, Yost B (1984): Calmodulin-binding proteins: visualization by ^{125}I-calmodulin overlay on blots quenched with Tween 20 or bovine serum albumin and poly(ethylene oxide). *Anal Biochem* 140:510–519
Gershoni JM (1988): Protein blotting: a manual. *Methods Biochem Anal* 33:1–58
Gershoni JM, Palade GE (1983): Protein blotting: principles and applications. *Anal Biochem* 131:1–15
Gershoni JM, Hawrot E, Lentz TL (1983): Binding of alpha-bungarotoxin to isolated alpha subunit of the acetylcholine receptor of *Torpedo californica*: quantitative analysis with protein blots. *Proc Natl Acad Sci USA* 80:4973–4977
Geysen J, DeLoof A, Vandesande F (1984): How to perform subsequent or "double" immunostaining of two different antigens on a single nitrocellulose blot within one day with an immunoperoxidase technique. *Electrophoresis* 15:129–131
Guichet A, Copeland JWR, Erdelyi M, Hlousek D, Zavorsky P, Ho J, Brown S, Percival-Smith A, Krause HM, Ephrussi A (1997): The nuclear receptor homologue Ftz-F1 and the homeodomain protein Ftz are mutually dependent cofactors. *Nature* 385:548–552
Hames BD, Rickwood D (1981): *Gel Electrophoresis of Proteins: A Practical Approach.* Oxford: IRL Press
Hancock K, Tsang VCW (1983): India ink staining of proteins on nitrocellulose paper. *Anal Biochem* 133:157–162
Handen JS, Rosenberg HF (1997): An improved method for Southwestern blotting. *Front Biosci* 2:c9–c11
Hawkes R, Niday E, Gordon J (1982): A dot-immunobinding assay for monoclonal and other antibodies. *Anal Biochem* 119:142–147
Hayman EG, Pierschbacher MD, Ohgren Y, Ruoslahti E (1983): Serum spreading factor (vitronectin) is present at the cell surface and in tissues. *Proc Natl Acad Sci USA* 80:4003–4007
Ibraghimov-Beskrovnaya O, Ervasti JM, Leveille CJ, Slaughter CA, Sernett SW, Campbell KP (1992): Primary structure of dystrophin-associated glycoproteins linking dystrophin to the extracellular matrix. *Nature* 355:696–702
Laskey RA (1980): The use of intensifying screens or organic scintillators for visualizing radioactive molecules resolved by gel electrophoresis. *Methods Enzymol* 65:363–371
Lis H, Sharon N (1984): Lectins: properties and applications to the study of complex carbohydrates in solution and on cell surfaces. In: *Biology of Carbohydrates, Vol. 2,* Ginsburg V, Robbins PW, eds. New York: John Wiley & Sons
Maillet C, Gupta RK, Schell MG, Brewton RG, Murphy CL, Wall JS, Mullin BC (2001): Enhanced capture of small histidine-containing polypeptides on membranes in the presence of $ZnCl_2$. *BioTechniques* 30:1224–1230

Montelaro R (1987): Protein antigen purification by preparative protein blotting. *Electrophoresis* 8:432–438

Parekh BS, Mehta HB, West MD, Montelaro RC (1985): Preparative elution of proteins from nitrocellulose membranes after separation by sodium dodecyl sulfate polyacrylamide gel electrophoresis. *Anal Biochem* 148:87–92

Peranen J (1992): Rapid affinity-purification and biotinylation of antibodies. *BioTechniques* 13:546–549

Roche PC, Ryan RJ (1989): Purification, characterization and amino-terminal sequence of rat ovarian receptor for luteinizing hormone/human choriogonadotropin. *J Biol Chem* 264:4636–4641

Sheng S, Schuster SM (1992): Simple modifications of a protein immunoblotting protocol to reduce nonspecific background. *BioTechniques* 13:704–708

Shibuya N, Goldstein IJ, Broekaert WF, Nsimba-Lubaki M, Peeters B, Peumans WJ (1987): The Elderberry (*Sambucus nigra L.*) bark lectin recognizes the Neu5Ac(α2-6)Gal/GalNAc sequence. *J Biol Chem* 262:1596–1601

Simpson RJ, Ward LD, Reid GE, Batterham MP, Moritz R (1989): Peptide-mapping and internal sequencing of proteins electroblotted from two-dimensional gels onto polyvinylidene difluoride membranes-a chromatographic procedure for separating proteins from detergents. *J Chromatogr* 476:345–361

Soutar AK, Wade DP (1989): Ligand blotting. In: *Protein Function: A Practical Approach*, Creighton TE, ed. Oxford: Oxford University Press

Stromquist M, Gruffman H (1992): Periodic acid/Schiff staining of glycoproteins immobilized on a blotting matrix. *BioTechniques* 13:744–749

Taki T, Ishikawa D (1997): TLC blotting: application to microscale analysis of lipids and as a new approach to lipid-protein interaction. *Anal Biochem* 251:135–143

Thompson D, Larson G (1992): Western blots using stained protein gels. *BioTechniques* 12:656–658

Towbin J, Staehelin T, Gordon J (1979): Electrophoretic transfer of proteins from polyacrylamide gels to nitrocellulose sheets: procedure and some applications. *Proc Natl Acad Sci USA* 76:4350–4354

Wang W, Cummings RD (1988): The immobilized leukoagglutinin from the seeds of *Maakia amurensis* binds with high affinity to complex-type Asn-linked oligosaccharides containing terminal sialic acid-linked α-2,3 to penultimate galactose residues. *J Biol Chem* 263:4576–4585

Wilson PT, Gershoni JM, Hawrot E, Lentz TL (1984): Binding of alpha-bungarotoxin to proteolytic fragments of the alpha-subunit of Torpedo acetylcholine receptor analyzed by protein transfer on positively charged membrane filters. *Proc Natl Acad Sci USA* 81:2553–2557

Wordinger R, Miller G, Nicodemus D (1983): *Manual of Immunoperoxidase Techniques.* Chicago: American Society of Clinical Pathologists Press

Yuen S, Chui A, Wilson K, Yuan P (1989): Microanalysis of SDS-PAGE electroblotted proteins. *BioTechniques* 7:74–83

Zacharius RM, Zell TE, Morrison JH, Woodlock JJ (1969): Glycoprotein staining following electrophoresis on acrylamide gels. *Anal Biochem* 30:148–152

General References

Cummings RD (1994): Use of lectins in analysis of glycoconjugates. *Methods Enzymol* 230:66–86

Harlow E, Lane D (1988): *Antibodies: A Laboratory Manual.* Cold Spring Harbor, NY: Cold Spring Harbor Laboratory

8

Identification of the Target Protein

Introduction

This chapter consists of three related sections. The first part presents methods that are used to compare proteins based on their composition and structure. Some of these techniques can also be used in the middle section where the target protein is isolated and analyzed. The final section is an introduction to bioinformatics, and how the expanding discipline of *"in silico"* science is playing a major role in biomedical research.

A. Peptide Mapping

Peptide mapping is a powerful technique used to analyze and compare the structure and composition of proteins. Denatured proteins are digested to completion using a proteolytic enzyme and the peptides are resolved either on one-dimensional SDS-PAGE, reversed-phase high-performance liquid chromatography (RP-HPLC), or by two-dimensional separation on thin-layer cellulose (TLC) plates. The resulting patterns of bands, spots or peaks are referred to as a peptide map.

The most common applications of peptide mapping are: (1) to compare the primary structure of proteins suspected of being encoded by the same or related genes; (2) to determine the precise location of amino acid residues that are posttranslationally modified (Boyle et al, 1991); (3) to locate the binding sites of antibodies and functionally important sites of the target protein; and (4) to prepare individual peptides to determine amino acid composition and sequence.

Fractionation of proteins by SDS-PAGE is a high-resolution method. However, because it is based on molecular size alone, it is not possible to conclude that since two polypeptides have identical mobilities they must be related. Conversely, two proteins may be closely related but have different mobilities because of a precursor-product relationship. Peptide mapping is a method of demonstrating protein similarity. Modifications of classical two-dimensional tryptic fingerprinting have allowed the technique to be used with small amounts of protein that can be prepared from analytical SDS-PAGE. Several polypeptides can be compared simultaneously, and the only apparatus required is standard slab gel electrophoresis equipment.

If two proteins have the same primary structure, and have not been co- or posttranslationally modified (identical amino acid sequence), then performing identical cleavage reactions on them either enzymatically or chemically will theoretically yield identical peptide fragments. Conversely, if the proteins differ in primary structure, then the cleavage will yield dissimilar peptides. Following digestion and separation of the peptides on a high percentage SDS gel, the resultant peptide banding pattern is characteristic of the specific protein and the proteolytic reagent used. Relationships between polypeptides are inferred by comparing the peptide banding patterns. In this way, one can compare proteins by the similarity of the resultant peptides.

If only small numbers of fragments are produced after the specific cleavage of the target protein, maps will be easier to interpret. On the other hand, a larger number of fragments have the potential to produce maps with a broader scope.

Peptide mapping can be used to identify epitopes recognized by a specific antibody. For example, a tryptic digest of an isolated protein is followed by SDS-PAGE on a high percentage gel, transfer of the peptides to a membrane then performing an immunodetection reaction. Only the peptide containing the specific antigenic determinant recognized by the antibody will react. This technique can be extended to examine binding sites for associated proteins and for positional localization of covalent modifications that are detected by blotting techniques following cleavage as described in Chapter 7.

Peptide mapping can be performed on proteins isolated from SDS-PAGE. Since only small amounts of a protein are available for analysis, peptide mapping is usually carried out on proteins that have been radiolabeled. For example, the protein can be metabolically labeled prior to electrophoresis and cleaved directly in SDS-PAGE slices, or electroblotted onto nitrocellulose (NC) or polyvinylidene difluoride (PVDF) where the proteins can be cleaved.

Peptide maps can also be used to analyze proteins that are not radiolabeled. However, the quantity of the protein must be sufficient. The cleavage reagent should not interfere with the visualization of Coomassie brilliant blue or silver stained proteins.

A series of methods are described that are used to analyze peptides that have been generated by either thermal, chemical or enzymatic treatment of the target protein in a complex solution, directly from SDS gels or following transfer to PVDF membranes.

PROTOCOL 8.1 Thermal Denaturation

Proteins can be thermally denatured in solution prior to digestion and mass spectral peptide mapping. Sample preparation steps are minimal and purification and concentration steps are not required prior to mass spectral analysis (Park and Russell, 2000). Proteins that are resistant to enzymatic digestion become more susceptible to digestion, independent of protein size, following thermal denaturation. Mixture analysis does not require separation of the proteolytic peptides. Although some proteins aggregate upon thermal denaturation, the protein aggregates are easily digested by trypsin and generate sufficient numbers of digest peptides for protein identification. Protein aggregates of denatured proteins can have a more protease-labile structure than nondenatured proteins because more cleavage sites are exposed to the environment (Park and Russell, 2001).

Materials

Ammonium bicarbonate
Microfuge tubes
Heating block

1. Dissolve the protein in 50mM ammonium bicarbonate.
2. Incubate the protein solutions at 90°C for 20min in airtight microcentrifuge tubes.
3. Transfer the tubes to an ice-water bath to quench the denaturation process. The proteins are ready to be enzymatically digested.

Preparing the Target Protein for Digestion

Three strategies have been used to generate peptide fragments from proteins separated by SDS-PAGE: (i) elution of the intact protein from the gel, (ii) direct digestion of the protein in the gel, (iii) electrotransfer of the protein to a membrane (Patterson and Aebersold, 1995). Some features of these methods are mentioned below.

The oldest method for the recovery of gel-separated proteins is by passive-elution, which is sometimes associated with unacceptable protein loss, and by electroelution. Electroeluted proteins must be freed of salts and detergents prior to digestion. Removal of SDS will require more labor intensive techniques. An alternative approach is to digest the protein within the gel matrix and to recover the released peptides for further sequence analysis (Cleveland et al, 1977). An obvious advantage of in gel digestion is that blotting is not required. As long as the protease is small enough to permeate the gel matrix, digestion will take place (Jeno et al, 1995).

Proteins have been successfully cleaved within polyacrylamide gels with proteolytic enzymes (see Table 8.1), or by chemical procedures. These methods have proven to be highly effective for the comparison of the cleavage pattern of different proteins. The usefulness of these reagents for obtaining internal sequence information depends on the efficiency of peptide recovery.

Blotting onto a membrane has the advantage that the protein is obtained in a form that is compatible with enzymatic cleavage as most residual SDS can quickly be rinsed away prior to digestion. It can be speculated that a protease has only limited access to a protein embedded in a matrix of PVDF or nitrocellulose. A detailed study has estab-

TABLE 8.1 Proteolytic Enzymes for Protein Cleavage

Enzyme*	Site of Cleavage***	Conditions
α-Chymotrypsin	Phe-X, Trp-X, Leu-X, His-X, Met-X (possible exeption X = Pro)	NH_4HCOOH or Tris-HCl, 2–18 h; enzyme-to-protein ratio of 1:20 or 1:100, pH 8.5
Pepsin A	X-Phe, X-Leu	$CH_3COOH—H_2O$, pH 3.0
Trypsin	Arg-X, Lys-X	50 mM NH_4HCOOH, pH 8.5 same enzyme-to-protein ratio as chymotrypsin
V8 protease (*S. aureus*)**	Glu-X, Asp-X (possible exeption X = Pro)	50 mM NH_4HCOOH, pH 7.8, pH 4.0 for Glu-X cleavage or phosphates, pH 7.8 for both Glu-X and Asp-X cleavages; 25°C to prevent autolysis; enzyme-to-protein ratio of 1:20 to 1:100
Lys-C (endoproteinase Lys-C)	Lys-X (exeption X = Pro)	Similiar to chymotrypsin
Arg-C (endoproteinase Lys-C)	Arg-X	Similiar to chymotrypsin
Aspartyl protease (endoproteinase Asp-N)	X-Asp, X-Cysteic acid	Similiar to chymotrypsin exept 18 h, 37°C

*All enzyme stocks are prepared at 1 mg/ml in the appropriate buffer.
**Should only cleave Glu in ammonium acetate and in the Tris buffers used for electophoresis.
***X can be any amino acid.
Data from *Biochemica Information: A revised biochemical reference source, First Edition,* compiled and edited by J. Keesey. Copyright © 1987, Boehringer Mannheim Biochemicals, Indianapolis, IN.

lished the optimal conditions for the digestion and release of peptides from proteins electroblotted onto nitrocellulose (Lui et al, 1996).

B. Enzymatic Cleavage of Proteins

The specificity of the enzyme or the chemical reagent used to cleave the polypeptide is a critical consideration. Many proteolytic enzymes have been characterized (Barrett et al, 1998). Proteinases cleave peptide bonds at points within the protein. They are classified into metallo, serine, cysteine, and aspartic proteinases based on their catalytic mechanisms. The purer, and therefore the more specific the reagent, the more reliable the results will be. Not all proteins, gel separated or not, will be amenable to digestion with trypsin or any other specific protease. However, in the majority of cases, gel-separated proteins, due to their denatured state, are effectively digested.

Proteases can be chosen or eliminated based on the amino acid sequence of the target protein if this information is known. Sequence data may be helpful when comparing peptide maps of known proteins with an experimental, possibly related protein. Amino acid sequence information is also helpful in designing structure-function studies when cleavage at a specific location is desirable.

Currently, trypsin is the enzyme of choice. Trypsin is specific for cleavage at the C-terminal side of Arg and Lys. Partial digestion products may be generated when multiple Arg and Lys residues appear in tandem. In addition, trypsin does not cleave either Arg-Pro or Lys-Pro and does not efficiently cleave the sequences Arg-Asp/Glu and Lys-Asp/Glu. These sequences will lead to multiple partial digestion products.

Chymotrypsin cleaves at the C-terminal side of Phe, Tyr, Try, Leu, Met and more slowly His, Asn, and Gln. The digestion conditions are as for trypsin.

Staphylococcus aureus V-8 proteinase (endoproteinase Glu-C) is specific for the C-terminus of glutamate in Glu-Xaa bonds. This proteinase is also stable in 4M urea and up to 2% SDS. Extended reaction times may also be employed (12–48h at 25°C) to achieve optimal cleavage.

Thermolysin has a broad specificity and is often referred to as a "non-specific "protease. This metallo-proteinase cleaves peptide bonds at the N-terminal side of hydrophobic amino acids especially, and in descending order, Leu, Ile, Phe, Val, Ala, and Met. Although active at 50–60°C the buffer should contain 2–3mM calcium to enhance thermostability. Thermolysin is also active in 6M urea and 0.5% SDS.

A metallo-proteinase at the other end of the specificity spectrum, *Endoproteinase Asp-N* specifically cleaves peptide bonds at the N-terminal side of either aspartic acid or cysteic acid. Metallo-proteinases are rapidly inhibited by the addition of metal chelating agents.

Another useful proteinase that is specific for the C-terminal side of lysine bonds is *Lysobacter* endoproteinase Lys-C. The endoproteinase LysC retains full activity in the presence of 0.1% SDS. Many proteases are inhibited to various extents by SDS, with trypsin being one of the most sensitive. LysC also generates longer peptide fragments than trypsin.

Arg-C, also known as Clostripain, requires the presence of sulfhydryl reagent for activity. Arg-C/clostripain selectively hydrolyzes the peptide bond C-terminal to Arg.

PROTOCOL 8.2 Peptide Mapping by Proteolysis and Analysis by Electrophoresis

Partial digestion of denatured proteins is often referred to as protein fingerprinting or Cleveland digests (Cleveland et al, 1977). The technique is used to study the relatedness of two proteins or compare epitopic or modification sites. This method can be performed on radiolabeled or unlabeled polypeptides. When the protein is not labeled, one should start with 10µg per band. **Good results require careful sample preparation.**

Materials

Equilibration buffer: 0.125M Tris-HCl, pH 6.8, 0.1% SDS, 1mM EDTA
Weighing spatula
10% glycerol solution
Parafilm
Razor blade
1 × SDS sample buffer: See Appendix C

1. Excise the band of interest from the gel (see Protocol 4.14) and soak it in equilibration buffer for 30min with occasional mixing.
2. If the band of interest is radiolabeled, and the gel has been dried onto paper, line up the dried gel with the autoradiogram to positively identify the band. To remove the paper backing and rehydrate the gel slice, place the excised band on a piece of Parafilm and soak it in a small volume of equilibration buffer. Peel off the paper and soak the gel slice as in step 1.
3. Cast an SDS-gel (see Protocol 4.2) with an acrylamide concentration between 15–20% and a stacking gel slightly longer than usual. The optimal acrylamide concentration (see Table 4.1) will be determined by the sizes of the peptides generated, which will vary with protein and protease. Limited proteolysis may generate large fragments.
4. Fill the sample wells with equilibration buffer and place each gel slice in the bottom of a well with a spatula. If there are extra wells, add 1 × SDS sample buffer.
5. Add equilibration buffer containing 10% glycerol and a given amount of a protease (see Table 8.1) to each well. This solution should displace the previous solution.
6. Electrophoresis is performed normally except that to allow further digestion, the power is turned off for 30min when the bromophenol blue reaches the bottom of the stacking gel. Turn the current on and complete the run.

Comments: A protease found useful for digests using the conditions described above is *Staphylococcus aureus* V8 protease (V8), also called GluC, which cleaves at the COOH-terminal side of aspartic and glutamic acid residues in most buffers and only at Glu in the absence of PO_4. Use three concentrations of V8 when comparing different proteins. A useful range is 0.5, 5.0, and 50μg/ml protease.

The technique is versatile and easy to use. Several patterns can be obtained simultaneously on the same gel, allowing easy, unambiguous comparison.

If the band to be digested has not been radiolabeled, then a control lane should be run containing only the protease so that the protease band and autolytic products can be identified and will not be confused with the specific product.

Do not attempt Cleveland digests on gels that have been soaked in a fluorographic signal enhancing solution prior to drying as this treatment may adversely affect the digest.

Cleavage of Proteins Transferred to PVDF or NC Membranes

Proteins are separated by SDS-PAGE, transferred to a PVDF membrane and stained with Ponceau S (Protocol 7.5). The band of interest containing the isolated target protein is then excised. Two protocols are presented. Protocol 8.3 describes digestion of the protein on the membrane. Protocol 8.4 describes elution of the target protein from the membrane which is then digested enzymatically.

PROTOCOL 8.3 Cleavage of Proteins Immobilized on the Membrane

Materials

PVDF destain solution: 0.2M NaOH in 20% acetonitrile

0.2M NaOH

Polyvinyl pyrolidone 40 (PVP-40) solution: 0.2% PVP-40 (w/v) in methanol

Methanol

Digestion buffer: 1% dehydrogenated Triton X-100 (RTX-100), 10% acetonitrile, 100mM Tris-HCl, pH 8.0

Trifluoroacetic acid (TFA): 0.1%

Sonicator equipped with a microtip

Speedvac

Milli-Q water

Cleavage enzyme

1. Excise the Ponceau S stained protein band and wash with Milli-Q water. If using a PVDF membrane, destain for 1 min with 0.5 ml of PVDF destain solution. For NC membranes use 0.5 ml of 0.2M NaOH for 1 min followed by one wash with Milli-Q water.
2. For PVDF membrane, incubate the band in 0.5 ml of PVP-40 solution for 30 min at room temperature, then add 0.5 ml of Milli-Q water. For NC, use 1.0 ml of 0.5% PVP-40 in 0.1M acetic acid (w/v) at 37°C for 30 min.
3. Wash the strips extensively six to ten times with Milli-Q water to remove any traces of PVP-40.
4. Cut the strips into small squares and return them to the same microfuge tube.
5. Add 50 μl of digestion buffer followed by 5 μl of trypsin (0.1 μg of enzyme/estimated μg of protein), endoproteinase Glu-C (0.1 μg of enzyme/estimated μg of protein), or endoproteinase Lys-C (0.0075 U/estimated μg of protein). The volume of the reaction should be enough to cover the membrane completely.
6. Digest at 37°C for 24 h.
7. Following digestion, sonicate the samples for 5 min then centrifuge in a microfuge at maximum speed for 5 min. Transfer the supernatant to a new microfuge tube.

8. Wash the membrane pieces with 50 µl of digestion buffer and sonicate and centrifuge as described above. Wash the membrane twice with 50 µl 0.1% TFA. Pool all supernatants for a total volume of 200 µl.
9. Dry the contents of the microfuge tube on a Speedvac. Resuspend the peptides in H_2O and analyze the sample by electrophoresis or RP-HPLC.

Comments: Enzymatic digestion of proteins on the membrane requires that permanent absorption of the proteolytic enzyme be prevented and that peptide fragments be released into solution. Enzyme absorption is quenched by treating the protein band with PVP-40 prior to the addition of the protease. Peptide recovery is enhanced by the addition of organic solvents like acetonitrile, which increases peptide solubility.

RTX-100 was reported to increase the yield from digests conducted on either nitrocellulose or high retention PVDF (Fernendez et al, 1992). However, the detergent caused significant problems during LC-MS, including signal suppression and obscuring of peptide signals co-eluting with RTX species (Loo et al, 1994). Similar problems were observed during MALDI-MS of RTX-100 containing peptide samples.

PROTOCOL 8.4 Tryptic Cleavage of Protein Eluted from PVDF Membrane

Following SDS-PAGE and transfer to a PVDF membrane, the method below describes elution of the target protein from the PVDF membrane followed by digestion.

Materials

Acetonitrile: 40%

Re-extraction buffer: 40% acetonitrile, 0.05% Trifluoroacetic acid

Resuspension buffer: 8 M urea, 0.4 M NH_4HCO_3, pH 8.0

Dithiothreitol (DTT): 45 mM

Iodoacetamide: 100 mM

Trypsin: TPCK treated

Speedvac

1. After extensive washing with water, excise the Coomassie brilliant blue stained protein band from the PVDF membrane and incubate it in 200 µl of 40% acetonitrile for 3 h at 37°C.
2. Remove the supernatant and incubate the membrane with reextraction buffer for 20 min at 50°C.
3. Combine both supernatants and evaporate to dryness on a Speedvac.

4. Redissolve the residue in 50 µl Resuspension buffer.
5. Reduce the S-S bonds by adding 5 µl of 45 mM DTT and incubate for 15 min at 50°C.
6. Cool to room temperature and add 5 µl of 100 mM iodoacetamide. Incubate for 15 min at room temperature then add 140 µl of water.
7. Add trypsin (TPCK-treated) in a volume of 5 µl at an enzyme/substrate (w/w) ratio of 1:25, and incubate the reaction for 24 h at 37°C. Stop the digest by freezing or analyze directly by SDS-PAGE or HPLC.

Comments: Some common reasons why tryptic digests are not successful are: (1) the protein concentration is less than estimated; (2) too much residual SDS present as > 0.005%, (10 µg/200 µl digestion volume) decreases the efficiency of the trypsin digestion; and (3) complete denaturation is essential.

Treatment of trypsin with L-1-tosylamido-2-phenylethyl chloromethyl ketone (TPCK) is commonly performed to inhibit contaminating chymotrypsin. TPCK has been reported to inhibit chymotrypsin activity without affecting trypsin (Kostka and Carpenter, 1964).

C. Chemical Cleavage

Chemical cleavage of proteins immobilized on PVDF membranes has some advantages over proteolytic cleavage. Cleavage reagents are highly specific and the fragments are generally larger and fewer in number than those generated proteolytically. PVDF is inert and mechanically stable in the presence of a number of cleavage reagents. Some of the more widely used reagents are listed in Table 8.2.

Solubilized pure proteins can also be chemically cleaved.

PROTOCOL 8.5 Cyanogen Bromide Cleavage of Proteins on PVDF Membrane

CNBr is an extremely dangerous reagent. Exercise care when handling it. Use a fume hood. Keep contact with CNBr to a minimum. To prepare the CNBr solution, weigh an empty microfuge tube and then add the CNBr directly to it in a fume hood. Reweigh the tube and

TABLE 8.2 Chemical Cleavage of Proteins

Reagent	Site	Reference
Formic Acid	Between Asp-Pro	Miczka and Kula (1989)
Trifluoroacetic acid (TFA)	amino side of Ser and Thr	Hulmes et al. (1989)
Cyanogen Bromide (CNBr)	Carboxy side of Met	Scott et al. (1988); Choli et al. (1989)
N-Chlorosuccinimide (NCS)	Tryptophanyl peptide bonds	Lischwe and Sung (1977)
Hydroxylamine	Between Asn-Gly	
BNPS-skatole	Carboxy side of Trp	Crimmins et al. (1990)
2-Nitro-5-Thiocyanobenzoate	Cys	
5-5′-dithio-bis-(2-nitro-benzoic acid) (DTNB)	Amino side of Cys	Degani and Patchornik (1971)
o-iodosobenzoic acid	Carboxy side of Trp	Aitken (1990)

then add the appropriate amount of solvent to reach the desired final concentration.

Following cleavage, the product should still be associated with the membrane. In order to be sure, collect the CNBr solution and further dilute it with water, lyophilize and prepare for electrophoresis. Figure 8.1 illustrates the results that can be obtained when CNBr cleavage is used with the Tricine SDS system described in Protocol 4.3.

Materials

0.5 ml microfuge tubes

Cyanogen bromide (CNBr) solution: 0.15 M CNBr in 70% formic acid (v/v)

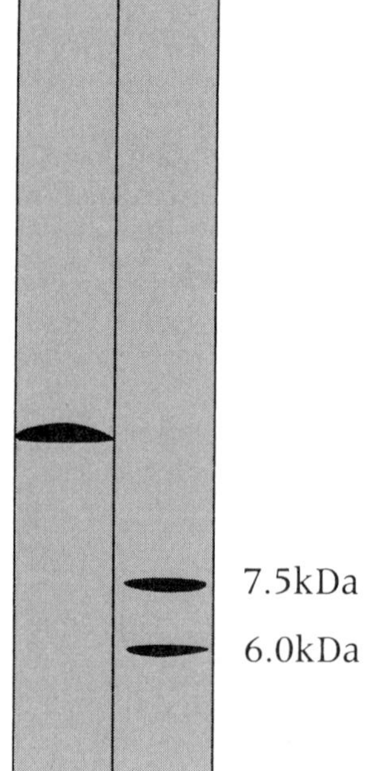

FIGURE 8.1 CNBr cleavage of bovine uroplakin II. Lanes 1 and 2 show gel purified uroplakin II before and after cleavage with CNBr. The products are resolved on a Tricine SDS-gel. The two major fragments are 7.5 and 6.0 kDa. (Adapted with permission from Lin et al, 1994.)

Elution buffer: 2% SDS, 1% Triton X-100, 50 mM Tris, pH 9.2–9.5
Speedvac
4 × SDS sample buffer: see Appendix C

1. Stain the PVDF membrane with 0.1% Ponceau S and identify the band of interest.
2. Excise the band and cut the membrane into small pieces and place them in a 0.5 ml microfuge tube.
3. Add 150 µl of the CNBr solution and incubate overnight in the dark at room temperature with occasional agitation.
4. Following cleavage, remove the CNBr solution and wash the pieces of membrane with water. Dry the membrane in a Speedvac to remove traces of the formic acid as low pH will inhibit subsequent elution of the peptides from the membrane.
5. Resuspend the washed membrane in 50–75 µl of elution buffer for 90 min at room temperature with frequent agitation so that all surfaces are exposed to the elution buffer.
6. Add 1/4 of the volume of 4 × SDS sample buffer and isolate the digestion products on a 15% SDS-polyacrylamide gel. Transfer to PVDF (Protocol 7.1) and stain the blot with Coomassie brilliant blue.
7. Cut out the band of interest and submit it for mass spectrometric analysis.

Comments: CNBr digestion will fail if the protein is oxidized, and CNBr activity is reduced in the presence of reducing agents like DTT and 2-ME. The immobilized protein can be reduced by DTT and then the membrane washed prior to CNBr treatment. Incubate soluble proteins in buffer containing 10 mM DTT then TCA precipitate the protein (Protocol 4.16), wash and resuspend in the CNBr 70% formic acid solution.

PROTOCOL 8.6 N-Chlorosuccinimide (NCS) Mapping

N-chlorosuccinimide (NCS) specifically cleaves tryptophanyl peptide bonds (Lischwe and Sung, 1977; Vorburger et al, 1989).

Materials

Equilibration buffer: 1 g urea, 1 ml water, 1 ml acetic acid
N-Chlorosuccinimide (NCS) solution: 40 mM NCS in equilibration buffer
Wash buffer: 10 mM Tris-HCl, pH 8.0
1 × SDS sample buffer: see Appendix C
X-ray film

1. Following electrophoresis of radiolabeled proteins and gel drying, expose the gel to X-ray film for several hours at room temperature.
2. Excise the band of interest from the dried gel and rinse it first with 2 ml of water for 20 min, with one change, and then with equilibration buffer for 20 min with one change.

3. Incubate the gel slice for 1 h in NCS solution.
4. Rinse the gel slice 5 times with wash buffer (5 min per wash) then equilibrate the slice in 1 × SDS sample buffer for 1.5 h with three changes.
5. Place the piece of gel at the bottom of the well in the stacking gel and fill the well with 1 × SDS sample buffer. Run the gel and analyze the cleavage products by autoradiography.

Comments: Cleavage does not always go to completion. Therefore, partial cleavage products should be expected. This method may also be performed with proteins that are not radiolabeled.

It has been observed that the apparent molecular weight of fragments increased following NCS treatment (Kilic and Ball, 1991). This may be due to a side reaction with NCS resulting in a product which binds less SDS.

PROTOCOL 8.7 Hydroxylamine Cleavage of Proteins in Polyacrylamide Gels

Peptides can be generated from proteins that are immobilized in polyacrylamide gels using hydroxylamine which cleaves specifically at Asn-Gly (Saris et al, 1983). The cleavage is performed on individual protein bands that have been excised from dried slab gels after identification of the target band by staining, fluorography, or autoradiography. The procedure can be combined with other techniques to further characterize the cleavage products.

Materials

Razor blade

5% methanol

40% methanol

Vacuum dessicator

Hydroxylamine solution: Dissolve 1.1 g hydroxylamine hydrochloride (final concentration 2 M), 4.6 g guanidine hydrochloride (final concentration 6 M), and 15 mg Tris base (final concentration 15 mM) in 4.0 M lithium hydroxide to give a pH of 9.0 and a final volume of 8 ml.

Vehicle solution: Dissolve 4.6 g guanidine hydrochloride (final concentration 6 M), and 15 mg Tris base (final concentration 15 mM) in 4.0 M lithium hydroxide to give a pH of 9.0 and a final volume of 8 ml.

15 ml conical polypropylene centrifuge tubes

1. Following electrophoresis and detection of the target protein, excise the band of interest.
2. Swell the gel pieces in 5% methanol. Peel off the paper backing and remove free SDS by washing at 4°C in four changes of 3 ml 5% methanol. Pieces cut out of a low percentage gel (<10%) are swelled in 40% methanol, 5% acetic acid prior to washing in 5% methanol.
3. The gel pieces are brought to near dryness in a vacuum desiccator.
4. Submerge the gel pieces in hydroxylamine solution or vehicle solution (50–200 µl of reaction mixture to completely cover the gel pieces) and incubate for 3 h at 45°C in closed polypropylene tubes.

5. After the reaction, blot the gel pieces on tissue paper and wash four times with 3ml of 5% methanol. Briefly wash (5min) low percentage gels in 40% methanol, 5% acetic acid prior to the methanol washes to prevent elution of small fragments.
6. Bring the gel pieces to near-dryness then submerge them in SDS-sample buffer and incubate at 37°C for 2h then at 95°C for 5min.
7. Analyze the cleavage products by SDS-PAGE. Place the gel pieces at the bottom of the sample well in the stacking gel. Fill the appropriate slot with the sample buffer used to boil the gel pieces.

Comments: Cleavage occurs specifically but partially, at a limited number of sites (Asn-Gly). The target protein can be purified by other methods rather than SDS-PAGE. In this case the protein is dissolved in 6M guanidine-HCl and incubated with 2M hydroxylamine-HCl (adjusted to pH 9.0 with 4M LiOH) for 4h at 25°C.

Manipulation of gel pieces is facilitated by using spacers that are slightly wider than those used for the first gel making it easier to insert the strip between the plates.

Controls should be included and run in parallel with the hydroxylamine cleavage products. A comparison will show which bands are hydroxylamine specific.

PROTOCOL 8.8 Formic Acid Cleavage

Conditions for the selective cleavage of proteins at the carboxyl side of aspartic acid residues using formic acid have been reported (Inglis, 1983). Limited acid hydrolysis is usually not complete, generating overlapping peptides, but so do most protein fragmentation techniques. The following method describes limited acid hydrolysis of proteins in the polyacrylamide gel matrix (Vanfleteren et al, 1992).

Materials

Ponceau S solution: 0.1% Ponceau S in 1% acetic acid
NaOH: 1mM
Glass digestion tubes (Pierce)
HPLC grade water
Formic acid: 20% in HPLC grade water
Formic acid: 2% in HPLC grade water
100ml Duran flask
Shaker
Acetonitrile: HPLC grade

1. Following electrophoresis, stain the gel with Ponceau S.
2. Excise the band containing the protein of interest and cut it into transverse slices. The volumes that follow are applicable for a gel slice loaded on a 10mm wide slot.
3. Destain the gel slices by soaking them in at least 1000vol of HPLC grade H_2O for 1h. Transfer the slices to glass digestion tubes containing 0.9ml

of 1 mM NaOH. Tape the tubes onto a gyrotory shaker and incubate for 1 h at 300 oscillations/min. This should be sufficient to completely destain the protein.

4. Add 0.1 ml of a 20% solution of formic acid in HPLC grade H_2O and allow the gel slices to equilibrate for no longer than 10–15 min.
5. Transfer the contents of the digestion tube to a 100 ml Duran flask containing 3 ml of 2% formic acid, pH 2.0, to avoid excess evaporation.
6. Tightly close the flask and incubate at 112°C for 4 h.
7. After completion of the acid hydrolysis, add 0.1 ml of HPLC grade acetonitrile. The released peptides diffuse out of the gel at the elevated temperature and can be separated by RP-HPLC using a Delta-Pak 3.9 × 150 mm C_{18} column (Waters). See chapter 10 for more details.

Comments: Limited acid hydrolysis can be performed on proteins that are in the gel matrix.

Staining with Ponceau S has two main advantages. The dye dissociates rapidly from the protein in weak alkaline solution. There is no need to extract Ponceau S prior to the acid hydrolysis treatment, which would lead to protein loss since the dye is eluted before the peptides from a reversed-phase support. The drawback to Ponceau S is that it is five to ten times less sensitive than Coomassie staining. Therefore, if you can visualize the band of interest with Ponceau S staining, there should be enough material for a successful analysis and product recovery.

PROTOCOL 8.9 Chemical Cleavage at Cysteine Residues with DTNB

The cysteine specific reagent 5-5′-dithio-bis-(2-nitrobenzoic acid) (DTNB), (Degani and Patchornik, 1971), cleaves the polypeptide chain on the amino side of cysteine residues.

The specificity of the reaction can be checked by first blocking the sulfhydryl groups with N-ethylmaleimide. Under these conditions, no fragments should be generated. A representative digest is shown in Figure 8.2. For ease of analysis the fragments were assigned numbers starting from the largest to the smallest.

The method described below (Kilic and Ball, 1991) was used to cleave the cytoskeletal protein vinculin.

Materials

Dithiothreitol (DTT): 1 M

Reduction buffer: 100 mM Tris-HCl, pH 8.0, 8 M urea, 5 mM DTT

Cleavage buffer: 100 mM Na borate, 500 mM glycylglycine, 8 M urea, pH 9.0, 1 mM KCN

5-5′-dithio-bis-(2-nitrobenzoic acid) (DTNB)

2 × SDS sample buffer: see Appendix C

1. Dissolve 60 μg of purified protein in reduction buffer and incubate at room temperature for 20 min. This step keeps cysteine residues in the reduced state.

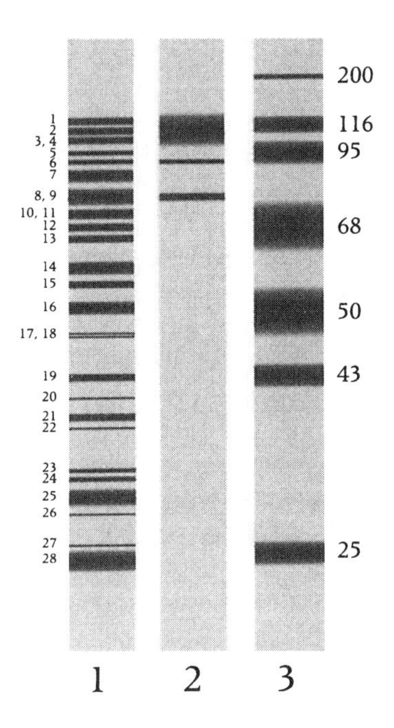

FIGURE 8.2 Cleavage of chicken gizzard vinculin with DTNB. Chicken gizzard vinculin (60 μg) was cleaved with DTNB. The reaction products were run on a 6–15% gradient SDS-gel and stained with Coomassie blue (lane 1); intact vinculin (lane 2); molecular mass marker proteins (lane 3). This digestion resulted in roughly 20 major digestion products which were distributed evenly throughout the gel. (Adapted with permission from Kilic and Ball, 1991.)

2. Make the solution 15 mM with DTNB and incubate for 15 min. If necessary, adjust the pH to 8.0 with NaOH.
3. To cleave the DTNB-modified protein, precipitate with trichloroacetic acid as described in Protocol 4.16. Dissolve the precipitated protein in cleavage buffer and incubate at 40°C for 3 h. For cleavage with radiolabeling, use the same conditions substituting 1 mM $Na^{14}CN$ for the KCN (Nefsky and Bretscher, 1989). ☢
4. Make the solution 100 mM with DTT and incubate at 40°C for an additional 15 min.
5. Add an equal volume of 2 × SDS sample buffer, boil for 3 min and analyze the peptide products of the cleavage reaction by SDS PAGE.

D. Microsequencing from PVDF Membranes

The identification of a protein assigns a name and database accession code to a spot on a 2-D gel or a band on an SDS-PAGE gel. The amino acid sequence of the target protein is linked to a corresponding DNA sequence thereby connecting the proteome to the genome. The identification of the target protein is the first step toward studies detailing its co- and posttranslational modifications, ultimately leading to the elucidation of its function.

The challenge to the researcher is how the target protein can be identified with a minimum of effort in a cost- and time- effective manner.

One way to identify a protein is to query a database with all the attributes that are known about the target protein. The more information available, the greater the likelihood that the unknown protein can be identified by matching its characteristics with those of a protein in a database. This section will examine various attributes of a protein and how they can be used by themselves and combined to identify the object protein. Attributes of a protein that have been incorporated into databases are species of origin, mass, isoelectric point, protein N- and C-terminal sequence tags, peptide mass fingerprinting, amino acid composition and amino acid sequence. Properties of the target protein are matched against protein databases to identify the unknown protein.

During the last several years there has been a radical shift from chemical based protein sequencing that relied almost exclusively on the Edman degradation reaction (Edman and Begg, 1967), to mass spectrometric based methods of protein identification coupled with *in silico* analysis. Newer methodology that is being introduced is rapidly becoming routine in many core facilities. Therefore, some of the techniques used for isolating and preparing samples for subsequent analyses are included here.

The 1980s saw a gradual shift away from classical brute force protein purification where the goal was to purify enough of the target protein so that it could be sequenced. Technological advances made it possible and practical for small amounts of a specific protein to be isolated by SDS-PAGE and subsequently identified. As a ballpark estimation, the average amount of protein required to obtain 15 amino acids of sequence from the NH_2 terminus of an intact protein by Edman degradation was 150 pmol. Eight times this amount, roughly 1.2 nmol was needeed if the protein required trypsin digestion and then one of the resulting HPLC purified peptides was subjected to Edman degradation to obtain 15 amino acids. For reference,

$$10 \text{ pmol of a } 100 \text{ kDa protein} = 1\,\mu g.$$

Protein sequencing in the mid eighties was performed by the stepwise N-terminal chemical degradation of the intact polypeptide isolated from SDS-PAGE. Proteins were recovered from an SDS-PAGE gel by diffusion and sequenced manually by dansyl-Edman degradation at the nanomole level. The sensitivity was increased by using electroelution to increase the recovery of proteins. Drawbacks to these methods were low yields, free amino acids in the protein samples and NH_2-terminal blocking.

Protein-sequencing instruments and methodology evolved considerably in the late 80s. With the use of automated gas-phase sequencing instruments and on-line HPLC systems, sequence data could routinely be obtained with less than 10 pmol of protein or peptide.

The advent of sensitive microsequencing instrumentation coupled with SDS-PAGE and electrophoretic transfer to PVDF made amino-terminal sequence analysis possible for proteins not amenable to purification by other means. Proteins that are electro-transferred onto chemically inert membranes could be sequenced directly without any additional manipulations. This direct sequencing procedure generally yielded NH_2-terminal sequences of 10–40 residues. Blotted proteins that are weakly stained with Coomassie blue represent 50–200 ng of material. Practically speaking, a good test to determine whether there is enough material to obtain sequence is to stain the blot with Coomassie and photocopy the blot. If the band of interest is visible on the copy, then there should be enough material for sequencing.

Membranes suitable for direct gas and liquid phase sequence analysis include polybrene-coated, chemically derivatized and siliconized glass fiber filters. A popular membrane used frequently is polyvinylidene difluoride (PVDF). The binding capacity of this membrane is roughly 170μg/cm^2 for BSA.

If the target protein has been purified, there is no need to expose it to an additional electrophoresis step. Sample preparation cartridges (ProSpin®, Applied Biosystems) are commercially available that provide a simple one-step method for collecting, concentrating and desalting the protein solution. The method is based on a gentle centrifugation through a PVDF membrane which will retain the target protein and wash away the interfering substances. After the centrifugation, simply remove the PVDF membrane and insert it directly into the microsequencer.

The presence of a blocked N-terminus, or when N-terminal data provided little discrimination between close members of a family of proteins having extensive N-terminal homology or the need for two or more segments of protein sequence data for successful cloning of cDNA often necessitated the generation of internal peptide fragments. When it has been proven that the protein of interest is present in sufficient quantity but is NH_2-terminally blocked, the next step was to generate peptides derived from the target protein by either enzymatic or chemical cleavage.

Transferring proteins to a membrane is an effective way to isolate the protein of interest from the majority of contaminants. If the protocol calls for proteolytic digestion of the protein followed by reverse phase HPLC to isolate the peptides, the transfer can be made to either NC or PVDF. The immobilized protein is subjected to exhaustive proteolytic digestion. Proteases that have been used successfully for this are trypsin, chymotrypsin, Staph V-8 protease and pepsin. Protease digestion generates small peptides that elute from the membrane into the digestion buffer. However, membrane-immobilized proteins are not homogeneously digested. The number of peptides generated is less than that theoretically expected from the number of cleavage sites. This is advantageous because when the peptides are subsequently purified by reversed-phase HPLC the resulting peptide chromatogram is less complex. Isolated peptides can now be sequenced individually. Alternatively, the RP-HPLC isolated peptide fragments can be electrophoresed, transferred to a suitable membrane and then sequenced.

When, and In What Form Do You Submit the Target Protein to the Protein Sequencing Core?

It has become routine to obtain sequence data from less than 5pmol of gel purified target protein, which, for a polypeptide of 50kDa, is 0.25μg. If the band of interest can be visualized following silver staining, indicating ~5ng/band, then most sequencing cores should be able to identify the target protein.

If your lab is not set up for protein identification by mass spectrometry you will have to submit the unknown protein to a core facility for analysis. Different facilities prefer to receive a band of SDS-PAGE gel or a piece of PVDF or nitrocellulose membrane. When you are getting ready to run the gel to enrich for the target protein, contact your core facility and ask how they would prefer to receive your precious sample.

PROTOCOL 8.10 Transferring Spots from 2-D Gels to PVDF Membranes

Microsequencing can be performed on protein spots collected from 2-D gels that have been stained with Coomassie brilliant blue, dried and stored. Proteins are recovered from the collected gel pieces by a protein-elution-concentration method combined with electrophoresis and electroblotting (Rasmussen et al, 1991).

Materials

1 × SDS sample buffer: see Appendix C

Sephadex G-10 equilibrated in sample buffer

1. Combine identical spots collected from 2-D gels and allow them to swell in 1 × SDS sample buffer in a total volume of 1.5 ml.
2. Load the gel pieces along with the supernatant into a wide (at least 7 mm) sample well made in a new gel.
3. Fill the slot with Sephadex G-10 that has been equilibrated in sample buffer.
4. During electrophoresis most of the electric current passes on the sides of the gel slot resulting in a horizontal contraction of the protein band. In this manner, the protein is efficiently eluted from the gel pieces and concentrated from a large volume into a narrow spot.
5. Following the run, transfer the proteins to a PVDF membrane (Protocol 7.1). Stain the blot with CBB (Protocol 7.6) to locate the target protein and proceed with microsequencing. Alternatively, the protein can be digested on the membrane with trypsin. The peptides that are generated during the digestion elute from the membrane into the supernatant which can be separated by narrow bore RP-HPLC (Chapter 10) and collected individually for sequence analysis.

Sequencing Glycopeptides

In many studies, the fine structure of the oligosaccharides responsible for the separation of protein isoforms by 2-D PAGE may not be necessary. However, if they are to be characterized, the best approach is to release the sugars from the separated protein spots. A detailed structural analysis could involve the determination of: a) the monosaccharide composition, b) the attachment sites to the protein, c) the sequence, branching and linkage positions and d) the anomeric configuration of the monosaccharides.

Sequencing of an *N*-glycosylasparaginyl-linked peptide proceeds normally through this residue but a blank appears at the position of the glycosylated asparagine residue (Evans et al, 1988). Subsequent residues are identified normally, but in many instances the yield of phenylthiohydantoin-derivatives is considerably reduced. Most likely, the large mass of carbohydrate is the cause of the problem. This has been overcome by the removal of the N-linked sugar with specific glycosidases (Deshpande et al, 1987) and by β-elimination of the labile O-linked carbohydrate (Downs et al, 1977).

In order to obtain information on the site of glycan attachment it is almost always necessary to first release N-linked glycans with peptide-N^4-(*N*-acetyl-β-glucosaminyl) asparagine amidase F (PNGase F). After deglycosylation, the formerly N-glycosylated asparagine residue is identified as aspartic acid. O-glycosylation sites can be determined following reductive—β-elimination of an O-linked glycopeptide simultaneously releasing the carbohydrate chain and generating a modified amino acid. After deglycosylation, the modified serine is identified as dehydroalanine.

PROTOCOL 8.11 Protein Hydrolysis: Total Amino Acid Composition of the Target Protein

Protein amino acid composition refers to the number of each amino acid present in the protein expressed as a percentage. The amino acid composition is sequence independent. Theoretically proteins can be identified by amino acid composition by comparing the experimentally determined result with a database. Under ideal conditions, about 50% of proteins from species with sequenced genomes can be identified with high confidence by amino acid composition, when accompanied by pI and mass (Wilkins et al, 1996).

There are at least two programs that can use amino acid composition for identification purposes: AACompIdent at www.expasy.ch/ch2d/aacompi.html. and PropSearch at www.emblheidelberg.de/aaa.html.

Proteins require some form of chemical treatment before their component amino acids are suitable for analysis. Protein and peptide samples must be hydrolyzed to free amino acids from peptide linkages. Hydrochloric acid is the most widely used agent for hydrolyzing proteins. A simplified hydrolysis procedure involves refluxing the protein with excess HCl then removing the excess HCl in vacuum (Moore and Stein, 1963).

Materials

Glass test tubes (Pyrex) 13 × 100 mm

HCl: 6N

Acetylene torch

Dry ice

Acetone

1. Perform the hydrolysis on ~0.2–0.5 mg of protein in 13 × 100 Pyrex test tubes which have been thoroughly cleaned and are free of trace metals and oxidants.
2. Resuspend the lyophilized protein in 6N HCl and add it to the tube. Constrict the tube using a torch.
3. Freeze the contents of the tube using a bath of dry ice and acetone. Evacuate the tube under vacuum and seal it at the constricted area.
4. Hydrolyze the protein at 110°C for 20–70 h. Remove the residual HCl by rotary evaporation.
5. Dissolve the residue in the appropriate buffer for amino acid analysis.

Comments: A Protein Hydrolysis Kit complete with a dry block heating module is available from Pierce.

Identifying Proteins Using Mass Spectrometry

Over the last several years, mass spectrometry has become the core technology for analyzing protein covalent structure. Once regarded as an analytical chemistry technique for analyzing small molecules, MS has evolved into a vital tool for the biologist. Mass spectrometry is used to accurately measure the molecular mass of large and small biological molecules. In essence, the mass spectrometer is a very sensitive balance for weighing molecules. Accurate mass assignment permits the unambiguous identification of proteins.

Mass spectrometry has emerged as a focal technology for the identification of proteins in a complex mixture. It provides sensitivity and high throughput using two approaches: (i) mass mapping, also referred to as fingerprinting of peptide fragments using matrix-assisted laser desorption/ionization (MALDI) (Berndt et al, 1999) and (ii) tandem mass spectrometry of peptide fragments to obtain specific sequence information using electrospray ionization (ES-MS), and by MALDI quadrapole time-of-flight mass spectrometry (MALDI-TOF-MS) (Shevchenko et al, 2000). The success of protein identification by peptide mapping is based on the properties of individual proteins; the limited number of proteins for each organism, the large difference in amino acid sequence, and the large mass difference between different amino acids.

An in depth treatment of MS as it relates to the analysis of proteins is beyond the scope of this book. However, the technique is so powerful, and is used so frequently that it deserves to be presented here. The aim of this section is to introduce some of the fundamentals of MS and demonstrate how they are used to identify proteins.

In practice, the identification of proteins by MS reduces down to the identification of peptides. Sequencing peptides by MS has several advantages over classical Edman degradation. Sequence analysis can be performed on a peptide within a mixture because individual peptides can be selected by the mass spectrometer. In addition, the fragmentation of a molecular ion is not affected by the presence of an N-terminal modification, a significant drawback with Edman degradation.

MS sequencing strategies analyze peptides rather than their full-length proteins. Peptides can be introduced into the mass spectrometer by several methods. Proteins that are in the denatured state are routinely digested by a proteolytic enzyme. Trypsin is frequently used because the resultant peptides produce spectra that are more amenable to interpretation.

To allow the mass spectrometer to analyse the peptides, gas-phase ions must be formed through electrospray ionization or matrix-assisted laser desorption ionization. A mass analyzer then scans through the m/z (ratio of mass to charge) range, and a detector records the ions, producing a mass spectrum (Tabb et al, 2001).

Preparation of Proteins for MS Analysis

Peptides can be produced from a protein band on a gel by three methods; elution of the intact protein from the gel followed by digestion, digestion of the protein within the gel followed by elution of the peptides from the gel, and electroblotting the protein onto a membrane followed by digestion on the membrane using conditions that release the peptides into solution.

Partial Proteolysis

Limited digestion of isolated proteins will produce relatively large fragments for automated microsequencing. Proteolysis is usually performed around pH 8. Volatile buffers are desirable since they may be easily removed by lyophilization. Ammonium bicarbonate, which gives a pH of ~7.8 when dissolved, and N-ethylmorpholine acetate are buffers of choice. SDS may be used if the solubility of the substrate protein is a problem since most proteases are active in the presence of 0.05–0.5% SDS. In addition, proteolysis is more efficient when the substrate is denatured. Digestion in the presence of SDS will yield patterns from unfolded proteins. Digestion under native conditions may be useful in structure-function studies of proteins. Specific domains may be identified. It is also helpful to add aliquots of enzyme over the total digestion period without exceeding an enzyme to substrate ratio of 1/20 (Fernandez et al, 1992).

Note that different digestion patterns may arise depending on buffer (e.g. Tris vs. PO_4), pH, temperature, and time of digestion. Different lots of protease and protease from different vendors may also yield different results.

According to a literature analysis (Keil and Tong, 1988), trypsin has been used in 30% of digests; chymotrypsin 22%; endoproteinase Glu-C (*S. aureus V8*) 12%; thermolysin 11%; pepsin 6%; subtilisin, papain, clostripain and *Lysobacter* Lys-specific proteinase ~2% each; thrombin, elastase, pronase, submaxillary Arg-specific, and *A. mellea* proteinases ~1% each.

Currently, trypsin is the enzyme of choice. Trypsin is specific for cleavage at the C-terminal side of Arg and Lys. This means that the peptide will almost always have the charge contributed by the Arg or Lys residue. In addition, upon analysis of the tryptic peptides, Arg and

Lys residues function as landmarks. A tryptic peptide sequence that does not end in an Arg or Lys suggests that it is the C terminus of the protein. Extending this farther, in a database search this sequence would align just upstream of the stop codon, thereby establishing the reading frame as well as the translational stop site (Mann and Pandey, 2001). Partial digestion products may be generated when multiple Arg and Lys residues appear in tandem. Trypsin does not cleave either Arg-Pro or Lys-Pro and does not efficiently cleave the sequences Arg-Asp/Glu and Lys-Asp/Glu. These sequences will lead to multiple partial digestion products.

In general, the smaller the protein, the fewer tryptic peptides that can possibly be generated within the mass-to-charge ratio of the mass spectrometer. Furthermore, proteins with pIs < 4.3 have fewer lysine and arginine residues that can be targeted during a digestion with trypsin and Lys-C (Washburn et al, 2001).

PROTOCOL 8.12 In-gel Tryptic Digestion

In-gel digestion, followed by extraction of the resulting peptides from the gel and MS analysis has become well established (van Montfort et al, 2002).

Materials

Destaining buffer: 50 mM ammonium bicarbonate in 40% ethanol
Wash buffer: 25 mM ammonium bicarbonate
SpeedVac (Savant)

1. After visualization, excise the protein band from the gel and destain it completely with destain buffer.
2. Wash the gel pieces with 200 µl of wash buffer three times for 15 min each.
3. Cut the band into pieces <1 mm^3.
4. Dehydrate the gel pieces by incubating them in 100 µl of acetonitrile three times for 10 min each.
5. Dry the gel pieces using a SpeedVac.
6. For tryptic digestion, add 5 µl of 75 ng/µl trypsin in 25 mM ammonium bicarbonate to the dried gel pieces.
7. Cover the gel pieces with ~20 µl of 25 mM ammonium bicarbonate making sure that the gel pieces are immersed throughout the digestion.
8. Digest the protein for at least 14 h at 30°C.

Comments: Chemical cleavages can also be performed with proteins in-gel. For CNBr cleavages, add 25 µl of a 1 M CNBr solution in 70% TFA to the dried gel pieces. Digest for at least 14 h in the dark at 25°C. The in-gel digestion technique has been automated and proteins present in the low femtomolar range can be confidently identified (Nadler et al, 2000).

In contrast to trypsin cleavage sites, which are rarely found in transmembrane segments, CNBr cleavage sites are also present in membrane spanning regions. Consequently, a transmembrane segment can be cleaved into smaller hydrophobic parts, which may contain a solubilizing hydrophilic tail.

In-gel digestion combined with MS has some drawbacks. Low abundance proteins may not be detected unless large amounts of starting material are used. (However, the quantity of target protein will influence any method.) Only a few peptide fragments may be eluted from the gel. This is especially true for polypeptides <10 kDa. For larger proteins, poor sequence coverage may result due to difficulty in extracting hydrophobic peptides from the gel. Low digestion efficiency may occur because sufficient trypsin may fail to penetrate the gel matrix.

Membrane proteins pose problems inherent to their chemical nature. Membrane-spanning segments are either not readily accessible to proteolytic enzymes or lack the specific proteolytic cleavage site. These difficulties can sometimes be overcome with the use of a combination of proteases and chemical cleavage methods (Van Montfort et al, 2002).

PROTOCOL 8.13 Extraction of Peptides from Gel Pieces Containing Integral Membrane Proteins

Lysine and arginine residues are seldom found in the membrane-spanning regions. Therefore, tryptic cleavage will generate large hydrophobic fragments which are poorly soluble and do not contribute to the sequence coverage. β-octyl glucoside (βOG), a MALDI-MS-compatible detergent, increases the total number of extractable peptides when included in the extraction buffer (van Montfort et al, 2002).

Materials

Extraction buffer: 60% acetonitrile, 1% TFA, ±0.1% β-octyl glucoside
β-octyl glucoside (βOG): 10%
1% TFA

1. Sonicate the digested gel pieces for 5 min in the presence or absence of 0.1% βOG, adding the appropriate amount of the 10% stock.
2. Collect the sonicate in a clean tube.
3. Reextract the gel pieces twice more by sonication for 5 min each in 30 μl extraction buffer.
4. Pool the sonicated extracts and dry in a SpeedVac.
5. To remove the last traces of ammonium bicarbonate, add 10 μl of 1% TFA and dry in the SpeedVac.

Comments: This method is compatible with MALDI-MS and it has been shown to increase the recovery of hydrophobic tryptic fragments.

Elution of Target Protein from SDS-PAGE

An alternative to in-gel digestion is elution of the undigested protein from the gel. This normally requires picomoles of material and is performed by direct electroelution or passive solvent extraction with a solution of formic acid/water/2-propanol (Cohen and Chait, 1997). Enzymatic digestion of proteins in solution following elution from the gel has the potential of increasing the sequence coverage, ensuring that the whole protein is accessible to the enzyme. Cross-linkers and harsh oxidizing agents should be avoided as they interfere with extraction of peptides from the gel or may chemically modify the peptides (Shevchenko et al, 1996).

Protein Transfer to a Membrane

Gel separated proteins may be transferred to a nitrocellulose or PVDF membrane, with nitrocellulose being preferred because of its high solubility in most organic solvents used and compatability with MALDI-MS. Proteins in solution or immobilized on a membrane can be proteolytically digested, avoiding the problem of selective retention of the proteolytic fragments in the gel.

Considerations

Silver stained amounts necessary for successful MS identification of proteins are 5–50 ng or 0.1–1 pmol for a 50 kDa protein. It is important to minimize contamination from keratins, which are introduced by dust, chemicals, fur from small animals, and handling without gloves. The keratin peptides can dominate the spectra of low abundance proteins. Although target proteins can be identified by MS it is difficult to achieve 100% coverage over the complete protein. This is because: (i) some of the peptides derived from a given protein might not ionize well; (ii) the peptides might be too large or too small for analysis; or (iii) the peptides might yield complex fragmentation spectra that are difficult to interpret.

MS Basics

In the early 1990s two technical advances, matrix-assisted laser desorption ionization (MALDI) time of flight (TOF) and electrospray ionization (ESI) replaced the slower and less-sensitive chemical degradation methods. Mass spectrometry has become the method of choice for identifying proteins. Commercially available instruments can provide high quality sequence spectra in the low femtomole region (Morris et al, 1996). Experimental research instruments can attain low-attomole sensitivities (Valaskovic et al, 1996). It is now no longer necessary to isolate the target on SDS-PAGE. MS can identify the target protein which can even be present in a mixture, precluding the need to purify the target protein to homogeneity. MS is not affected by a blocked N terminus, greatly simplifying the procedure.

Protein identification is accomplished using peptide-mass fingerprinting by MALDI-TOF MS and nano-ESI tandem mass spectrometry (MS/MS). Tandem mass spectra are searched against theoretical tandem mass spectra of peptide sequences contained in a database using a variety of search algorithms (Beavis and Fenyo, 2000), the pioneering program being SEQUEST (Eng et al, 1994).

The two forms of MS that will be discussed in this manual are ESI-MS and MALDI-MS. Both are considered "soft" ionization methods in that they allow mass analysis in the gas phase with minimal ionization induced fragmentation. Both methods have been implemented in two-stage mass spectrometers with the possibility of inducing fragmentation of the peptide bonds.

Mass spectrometers have three main components, an ion source for the production of ions with subsequent introduction into the gas phase, a mass analyzer separating ions and measuring them according to their mass/charge (m/z) ratios and a detector for counting the ions (usually a photomultiplier connected to a computer). The most common type of ion source uses a pulsed nitrogen laser at 337nm. Ions are usually detected by electron multipliers.

Ions from a pure polypeptide sample are detected as a series of peaks of different *m/z* values. In many cases, the ions recorded in a mass spectrum will be singly charged, $z = 1$. Under these circumstances, the m/z ratio and the mass are identical so that the m/z is numerically equal to its molecular weight. From the series of measured *m/z* values, the molecular mass of a peptide is easily calculated by a process known as deconvolution.

Electrospray and Tandem Mass Spectrometry

As the name suggests, tandem MS is a two-step process, coupling two mass-selective devices capable of producing rapid and unambiguous protein identification. In electrospray ionization mass spectrometry (ESI-MS) the ions are formed at atmospheric pressure which makes this mode conducive to coupling with liquid chromatography and capillary electrophoresis. ESI-MS is effectively performed on-line with the HPLC separation by splitting the column effluent, allowing spontaneous mass measurement and fraction collection. Preferred conditions for ES ionization are similar to conditions used for RP-HPLC. It is therefore no surprise that ESI-MS is often used as a detector on HPLC systems and is referred to as LC-MS. First, an unseparated mixture of peptides is ionized by "electrospray ionization" directly from the liquid phase and is applied to a low-flow device called nanoelectrospray. The peptide mixture is then sprayed through a narrow orifice into the mass spectrometer where individual peptides are isolated. In the second step, an individual isolated peptide is dissociated into amino-or carboxy-terminal containing fragments, that when assembled should permit the elucidation of the amino acid sequence of the peptide (Wilm and Mann, 1996).

Tandem MS is currently the method of choice for sequencing peptides. Valuable information can be obtained by subjecting a precursor ion that has been isolated in an initial stage of MS to collision induced dissociation (CID) (also referred to as collision activated dissociation CAD) to a second stage of mass analysis on the product ion. CID is useful in the analysis of peptides because fragmentation frequently occurs at amide bonds, leading to a set of ions that is characteristic of the peptide's sequence. The data obtained by tandem mass spectrometry (MS/MS) generally comprise a partial sequence of a peptide that is referred to as a "peptide sequence tag". It usually contains enough information to identify the protein from databases. If a peptide match for a given sequence is not found it might mean that the peptide belongs to an unknown protein or that the peptide is modified.

MS/MS requires the detection of peptide ions followed by the isolation and fragmentation of a selected peptide by CID generating a MS/MS spectrum from which partial or complete peptide sequence information can be extracted. Protein identification based on CID spectra has been accelerated by the availability of sequence databases and the development of computer algorithms that correlate the information contained in CID spectra with sequence databases.

In CID, the breaks in the peptide chain commonly occur at the peptide bonds, along the peptide backbone as shown in Figure 8.3. The most common ion types are the b and the y ions, which denote fragmentation at the amide bond with charge retention on the N or C terminus, respectively. If the positive charge of the peptide ion remains on the N-terminus of the fragment ion, the ion is referred to as a b series ion, with a subscript reflecting the number of amino acid residues present on the fragment ion, counted from the N-terminus. Therefore, b ions represent the fragments with C-terminal deletions and an intact N-terminus. The b ion including only the first residue of the peptide is referred to as b_1. The remaining b ions are numbered progressively higher as one progresses toward the C-terminus. Y ions are numbered similarly but start at the C-terminus and progress towards the N-terminus. If the charge remains on the C-terminus, the ion is referred to as a y series ion (with the same subscript numbering as for the b series ion, but counting from the C-terminus. Therefore, y ions represent the fragments with an intact C-terminal (Tabb et al, 2000). On a spectrum, the b and y ions are not distinguishable a priori.

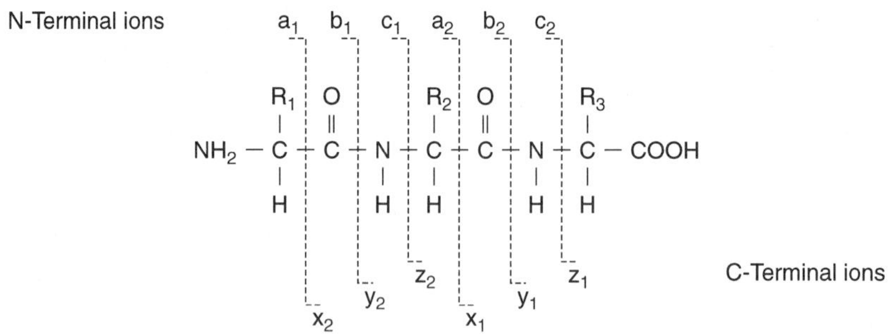

N-Terminal ions
a_1 b_1 c_1 a_2 b_2 c_2
R_1 O R_2 O R_3
NH_2 — C — C — N — C — C — N — C — COOH
H H H H H
z_2 z_1
y_2 y_1
x_2 x_1
C-Terminal ions

FIGURE 8.3 Fragmentation of the amide backbone. A peptide of three amino acids can fragment into a series of daughter ions. Fragmentation at the peptide bond will produce b series ions if the peptide positive charge remains at the peptide n-terminus, or y series ions if the charge remains at the peptide C-terminus. Only the charged portion of the peptide will be detected after fragmentation.

Most proteomics experiments are performed with tryptic peptides, which have arginyl or lysyl residues as their C-terminal residues. In this case, the formation of y ions is favored due to the basic amino acid at the C-terminus of the tryptic peptide (Mann et al, 2001).

MALDI and Peptide-Mass Mapping

Laser desorption ionization (LDI) of soluble compounds is based on air drying the solution containing the protein of interest on a metal target and then forming ions by using a UV laser pulse. The ions are detected by time-of-flight (TOF) mass analysis. For a variety of reasons this approach was not suited for peptide mass analysis. In matrix-assisted laser-desorption ionization time-of-flight (MALDI)-MS ions are formed from a solid state. The sample to be analyzed is deposited on a probe by co-crystalization with an excess of inert matrix which routinely consists of a weak aromatic acid that is readily soluble in solvents used regularly for polypeptides and strongly absorbs the light of the laser used for ionization. Small molecules like salts are excluded from the crystals. These crystals are introduced into the ionization chamber, which is under vacuum and then irradiated by short pulses of laser light focused on the sample probe resulting in the ionization of the solid matrix followed by an energy transfer to the polypeptide. The laser pulse triggers the clock used to measure the time of flight of the ions to the detector. Peptides generated by MALDI tend to have low charge states, i.e., most peptides that are ionized in positive ionization mode are observed as their singly protonated molecular ion making the data easier to interpret (Patterson and Aebersold, 1995). A portion of the tryptic peptide mixture is analyzed by MALDI mass spectrometry. The masses of tryptic peptides can be predicted *in silico* for any entry in a protein sequence database. The predicted peptide masses are then compared with the experimentally derived data obtained by MALDI analysis. The target protein can be identified if there are a sufficient number of matches with a protein in the database. MALDI identification by peptide-mass fingerprinting requires that the full-length gene be present in the database (Pandey and Mann, 2000).

Two characteristics of MALDI make it well suited for the analysis of digest mixtures. MALDI is tolerant of low levels of contaminants such as buffer salts and stain residue, and discrimination between components of mixtures is relatively low.

Post-Source Decay (PSD) MALDI-MS

To improve the confidence in the results of a peptide mass search, additional primary structural information about the peptides is often necessary. The peptide may be isolated in the gas phase of the mass spectrometer allowing the masses of the daughter ions to be measured. MALDI-MS lacks a designated collision cell for the generation of fragment ions. However, the intact molecular ions generated by MALDI ionization undergo significant "metastable fragmentation" termed post-source decay (PSD). The term "PSD" was introduced to summarize various mechanisms that together lead to fragmentation events (Spengler, 2001). PSD ion fragmentation is the result of internal energy gathered by electronic, vibrational excitation of the molecular ions during the desorption or ionization process. The fragmentation is often sequence specific generating the same types of ions observed with ESI-MS/MS. Therefore, PSD spectra have been used to determine peptide sequences.

Spectral interpretation for peptide analysis is based on fundamental investigations of fragmentation behavior and on accumulated PSD data from known peptide sequences. Software tools are available to aid in the interpretation of PSD spectra.

One of the drawbacks of peptide sequencing by MALDI-PSD is the relatively complex interpretation of fragment ion spectra.

Protein Identification by MS

The idea behind a protein identification strategy is to progress from an automated situation to more labor-intensive, time consuming analyses. First, SEQUEST is used to automatically search databases. SEQUEST searches the database for candidate peptides on the basis of mass alone. It creates a virtual spectrum for each of these peptides and checks to see how well it matches the observed spectrum. A new software package DTASelect assembles SEQUEST identifications and highlights the most significant matches. Developed to assemble and evaluate shotgun proteomics data, the DTASelect package applies advanced filtering to SEQUEST search results (Tabb et al, 2002). If results are inconclusive, sequence tags can be manually generated and used with various database-searching programs such as: http://www.matrixscience.com for links to the mascot program (Mann and Wilm, 1994).

Peptide Mass Fingerprint Analysis

The first phase of a spectrometric identification of an unknown protein is usually based on the masses of proteolytically generated fragments. Mass spectrometry can identify a protein by analyzing the masses of the peptides derived from proteolytic or chemical cleavage of the protein. Peptide mass fingerprinting takes advantage of reproducible peptide fragmentation patterns without explicitly interpreting the complete amino acid sequence. The collective peptide mass information obtained by MS analysis of a protein digest provides a unique fingerprint for that protein. Identification is based on comparisons of MS determined peptide molecular masses of the target protein with a theoretical peptide mass database. Protein databases and DNA sequence databases translated into protein sequences are assembled into a nonredundant database. Peptide fragment masses (using specific enzyme cleavage rules) are computed from this database to generate a peptide mass database that is queried with the experimentally derived masses. A ranked output is generated based on matches. The highest ranked is most likely the identity of the target protein. This strategy will only be successful if the target protein is in the database. When high mass accuracy is achieved (10 to 50 ppm), as a rule, at least five peptide masses need to be matched to the protein and 15% of the protein sequence needs to be covered for an unambiguous identification. Proteins from organisms with fully sequenced genomes can be identified with a 50–90% success rate when at least a few hundred femtomoles of gel-separated protein are present (Mann et al, 2001).

The following sites should aid in the analysis. They all supply instructions on-line how to use them.

1. PepSea: http://pepsea.protana.com/PA_PepSeaForm.html (Mann et al, 1993)
2. PeptIdent/MultiIdent: http://expasy.ch/tools/peptident.html (Wilkins et al, 1999)
3. MassSearch from CBRG, ETHZ, Zurich: http://cbrg.inf.ethz.ch/Server/MassSearch.html
4. MOWSE search program from SEQNET, Daresbury: http://www.dl.ac.uk/SEQNET/mowse.html. (Pappin et al, 1993)
5. ProFound (and PepFrag) search program from Rockefeller University, New York: http://chait-sgi.rockefeller.edu/cgi-bin/prot-id/1/1. (Zhang and Chait, 2000)
6. MS-Fit (and MS-Tag) http://rafael.ucsf.edu/MS-Fit.html.
7. The Darwin suite of programs at http://cbrg.inf.ethz.ch/ (James et al, 1994)

Protein identification via accurate mass measurements of enzymatically derived peptides is thus a comparison of experimental data with data deposited in sequence databases. Peptide mass fingerprinting does not work very well on complex mixtures of proteins. This makes gel electrophoresis a good choice for sample isolation.

Peptide Fragmentation

In contrast to MS analysis of peptide maps, as performed for fingerprint analysis, peptide fragmentation mass spectra yield information from individual peptides. The information present in a single peptide is sufficient to enable the identification of a protein. The use of peptide fragmentation MS is the method of choice for identifying proteins from complex mixtures. MS/MS data accumulation and database searching has become fully automated and is almost routine. The distribution of charge plays an important role in determining the relative amounts of b- and y- type ions that are detected.

Proton transfer to the residue with the highest basicity can occur, which explains the relative dominance of y-type ions in tryptic peptides that usually have Arg or Lys as the C-terminal residue.

The presence of a fixed charge at each terminus gives a dominant y-type series and a weak b-type series.

In practice, not all fragment ions are present at detectable levels and fragments can also arise by double fragmentation of the backbone. Therefore, it is often possible to interpret only part of the sequence with confidence.

Peptide Ladder Sequencing

MALDI-MS is used in protein ladder sequencing for the analysis of peptides (Bergman, 2000). The approach uses classical Edman chemistry to release one amino acid per cycle. MALDI-MS analyzes the remaining peptide rather than the released amino acid. The amino acid sequence is deduced from the mass differences between consecutive fragments. An example is shown in Figure 8.4. Subsequent advances showed that ladders could be generated by incomplete Edman degradation, relying on software to accurately analyze the products. Peptides can be fragmented within mass spectrometers using a variety of techniques. In all methods, the aim is to produce spectra containing a series of peptide peaks differing by the mass of single amino acid residues, thereby allowing stretches of peptide sequence to be elucidated.

Peptide Sequence Tag

The peptide sequence tag method takes advantage of the fact that nearly every tandem mass spectrum contains at least a short run of fragment ions that unambiguously specifies a short amino acid sequence. This partial sequence divides the peptide into three parts; region 1, the partial amino acid sequence of region 2, and the added mass of region 3 (Mann and Wilm, 1994). Together, these three regions constitute the peptide sequence tag, which along with the mass of the parent peptide constitute a highly specific identifier of the peptide (Fenyö, 2000). Peptide sequence tags can be used to query MS-Tag at: http://prospector.ucsf.edu/

Identification of a Gene Product

One peptide is often sufficient for the identification of the gene encoding the investigated protein. A stretch of seven amino acids is fre-

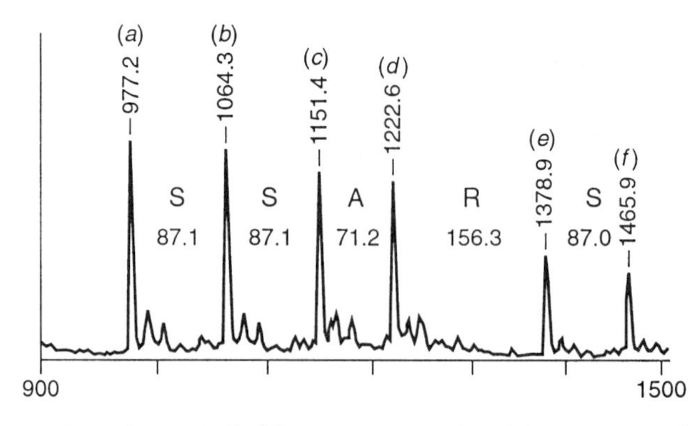

FIGURE 8.4 Partial protein ladder-sequencing data. The amino acid identities are determined by measuring the mass differences between consecutive peaks (e.g., [a] and [b], [b] and [c] etc.) and comparing these mass differences with the amino acid residue masses. The amino acid sequence is determined by the order of residues from the high-mass to low-mass end of the spectrum (e.g., . . . SRASS). (From Wang and Chait.)

quently unique to a specific protein in the human genome. A peptide that is too short will not be as useful in searches of large databases since too many proteins will have the same or similar sequence (Courchesne and Patterson, 1999). Several peptides from a target protein will identify the gene with relative certainty. Still, it is not necessarily known which alternatively spliced form or which modification is present. So for accuracy sake, one can state that the product of a certain gene has been identified rather than a certain protein (Rappsilber and Mann, 2002).

Selective tagging effectively reduces the sample complexity, greatly increasing the representation of selected peptides. The labeling of proteins at specific sites followed by proteolysis and selective purification of the labeled peptide fragments has proven to be an effective method for identifying low abundance proteins. The thiol group of cysteines is an attractive target for labeling because of its unique chemistry. Also, the frequency of cysteine residues in proteins (~10% of all tryptic peptides in the yeast *S. cerevisiae* contain cysteine) (Fenyo et al, 1998) make it a good candidate for tagging. Selective labeling and capture of cysteine containing peptides formed the basis of the isotope-coded affinity tag (ICAT) method (Gygi et al, 1999). In another approach cysteine side chains were labeled using a cleavable, biotinylated reagent. Peptides were then purified using immobilized avidin and identified by liquid chromatography coupled to mass spectrometry (Spahr et al, 2000).

Posttranslational Modifications and MS

Posttranslational modifications (PTMs) are covalent processing events that change the properties of the protein by proteolytic cleavage or by the addition of a modifying group to one or more amino acids (Mann and Jensen, 2003). Protein modifications do not present significant obstacles to identifying the target. A typical 50kDa protein that is digested with trypsin will produce about 50 peptides. However, only a few of them will contain modified amino acids. Since only a small number of peptides are required for unique matching to a database entry, extensive protein modification will only marginally increase the difficulty of protein identification (Mann et al, 2001).

Posttranslational modifications themselves remain difficult to analyze. All modifications cause a shift in mass and in principle, are detectable by MS. The mass of the modified peptide is usually not sufficient to determine the nature of the modification. Peptides are fragmented by MS to identify the peptide and localize its modification. In "tandem mass spectrometry" (MS/MS) experiments, peptide ions are collided with an inert gas, leading to fragmentation, which usually occurs at the peptide bonds. Some modified amino acids remain intact during this process. In this case, the fragmentation pattern is similar to the unmodified peptide with the difference being that the location of the modified amino acid is revealed by its mass increment. Often times, the modification is labile and will be lost before the peptide itself fragments. In this case the peptide can still be sequenced and identi-

TABLE 8.3 Some common and important posttranslational modifications

PTM type	ΔMass[a] (Da)	Stability[b]
Phosphorylation, pTyr	+80	+++
Phosphorylation, pSer, pThr	+80	+/++
Acetylation	+42	+++
Methylation	+14	+++
Farnesylation	+204	+++
Myristoylation	+210	+++
Palmitoylation	+238	+/++
Glycosylation N-linked	>800	+/++
Glycosylation O-linked	203, >800	+/++
GPI anchor	>1,000	++
Hydroxyproline	+16	+++
Sulfation (sTyr)	+80	+
Disulfide bond formation	−2	++
Deamidation	+1	+++
Pyroglutamic acid	−17	+++
Ubiquitination	>1,000	+/++
Nitration of tyrosine	+45	+/++

[a]A more comprehensive list of PTM Δmass values can be found at: http://www.abrf.org/index.cfm/dm.home.
[b]Stability: + labile in tandem mass spectrometry, ++ moderately stable; +++ stable.
Table modified from Mann and Jensen, 2003.

fied, but only the mass increment, not the location of the modification can be determined. A list of posttranslational modifications along with their relative stabilities and masses is presented in Table 8.3 (Mann and Jensen, 2003).

Peptides are identified in databases by their tandem mass spectra, using pattern matching with a variety of different algorithms such as SEQUEST or Mascot (Perkins et al, 1999). These algorithms have been extended to examine modified peptides. An error tolerant mode for database matching of uninterpreted tandem mass spectrometry data has been described (Creasey and Cottrell, 2002). If the peptide sequence is not in the database it could mean that the exact sequence is not in the database but a closely related sequence is. Using Mascot software found at www.matrixscience.com, uninterpreted spectra can be tested serially for a wide variety of posttranslational modifications. For example, tyrosine can be phosphorylated, increasing the molecular weight by 80 Da. Therefore, the peptides that contain a phosphorylated tyrosine will reflect the increased molecular weight. For each posttranslational modification, if there are multiple modifiable residues in a peptide, all permutations will be tested.

The N terminus of a protein is indicated by an acetylated peptide. Therefore, the N-terminal amino acid in the N-terminal peptide is 42 Da greater than expected. Similarly, myristoylation causes an increase of 210 Da, and GalNac or GlcNAc sugars increase the mass of peptides with these modifications by 203 Da (Mann and Pandey, 2001).

Phosphopeptides are generally difficult to analyze by MS for several reasons. They are negatively charged whereas electrospray is generally performed in the positive mode. Being hydrophilic, they do not bind well to columns that have been routinely used for peptide purification prior to MS analysis. Phosphopeptides are not observed as intense peaks, especially in the presence of other nonphosphorylated peptides due to ionic suppression. The presence of isobaric peptides in the sample can also complicate the analysis. If the protease produces peptide fragments that are too large or too small, the peptides might not be observed in the mass spectrum at all. Phosphoserine and phosphothreonine residues are labile whereas phosphotyrosine residues are relatively more stable.

MALDI-TOF MS is especially useful if performed on peptide mixtures that have first been purified on IMAC columns to enrich for phosphorylated peptides (Mann et al, 2002). MALDI-TOF MS has been used in combination with phosphatase treatment to specifically identify phosphopeptides that are identified based on a characteristic mass shift due to the loss of phosphate (80 Da or multiples).

Caveats When Using MS

In a proteomics project aimed at identifying an unknown target protein it sometimes happens that the MS data cannot be correlated with any sequence in the searched database. Some possible reasons for this are:

- Absence of the sequence in the database searched
- The presence of unexpected co-or posttranslational modifications
- Artifacts of sample handling
- The peptide is produced by an unexpected or nonspecific cleavage
- Technical difficulties- Either the quality of the spectrum is poor or the spectrum is that of a non-peptide contaminant. Frequently, the quality of the spectrum can be improved by purifying and concentrating the peptide samples prior to MS analysis.

There are also some potential pitfalls in the interpretation of MS. Various hexoses, glucose, mannose and galactose all have the same monoisotopic residue mass (162.058 Da). There is close mass similarity between sulfate 79.957 Da and phosphate 79.663 Da and sugars with quite different functions can have identical mass (e.g. the N-acetylhexosamines GalNAc and GlcNAc 203.079 Da).

The target protein can undergo covalent modifications that can alter the molecular weight. If the protein is digested with CNBr which cleaves at methionine residues, either homoserine (Hse) or Hse lactone will be produced. Therefore, to take this into account, the search should be modified by –30 for Hse and –48 for Hse lactone (Washburn et al, 2001).

SDS can only be tolerated in trace amounts. Other problems can arise from acrylamide monomer and many of the common protein stains (Cottrell and Sutton, 1996).

Integral membrane proteins are more difficult to identify than cytoplasmic proteins. Since lysines and arginines are infrequently found in membrane-spanning regions, complete tryptic cleavage will generate large hydrophobic fragments, due to cleavage at the borders of membrane-spanning regions which are difficult to solubilize. Peptides detected and identified from integral membrane proteins rarely contained any portion of a predicted transmembrane domain (Washburn et al, 2001).

Protein Database Searches

When a partial or complete amino acid sequence of the target protein is known, computer-assisted searches should be performed. Compare the newly obtained sequence data to a database of known sequences. This will reveal whether there is a protein of known structure or a homologous sequence that is sufficiently similar to the sequence of the target protein to suggest a familial relationship and a possible function. There are numerous examples of important discoveries resulting from database searches, for example, the similarity between platelet-derived growth factor and the v-sis oncogene product (Doolittle et al, 1983). A great deal of biology is learned from these searches, which should be performed routinely to create or testhypotheses about the function of a protein or the membership of a newly described protein in a family.

There are two mathematical aspects to database searches: the algorithm used to find sequence similarities, and the method used to determine which similarities are statistically significant and therefore, potentially interesting. For a mathematical description of database searches, see Waterman and Vingron, (1994). Two popular algorithms used for rapid searches of databases are FASTA (Wilbur and Lipman, 1983; Lipman and Pearson, 1985) and BLAST (Altschul et al, 1990) which look for intervals or segments of good matching between sequences.

If database screening based on overall sequence similarity does not identify the target protein as a family member, a short sequence of the target protein may be recognized as a commonly occurring structural motif, which may be helpful assigning a function.

Bioinformatics

Bioinformatics is an interdisciplinary research area bringing together the biological sciences with computational technology. It involves the solution of complex biological problems using computational tools and systems. It also includes the collection, organization, storage and retrieval of biological information from databases. Bioinformatics can be used to analyze the information content of biological sequence data, and subsequently process and use this information to predict the function of the corresponding molecules (Ibba, 2002).

Bioinformatics can be divided into three sub-disciplines:

- Development of new algorithms and statistics to probe the relationships among the members of the data sets
- Analysis and interpretation of data from nucleotide and amino acid sequences, protein domains and protein structures
- Development of new tools that will enable efficient access and management of different types of information. (www.ncbi.nlm.gov/Education/)

The Rise of Biological Databases

The term, "*in silico* science" refers to activities that were once carried out *in vitro* or *in vivo* that are now performed with a computer. The genomics revolution, fueled by the race to sequence the human genome and the rise of the Internet, has changed the way that researchers perform scientific investigation. Today, the latest research is just a click away. Online biological databases collect, organize and disseminate the latest research at no cost!

The thriving popularity of databases has demonstrated, perhaps to the surprise of the scientific community on the cutting edge, that the free flow of shared information which all of the databases encourage, benefits science as a whole and is a demonstration of the community spirit that still dominates scientific research (Wilson, 2002).

One major objective of this section is to describe resources that are available free of charge through the World Wide Web (WWW) a graphical interface based on hypertext through which both text and graphics can be displayed and highlighted. Each highlighted element is a link to another document, which can reside on any Internet host computer.

Biological databases are assembled by different teams in different geographical locations. They are usually compiled for different purposes and tend to focus on different subjects. Connecting the data they contain is not straightforward. One approach to database integration is a browser like "Entrez" (Schuler et al, 1996) which can maneuver easily within a series of linked databases. Other popular browsers are the Integrated Genomic Database (IGD) (Ritter et al, 1994), the ExPASy WWW server and the Sequence Retrieval System (SRS).

An up-to-date listing of addresses of databases along with brief summaries and updates for each database is included in the molecular biology database collection which can be accessed through the Nucleic Acids Research Web site at: http://nar.oupjournals.org.

The following resources and others like them are becoming increasingly important and useful research tools. They are worth spending the time to learn to use effectively and more than justify infrastructure investments to provide adequate Internet connections in the laboratory.

Search Engines

Web search engines are software tools that can locate information on the WWW. Search engines like the popular www.yahoo.com/, for example, are widely utilized by the majority of internet users. Search engines have a large number of documents available which may be queried by keywords. The drawback to the all-purpose search engines is that queries will often return too many unwanted results.

A solution to the problem is to access a dedicated search engine, which has a limited field, thereby limiting searches to potentially interesting areas. An example that provides links to documents and databases related to electrophoresis in general and 2-D in particular is: www.expasy.ch/ch2d/2DHunt/

Entrez
http://www.ncbi.nlm.nih.gov/Entrez

Entrez, a search and retrieval system produced by the NCBI, a division of the NIH, is a program that integrates DNA and protein sequence data with its associated abstracts and citations contained in the MEDLINE database. Using ENTREZ one can search databases by bibliographic associations (e.g. author name, gene family, or map location) as well as by sequence homology.

Entrez allows the user to access three databases: the National Library of Medicine Medline database, the NCBI protein database and the NCBI nucleotide database (GenBank). The Medline database is compiled from journal publications. The genetic sequences in this database are linked to the abstract of the journal article that first reported them. The protein and nucleotide entries in Entrez come from several databases: GenBank, EMBL, DDBJ, Protein Information Resource (PIR), SWISS-PROT, Protein Research Foundation (PRF) and Brookhaven Protein Data Bank (PDB). Their databases available for searching through the NCBI site include: Database of Expressed Sequence Tags (dbEST), Database of Sequence Tagged Sites (dbSTS) and the Molecular Modeling Database (MMDB).

Databases

Collections of protein sequences date back to the 1960s, preceding GenBank by almost 20 years (Dayhoff and Eck, 1966). The utilitarian aims of protein databases are minimal redundancy, maximal annotation and integration with other databases. The most comprehensive source of information about proteins is found in protein sequence databases, of which there are two types, universal and specialized. Universal databases store information on all types of proteins from all organisms. They can be simply repositories of sequence data or they can be annotated with relevant references and commentary such as the function of the protein, posttranslational modifications, and variants. The problem that investigators encounter with the larger databases is that they do not often contain specialized information that would be of interest to specific groups. To fill the void, specialized databases have been developed. Data in smaller databases tend to be curated by experts in a particular field, and are often experimentally verified and cater to a specific group of proteins or proteins produced by a specific organism. In addition to sequence information, they may provide data such as phenotypes, experimental conditions and in general information that may not be strictly sequence based.

In order to query a database the researcher can either type (or cut and paste) the sequence data (either nucleotide single letter or protein single letter code) directly into electronic fill-in-the-blank forms in the browser window and then click on a submit button. On-site instructions and documentation make this a user friendly experience.

Presently there are three major publicly accessible sequence database facilities. The National Center for Biotechnology information (NCBI) maintains the sequence database GenBank, which consists of nucleotide and protein sequences including those from EBI and DDBJ (see below). All three data banks exchange data nightly and therefore contain the same data.

National Center for Biotechnology Information (NCBI)
http://www.ncbi.nlm.nih.gov

The NCBI site contains several major resources, the most well known of these is probably GenBank, described below. Other information, such as links to literature databases (PubMed) and tools for data mining, (such as BLAST) can be accessed through this site.

PubMed is a service of the National Library of Medicine, which provides access to over 12 million MEDLINE citations dating back to the mid-1960's and additional life science journals. PubMed includes links to many sites providing full text articles and other related resources.

GenBank
www.ncbi.nlm.nih.gov/GenBank/

GenBank is a comprehensive annotated sequence database that contains sequences and related information that have been reported in journals, patents or directly submitted. It is managed by the National Institute of Health (NIH) and can be searched via several Internet sites free of charge. A GenBank report will include the nucleotide or protein sequence along with other information. Especially important is the accession number, unique to a particular protein. If available, searching by accession number is the quickest method to retrieve a sequence. Some additional information includes the source (the organism), bibliographic references, and the date and person responsible for the submission.

DNA Data Bank of Japan (DDBJ)
http://www.ddbj.ni.ac.jp/

The DNA data bank of Japan (DDBJ) is the sole DNA data bank in Japan and also contains sequences from GenBank and the EMBL Data Library. It provides similar tools for data retrieval as EBI and NCBI.

European Bioinformatics Institute (EBI)
http://www.ebi.ac.uk/

EBI, an EMBL subsidiary, is a comprehensive database of DNA sequences extracted from the scientific literature, patent applications, or through direct submission from researchers worldwide. The EMBL Nucleotide Sequence Database is maintained at the European Bioinformatics Institute (EBI), an EMBL outstation. In addition to sequences from GenBank, DDBJ, translated protein sequences from EMBL (TrEMBL) and SWISS-PROT, it also contains additional information, Macromolecular Structure Database (MSD), and links to protein domain and motif database tools (InterPro), database searching (SRS) and analysis tools such as ClustalW.

SWISS-PROT
http://ca.expasy.org/sprot/

The SWISS-PROT protein sequence database and its supplement TrEMBL (Bairoch and Apweiler, 1997) is an annotated, cross-referenced, protein sequence database that contains protein sequence data from various sources including translation of entries from the EMBL/GenBank/DDBJ database, from the Protein Identification Resource (PIR) and from the literature. The high level of annotation provides information on the function of a protein, its domain structure, post-translational modifications and variants. A SWISS-PROT entry is composed of line-by-line out-put where each line begins with a two-character code indicating what type of data is contained on the line. The first line of each SWISS-PROT entry is the ID, identification line. This contains the entry name. More useful is the second line, the AC, accession number line. This line contains the accession number of the protein, which can be used to probe other databases. A detailed description of the format and an in-depth line-by-line explanation is provided (Bairoch, 1997).

TRanslation of EMBL nucleotide sequence database: (TrEMBL)

TrEMBL is a protein sequence database that supplements SWISS-PROT. Its format is similar to SWISS-PROT and consists of the translation of coding sequences (CDS) in the EMBL nucleotide sequence database and is considered preliminary before being manually annotated and moved to the main part of the SWISS-PROT database.

TIGR: The Institute for Genomic Research
http://www.tigr.org

TIGR Database (TDB) is a collection of curated databases containing DNA and protein sequence, gene expression, cellular role, protein family and taxonomic data for microbes, plants and humans.

Identifying a Target Protein

TagIdent
http://us.expasy.org/tools/tagident.html

TagIdent is a tool that allows the generation of a list of proteins close to a given pI and Mw. Query proteins can be identified by matching a short sequence tag of up to 6 amino acids against proteins in the Swiss-Prot/TrEMBL databases. The search can also be limited to a specific organism. TagIdent removes signal sequences and /or propeptides before computing pI and Mw (Wilkins et al, 1998).

Gene Analysis Tools

BLAST
http://www.ncbi.nlm.nih.gov/BLAST/

The Basic Local Alignment Search Tool (BLAST) is a family of sequence alignment algorithms that allows the user to search protein or DNA databases regardless of whether the query sequence is amino acid or nucleic acid (Henikoff et al, 2000). Using BLAST one can discern whether a given sequence is novel or homologous to a given sequence. BLAST searches can be performed at a basic or advanced level. There are five different BLAST programs available, allowing everything from a simple sequence homology search to complex comparisons of the six frame translations of a nucleotide query sequence against the six-frame translations of a nucleotide sequence database.

Blastp compares an amino acid query sequence against a protein sequence database.
Blastn compares a nucleotide query sequence against a nucleotide sequence database.
Blastx compares a nucleotide query sequence translated in all reading frames against a protein sequence database.
Tblastn compares a protein query sequence against a nucleotide sequence database dynamically translated in all six reading frames.
Tblastx compares the six-frame translations of a nucleotide query sequence against the six-frame translations of a nucleotide sequence database.

A site for BLAST tutorials can be located at:

NCBI BLAST Tutorial
www.rickhershberger.com/darwin2000/blast

Ensembl
http://www.ensembl.org

The Ensembl program is a project launched jointly by EMBL-EBI and the Sanger Center. It assembles DNA fragments, predicts genes, and identifies single nucleotide polymorphisms, repeat regions and regions homologous to other sequences in public databases.

Open Reading Frame (ORF) Finder
http://www.ncbi.nlm.nih.gov/gorf/gorf.html

An Open Reading Frame may be defined as the region between two in-frame stop codons that may contain a putative start codon regardless of whether there is biological proof that the putative encoded protein even exists. The ORF Finder tool identifies all possible ORFs in a DNA sequence by locating the standard and alternative stop codons in all six reading frames of the corresponding translated amino acid sequence.

Subcellular Localization

http://psort.nibb.ac.jp/

PSORT is a program for predicting the subcellular localization of proteins in bacteria. This program predicts the location of a protein into one of four subcellular regions, cytoplasm, inner membrane, periplasm, or outer membrane.

ProtLock is a useful software program for localizing the target gene product. It predicts the localization of the protein in one of five subcellular locations, integral membrane, anchored membrane, extracellular, intracellular and nuclear (Cedano et al, 1997).

Protein Domain Families

It has become increasingly apparent that protein domains are the basic units of protein structure and function. Detecting the presence of a particular cluster of residues in a sequence may reveal relationships between proteins. This pattern has been referred to as a motif, signature, or domain.

PROSITE
http://us.expasy.org/prosite/

PROSITE (Bairoch et al, 1997) is a database that describes protein families and domains which are detected using specific amino acid sequence patterns. Each domain in PROSITE is defined as a simple pattern. A list of proteins known to contain that domain constitutes a protein family.

Pfam
http://www.sanger.ac.uk/Software/Pfam/
In the US at http://pfam.wustl.edu/

Pfam is a human-curated collection of protein sequences that specializes in classifying protein sequences into families and domains (Bateman et al, 2002). The latest version (6.6) contains 3071 families, which match 69% of proteins in SWISS-PROT 39 and TrEMBL14. The database is divided into two parts: Pfam-A and Pfam-B. Pfam-A is the curated part that contains 3,300 protein families. Pfam-B makes the database more comprehensive by listing a large number of lower quality, small protein families that do not overlap with Pfam-A.

Researchers can access Pfam to look at multiple alignments, view protein domain architectures, view known protein structures, and follow links to other databases. Pfam can alert the investigator that a particular protein is similar to a large number of other proteins that may have already been extensively characterized.

The Protein Information Resource (PIR)
http://www-nbrf.georgetown.edu/pir/

The Protein Information Resource is a site for the identification and analysis of protein sequences. It also maintains the PIR International Protein Sequence Database.

Protein composition and secondary structure
http://bioinformatics.weizmann.ac.il/hydroph/.

If the target protein does not have a known 3-D structure and the amino acid sequence is known, then some basic biochemical properties can be determined from the amino acids. Molecular weight and isoelectric point of the unmodified protein can be calculated easily from the primary sequence. Hydrophobicity, which is a property of groups of amino acids, best viewed in a hydrophobicity plot, will indicate regions that are hydrophilic and also identify membrane spanning regions.

Structural Classification of Proteins

SCOP
http://scop.mrc-lmb.cam.ac.uk/scop/

SCOP is a database of structural domains arranged in a hierarchical manner according to structure and evolutionary relatedness.

SMART
http://smart.embl-heidelberg.de

The SMART database (Shultz et al, 2000), which is extensively annotated, focuses on classifying protein families into domains. It is also used for comparative protein domain architecture analysis scanning for domain composition and features such as transmembrane domains. Currently, the database includes information on more than 500 domains encompassing more than 54,000 proteins. Details such as functional classes, tertiary structures and functionally important residues are annotated.

Protein Explorer (PE) for macromolecular visualization
www.proteinexplorer.org

Protein Explorer (Martz, 2002) is geared for non-specialists to easily visualize macromolecular structures. It also offers advanced capabilities for specialists. Great attention has been paid to make this easy to use, with extensive built-in instructions. It's free of charge, and has an extensive tutorial. The first image of a molecule shown in PE is designed to be maximally informative and is presented on the First View page.

Motif
http://motif.genome.ad.jp

Motif provides access for researchers to analyze protein structural motifs in sequence databases. Users can specify various structural parameters as defined by other protein-structure databases, including Pfam, BLOCKS, Prints, and ProSite. Searches are easy to perform, just paste a protein sequence into a window and specify a few search parameters.

Atomic Coordinates
www.rcsb.org

The knowledge of a protein's atomic coordinates will allow one to view the protein in 3-D interactively. This can be performed with Protein Explorer. The Brookhaven Protein Data Bank is the primary repository of experimentally determined atomic coordinates of proteins. Each coordinate set has a unique identification code that can be retrieved together with the coordinates and information about the protein such as structure, including interactive images, by searching for the protein's name or publication details (Berman et al., 2000).

DeepView
www.expasy.ch/spdbv

DeepView (also known as SwissPDBViewer) is a program designed to examine modeling, changing of protein conformation, mutation of residues, and homology modeling.

Genomic Organization

GENSCAN
http://genes.mit.edu/GENSCAN.html

The GENSCAN program predicts the exon/intron boundaries of a sequence as well as the 5′ and 3′UTR sequences and poly-A tail sequences.

Additional Useful Sites on the Internet

www.public.iastate.edu/~pedro/research_tools.html
Pedro's Biomolecular Research Tools site is a collection of hundreds of sites of interest to molecular biologists and a good place to start a search.

www.protocol-online.net
A collection of useful laboratory protocols

htwww.ceolas.org/VL/fly/protocols.html
The Drosophila protocols page of the WWW Virtual Library

www.spotfire.com
This site contains links to protocols from a variety of labs. There are also many resources and applications for microarray data analysis

tp://123genomics.com/
This site is a collection of well organized links which guides the user from basic tutorial information about biotechnology through databases and sequence analysis and much more.

www.genebrowser.com
A site for biotechnology.

www.sequenceanalysis.com
As the name suggests, this site is a guide for molecular sequence analysis.

PCR Primer Design Programs

There are many different software packages that can be used to design PCR primers. They all provide essentially the same set of features and use similar algorithms. Three free programs are listed below.

www.genomeweb.com
PCR primer selection

http://www-genome.wi.mit.edu/cgi-bin/primer/primer3_www.cgi
Primer 3 at MIT is a powerful PCR primer design program permitting one considerable control over the nature of the primers, including size of product desired, primer size and Tm range, and presence/absence of a 3′-GC clamp.

http://alces.med.umn.edu/rawprimer.html
PCR primer selection program at U. Minnesota

www.microbiology.adelaide.edu.au/learn/oligcalc.htm
Oligo Calculator (Adelaide, AU)

www.biochem.ucl.ac.uk/bsm/sidechains/index.html#
According to the description at the site, the Atlas of Protein Side-Chain Interactions "depicts how amino acid side-chains pack against one another within the known protein structures".

Microarrays and Data Mining—the Challenge of Data Analysis

http://cmgm.stanford.edu/pbrown/mguide/index.html
Microarrays yield expression levels of thousands of genes; in some experimental systems it can yield the complete set of recognized genes in a sequenced genome. These expression levels are simultaneously measured in a single hybridization experiment. General information about making and using microarrays is available from several sources and Web sites. Patrick Brown's lab page includes a microarrayer guide.

www.unil.ch/ibpv/microarrays.htm
Information about making microarrays, from technical "how to" details to links to information about equipment is provided at this site.

www.uib.no.aasland
This site provides an overview of bioinformatics and contains a well-organized, in-depth collection of links to online DNA and protein analysis tools.

http://members.aol.com/johnp71/javastat.html
A site that will perform statistical calculations

www.jeanweber.com/index.htm
A site that will assist in editing and writing

www.allseachengines.com
This site contains a general collection of databases. Here one can find search engines for education, infringed patents, books and much more.

http://images.google.com
This site provides useful illustrations and figures which can be useful when preparing a lecture or presentation. Copyright is a consideration so permission is necessary.

References

Aitken A (1990): *Identification of Protein Consensus Sequences.* Chichester, UK: Ellis Horwood

Altschul SF, Gish W, Miller W, Myers EW, Lipman DJ (1990): Basic local alignment search tool. *J Mol Biol* 215:403–410

Bairoch A (1997): Proteome databases. In: *Proteome Research: New Frontiers in Functional Genomics*, Wilkins MR, Williams KL, Appel RD, Hochstrasser DF, eds. Berlin: Springer-Verlag

Bairoch A, Apweiler R (1997): The SWISS-PROT protein sequence data bank and its supplement TrEMBL. *Nucleic Acids Res* 25:31–36

Bairoch A, Bucher P, Hofmann K (1997): The PROSITE database, its status in 1997. *Nucleic Acids Res* 25:217–221

Barrett AJ et al ed. (1998): *Handbook of Proteolytic Enzymes CD-ROM*. San Diego, CA: Academic Press

Bateman A, Birney E, Cerruti L, Durbin R, Etwiller L, Eddy SR, Griffiths-Jones S, Howe KL, Marshall M, Sonhammer EL (2000): The Pfam protein families database. *Nucleic Acids Res* 30:276–280

Beavis RC, Fenyo D (2000): In: *Proteomics: A Trends Guide*, Mann M, Blackenstock W, eds. London: Elsevier, pp. 12–17

Bergman HM, Westbrook J, Feng Z, Gilliland G, Bhat TN, Weissig H, Shindyalov IN, Bourne PE (2000): The protein data bank. *Nucleic Acids Rec* 28:235–242

Bergman T (2000): Ladder sequencing. *Experientia Supplementa* 88:133–144

Berndt P, Hobohm U, Langen H (1999): Reliable automatic protein identification from matrix-assisted laser desorption/ionization mass spectrometric peptide fingerprints. *Electrophoresis* 20:3521–3526

Boyle WJ, Van der Geer P, Hunter T (1991): Phosphopeptide mapping and phosphoamino acid analysis by two-dimensional separation on thin-layer cellulose plates. *Methods Enzymol* 201:110–111

Cedano J, Aloy P, Perez-Pons JA, Querol E (1997): Relation between amino acid composition and cellular location of proteins. *J Mol Biol* 266:594–600

Celis JE, Rasmussen HH, Leffers H, Madsen P, Honore B, Gesser B, Dejgaard K, Vandekerckhove J (1991): Human cellular protein patterns and their link to genome DNA sequence data: Usefulness of two-dimensional gel electrophoresis and microsequencing. *FASEB J* 5:2200–2208

Choli T, Kapp U, Wittmann-Liebold B (1989): Blotting of proteins onto immobilon membranes: In situ characterization and comparison with high performance liquid chromatography. *J Chromatogr* 476:59–72

Cleveland DW, Fischer SG, Kirschner MW, Laemmli UK (1977): Peptide mapping by limited proteolysis in SDS by gel electrophoresis. *J Biol Chem* 252:1102–1106

Cohen SL, Chait BT (1997): Mass spectrometry of whole proteins eluted from SDS-PAGE gels. *Anal Biochem* 247:257–267

Cottrell JS, Sutton CW (1996): The identification of electrophoretically separated proteins by peptide mass fingerprinting. In: *Methods in Molecular Biology Vol. 61: Protein and Peptide Analysis by Mass Spectrometry*, Chapman JR, ed. Totowa, NJ: Humana Press Inc.

Courchesne PL, Patterson SD (1999): Identification of proteins by matrix-assisted laser desorption/ionization mass spectrometry using peptide and fragment ion masses. *Methods Mol Biol* 112:487–511

Creasy DM, Cottrell, JS (2002): Error tolerant searching of uninterpreted tandem mass spectrometry data. *Proteomics* 2:1426–1434

Crimmins DL, McCourt DW, Thoma TS, Scott MG, Macke K, Schwartz BD (1990): In situ chemical cleavage of proteins immobilized to glass-fiber and polyvinylidenedifluoride membranes: Cleavage at tryptophan residues with 2-(2′-nitrophenylsulfenyl)-3-methyl-3′-bromoindolenine to obtain internal amino acid sequence. *Anal Biochem* 187:27–38

Dayhoff MO, Eck RV (1966): *Atlas of Protein Sequence and Structure.* Silver Spring, MD: National Biomedical Research Foundation

Degani Y, Patchornik A (1971): Selective cyanylation of sulfhydryl groups. II. On the synthesis of 2-Nitro-5-thiocyanatobenzoic acid. *J Org Chem* 36:2727

Deshpande KL, Fried VA, Ando M, Webster G (1987): Glycosylation affects cleavage of an H5N2 influenza virus hemagglutinin and regulates virulence. *Proc Natl Acad Sci USA* 84:36–40

Doolittle RF, Hunkapiller MW, Hood LE, Devare SG, Robbins KC, Aaronson SA, Antoniades HN (1983): Simian sarcoma virus oncgene, r-*sis*, is derived from the gene (or genes) encoding a platelet-derived growth factor. *Science* 221:275–276

Downs F, Peterson C, Murty VLN, Pigman W (1977): Quantitation of the beta-elimination reaction as used on glycoproteins. *Int J Peptide Protein Res* 10:315–322

Edman P, Begg G (1967): A protein sequenator. *Eur J Biochem* 1:80–91

Eng J, McCormack AL, Yates JR III (1994): *J Am Soc Mass Spectrom* 5:976–989

Evans RW, Aitken A, Patel KJ (1988): Evidence for a single glycan moiety in rabbit serum transferrin and location of the glycan within the polypeptide chain. *FEBS Lett* 238:39–42

Fenyö D (2000): Identifying the proteome: software tools. *Curr Opin Biotechnol* 11:391–395

Fenyö D, Qin J, Chait BT (1998): Protein identification using mass spectrometric information. *Electrophoresis* 19:998–1005

Fernandez J, DeMott M, Atherton D, Mische SM (1992): Internal protein sequence analysis: enzymatic digestion for less than 10 micrograms of protein bound to polyvinylidene difluoride or nitrocellulose membranes. *Anal Biochem* 201:255–264

Gygi SP, Rist B, Gerber SA, Turecek F, Gelb MH, Aebersold R (1999): Quantitative analysis of complex protein mixtures using isotope-coded affinity tags. *Nat Biotechnol* 17:994–999

Hearn MTW, Aguilar MI (1988): Reversed phase high performance liquid chromatography of peptides and proteins. In: *Modern Physical Methods in Biochemistry*, Neuberger A, Van Deenen LLM, eds. Amsterdam: Elsevier

Henikoff JG, Greene EA, Pietrokovski S, Henikoff S (2000): Increased coverage of protein families with the blocks database servers. *Nucleic Acids Res* 28:228–230

Hochstrasser DF, Patchornik A, Merril CR (1988): Development of polyacrylamide gels that improve the separation of proteins and their detection by silver staining. *Anal Biochem* 173:412–423

Hulmes JD, Miedel MC, Pan Y CE (1989): Strategies for microcharacterization of proteins using direct chemistry on sequencer supports. In: *Techniques in Protein Chemistry*, Hugh TE, ed. San Diego: Academic Press

Ibba M (2002): Biochemistry and bioinformatics: when worlds collide. *Trends Biochem Sci* 27:64

James P, Quadroni M, Carafoli E, Gonnet G (1994): Protein identification in DNA databases by peptide mass fingerprinting. *Protein Sci* 3: 1347–1350

Jeno P, Mini T, Moes S, Hintermann E, Horst M (1995): Internal sequences from proteins digested in polyacrylamide gels. *Anal Biochem* 224:75–82

Inglis AS (1983): Cleavage at aspartic acid. *Methods Enzymol* 91:324–332

Keil B, Tong NT (1988): Database lysis: computer-assisted investigation of cleavage sites in proteins. In: *Methods in Protein Sequence Analysis*, Wittman-Liebold B, ed. Heidelberg: Springer-Verlag

Kilic F, Ball EH (1991): Partial cleavage mapping of the cytoskeletal protein vinculin. *J Biol Chem* 266:8734–8740

Kostka V, Carpenter FH (1964): Inhibition of chymotrypsin activity in crystalline trypsin preparations. *J Biol Chem* 239:1799–1803

Lin J-H, Wu X-R, Kreibich G, Sun T-T (1994): Precursor sequence, processing and urothelium-specific expression of a major 15-kDa protein subunit of asymmetric unit membrane. *J Biol Chem* 269:1775–1784

Lipman DJ, Pearson WR (1985): Rapid and sensitive protein similarity searches. *Science* 227:1435–1441

Lischwe MA, Sung MA (1977): Use of N-chlorosuccinimide/urea for the selective cleavage of tryptophanyl peptide bonds in proteins. *J Biol Chem* 252:4976–4980

Loo RR, Dales N, Andrews PC (1994): Surfactant effects on protein structure examined by electrospray ionization mass spectrometry. *Protein Sci* 3: 1975–1983

Lui M, Tempst P, Erdjument-Bromage H (1996): Methodical analysis of protein-nitrocellulose interactions to design a refined digestion protocol. *Anal Biochem* 241:156–166

Mann M, Hendrickson RC, Pandey A (2001): Analysis of proteins and proteomes by mass spectrometry. *Annu Rev Biochem* 70:437–473

Mann M, Hojrup P, Roepstoff P (1993): Use of mass spectrometric molecular weight information to identify proteins in sequence databases. *Biol Mass Spectrom* 22:338–345

Mann M, Jensen ON (2003): Proteomic analysis of post-translational modifications. *Nature Biotechnol* 21:255–261

Mann M, Ong S-E, Gronborg M, Steen H, Jensen ON, Pandey A (2002): Analysis of protein phosphorylation using mass spectreometry: deciphering the phosphoproteome. *TRENDS Biotechnol* 20:261–268

Mann M, Pandey A (2001): Use of mass spectrometry-derived data to annotate nucleotide and protein sequence databases. *TRENDS Bio Sci* 26:54–61

Mann M, Wilm M (1994): Error-tolerant identification of peptides in sequence databases by peptide sequence tags. *Anal Chem* 66:4390–4399

Martz E (2002): Protein Explorer: easy yet powerful macromolecular visualization. *TRENDS Bio Sci* 27:107–109

Miczka G, Kula MR (1989): The use of polyvinylidene difluoride membranes as blotting matrix in combination with sequencing; applications to pyruvate decarboxylase from *Zymomonas mobilis*. *Anal Lett* 22:2771–2782

Morris HR, Paxton T, Dell A, Langhorne J, Bordoli RS, Hoyes J, Bateman RH (1996): High sensitivity collisionally activated decomposition tandem mass spectrometry on a novel quadrapole/orthogonal-acceleration Time-of-Flight mass spectrometer. *Rapid Commun Mass Spectrom* 10:889–896

Moore S, Stein WH (1963): Chromatographic determination of amino acids by the use of automatic recording equipment. *Methods Enzymol* 6: 819–831

Moos M, Nguyen NY, Liu T-Y (1988): Reproducible, high yield sequencing of proteins electrophoretically separated and transferred to an inert support. *J Biol Chem* 263:6005–6008

Nadler T, Parker K, Huang Y, Degnore J, Wolf B, Anderson L, Anderson N, Lennon J, Bappanad D, McGrath A (2000): Automation of gel slice extraction and MALDI-TOF-MS sample preparation on a robotic platform.in: Proceedings of the 48th ASMS Conference on Mass Spectrometry and Allied Topics. Long Beach, CA, pp. 345–346

Nefsky B, Bretscher A (1989): Landmark mapping: a general method for localizing cysteine residues within a protein. *Proc Natl Acad Sci USA* 86: 3549–3553

Pandey A, Mann M (2000): Proteomics to study genes and genomes. *Nature* 405:837–846

Pappin DJC, Hojrup P, Bleasby AJ (1993): Rapid identification of proteins by peptide-mass fingerprinting. *Curr Biol* 3:327–332

Park Z-Y, Russell DH (2000): Thermal denaturation: a useful technique in peptide mass mapping. *Anal Chem* 72:2667–2670

Park Z-Y, Russell DH (2001): Identification of individual proteins in complex protein mixtures by high-resolution, high-mass-accuracy MALDI TOF-mass spectrometry analysis of in-solution thermal denaturation/enzymatic digestion. *Anal Chem* 73:2558–2564

Patterson SD, Aebersold R (1995): Mass spectrometric approaches for the identification of gel-separated proteins. *Electrophoesis* 16:1791–1814

Perkins DN, Pappin DJ, Creasey DM, Cottrell (1999): Probability-based protein identification by searching sequence databases using mass spectrometry data. *Electrophoresis* 20:3551–3567

Rappsilber J, Mann M (2002): What does it mean to identify a protein in proteomics? *TRENDS Biochem Sci* 27:74–78

Rasmussen HH, Van Damme J, Bauw G, Puype M, Gesser B, Celis JE, Vandekerckhove J (1991): Protein electroblotting and microsequencing in establishing integrated human protein databases. In: *Methods in Protein Sequence Analysis*, Jörnvall H, Höög JO, eds. Basel: Birkhäuser Verlag

Saris CJM, van Eenbergen J, Jenks BG, Bloemers HPJ (1983): Hydroxylamine cleavage of proteins in polyacrylamide gels. *Anal Biochem* 132:54–67

Scott MG, Crimmins DL, McCourt DW, Tarrand JJ, Eyerman MC, Nahm, MH (1988): A simple *in situ* cyanogen bromide cleavage method to obtain internal amino acid sequence of proteins electroblotted to polyvinyldifluoride membranes. *Biochem Biophys Res Commun* 155:1353–1359

Shevchenko A, Loboda A, Shevchenko A, Ens W, Standing KG (2000): MALDI quadrupole time-of-flight mass spectrometry: a powerful tool for proteomic research. *Anal Chem* 72:2132–2141

Shevchenko A, Wilm M, Vorm O, Mann M (1996): *Anal Chem* 68:850–858

Shultz J et al (2002): SMART: A web-based tool for the study of genetically mobile domains. *Nucleic Acids Res* 28:231–234

Spahr CS, Susin SA, Bures EJ, Robinson JH, Davis MT, McGinley MD, Kroemer G, Patterson SD (2000): Simplification of complex peptide mixtures for proteomic analysis: reversible biotinylation of cysteinyl peptides. *Electrophoresis* 21:1635–1650

Spengler B (2001): The basics of Matrix-assisted laser desorption, ionization time-of-flight mass spectrometry and post-source decay analysis. In: *Proteome Research: Mass Spectrometry Principles and Practice*, James P, ed. Berlin: Springer-Verlag

Stone KL, LoPresti MB, Crawford JM, DeAngelis R, Williams KR (1989): Reverse phase HPLC separation of sub-nanomole amounts of peptides obtained from enzymatic digests. In: *HPLC of Peptides and Proteins: Separation, Analysis and Conformation*, Hodges RS, ed. Boca Raton, FL: CRC Press

Tabb DL, Eng JK, Yates JR III (2001): Protein identification by SEQUEST. In: Proteome Research: Mass Spectrometry, James P, ed. Berlin: Springer-Verlag

Tabb DL, McDonald WH, Yates JR (2002): DTASelect and contrast: tools for assembling and comparing protein identifications from shotgun proteomics. *J Proteome Res* 1:21–26

Vanfleteren JR, Raymackers JG, Van Bun SM, Meheus LA (1992): Peptide mapping and microsequencing of proteins separated by SDS-PAGE after limited *in situ* acid hydrolysis. *BioTechniques* 12:551–557

Van Montfort BA, Canas B, Duurkens R, Godavac-Zimmermann, Robillard GT (2002): Improved in-gel approaches to generate peptide maps of integral membrane proteins with matrix-assisted laser desorption/ionization time-of-flight mass spectrometry. *J Mass Spectrometry* 37:322–330

Valaskovic GA, Kelleher NL, McLafferty FW (1996): Attomole protein characterization by capillary electrophoresis-mass spectrometry. *Science* 273:1199–1202

Vorburger K, Kitten GT, Nigg EA (1989): Modification of nuclear lamin proteins by a mevalonic acid derivative occurs in reticulocyte lysates and requires the cysteine residue of the C-terminal CXXM motif. *EMBO J* 8:4007–4013

Waterman MS, Vingron M (1994): Rapid and accurate estimates of statistical significance for sequence data base searches. *Proc Natl Acad Sci USA* 91:4625–4628

Weber K, Osborn M (1975): Proteins and sodium dodecyl sulfate: Molecular weight determination on polyacrylamide gels and related procedures. In: *The Proteins, Vol. I,* Neurath H, Hill RL, eds. New York: Academic Press

Wilbur WJ, Lipman DJ (1983): Rapid similarity searches of nucleic acid and protein data banks. *Proc Natl Acad Sci USA* 80:726–730

Wilkins MR, Pasquali C, Appel RD, Ou K, Golaz O, Sanchez J-C, Yan JX, Gooley AA, Hughes G, Humphery-Smith I, Williams KL, Hochstrasser DF (1996): From proteins to proteomes: Large scale protein identification by two-dimensional electrophoresis and amino acid analysis. *Biotechnology* 14:61–65

Wilkins MR, Gasteiger E, Bairoch A, Sanchez J-C, Williams KL, Appel RD, Hochstrasser DF (1998): Protein identification and analysis tools in the ExPASy server. In: *2-D Proteome Analysis Protocols,* AJ Link, ed. New Jersey: Humana Press

Wilkins MR, Gasteiger E, Bairoch A, Sanchez J-C, Williams KL, Appel, RD, Hochstrasser DF (1999): Protein identification and analysis tools in the ExPASy server. *Methods Mol Biol* 112:531–552

Wilm M, Mann M (1996): Analytical properties of the nanoelectrospray ion source. *Anal Chem* 68:1–8

Wilson JF (2002): The rise of biological databases. *The Scientist* Mar 18 p. 34–35

Zhang W, Chait BT (2000): ProFound-an expert system for protein identification using mass spectrometric peptide mapping information. *Anal Chem* 72:2482–24

9

Identifying and Analyzing Posttranslational Modifications

Introduction

The aim of this chapter is to present an overview of bench-top methodology for the structural elucidation of modified proteins and peptides. The techniques covered in previous chapters provide most of the basics for studying protein modifications.

Protein function is controlled by many posttranslational modifications. As part of a cellular response to external stimuli, proteins may be altered by chemical modification or change subcellular localization. Protein maturation and degradation are dynamic processes that alter the final amount of active protein, independent of the mRNA level.

Release of the completed polypeptide chain from the ribosome is not necessarily the final step in the formation of a biologically active protein. Covalent modification of proteins can occur during (cotranslationally) and after protein synthesis (posttranslationally). Consequently, the primary structure may be altered, and novel amino acid side chains may be introduced which can influence a protein's charge, hydrophobicity, conformation and/or stability. Hence, the "one-gene-one-polypeptide" paradigm is now outdated. In both eukaryotes and prokaryotes the peptide translation of many genes is modified to create multiple gene products from a single DNA sequence (Gooley and Packer, 1997).

There are at least 200 known posttranslational modifications of proteins (Yan et al, 1989. Many are easy to overlook. Some modifications are common, others extremely rare. There are many ways that proteins can be modified after they have been synthesized. Amino-terminal acylation, cyclization of amino terminal Glu to yield pyroglutamic acid, carboxy-terminal amidation, sulfation of Tyr residues, hydroxylation of Lys and Pro residues as well as methylation of various amino acids have all been described. It is beyond the scope of this manual to describe them all, or even the majority of them for that matter.

A case could be made that all eukaryotic proteins are posttranslationally modified in some way (e.g. truncation at the N- or C-terminus, addition of various substituents such as carbohydrate, phosphate, sulfate, methyl, acetyl, or lipid groups). Some modifications occur while the polypeptide is being assembled; others are manifestations of old age. Many modifications are products of specific enzyme catalyzed reactions, while others result from nonenzymatic chemical modifications. Some modifications are reversible. Some are physiologically important; others may arise artifactually during isolation and purification.

Modifications to the linear amino acid sequence vary widely in chemical structure and in their attachment to different amino acids. Protein modifications have been classified into different types according to which type of amino acid side chain is modified. Alternatively, as will be the case with the modifications discussed here, they are grouped on the basis of the chemical nature of the attached group.

Protein modifications play important roles in cell-cell interactions and intracellular regulation, and they are often vital for cell functions. For example, protein phosphorylation and dephosphorylation play important roles in regulation. Proteins are frequently modified by more than one type of posttranslational modification, e.g., where proteins are acylated as well as phosphorylated. The presence of multiple covalent modifications on a single protein can greatly expand the possibilities for precisely controlling metabolic pathways. An important point to keep in mind is that many co- and posttranslational modifications described below do not occur when recombinant proteins are expressed in bacterial cells or even in a specific eukaryotic expression system because the enzymes needed to produce the proper processing may not be present.

Characterization of co-and posttranslationally modified proteins isolated from gels or blots can be approached in two ways. Firstly, it is possible to study the modification while still attached to the target protein. Alternatively, the modification can be studied following its release from the target protein. Methods for studying the precise chemical nature of protein modifications include mass spectrometry (fast bombardment and direct chemical ionization), and high-sensitivity gas-phase sequencing. The modification affecting the primary amino acid sequence often becomes apparent when unexpected masses for known peptides are obtained. All modifications cause a shift in mass and therefore, should be detectable by mass spectrometry. For example, the N terminus of a protein is indicated by an acetylated peptide. In this case, the N-terminal amino acid contained within the peptide is 42 Da greater than expected (Mann and Pandey, 2001). Using MS, the determination of a peptide that is acetylated at the N terminus establishes it as an N-terminal peptide distinguishing it from other internal peptides.

In many cases proteins are modified in transit through a compartment. Most cell-surface proteins that transit through the Golgi contain one or more *N*-glycosylation site and several *O*-linked glycosylation sites. Gas chromatography coupled to mass spectrometry has been used to identify *O*-linked oligosaccharides released with alkali/borohydride treatment (Karlsson et al, 1987). Nuclear magnetic resonance (NMR) has also been widely used, particularly for the determination of the structure and configuration of the carbohydrate involved in glycosylation and the stereochemical configuration of the isoprenoid alkyl modification. The complete structure of complex polysaccharides is beyond the scope of this manual.

In this chapter, strategies for the identification of commonly encountered protein modifications are presented. As analytical techniques have improved, the researcher now has the bench-top capability of probing the structure of the target molecule. After isolation of the modified amino acids, analysis and identification can be performed by one or more techniques. Obtaining structural data that would have been possible only with highly specialized, sophisticated (expensive) equipment that is usually not readily available is now a reality.

A. Glycosylation

Carbohydrates are information rich molecules vital to recognition processes ranging from host-pathogen interactions to the recruitment of neutrophils to the sites of tissue damage. Carbohydrates exist in diverse forms. Some are attached to protein cores, as in glycoproteins and proteoglycans while others appear as glycolipids, as in microbial lipopolysaccharides. Still others are found as polysaccharides, as in glycosaminoglycans and bacterial capsular polysaccharides. Carbohydrates can achieve additional complexity by branching where multiple glycosidic linkages are made to a single residue (Kiessling and Cairo, 2002).

Protein glycosylation is influenced by the local protein structure and by the available repertoire of glycosylating enzymes in the cell. The enzyme reactions do not always go to completion; therefore, a single glycoprotein normally leaves the pathway as a collection of glycosylated variants of the same amino acid sequence.

Protein glycosylation is the most complex posttranslational modification. The number of different components, some of which may be modified themselves, the different anomeric forms (a or b), chain lengths and position of linkages and branching points translates into a large number of possible structures that can occupy a single glycosylation site (Lis and Sharon, 1993).

Glycoproteins generally exist as populations of glycosylated variants, also referred to as glycoforms, of a single homogeneous polypeptide. Although the same glycosylation machinery is available to all proteins that enter the secretory pathway, most glycoproteins emerge with characteristic glycosylation patterns and heterogeneous population of glycans at each glycosylation site. It is not uncommon for a glycoprotein to be processed with more than 100 alternative glycans at a single glycosylation site (Rudd et al, 1997).

This section attempts to answer two fundamental questions. Is the target protein glycosylated, and if so, how are the oligosaccharides linked? Protein glycosylation is a major covalent modification and perhaps the one most well studied. There are two main modes of linkage of the sugar chains to the polypeptide backbone. Glycosylation

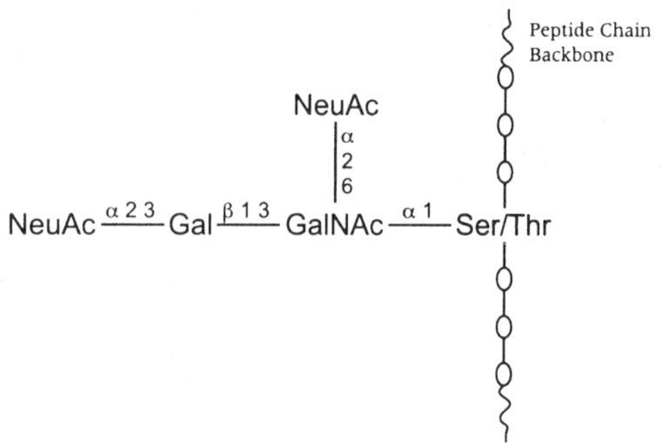

Peptide Chain Backbone

NeuAc α2 6

NeuAc —α2 3— Gal —β1 3— GalNAc —α1— Ser/Thr

FIGURE 9.1 Mucin-type *O*-linked oligosaccharide.

of proteins and peptides may occur through *O*-glycosyl- or N-glycosyl-linkages. *O*-Glycosidic bonds are links between N-acetylgalactosamine (GalNAc) and the hydroxyl groups of serine and threonine as exemplified by mucin shown in Figure 9.1.

In contrast, N-linked carbohydrate chains are formed through N-acetylglucosamine (GlcNAc) to asparagine residues. N-linked oligosaccharides may be required for proper processing, correct membrane targeting, intrinsic biological activity, or protein stability. N-linked carbohydrate units all start out with a common structure, the product of the dolichol pathway, shown in Figure 9.2, consisting of two GlcNAc residues, three glucose (Glc) and nine mannose (Man) monosaccharides which is transferred as a presynthesized oligosaccharide from a lipid precursor to an asparagine residue in the nascent protein chain. This step occurs cotranslationally in the rough endoplasmic reticulum. Subsequent processing of the carbohydrate chain begins in the rough endoplasmic reticulum and continues in the Golgi, catalyzed by glycosidases and glycosyltransferases, which sequentially remove the Glc residues and trim the mannose residues to produce the core structure shown in Figure 9.2 as two shaded GlcNAc residues and three shaded mannose residues.

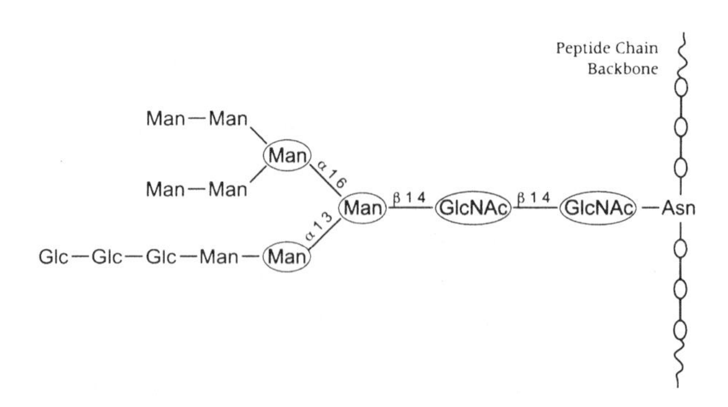

FIGURE 9.2 The common intermediate for the formation of all N-linked oligosaccharides which is the product of the dolichol pathway. The common core pentasaccharide sequence is shaded.

The Golgi apparatus is the central organelle of the exocytotic pathway, composed of membrane-bound tubules, cisternae, and vesicles. It is divided into subcompartments, each having a distinct set of resident enzymes to carry out sequential modification of carbohydrate moieties on proteins as they transit through the Golgi apparatus (Dunphy and Rothman, 1983). A large variety of monosaccharide units are added to the core structure as the protein goes through the Golgi network. The terminal glycosylations are tissue-specific and cell-type-specific (Paulson, 1989), often contributing to the function of the glycoprotein.

Three major types of asparagine-linked oligosaccharides (shown in Figure 9.3) are found: high-mannose, complex, and hybrid. In high mannose oligosaccharides, all residues attached to the core are mannose residues, with the total number being 5–9 (including the three in the core). If both the α1–3 and α1–6 mannose residues in the core have GlcNAc attached, the oligosaccharide referred to is a complex oligosaccharide. In complex oligosaccharides, branching structures originating from the GlcNAc residues are known as antennae. Since each core mannose can have several GlcNAc residues attached to it, there can be more than two antennae on complex oligosaccharides. The

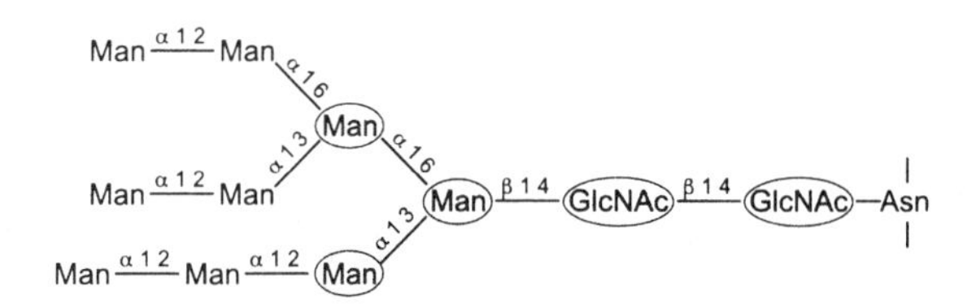

High Mannose

Man α16 Man α13 α16 Man Man β14 GlcNAc β14 GlcNAc —Asn α13 Gal β14 GlcNAc β12 Man β14 GlcNAc

Hybrid

SA α 2,3 or 6 3 Gal β14 GlcNAc β12 Man α16 Man β14 GlcNAc β14 GlcNAc —Asn α13 SA α 2,3 or 6 3 Gal β14 GlcNAc β12 Man β14 GlcNAc α16 Fuc

Complex

FIGURE 9.3 Some typical structures of N-linked oligosaccharide chains found attached to proteins. The shaded residues constitute the pentasaccharide core that is common to all three types.

resulting structures are termed biantennary (2 GlcNAc antennae), triantennary (3 GlcNAc antennae), tetraantennary (4 GlcNAc antennae), etc. (In an apparent nomenclatural inconsistency, the term antenna is usually applied only to the branches of a hybrid or complex structure. Even though high-mannose structures are branched, these structures are not referred to as having antenna.) The third major type of asparagine-linked oligosaccharide, the hybrid type, is composed of structural elements of the other two types. In the hybrid structure, the α1–6-linked core mannose has only mannose residues attached to it (like the high-mannose structure). The α1–3-linked core mannose has one or more GlcNAc-initiated antennae attached to it (like a complex structure). In another nomenclatural irregularity, the mannose-containing arm of the hybrid structure is termed a single antenna, even if it has branch points. The most common hybrid structures are the biantennary and the triantennarry hybrids.

Preliminary analysis of the type of oligosaccharides present on a glycoprotein can be obtained if the carbohydrate units are radioactively labeled. One way to introduce radioactivity into the saccharide moiety of a glycoprotein is to use radiolabeled sugar nucleotides and purified glycosyltransferases that will transfer the labeled monosaccharides to a terminal position on the glycoprotein oligosaccharides (Whiteheart et al, 1989). A second general method of introducing radioactivity into glycoprotein oligosaccharides is by metabolic labeling with radioactive monosaccharide precursors (Shen and Ginsburg, 1967; Yurchenco et al, 1978).

Sugar nucleotides are the immediate donors for glycosylation reactions. However, they are not internalized by cells. Therefore, metabolic labeling is performed with radiolabeled monosaccharide precursors. These molecules are internalized and modified to labeled sugar nucleotides, then transported to the Golgi apparatus where luminally oriented transferases add the monosaccharides to luminally oriented acceptors.

Labeling with radioactive sugars is not as efficient as metabolically labeling proteins with radioactive amino acids. Therefore, to optimize incorporation into the glycoprotein of interest the following factors should be considered: (1) amount and concentration of label in the media; (2) the number of cells; (3) the duration of labeling. The most specific factor affecting uptake and incorporation of a radioactive monosaccharide is whether or not it competes with glucose for uptake. Glucosamine, galactosamine, galactose and mannose are known to compete with glucose for uptake into most cells while N-acetylglucosamine, N-acetylmannosamine, mannosamine, fucose and xylose do not (Yurchenco et al, 1978). Therefore, depending on the monosaccharide, the medium can contain normal or reduced levels of glucose.

The labeling strategy can be long term (equilibrium) or short term (pulse-chase analysis, see Protocol 3.3). The labeled glycoprotein of interest is captured by immunoprecipitation (3.11) and identified by Western blotting (7.1). Structural analysis can then be performed on the isolated target glycoprotein.

Cleavage with endoglycosidases followed by SDS-PAGE analysis reveals the general types of oligosaccharides present and also examines the contribution of the oligosaccharides to the apparent molecular weight of the glycoprotein.

Lectin binding, described in Chapter 7, provides some indication as to the types of terminal sugars present.

Pulse-chase analysis coupled with endoglycosidase treatments yields information on the time of saccharide addition and evidence for processing of oligosaccharides. This line of experimentation provides data on the general types of oligosaccharides present and also the contribution of the oligosaccharide to the apparent molecular weight of the glycoconjugate.

Glycosidase inhibitors have been used to study how alterations in oligosaccharide structure affect the function of specific N-linked glycoproteins. The advantage of using processing inhibitors to cause modifications in the structure of the N-linked oligosaccharides, rather than a glycosylation inhibitor such as tunicamycin which completely prevents glycosylation, is that many glycoproteins require carbohydrate for proper conformation or solubility, and they also participate in molecular and cell recognition. Having the carbohydrate added intact to the protein as the core oligosaccharide, shown in Figure 9.2, and then causing modifications in the processing by blocking the modification at different steps, thus altering the structure of the mature molecule, would appear to be a better approach to learning about the role of certain carbohydrate structures in glycoprotein function (Elbein, 1991). A number of compounds have been identified that specifically inhibit glucosidases and mannosidases involved in glycoprotein processing. Some inhibitors of glycoprotein processing and their sites of action are shown in Figure 9.4.

The complete structure of complex polysaccharides is beyond the scope of this book. Glyco, Inc. and Roche Diagnostics Corp., (Glycan Differentiation Kit) market reagents that determine the linkages of the oligosaccharides on the target protein as well as the specific type of monosaccharides that are present. You will need 50–100 μg of target glycoprotein for the analysis.

Fast atom bombardment mass spectrometry (FAB-MS) is a very powerful technique that has been used to uncover glycan structures in glycoproteins. Although this methodology will not be covered in this manual, interested readers are encouraged to read Dell et al (1992), for an introductory description of the many areas where FAB-MS has been applied to the structure and biosynthesis of various oligosaccharide moieties in glycoproteins.

PROTOCOL 9.1 Chemical Deglycosylation Using Trifluoromethanesulfonic Acid (TFMS)

TFMS breaks glucosidic bonds between adjacent monosaccharides as well as *O*-glycosidic linkages between carbohydrate moieties and

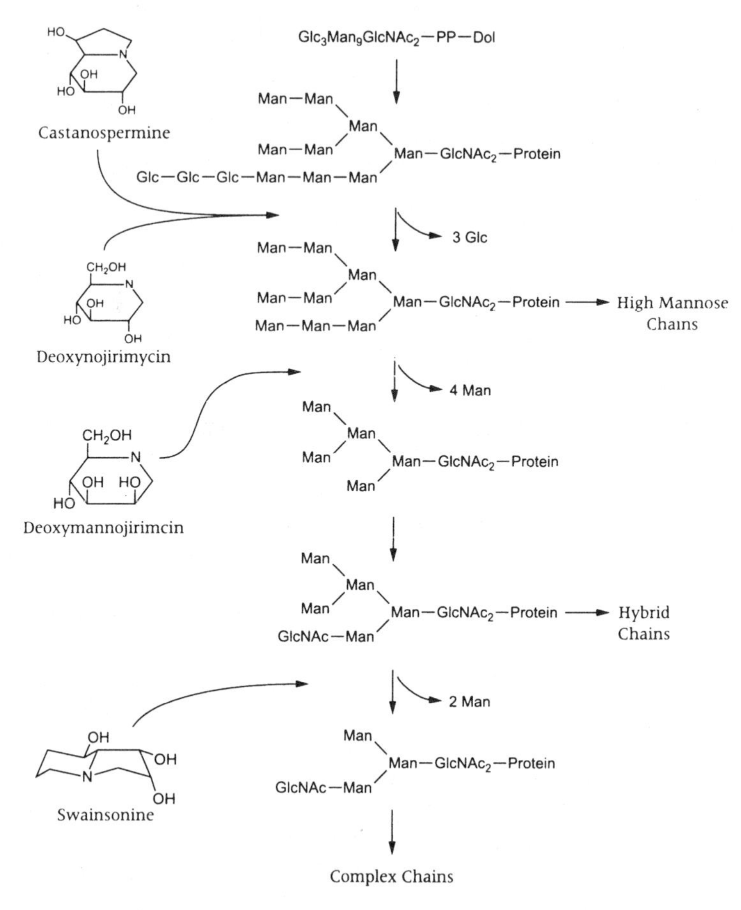

FIGURE 9.4 Inhibitors of glycoprotein processing. The site of action of selected glycoprotein-processing inhibitors. Castanospermine and deoxynojirimycin are inhibitors of glucosidase I; deoxymannojirimycin inhibits mannosidase I. Swainsonine inhibits mannosidase II. (Adapted with permission from Elbein, 1991.)

amino acids serine and threonine but not asparaginyl: N-acetylglucosaminyl amide linkages (Burgess and Norman, 1988). Glycoproteins treated with TFMS are virtually denuded of carbohydrate with only N-acetylglucosamine remaining bound to the asparagine residue (Florman and Wassarman, 1985).

Materials

Reactivials: 5 ml (Pierce)
Nitrogen tank
Trifluoromethanesulfonic acid (TFMS) (Sigma)
Anisole
Pyridine solution: pyridine: H_2O (3:5 v/v)
Ether
1 × SDS sample buffer: see Appendix C

1. Lyophilize the extract to be treated in a 5 ml Reactivial.
2. Incubate the sample under nitrogen for 4 h at 4°C with 0.39 ml anisole and 0.59 ml TFMS.
3. Terminate the reaction by adding 1.57 ml of ice-cold pyridine solution and dialyze overnight at 4°C against H_2O.
4. Extract the dialyzed sample with anhydrous ether, lyophilize the aqueous phase, dissolve it in 1 × SDS sample buffer and analyze by SDS-PAGE.

Comments: Deglycosylation can be followed by SDS-PAGE, Western blotting or probing with a ligand that recognizes the complete molecule. The treated sample is compared to the untreated material. In this way, the contribution of the carbohydrate to antigenicity and biological activity can be examined.

N-Glycosylation

N-linked glycosylation influences many properties of proteins, including intracellular transport, biological activity, stability, and antigenicity. Glycosylation of asparagine residues (N-glycosylation) occurs only at the sequence Asn-X-Ser/Thr, where X can be any amino acid except proline. However, not every Asn-X Ser/Thr motif is glycosylated. Steric hindrance or inaccessability in the folded protein is just as important in directing the covalent modification.

An acceptable method for determining whether a candidate glycoprotein is N-glycosylated is to treat the protein with N-glycanase and examine the product by SDS-PAGE. If the product of the N-glycanase reaction has a decreased molecular weight in comparison to the undigested glycoprotein, then the reduction in molecular weight was due to the removal of N-linked carbohydrate.

N-glycanase, Peptide-N^4-(N-acetyl-B-glucosaminyl) asparagine amidase, or PNGase F cleaves the β-aspartylglycosylamine linkage of N-linked sugar chains, releasing oligosaccharide-glycosylamine and peptide. This reaction differs from those catalyzed by the endo-β-N-acetylglucosaminidases such as the endoglycosylases D, H, and F that cleave the glycosidic bond within the chitobiose core structure. N-glycanase releases all common classes of Asn-linked oligosaccharides (high-mannose, complex and hybrid), including sialylated, phosphorylated and sulfated sugar chains (Maley et al, 1989). However, N-glycanase will not release oligosaccharides that are linked to an amino- or carboxy-terminal Asn residue (Elder and Alexander, 1982).

PROTOCOL 9.2 Removal of the Oligosaccharide from the Glycoprotein with N-Glycanase

N-Glycanase and peptide N-glycosidase F (Glycopeptidase F, GPase F) cleave an intact oligosaccharide from native or denatured glycoprotein, removing N-linked oligosaccharides from the polypeptide backbone. The cleavage occurs between the innermost residue of the oligosac-

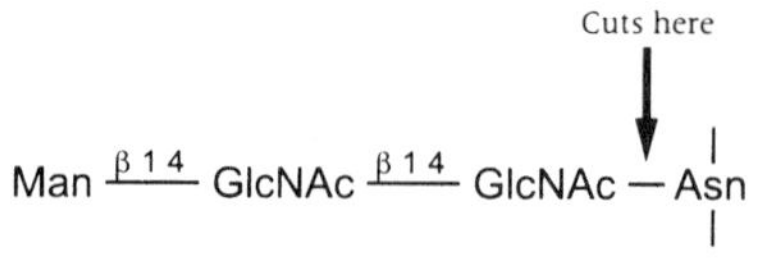

FIGURE 9.5 Glycopeptidase F specificity.

charide and the asparagine residue to which the oligosaccharide is linked as shown in Figure 9.5. Asn-linked high mannose and hybrid and complex oligosaccharides are good substrates. Deglycosylation is routinely monitored by increases in electrophoretic mobility and loss of lectin staining compared to the untreated glycoprotein.

Materials

NP-40

SDS: 10% (w/v): see Appendix C

2-ME: 0.1M

Phosphate buffer: 0.55M sodium phosphate, pH 8.6, 7.5% NP-40

N-glycanase: Genzyme

4 × SDS sample buffer: see Appendix C

1. Boil the sample (which can be immunoprecipitated material) for 3min in the presence of 0.5% SDS, 0.1M 2-ME.
2. Divide the sample into two equal fractions, one to be digested with N-glycanase and the other to be a mock digested control. Running digested and undigested samples side by side will rule out the contribution of any proteases that may be present in the sample that could digest the sample producing extra bands. (However, this control will not rule out the possibility of a protease contaminant in the N-glycanase, although most enzymes are assayed for the presence of contaminating proteases.)
3. Dilute the sample with phosphate buffer to reach a final concentration of 0.2M sodium phosphate and at least a sevenfold excess of NP-40 to SDS.
4. Add N-glycanase (final concentration 10 units/ml) and incubate the reaction at 37°C overnight.
5. Add 1/4 of the total reaction volume of 4 × SDS sample buffer and analyze the product by SDS-PAGE, running the digested and mock digested samples side-by-side.

Comments: For native glycoproteins, 20 μg of sample usually requires a minimum of 1.8 units of enzyme (60 units/ml in the digest). SDS denatured samples require 0.03–0.3 units (1–10 units/ml). Enzyme concentrations can be reduced for radiolabelled samples that contain less protein. When using this enzyme for the first time, run a control digestion on a protein with known N-linked carbohydrate like ovalbumin, fetuin, or α-acid-1-glycoprotein.

PROTOCOL 9.3 N-glycosidase F (GPase F) Treatment of Glycoproteins in Immunoprecipitates

This reaction is identical to the one described in the previous protocol using N-Glycanases however, the enzymes and the buffers differ (Elder and Alexander, 1982). GPase F should not be confused with Endo F.

Materials

Reaction buffer: 0.25M Na_2HPO_4, pH 7.8, 0.5% Triton X-100, 10mM EDTA, 10mM 2ME

N-glycosidase F (Boehringer)

4 × SDS sample buffer: see Appendix C

1. Isolate the target glycoprotein by immunoprecipitation (Protocol 3.11).
2. After the last wash, resuspend the Sepharose or agarose beads in 30µl of reaction buffer.
3. Heat the mixture for 3min at 100°C and separate the sample into 2 fractions.
4. To one fraction add 15µl of reaction buffer plus 0.06–6 units of N-glycosidase F. To the other fraction add 15µl of reaction buffer without enzyme.
5. Incubate the reaction mixtures overnight at 37°C.
6. Stop the reaction by adding 4 × SDS sample buffer diluted to a final 1 × concentration.
7. Heat for 5min at 100°C and load the samples on an SDS polyacrylamide gel.

Comments: As a positive control for N-glycosidase F, use 25µg of α-acid-glycoprotein or ovalbumin and digest as described above.

GPase F is sensitive to SDS. The final concentration of SDS in the reaction should not exceed 1mg/ml. If necessary, add concentrated Triton X-100 to the incubation mixture to reach a concentration of 6–7mg/ml prior to adding the GPase F.

GPase F will not cleave oligosaccharides attached to an N- or C-terminal Asn residue (i.e., a free amino or carboxy terminus) nor will it cleave *O*-linked oligosaccharides.

PROTOCOL 9.4 Tunicamycin

Tunicamycin prevents the formation of the asparagine-linked precursor oligosaccharide from which all N-linked oligosaccharides are derived by blocking the first step in the pathway to dolichol pyrophosphate oligosaccharide (Elbein, 1981). Therefore, glycoproteins synthesized in the presence of tunicamycin do not carry any N-linked oligosaccharides. However, they may still carry *O*-linked oligosaccharides, the formation of which is not blocked by tunicamycin.

Materials

Tunicamycin dissolved in DMSO. **Handle tunicamycin with care as it is a highly toxic substance.**

1. Use tunicamycin over a concentration range of 0–5.0µg/ml. (Usually, two concentrations in addition to an untreated control will be adequate: 0, 2.5, and 5.0µg/ml.)
2. Treat cells with tunicamycin for 1.5h prior to and during a 2h metabolic labeling with [^{35}S]methionine (100µCi/ml). See Protocols 3.1 and 3.2.
3. Analyze the products by immunoprecipitation or Western blotting.

Comments: Compare the pattern of bands obtained from Tunicamycin treated cells vs. that of the untreated cells. The Tunicamycin treated cells should contain specific bands of higher electrophoretic mobility indicating a lower molecular weight due to the absence of the N-linked oligosaccharides.

O-Glycosylation

PROTOCOL 9.5 Identification of *O*-Glycosylated Amino Acids by Alkaline β-Elimination

Reductive β elimination releases *O*-linked oligosaccharides from glycosylated serine and threonine residues, converting glycosylated serine residues to alanine and glycosylated threonine residues to β-aminobutyric acid. *O*-Glycosylation sites can be identified by comparing the amino acid sequence of the target protein before and after reductive β elimination. Unglycosylated serine and threonine residues are unaffected by reductive β elimination. Amino acid sequence analysis (Protocol 8.10) will therefore show a conversion of glycosylated serine and threonine residues to alanine and β-aminobutyric acid, respectively, after reductive β elimination. It should be noted that reductive β elimination can also cause some non-specific peptide bond cleavage (Roquemore et al, 1994).

Materials

β-elimination buffer: 1 M $NaBH_4$, 0.1 M NaOH
15 ml plastic conical centrifuge tubes
Acetic acid: 4 M
Sephadex G-50 (coarse) column (1 × 30 cm)
Sephadex column buffer: 50 mM ammonia, 0.1% (w/v) SDS
pH paper

1. Resuspend acetone-precipitated proteins (Protocol 5.17) in 500 μl of fresh β elimination buffer and incubate at 37°C for 18–48 h. Check the pH of the reaction mixture after several hours of incubation. If it is less than 13, increase the alkalinity by adding more β elimination buffer.
2. Transfer the sample to a 15 ml plastic tube on ice.
3. Add ice-cold acetic acid dropwise while vortexing over the course of 1 h, until the pH is between 6 and 7.
4. Resolve the reaction products by gel filtration through the Sephadex G-50 column (Chapter 10). The released carbohydrate should be in the included volume. Lyophilize to dryness.
5. Analyze the β-eliminated products by chromatography (Chapter 10) and mass spectrometry (Chapter 8).

Comments: In order to achieve β elimination, various strengths of alkali (0.05–0.5 N NaOH), temperatures (0–45°C) and length of incubation (15–216 h) have been used. Conditions should be established separately for each glycoprotein.

PROTOCOL 9.6 β–Elimination of *O*-Glycans from Glycoproteins Immobilized on Blots

The linkage between GalNAc and Ser/Thr residues is alkali-labile and can be cleaved under mild alkaline conditions by the β-elimination reaction. The following method describes the efficient release of *O*-glycans from electrotransferred glycoproteins, without release or destruction of polypeptide chains and *N*-glycans.

Materials

0.025M sulfuric acid

0.055M NaOH

1. Separate proteins by SDS-PAGE as described in Protocol 4.6.
2. Electrophoretically transfer proteins to PVDF membranes.
3. To first desialylate glycoproteins, incubate the blot in 0.025M sulfuric acid for 1h at 80°C then wash with water.
4. Incubate the blot in 0.055M NaOH for 16h at 40°C.
5. Wash three times with water.
6. Perform blocking step and probe the blot with antibody or lectin.

Comments: Do not use nitrocellulose as it will be destroyed by alkaline treatment. Sodium borohydride used in the β–elimination reaction described in Protocol 9.5 is omitted since reduction of released *O*-glycans is not necessary.

PROTOCOL 9.7 *O*-Glycosidase

O-glycosidase cleaves Ser/Thr-linked disaccharide galβ1–3GalNAc units unless the saccharide contains terminal sialic acid residues (Ervasti and Campbell, 1991). To make the glycoprotein susceptible to *O*-glycosidase digestion, it will be necessary to remove terminal sialic acid residues if they are present. This is accomplished in the early steps of the protocol.

Materials

Neuraminidase: *Vibrio cholera* neuraminidase (Roche)

Diplococcus pneumoniae O-glycosidase (Roche)

10% (w/v) SDS: see Appendix C

Citrate buffer: 20mM citrate, pH 6.1.

4 × SDS sample buffer: see Appendix C

Triton X-100: 10% (w/v)

Sodium phosphate buffer: 0.5M, pH 7.4

1. Following the final wash of the immunoprecipitated material (Protocol 3.11), equilibrate the beads in 50μl of citrate buffer.
2. Incubate the samples at 37°C with 0.1U/ml *Vibrio cholera* neuraminidase.
3. Add SDS to the sample (1/10 of the total volume of the 10% stock solution) to a final concentration of 1% and boil for 5min.

4. Dilute the reaction mixture to final concentrations of 50mM sodium phosphate, pH 7.4, 1% Triton X-100, 0.1% SDS. Add 17mU/ml *Diplococcus pneumoniae O*-glycosidase and incubate at 37°C for 2h.
5. Add 1/4 of the total reaction volume of SDS sample buffer and analyze by SDS-PAGE.

PROTOCOL 9.8 *O*-Glycanase

O-glycanase, like *O*-glycosidase, is an endo-α-N-acetylgalactosaminidase. Like *O*-glycosidase, *O*-glycanase will only cleave desialylated *O*-linked sugars.

Materials

Citrate buffer: 20mM, pH 6.1.

Neuraminidase: *Vibrio cholera* neuraminidase (Roche)

O-glycanase

1. Following the final wash of the immunoprecipitated material (Protocol 3.11), equilibrate the beads in 50μl of citrate buffer.
2. Incubate the samples at 37°C with 0.1U/ml *Vibrio cholera* neuraminidase.
3. Divide the sample into two equal fractions.
4. Add *O*-Glycanase (0.06 milliunits) to one fraction and an equal volume of buffer only to the other fraction. Incubate for 18h at room temperature.
5. Analyze the digested and mock digest fractions by SDS-PAGE.

Comments: O-glycanase from *D. pneumoniae* removes Galβ1–3GalNAc-units linked to Ser/Thr residues, but it does not reduce more complex *O*-glycans (Umemoto et al, 1978).

Combined Use of N-Glycanase and *O*-Glycanase

If sequential digestions of a sample with both N-Glycanase and *O*-Glycanase are desired, perform N-Glycanase followed by *O*-Glycanase. The recommended buffer is 15–20mM sodium phosphate, pH 7–7.5 that represents a compromise and should produce acceptable results. Alternatively, a pH of 7–8.6 is used for the N-Glycanase reaction, then the pH is adjusted to pH 6–7 with an appropriate acid buffer for subsequent digests.

PROTOCOL 9.9 Endoglycosidase H

Endo-β-N-acetylglucosaminidase H (Endo H) selectively cleaves N-linked oligosaccaride of the high mannose and hybrid types from a glycoprotein leaving an N-acetylglucosamine (GlcNAc) residue attached to the polypeptide as shown in Figure 9.6. Acquisition of Endo H resistance is used as a measure of the time taken by secretory proteins to

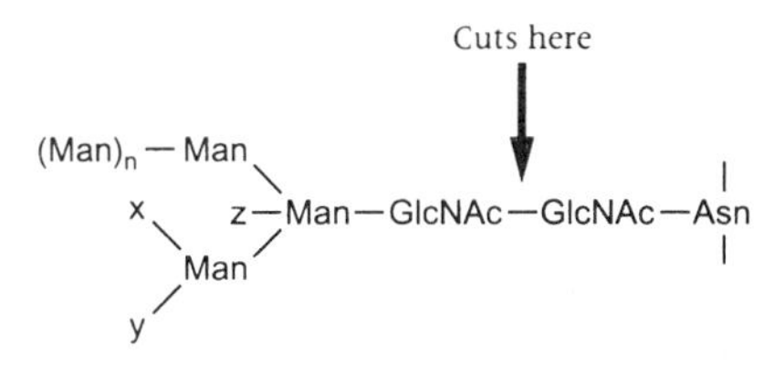

FIGURE 9.6 Endoglycosidase H specificity. Endo H is active on N-linked oligosaccharides of glycoproteins. Endo H will cleave only the following structures: High mannose where n = 2–150, x = (Man)1–2, y and z = H; Hybrid structures where n = 2, x and/or y = SA-Gal-GlcNAc or similar and z = H or GlcNAc. Endo H does not cleave sulfated high mannose structures or any complex structure.

move from the endoplasmic reticulum to the Golgi complex (Strous and Lodish, 1980). This reaction provides a useful tool for localizing glycoproteins to the rough endoplasmic reticulum before they are translocated across the Golgi where they acquire complex types of carbohydrates which are now resistant to Endo H. Incubation times can be varied and one can use Endo H resistance as a measure of the time taken by secretory proteins for transport from the ER to the Golgi.

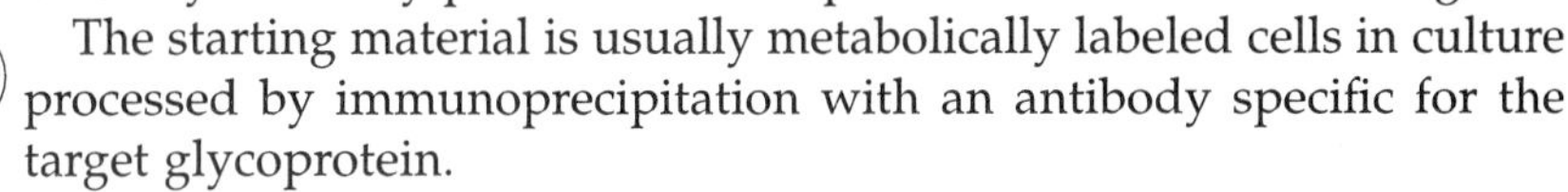

The starting material is usually metabolically labeled cells in culture processed by immunoprecipitation with an antibody specific for the target glycoprotein.

Materials

Reaction buffer: 50 mM sodium citrate, pH 5.5, 0.1% SDS, 50 mM 2-ME
Endo H
4 × SDS sample buffer: See Appendix C

1. Following the final wash of the immunoprecipitated complex (Protocol 3.11), add 30 μl of reaction buffer to the immunoprecipitate, and boil for 5 min.
2. Divide the supernatant into two equal fractions.
3. Add Endo H (1.3–2.0 mU) to one fraction. Add only reaction buffer to the second fraction (mock digest). Digest for 20–24 h at 37°C with shaking.
4. Terminate the reaction by adding 1/4 of the reaction volume of 4 × SDS sample buffer and analyze by SDS-PAGE followed by fluorography or phosphorimaging.

Neuraminidase (NA)

The term "sialic acid" first appeared in the literature in 1952 to describe an unusual acidic aminosugar present in gangliosides and submaxillary mucin (Blix et al, 1952). Nomenclature designated "neuraminic acid" as the unsubstituted parent structure and "sialic acid" as the generic term for the family of related derivatives having an acyl group on the amino nitrogen and frequently other substituents elsewhere. The most common of the sialic acids is N-acetylneuraminic acid (NANA) shown in Figure 9.7.

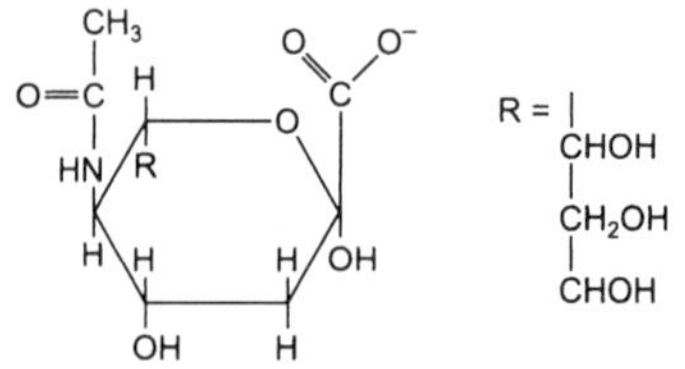

FIGURE 9.7 A schematic representation of N-acetylneuraminic acid.

Sialylation is a posttranslational modification of glycoproteins implicated in the regulation of processes as diverse as receptor-mediated endocytosis, protein targeting, cell adhesion, virus-host recognition, and hormone signal transduction (Ashwell and Harford, 1982; Rutishauser et al, 1988; Lasky, 1992; Stockell Hartree and Renwick, 1992). Sialic acids occur on glycoproteins in a wide variety of structures, the most common linkages are Siaα2—6Gal, Siaα2—3Gal, Siaα2—6GAlNac, and Siaα2—8Sia. Sialic acids are added to membrane and secretory glycoproteins during their posttranslational processing in the Golgi complex to become the terminal sugars on N- and *O*-linked oligosaccharides (Kornfeld and Kornfeld, 1985). Two sialyltransferases are responsible for adding sialic acids in a linkage specific manner to the Gal residues of nascent complex-type N-linked oligosaccharides: the β-galactoside α2,3-sialyltransferase and the β-galactoside α2,6-sialyltransferase (Weinstein et al, 1987; Wen et al, 1992). It is important to keep in mind that not all mammalian cell types produce all of the modifying enzymes. For example, Chinese hamster ovary cells do not have the capacity for an α-2,6-sialylation (Lee et al, 1989). A protocol for sialyltransferase assay is presented by Fayos and Bartles, (1994).

Neuraminidase (sialidase) hydrolyzes terminal α2,3-, α2,6- and α2,8-ketosidic bonds which join sialic acid to oligosaccharides. The enzyme from *Clostridium perfringens* hydrolyzes α2,3 bonds most rapidly, while the enzyme produced by *Arthrobacter ureafaciens* prefers α2,6 bonds. Both enzymes hydrolyze α2,8 bonds more slowly than α2,3 bonds. Either enzyme will hydrolyze terminal links between NANA and Gal, NANA and GlcNAc or other N-acetylhexosamine and N-acetylneuraminic acid.

PROTOCOL 9.10 Desialylation with *Clostridium perfringens* Neuraminidase

The method described below removes sialic acid residues from glycoproteins so that parameters of the glycoprotein can be compared in the sialylated versus the desialylated form of the glycoprotein.

Materials

Neuraminidase: *Clostridium perfringens* Neuraminidase (Calbiochem)

Neuraminidase buffer: 50 mM sodium acetate, pH 5.5, 10 mM calcium acetate, 3 mM sodium azide

Microcentrifuge

1. Isolate the target protein by immunoprecipation (Protocol 3.11).
2. Resuspend the immunoprecipitated complex in 1 ml of neuraminidase buffer and divide it into two equal fractions.
3. Centrifuge for 1 min at maximum speed and remove the neuraminidase buffer.
4. Incubate one fraction with 2 units/ml of *C. perfringens* neuraminidase in 50 μl of neuraminidase buffer at 37°C for 4 h, and in parallel, incubate the second fraction (mock digest) in neuraminidase buffer only.
5. Analyze by SDS-PAGE followed by lectin blotting (Protocol 7.14).

PROTOCOL 9.11 Desialylation with *Arthrobacter ureafaciens* Neuraminidase

Neuraminidase can be used to remove sialic acids from target proteins present in a complex mixture like tissue culture conditioned media. Glycoproteins are metabolically labeled (Protocol 3.1) and the neuraminidase reaction is performed prior to immunoprecipitation. Alternatively, neuraminidase digestion can be performed immediately following immunoprecipatation. Both protocols (A and B) are described below.

Materials

Neuraminidase: *Arthrobacter ureafaciens* neuraminidase (81.6 units/mg) (Calbiochem)

SDS: 10% (w/v): see Appendix C

Neuraminidase buffer: 20 mM sodium acetate, pH 6.0, 2.5% (v/v) NP-40, 0.16% SDS, 3 mM sodium azide

4 × SDS sample buffer: see Appendix C

A: Desialylation of Conditioned Media

1. Add 0.005 units of neuraminidase to 100 μl of conditioned media. Add an equivalent volume (5 μl) of water to the mock digest controls. Incubate at 37°C for 1 h.
2. Isolate the target molecule by immunoprecipitation and analyze the products by SDS-PAGE and fluorography or phosphorimaging.

B: Desialylation of Immunoprecipitated Material

1. Isolate the target protein by immunoprecipitation (Protocol 3.11).
2. Elute the product by boiling the beads for 3 min in 0.5% SDS.
3. Divide the immunoprecipitated material equally and treat one fraction with 2 units/ml of neuraminidase in neuraminidase buffer for 18 h at 37°C and buffer treat (mock digest) the other.
4. Stop the reaction by adding 1/4 of the total volume of 4 × SDS sample buffer.
5. Analyze the products by SDS-PAGE followed by fluorography running the digested and mock samples side by side.

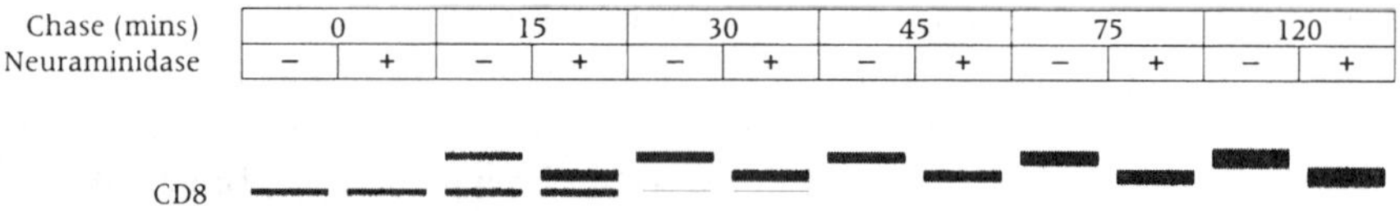

Chase (mins)	0		15		30		45		75		120	
Neuraminidase	–	+	–	+	–	+	–	+	–	+	–	+

CD8

FIGURE 9.8 Neuraminidase digestions of pulse-chased HeLa cells transfected with CD8. Transfected HeLa cells were labeled with [^{35}S]methionine/cysteine for 20 min and chased for 0, 15, 30, 45, 75, or 120 min. Proteins were immunoprecipitated with OKT8 (anti-CD8) antibody, and immunoprecipitates were mock-digested (–) or digested (+) with neuraminidase and analyzed by SDS-PAGE and fluorography. (Adapted with permission from Ponnambalam et al, 1994.)

Comments: An example of this experimental approach is shown in Figure 9.8 using the processing of CD8 (Ponnanbalam et al, 1994).

Lectins as Tools for Carbohydrate Analysis

Lectins are carbohydrate-binding proteins that possess characteristic, high-affinity binding specificity for various sugar residues. Due to their affinity for sugars, they can be used to discriminate between glycosylated and nonglycosylated or deglycosylated proteins. On this basis, they are useful for evaluating the removal of sugar chains from glycoproteins by enzymes like N-Glycanase, Endo H, Endo F, neuraminidase, or other similar enzymes.

Neuraminidase treatment is routinely followed by SDS-PAGE and lectin blotting to further analyze the oligosaccharide structure. *Maackia amurensis* agglutinin (MAA) is a lectin specific for sialic acid linked α2–3 to galactose. Treatment with neuraminidase should abolish staining of glycoproteins with this type of sialic acid linkage. Following NA treatment, staining with *Sambucus nigra* agglutinin (SNA), a lectin specific for α2–6-linked sialic acid, and with wheat germ agglutinin (WGA), a lectin specific for both sialic acid and GlcNAc, should also be lost. However, neuraminidase treatment followed by blotting with peanut agglutinin (PNA), which binds only to the unsubstituted Ser/Thr linked oligosaccharide Galβ1–3 GalNAc unit should reveal glycoproteins with this linkage which were previously masked by the sialic acid residues. Subsequent treatment with *O*-glycosidase should remove PNA reactive disaccharides from the glycoprotein.

Erythrina cristagalli (ECA) binds specifically to the Galβ1–4GlcNAc moieties in N-linked glycans. Sialic acid substitution on this structure prevents its binding (Low et al, 1994). Upon removal of the sialic acid moieties, the glycoprotein should now be recognized by ECA.

Carboxypeptidase Y is a positive control for GNA (*Galanthus niralis*). Fetuin is a negative control for GNA and a positive control for SNA, MAA and DSA (*Dutura stramonium agglutinin*).

Probing a blot with a panel of lectins is described in Protocol 7.14.

Proteoglycans

Proteoglycans are abundant tissue components consisting of a core protein to which a variable number of glycosaminoglycan (GAG) chains are attached through a xyloside residue linked to a serine residue of the protein (Roden and Smith, 1966). The substituted serine residues are followed by a glycine residue. The significance of the Ser-Gly dipeptide in the glycosaminoglycan protein linkage is indicated by the fact that peptides containing this sequence can serve as biosynthetic acceptors for xylosyltransferase-catalyzed linkage of xyloside to peptide. An assay for xylosyltransferase is described in detail by Bourdon et al (1987). However, it is clear that additional signals must be involved in the biosynthetic recognition of the core proteins by the xylosyltransferase, since most proteins that contain Ser-Gly sequences are obviously not proteoglycans, and all Ser-Gly sequences of proteoglycan core proteins are not substituted.

The requirement for Gly to follow the Ser differs from the requirement for the attachment of *O*-glycosidically linked oligosaccharide units to proteins. In that case the linkage is also to serine or threonine, but there does not appear to be any requirement for a glycine next to it (Baenziger and Kornfeld, 1974).

There are four main forms of glycosaminoglycans: heparan sulfate and heparin; chondroitin sulfate and dermatan sulfate; keratan sulfate; and hyaluronic acid. The first three are protein-bound glycosaminoglycans in their natural form, and they all contain sulfate; hyaluronic acid is made as a free glycosaminoglycan and lacks sulfate (Ruoslahti and Yamaguchi, 1991). Each of the glycosaminoglycans, particularly the sulfated ones, has strong negative charges. Proteoglycans are involved in a variety of molecular interactions, which may be mediated by the GAG component or by the core protein.

PROTOCOL 9.12 Is the Target Protein a Proteoglycan?

This protocol is used to determine whether the target protein contains glycosaminoglycans (Barnea et al, 1994).

Materials

$[^{35}S]$methionine

$[^{35}S]$sulfate: (ICN)

Chondroitinase buffer: 100 mM Tris, 30 mM sodium acetate, adjusted to pH 8.3 with HCl

Chondroitinase ABC: Seikagaku America, Inc.

1. Metabolically label (Protocol 3.1) cells with $[^{35}S]$methionine (100 μCi/ml) for 5 h or with $[^{35}S]$sulfate (200 μCi/ml) for 20 h in sulfate-free medium.
2. Prepare cell lysates and perform the immunoprecipitation as described in Protocol 3.11.
3. Wash the immunoprecipitates twice with chondroitinase buffer.
4. Resuspend the immunoprecipitates in 100 μl of chondroitinase buffer. Divide the immunoprecipitates in half. Add protease-free chondroitinase

ABC to a final concentration of 0.5 milliunit/μl to one [^{35}S]sulfate immunoprecipitate and to one [^{35}S]methionine labeled immunoprecipitate. Incubate all four samples at 37°C for 1 h with agitation.

5. Analyze the samples by SDS-PAGE followed by fluorography or phosphorimaging.

Comments: Chondroitinase ABC is an enzyme that specifically depolymerizes chondroitin sulfate to disaccharides.

The results are analyzed by comparing the migration patterns of the treated samples. The [^{35}S]methionine labeled sample that was not chondroitinase treated should contain the intact proteoglycan which characteristically migrates as a broad band that does not efficiently enter the gel. When the [^{35}S]methionine sample is treated with chondroitinase ABC, a new band appears which migrates faster than the untreated sample band and with a more uniform mobility. Compare the [^{35}S]sulfate labeled lanes. [^{35}S]sulfate labels the sulfated glycosaminoglycan chains. If treatment with chondroitinase ABC results in the disappearance of the [^{35}S]sulfate labeled band, which should be present in the [^{35}S]sulfate undigested lane (which should be similar to the [^{35}S]methionine undigested band), the target protein is expressed in the form of a proteoglycan.

B. Phosphorylation

Although prokaryotes mainly phosphorylate histidine, glutamic acid and aspartic acid residues during the sensory transduction in bacterial chemotaxis (Sanders et al, 1989), phosphorylation of proteins on serine, threonine and tyrosine residues is recognized as a key mode of signal transduction in eukaryotic cells. Virtually any basic process of a eukaryotic cell is regulated at some point by the phosphorylation of one or more of its key protein components (Krebs, 1994). Only a few modifications have been shown to be reversible or to be of regulatory significance in biological processes. The best studied of these is protein phosphorylation. Two counteracting enzyme systems, kinases and phosphatases, modulate the level of protein phosphorylation (Gudepu and Wold, 1996).

Thousands of proteins are expressed in a typical mammalian cell, of which at least 30% are thought to contain covalently bound phosphate. (Ficarro et al, 2002). Protein kinases are coded by more than 2000 genes and thus constitute the largest single enzyme family in the human genome. Serine and threonine residues undergo phosphorylation more often than tyrosine. The phosphoamino acid content ratio (pSer:pThr:pTyr) of a vertebrate cell is 1800:200:1 (Mann et al, 2002).

The protein kinase family of enzymes can be classified into ~140 subfamilies based on their primary structures and substrate specificities. The majority of the protein kinases catalyze the phosphorylation of serine and threonine residues. However, there are at least 91 kinases that phosphorylate tyrosine residues. Tyrosine kinases can be subdivided into non-receptor tyrosine kinases and receptor tyrosine kinases, integral membrane proteins composed of an extracellular ligand binding domain, a membrane spanning region, and an intracellular kinase domain. Two examples of this class of molecule are the insulin receptor and epidermal growth factor receptor (Schlessinger, 2000).

The field of protein phosphorylation has grown exponentially in recent years as researchers from various disciplines have come to realize that key cellular functions are regulated by protein phosphorylation and dephosphorylation. The state of phosphorylation of a protein is generally highly dynamic and is modulated antagonistically by protein kinases and protein phosphatases. Protein phosphorylation on Ser, Thr and Tyr usually occurs at residues that are exposed on the surface of the protein, often on loops or turns, ensuring accessibility for the kinase.

The functional consequences of protein phosphorylation provide insights into how biological systems are controlled at the molecular level. Protein phosphorylation regulates many cellular functions including cell growth and differentiation, cell shape and locomotion, metabolism and intracellular signaling. Reversible protein phosphorylation is an essential protein modification which controls biological activities and functions by altering the catalytic activity and substrate specificity of enzymes and by modulating the stability and subcellular localization of regulatory proteins.

The reversible modification of protein function by phosphorylation and dephosphorylation reactions has been clearly established as a major component of both metabolic regulation and signal transduction pathways. The level of protein phosphorylation is controlled by modulation of both protein kinases, enzymes that phosphorylate proteins in a specific manner, and phosphatases, enzymes that specifically remove phosphate groups from proteins. Molecular biological techniques have fueled the growing interest in protein phosphorylation. The research community has reached the point where computer analysis of an amino acid sequence or cDNA leads to the realization that a molecular clone encodes a potential protein kinase (Sefton and Hunter, 1991).

Protein kinases have been unofficially classified based on the acceptor amino acid. The three major classifications are: (1) serine and threonine specific protein kinases; (2) tyrosine specific protein kinases; and (3) protein histidine kinases (Hunter, 1991).

Despite the importance and the widespread occurrence of protein phosphorylation, identifying sites of protein phosphorylation is still a challenge (Annan et al, 2001). Peptide and protein chips could be used for the identification and characterization of new substrates of protein kinases and for exploring mechanisms underlying the interactions among protein kinases (Schlessinger, 2002).

The aims of protein phosphorylation studies are threefold: (1) to determine the amino acid residues (sites) that are phosphorylated *in vivo* on a specific protein in a given cell in response to a given stimulus, (2) to identify the kinase responsible for the phosphorylation; and (3) to understand the functional consequences of the phosphorylation event.

Analytical strategies include, where possible, the generation of [^{32}P]-labeled proteins by incubation of cells, tissues, or cell free systems with [^{32}P]-orthophosphate, [γ ^{32}P] ATP, and [γ ^{32}P] GTP. The presence of a specific kinase and a protein substrate produces phosphorylated amino acid residues that can be traced by chromatographic methods described below. The phosphorylated proteins must be distinguished from other compounds such as phosphocholine, phosphoethanolamine, and phosphoinositols. The different reactivities exhibited by phosphorylated amino acids for hydrolysis with base or acid are useful for analysis.

Methods that are routinely used for the detection of phosphopeptides may also include some of the following strategies used either individually or in combination.

- Enrichment of phosphopeptides by immobilized metal affinity chromatography (IMAC).
- Treatment with alkaline phosphatase: Complex phosphopeptide mixtures can be analyzed by comparing alkaline phosphatase digested peptides with the same population of undigested peptides. A decrease in the mass of a peptide by 80 Da implies the presence of one phosphate group (Powell et al, 2000).

Although phosphorylation can occur to a variety of amino acid residues, by far the most common and important sites of phosphoylation in eukaryotes occur on serine, threonine and tyrosine residues. O-phosphates (O-phosphomonoesters) are formed by phosphorylation of the hydroxyamino acids serine, threonine and tyrosine. (2) N-phosphates (phosphoamidates) are generated by phosphorylation of the amino groups in arginine, lysine and histidine. (3) Acylphosphates (phosphate anhydrides) are produced by the phosphorylation of aspartic or glutamic acid. (4) S-phosphates (S-phosphothioesters) are formed by phosphorylation of cysteine (Sickmann and Meyer, 2001).

Analysis of phosphoproteins is not straightforward. The stoichiometry of phosphorylation is generally very low, only a small fraction of the available intracellular pool of a protein is phosphorylated at any given time. Phosphorylation sites on proteins may vary, most phosphoproteins undergo phosphorylation on more than one residue, suggesting that a phophorylated gene product is heterogenous (i.e. it exists in several different phosphorylated forms). Many phosphoproteins are present in low abundance, making enrichment a prerequisite prior to analysis.

PROTOCOL 9.13 Metabolic Labeling of Cells with [^{32}P]orthophosphate

Metabolic radiolabeling remains the method of choice for studies involving *in vivo* protein phosphorylation. Cells or tissue to be labeled are incubated with $^{32}PO_4$ for a time long enough to equilibrate the cellular ATP pool with ^{32}P. The radiolabeled ATP is then used by protein kinases to phosphorylate their substrates.

Cells can be metabolically labeled by adding [^{32}P]orthophosphate to the tissue culture media. In this fashion, proteins that are phosphorylated are now radiolabeled and their fates can be followed.

Materials

Phosphate-free Krebs-Ringer buffer plus 20 mM HEPES, pH 7.3, 0.1% BSA, and 0.2% glucose. Alternatively, any buffer or media that does not contain phosphate can be used, like phosphate free Dulbecco's modified Eagle medium with 2% dialyzed fetal calf serum.

[^{32}P]orthophosphate

1. Wash then starve (Protocol 3.1) subconfluent cultures (0.5–1 × 10^6 cells/35 mm dish) for 60 min in phosphate-free Krebs-Ringer buffer.
2. Label the cells by adding 0.5 mCi of [^{32}P]orthophosphate directly to the plate and incubate for 4 h at 37°C.
3. Analyze the products by SDS-PAGE and autoradiography (Protocol 7.22). All phosphorylated proteins will be radioactive.

PROTOCOL 9.14 Can the Target Protein Be Phosphorylated?

If you suspect that the protein of interest is, or can be phosphorylated, an *in vitro* phosphorylation reaction may be performed, preferably on the purified target protein (Robinson et al, 1993).

Materials

Reaction buffer: 30 mM Tris-HCl, pH 7.4, 1 mM $MgCl_2$, 1 mM EGTA, 40 μM ATP and 3.75 μCi[γ-^{32}P]ATP in the presence (where appropriate) of 10 μM cGMP or 200 μM Ca^{2+} plus 10 mg/L phosphatidylserine

Protein Kinases

4 × SDS sample buffer: see Appendix C

1. Measure out ~ 0.25 μg of the target protein.
2. Incubate the protein in 40 μl of reaction buffer.
3. To initiate the reactions, add protein kinases (20 ng). Try protein kinase C, cAMP dependent protein kinase, cGMP-dependent protein kinase, and casein kinase II. In parallel, set up a mock reaction containing all components except for the exogenously added kinase.
4. Perform the reaction at 30°C for 30 min.
5. Add 1/4 of the total reaction volume of 4 × SDS sample buffer, boil for 3 min and analyze the reaction products by SDS-PAGE followed by autoradiography (Protocol 7.20).

Comments: The choice of kinases need not be limited to the few that are mentioned.

Combinations of kinases can also be used. Reaction products can be analyzed further by phosphopeptide and phosphoamino acid analysis described below.

In vitro protein phosphorylation experiments are generally performed using [γ-^{32}P]ATP as the source of the radiolabel.

This is a point in the analysis where many conditions can be changed to examine the behavior of the target protein.

PROTOCOL 9.15 Determination of the Type of Phosphorylated Amino Acid-Immunoblotting with Anti-Phosphoamino Acid Antibodies

The use of antibodies specific for particular phosphoamino acids is a corroborative method to determine the type of phosphorylated amino acid. However, phosphoamino acid analysis remains the gold standard. Many companies have produced reliable anti-phosphotyrosine mouse monoclonal antibodies that can be used for immunoprecipitation and Western blotting. If you suspect that the target protein is phosphorylated on a tyrosine residue, performing this protocol is an important early analytical step.

Materials

Blocking buffer: 1% BSA in TTBS (see Appendix C)

Lysis buffer: 20 mM Tris-HCl, pH 7.8, 150 mM NaCl, 2 mM EDTA, 10 mM NaF, 100 µM Na orthovanadate, 10 µM leupeptin, 10 µM aprotinin, 2 mM PMSF

Anti-phosphotyrosine antibody: PY20 (ICN Biochemicals), 4G10 (UBI), or RC10 (Transduction Labs)

Donkey anti-mouse conjugated to horseradish peroxidase

Instant non-fat dried milk

1. Solubilize the cells producing the phosphotyrosine containing target protein.
2. Perform the immunoprecipitation (Protocol 3.11) on the lysate using specific antibody to your target protein.
3. Separate the proteins by SDS-PAGE. Transfer the proteins to a PVDF membrane (Protocol 7.1).
4. Block the membrane for at least 1 h in blocking buffer or overnight at 4°C.
5. Incubate the membrane in anti-phosphotyrosine antibody (1 µg/ml) for at least 1 h or overnight at 4°C.
6. Wash the membrane with 4 changes of TTBS for 5 min each.
7. Incubate the membrane in a 1:10,000 solution of donkey anti-mouse conjugated to horseradish peroxidase in TTBS.
8. Wash the membrane extensively with TTBS and visualize the protein bands by chemiluminescence (Protocol 7.20).

Comments: Phosphoserine and phosphothreonine specific antibodies have been described and are commercially available although not as reliable as the anti-phosphotyrosine antibodies.

These antibodies can also be used in the straight Western format.

PROTOCOL 9.16 Phosphorylation of Membrane Proteins with [γ ^{32}P]GTP

This protocol uses membranes prepared from the human leukemia cell line HL-60 to investigate phosphorylation of G proteins with the natural nucleotide GTP (Wieland et al, 1993). Note that the membrane preparation contains enzymes and substrates.

Materials

Reaction buffer: 50 mM triethanolamine, pH 7.4, 150 mM NaCl, 5 mM $MgCl_2$, 1 mM EDTA, 1 mM DTT

[γ ^{32}P]GTP: (~5000 Ci/mmol) Amersham

Preparation of Membrane resuspended in 10 mM triethanolamine HCl, pH 7.4, (Wieland et al, 1993)

4 × SDS sample buffer: see Appendix C

1. Prewarm 50 µl of reaction buffer to 30°C and add 1 µCi/tube of [γ-^{32}P]GTP.
2. Start the reaction by adding ~20 µg of membrane protein. Perform the reaction at 30°C for 30 min. The time interval of the reaction can be extended or shortened according to the results of a small scale pilot experiment.
3. Stop the reaction by adding 1/4 vol of 4 × SDS sample buffer. If concerned with phosphohistidine residues, do not heat the sample prior to electrophoresis.
4. Analyze the products by SDS-PAGE.

Enzymatic Dephosphorylation

Phosphate groups can be removed from phosphorylated proteins by the action of phosphatases. Two methods are presented to analyze phosphorylated proteins. The first method uses potato acid phosphatase (Mizutani et al, 1993) and the second technique uses calf intestinal phosphatase (Carmo-Fonseca et al, 1993) which is active at alkaline pH. The dephosphorylation reaction can be performed on the purified product following immunoprecipitation and immediately prior to SDS-PAGE or following transfer to a PVDF membrane (Harper et al, 1993). In Figure 9.9, whole cell extracts were either treated with acid phosphatase (Lane B) or untreated (Lane A). Following SDS-PAGE and transfer, the blot was probed with a specific antibody which no longer recognizes the dephosphorylated bands (Lane B) suggesting that the antibody recognizes only the phosphoprotein.

PROTOCOL 9.17 Potato Acid Phosphatase

In this Protocol, the protein of interest is isolated by immunoprecipitation (Protocol 3.11). The dephosphorylation reaction is performed on the immune complex. Following SDS-PAGE and transfer to PVDF, the membrane is probed with target specific antibody.

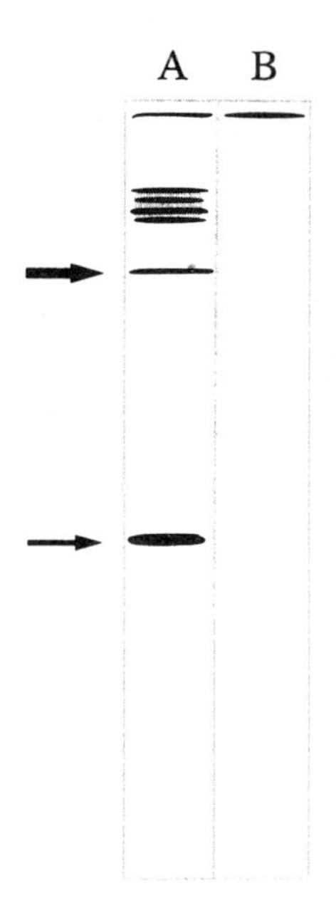

FIGURE 9.9 Acid phosphatase treatment of the blot prevents reactivity with specific antibody. Whole cell extracts of deflagellated *Chlamydomonas* were electrophoresed in duplicate. Lane A was incubated for 8h in phosphatase buffer. Lane B was incubated for 8h in phosphatase buffer containing 33U/ml acid phosphatase. Both lanes were then incubated with an antibody, MPM-2, which recognizes a 90 and a 34kDa protein. Treatment with acid phosphatase resulted in the loss of MPM-2 recognition of the 90kDa (thick arrow) and the 34kDa (thin arrow) proteins. (Adapted with permission from Harper et al, 1993.)

Materials

2 × SDS sample buffer: see Appendix C

Storage buffer: 10mM HEPES, pH 7.5, 0.5mM $MgCl_2$, 0.5mM DTT, 50% glycerol

Potato acid phosphatase: Potato acid phosphatase is usually supplied as an ammonium sulfate suspension. To prepare, mix the vial then measure out the desired amount of enzyme and centrifuge suspension for 10min at maximum speed in a microfuge. Dissolve the pellet in storage buffer then dialyze the acid phosphatase solution against storage buffer and store aliquotted at –80°C until use.

Reaction buffer: 50mM MOPS, pH 6.0, 1mM DTT, 10μg/ml leupeptin

RIPA buffer plus: RIPA buffer (see Appendix C) plus 200μM Na_3VO_4, 10mM β-glycerophosphate, 5mM NaF and 1mM p-nitrophenyl phosphate

Wash buffer: 10mM Tris-HCl, pH 7.4, 140mM NaCl

1. Isolate the target protein by immunoprecipation (Protocol 3.11).
2. Resuspend the beads in reaction buffer and divide them into two equal fractions.
3. Wash the fractions twice with reaction buffer and resuspend each in 50μl of reaction buffer containing 0 (mock control) and 2μg of potato acid phosphatase.
4. Incubate the reactions at 30°C for 15min then remove the acid phosphatase by washing the immunoprecipitates three times in RIPA buffer plus and once with wash buffer.
5. Add an equal volume of 2 × SDS sample buffer to the beads and boil for 3min. Analyze by SDS-PAGE followed by blotting with antibody to the target protein.

PROTOCOL 9.18 Alkaline Phosphatase

Materials

Calf intestinal phosphatase (CIP) buffer: 50 mM Tris-HCl, pH 9.0, 1 mM $MgCl_2$, 0.1 mM $ZnCl_2$, 1 mM spermidine, 0.1 mM PMSF and 10 μg/ml of each of the following protease inhibitors: chymostatin, leupeptin, antipain, and pepstatin

Alkaline phosphatase: (bovine intestinal mucosa, P-7915 Sigma)

β-glycerophosphate:1 M

Wash buffer: 10 mM Tris, pH 4.5

2 × SDS sample buffer: see Appendix C

1. Isolate the target protein by immunoprecipation (Protocol 3.11) then resuspend samples in 30 μl CIP buffer.
2. Add ~10 Units of alkaline phosphatase to 30 μl of resuspended immune complex Sepharose protein A slurry. As a control to be performed in parallel, add the enzyme to an immunoprecipitated sample in the presence of 100 mM β-glycerophosphate.
3. Incubate the samples for 30 min at 30°C. Wash the beads twice with 10 mM Tris pH 4.5. Add an equal volume of 2 × SDS sample buffer to the beads and boil for 3 min. Analyze by SDS-PAGE followed by blotting with antiphosphotyrosine antibody (Protocol 9.13 step 3). If the target protein was tyrosine phosphorylated and dephosphorylated, the band will be decreased or absent compared to the untreated control.

Comments: The dephosphorylation reaction can also be performed on proteins after they are transferred to a membrane.

Using alkaline phosphatase treatment the phosphopeptide can be identified by MALDI-TOF-MS as shown in Figure 9.10.

PROTOCOL 9.19 Immune Complex Kinase

If you suspect that your target protein is a kinase, you may perform a kinase reaction directly on the immunoprecipitated complex. The products are separated by SDS-PAGE and then analyzed by autoradiography (Protocol 7.20). This technique is also referred to as the *in vitro* kinase assay.

Materials

Lysis Buffer: 20 mM Tris-HCl, pH 7.5, 150 mM NaCl, 0.5% Triton X-100, 2 mM Na_3VO_4, 30 mM Na pyrophosphate, 50 mM NaF, 1 mM PMSF added just prior to use

Kinase Buffer: 20 mM HEPES pH 7.4, 100 mM NaCl, 15 mM $MnCl_2$, 0.1% Triton X-100

[γ^{32}P]ATP: 3000–6000 Ci/mmol

PBS: see Appendix C

2 × SDS sample buffer: see Appendix C

20 mM HEPES, pH 7.4

Microcentrifuge

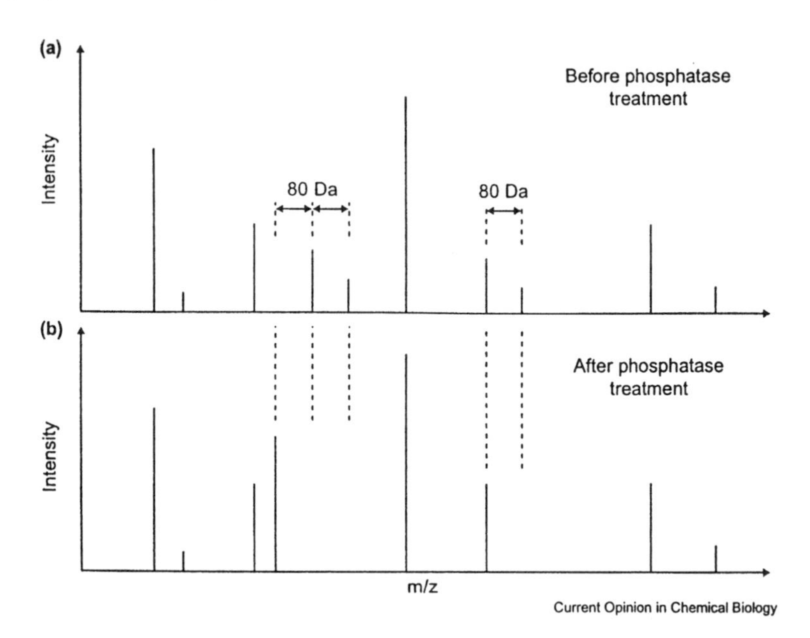

FIGURE 9.10 Phosphopeptide identification by MALDI-TOF-MS mapping combined with alkaline phosphatase treatment. (a) The MALDI-TOF-MS spectrum of a proteolytic digest. Phosphoproteins are indicated by peaks shifted by multiples of 80 Da (HPO_3 = 80 Da) relative to predicted unphosphorylated peptide masses. (b) The disappearance of such peaks upon treatment with a phosphatase confirms their identity as phosphopeptides (Adapted with permission from Mclachlin and Chait, 2001.)

KOH: 1 M

Fixative: 20% methanol, 10% acetic acid

X-ray film

1. Wash the cells 3 times with PBS. Add 1 ml of lysis buffer to a 10 cm dish. Collect the lysate in a 1.5 ml microcentrifuge tube and incubate at 4°C for 30 min. Centrifuge for 5 min at maximum speed at 4°C to remove nuclei and debris.
2. Do not preclear. Perform the immunoprecipitation (Protocol 3.11) with immune serum and protein A-Sepharose for 3 h, shaking in the cold.
3. Wash three times with lysis buffer, then twice with kinase buffer.
4. Incubate the beads in 60 μl of kinase buffer and add 10 μCi of [$\gamma^{32}P$]ATP and incubate at 25°C for 30 min.
5. Wash the beads twice with 20 mM HEPES pH 7.4 and transfer to a fresh tube.
6. Add an equal volume of 2 × SDS sample buffer and boil for 3 min. Separate the proteins by SDS-PAGE. Dry the gel (Protocol 4.12) and expose to X-ray film for 1–16 h at room temperature. The gel may crack if film is exposed at –70°C. Alternatively, place the dried gel in a phosphorimage cassette. If at this juncture you suspect that your target protein is Tyrosine phosphorylated, follow steps 7 through 9 below.
7. To remove P-Ser and P-Thr, incubate the dried gel in 1 M KOH for 2 h at 55°C. The gel will be very delicate. Care should be taken to avoid breakage.

8. Pour off the KOH solution carefully and incubate the gel in fixative for 30 min.
9. Dry the gel and expose it to X-ray film.

Comments: After performing the *in vitro* kinase labeling reaction, it is often desirable to further fractionate the products. The phosphoproteins can be bound to a column of anti-phosphotyrosine coupled to agarose beads (Chapter 10). The tyrosyl phosphorylated proteins are specifically eluted by incubating the complexes with 50 mM p-nitrophenyl phosphate. These proteins can either be analyzed by SDS-PAGE or reimmunoprecipitated.

After KOH hydrolysis the dried gels require an overnight exposure at –70°C. The resultant autoradiographic signal will be largely P-Tyr. However, in some cases, depending on the surrounding sequence, some P-Ser and P-Thr that are difficult to hydrolyze will be carried through. The P-Ser and P-Thr bands will either be removed completely or their intensities markedly decreased. The kinase buffer can be modified to include any other component that you might want to add (i.e., Mg^{2+}, dithiothreitol). The pH can also be changed and other buffers such as MES and Tris can also be used. In addition, the amount of [γ^{32}P]ATP can also be changed.

PROTOCOL 9.20 Renaturation of Immobilized Kinases on PVDF Membranes

Proteins from cell extracts are separated by SDS-PAGE and transferred to PVDF membranes where they are exposed to guanidine HCl and then allowed to renature (Ferrell and Martin, 1991). A phosphorylation reaction, likely autophosphorylation, is carried out by incubating the filter-bound proteins with [γ-^{32}P]ATP and kinase activity identified. The method takes advantage of the ability of SDS-denatured enzymes to regain at least partial activity after treatment with guanidine and a nonionic detergent. After locating phosphorylated proteins, the blot can be reblocked and probed with antibodies to determine the positions of known protein kinases. Tyrosine kinases renature very poorly, if at all, using this method (Grinstein et al, 1993).

Materials

Plastic containers

Lysis buffer: 50 mM Tris-HCl, pH 7.4, 100 mM NaCl, 1% NP-40,1% DOC, 0.1% SDS, 10 mM EDTA, protease inhibitors 10 µg/ml, leupeptin and aprotinin, 200 µM PMSF and 1 mM sodium orthovanadate.

Denaturation buffer: 50 mM Tris-HCl, pH 8.3, 7 M guanidine HCl, 50 mM DTT, 2 mM EDTA

TBS: 10 mM Tris HCl, pH 7.6, 150 mM NaCl

Renaturation buffer: 10 mM Tris-HCl, pH 7.4, 140 mM NaCl, 2 mM DTT, 2 mM EDTA, 1% (w/v) BSA, 0.1% NP-40

Blocking solution: 30 mM Tris-HCl, pH 7.4, 5% BSA

Reaction buffer: 30 mM Tris-HCl, pH 7.4, 10 mM $MgCl_2$, 2 mM $MnCl_2$, 100 μCi/ml [γ-^{32}P]ATP, (3000–6000 Ci/mmol)

Wash buffer A: 30 mM Tris-HCl, pH 7.4

Wash buffer B: 30 mM Tris-HCl, pH 7.4 + 0.05% NP-40

KOH: 1 M

4 × SDS sample buffer: see Appendix C

1. Lyse the tissue culture cells (growing either adherent or in suspension) with lysis buffer, 1 ml/10 cm dish.
2. Centrifuge the lysate for 5 min at maximum speed in a microfuge. Save the supernatant and determine the protein concentration (Chapter 5). Add 1/4 the volume of 4 × SDS sample buffer to 100 μg of protein.
3. Boil the sample for 3 min and load onto an SDS gel (Protocol 4.5).
4. Transfer the proteins to PVDF (Protocol 7.1) at 4°C for 90 min at 5 V/cm. For low molecular weight proteins, decrease the voltage. Do not soak the gel in transfer buffer prior to transfer.
5. Incubate the blot in denaturation buffer for 1 h at room temperature with gentle agitation.
6. Pour off the denaturation buffer and rinse the blot briefly with TBS.
7. Incubate the blot in renaturation buffer with gentle shaking overnight at 4°C.
8. Pour off the renaturation buffer and incubate the blot for 1 h at room temperature in blocking solution.
9. Pour off the blocking solution and cover the blot with reaction buffer and incubate with rocking at room temperature for 30 min.
10. Pour off the reaction mixture and **dispose of it according to your institution's guidelines for ^{32}P waste**. Quickly wash the blot with 50 ml of wash buffer A. Transfer the blot to a larger plastic container and follow the washing protocol, 10 min/wash, as follows:

    2 × 250 ml Wash buffer A

    1 × 250 ml Wash buffer B

    2 × 250 ml Wash buffer A

    1 × 250 ml 1 M KOH

    2 × 250 ml Wash buffer A
11. Let the blot air dry (except in cases where proteins will be excised for further analysis in which case do not dry) and expose it to X-ray film overnight at –70°C.

Comments: When the blot is exposed to denaturation buffer (step 5), it will become translucent. After rinsing with TBS (step 6), the blot will become opaque again.

Rubbermaid brand plastic containers are convenient for incubations and washes as they are easy to decontaminate.

A variety of proteins are phosphorylated using this protocol, including autophosphorylation, proteins present in the blocking solution, and proteins comigrating with the renaturable kinases. These products can be further analyzed by eluting them from the blot with 2% SDS, 1% NP-40 in 30 mM sodium phosphate, pH 8.6. Do not allow the blot to dry prior to protein elution.

Two controls should be included to ensure that the band on the film is due to kinase and not to ATP binding. Perform the kinase reaction in parallel with [α-^{32}P]ATP This should be background. In addition, the phosphorylated bands should be subjected to phosphoamino acid analysis as described later in Protocol 9.28.

PROTOCOL 9.21 Phosphorylation of Substrates in SDS-Gels

Protein kinase activity can be detected directly in the gel after SDS-PAGE. If the target protein exhibits kinase activity, the specificity of this activity can be examined by incubating the protein with putative substrates of various kinases. In the method described below (Kameshita and Fujisawa, 1989), kinase substrates are trapped within the matrix of the polyacrylamide gel. By including significant concentrations of non-radioactive ATP, the contribution of kinase autophosphorylation is minimized, revealing primarily the phosphorylation of the exogenous substrate trapped in the gel matrix.

Materials

Myelin basic protein (MBP)

BSA

Wash buffer: 50 mM Tris-HCl, pH 8.0, 20% isopropanol

Equilibration buffer: 50 mM Tris-HCl, pH 8.0, 5 mM 2-ME

Denaturation buffer: 50 mM Tris-HCl, pH 8.0, 5 mM 2-ME, 6 M guanidine HCl

Renaturation buffer: 50 mM Tris-HCl, pH 8.0, 5 mM 2-ME, 0.04% Tween 40

Reaction buffer: 40 mM HEPES-NaOH, pH 8.0, 2 mM DTT, 0.1 mM EGTA, 5 mM $Mg(C_2H_3O_2)_2 \cdot H_2O$, 0.15 mM $CaCl_2$, 14 μg/ml calmodulin

ATP

[γ^{32}P]ATP Fixation buffer: 5% (w/v) trichloroacetic acid (TCA), 1% (w/v) sodium pyrophosphate

Seal-A-meal bags: Kapak Corp, Minneapolis MN

Bag Sealer

X-ray film

1. Choose an appropriate concentration of acrylamide from Table 4.1. Prior to polymerization of the gel, add 0.5 mg of myelin basic protein to the acrylamide mixture. For a negative control, in another gel, add 1 mg/ml BSA.
2. Do not boil the sample prior to electrophoresis. Routinely, run SDS-polyacrylamide slab gels as described in Chapter 4 with a 3% stacking gel. If available, use 1.0 mm thick spacers. Perform a short run of 60–90 min at 25 mA which should produce adequate separation.
3. After electrophoresis, remove the SDS by rinsing the gel with two changes (30 min each) of 100 ml of wash buffer.
4. Incubate the gel for 1 h at room temperature in 250 ml of equilibration buffer.

5. Denature the enzyme by treating the gel first with 2 changes of 100 ml of denaturation buffer at room temperature for 1 h.
6. Renature the enzyme by washing the gel with five changes, 250 ml each of renaturation buffer over 16 h at 4°C.
7. Incubate the gel at room temperature for 30 min with 3.5 ml of reaction buffer in a Seal-a-Meal type of bag.
8. Initiate the phosphorylation reaction by incubating the gel at 22°C for 1 h in reaction buffer containing in addition 50 μM ATP and 50 μCi of [γ^{32}P]ATP.
9. Wash the gel with five changes of 500 ml each with fixation buffer. The radioactivity of the wash solution should become negligible.
10. Dry the gel and expose it to X-ray film or perform phosphorimaging.

Comments: The protein substrate added to the separating gel prior to polymerization appears to be fixed to the gel throughout the electrophoretic run and does not seem to affect the electrophoretic mobility of the protein samples.

Less than 1 μg of a purified kinase can be detected with this method.

Phosphopeptide and Phosphoamino Acid Analysis

By this stage of analysis it should become apparent that the target protein is phosphorylated. The knowledge of the type of amino acid(s) that are phosphorylated can reduce the number of potential phosphorylated sites, greatly simplifying the assignment of the phosphorylated residues within the peptide chain. The type of residue phosphorylated is commonly determined either by phosphoamino acid analysis or by phosphoamino acid-specific immunodetection.

After a phosphopeptide has been identified, one would then like to know which amino acid is modified. In some cases, the peptide will have only one serine, threonine or tyrosine making the assignment of the phosphorylation site easy. In most cases fragmentation of the peptide and characterization of the fragments is required.

Miniaturized immobilized metal affinity chromatography (IMAC) columns have been used for the enrichment of phosphopeptides. IMAC exploits the high affinity of phosphate groups for Fe^{3+} and Ga^{3+}. Phosphopeptides can be released using high pH or phosphate buffer, which would require desalting prior to MS analysis. IMAC has been coupled on-line to MS analysis directly or with intervening separation techniques such as HPLC (McLachlin and Chait, 2001).

Two newer methods have recently been described which chemically modify the phosphate group (Goshe et al, 2001; Zhou et al, 2001).

Peptide maps or fingerprints of proteolyzed proteins are usually obtained by separation on either one-dimensional SDS-PAGE or by two-dimensional separation on thin-layer cellulose (TLC) plates. Phosphopeptide mapping is used to analyze the degree of phosphorylation of the target protein and to identify individual phosphorylation sites. To determine whether treatments with specific stimuli affect phosphorylation, peptide maps from control and treated cells can be compared. The target protein, which can be the native, wild-type molecule or a genetically engineered mutant with an amino acid substitution for the residue suspected of being phosphorylated, is isolated and digested enzymatically into peptides. Analysis of the reaction can be performed by enzymatic or chemical cleavage followed by electrophoresis (Protocol 9.22) (Thomas et al, 1991), or in two dimensions by TLC in two steps (Protocols 9.22–9.25) (Mitsui et al, 1993). Both techniques are described below. These experiments can only be performed if an antibody capable of immunoprecipitating the target protein is available.

PROTOCOL 9.22 One-Dimensional Phosphopeptide Mapping

One dimensional phosphopeptide mapping is a combination of techniques that were previously described i.e., immunoprecipitation, chemical or enzymatic cleavage of proteins, and electrophoresis.

1. Immunoprecipitate the ^{32}P-labeled target protein, isolate it on SDS-PAGE and excise it from the gel (Protocol 4.14).
2. Cleave the target protein with cyanogen bromide as described in Protocol 8.5.
3. Alternatively, excise the band from the gel and digest it with 600 ng of staphlococcal V8 protease during electrophoresis as described in Protocol 8.2.
4. Run the products of the cleavage reaction on a 15% SDS polyacrylamide gel (see Chapter 8). Dry the gel and analyze by autoradiography or phosphorimaging.

Two-Dimensional Phosphopeptide Mapping

Two dimensional phosphopeptide mapping extends the one dimensional analysis, producing a two dimensional array of spots. In 2D-PP mapping, peptides are separated in a first dimension by electrophoresis on a thin-layer cellulose plate and in the second dimension by thin-layer chromatography on the same plate. Separated ^{32}P labeled phosphopeptides can be detected by autoradiography or phosphorimaging. The radiogram can yield a large amount of data. The number of spots provides an estimate of the maximum number of phosphorylated sites. The spot intensity provides an indication of the relative levels of phosphorylation between peptides. The number of spots does not necessarily correlate with the number of phosphorylation sites due to incomplete proteolytic digestion which is frequently observed with phosphoproteins.

Another advantage of 2D-PP mapping is that it produces purified phosphopeptides that can be analyzed by MS after extraction from the plate (Affolter et al, 1994).

The phosphorylated target protein is identified by SDS-PAGE followed by autoradiography (Protocol 7.22). It is then isolated and digested into peptide fragments either chemically or enzymatically. Phosphopeptides are identified by thin layer chromatography on cellulose plates using electrophoresis in the first dimension followed by liquid chromatography in the second dimension. Because cellulose is an inert substance, the peptide can be recovered for further analysis such as determining the presence of phosphoamino acid residues.

PROTOCOL 9.23 Isolation of Phospho-Proteins from SDS Gels: Preparation for Phosphopeptide Mapping

Tryptic phosphopeptide mapping is used to study the phosphorylation status of the target protein. The protein of interest is isolated from ^{32}P-labeled cells, usually by immunoprecipitation. Phosphopeptides are generated by tryptic digest. However, peptides can also be generated by other enzymes or chemically, as described in the previous section.

Materials

Hydration buffer: 50 mM NH_4HCO_3, 0.1% SDS, and 0.5% 2-ME

Disposable tissue grinder: Kontes

BSA: 1 mg/ml

TCA: 100% w/v (See Appendix C)

Ethanol: 95%

1. Excise gel pieces (Protocol 4.14) containing the ^{32}P-labeled protein from the dried SDS gel. Treating the gels with fluors is not recommended.
2. Peel the paper backing from the gel slice. Remove as much of the paper as possible without shaving down the gel slice.
3. Place the gel slice in a Kontes disposable tissue grinder tube. Add 500 µl of freshly prepared hydration buffer. Let the gel piece hydrate for 5 min then grind it. The ground gel should be fine enough to pass through a 200 µl pipette tip. Incubate on ice for 2–3 h. Centrifuge the gel fragments at maximum speed in a microfuge and save the eluate.
4. Add another 500 µl of hydration buffer and re-extract the gel slurry. The combined volume should be 1.0 ml. Incubate on ice for 2–3 h. Centrifuge the gel fragments at maximum speed in a microfuge and save the eluate.
5. Clear the combined eluate by centrifuging in a microfuge for 10 min at maximum speed.
6. Remove the supernatant and add 10 µg/ml of carrier protein (BSA) and precipitate the proteins by adding 250 µl of an ice cold 100% TCA solution (20% final TCA) and incubate on ice for 1 h.
7. Collect the precipitate by centrifugation for 10 min at maximum speed in a microfuge at 4°C. Wash the pellet once with ice cold 95% ethanol and air dry. A small white pellet at the bottom of the tube should be visible.

PROTOCOL 9.24 Tryptic Digestion of Isolated Phosphoproteins

This protocol generates peptides which will be further analyzed in Protocols 9.25–9.27

Materials

Trypsin (TPCK-treated): Prepare stocks of TPCK-treated trypsin and other enzymes in 0.1 mM HCl at a concentration of 1 mg/ml and store in small aliquots at –70°C.

Target protein: Previously labeled and extracted from SDS gels

Performic acid solution: Mix 900 µl of formic acid (98%) and 100 µl of 30% H_2O_2, incubate for 60 min at room temperature, then chill.

Reaction buffer: 50 mM NH_4HCO_3, pH 8.0

1. Dissolve the pellet generated from the previous protocol in 50 µl of cold performic acid solution. Incubate for 60 min on ice. Add 400 µl of water, freeze and lyophilize.
2. Resuspend the lyophilized material in 50 µl of reaction buffer. Add 10 µg trypsin and incubate at 37°C for 2–20 h.
3. Centrifuge the tube for 15 sec and add an additional 10 µg of trypsin and incubate again for 3–5 h at 37°C. Add 400 µl of H_2O, vortex, and lyophilize. Repeat the water addition and lyophilization.
4. Save the lyophilized pellet and proceed with Protocol 9.25.
5. Resuspend the pellet in 6 µl of electrophoresis buffer (pH 1.9 or 4.72. See Table 9.1). The samples are now ready for application onto the TLC plates. Alternatively, they can be stored at –20°C until used.

TABLE 9.1 Composition of Electrophoresis Buffers

Buffer Ingredients	Amount
pH 1.9 buffer	
formic acid (88% w/v)	50 ml
glacial acetic acid	156 ml
deionized water	1794 ml
pH 3.5 buffer	
glacial acetic acid	100 ml
pyridine	10 ml
deionized water	1890
pH 4.72 buffer	
n-butanol	100 ml
pyridine	50 ml
glacial acetic acid	50 ml
deionized water	1800 ml
pH 8.9 buffer	
ammonium carbonate	20 g
deionized water	2000 ml

PROTOCOL 9.25 Applying the Sample to the TLC Plate and Electrophoresis in the First Dimension

The first dimension of separation is the electrophoresis of the sample on a TLC plate. The second dimension is ascending thin-layer chromatography, described in Protocol 9.26. Optimal conditions are often determined empirically. As a starting point try the pH 1.9 buffer described in Table 9.1 because most peptides are soluble in it. Two-dimensional peptide analysis on TLC plates has been performed successfully using the Hunter thin-layer electrophoresis system (HTLE-7000; CBS Scientific, Del Mar, CA). Consult the manual for preparing your particular apparatus for electrophoresis. The Hunter Thin Layer Electrophoresis apparatus HTLE 7000 is recommended. In depth explanations for this equipment is provided by Boyle et al (1991) and van der Geer et al (1993). For a typical run using pH 1.9 buffer, electrophorese for 25 min at 1.0 kV (~21 mA).

Materials

TLC plates: EM Science

The Hunter Thin Layer Electrophoresis apparatus HTLE 7000

Green marker dye solution: 5 mg/ml ε-dinitrophenyl-lysine and 1 mg/ml xylene cyanol FF (blue) in 50% pH 4.72 buffer in deionized water

Electrophoresis buffer: Consult Table 9.1

1. Using an extra soft blunt-ended pencil, mark the sample and dye origins on the TLC plate according to Figure 9.11.
2. Resuspend the pellet in 6 μl of electrophoresis buffer (pH 1.9 or 4.72). The samples are now ready for application onto the TLC plates. Alternatively, they can be stored at −20°C.
3. Before spotting the sample on the origin (+), centrifuge for 5 min at 12,000 × *g*. Spot up to 5 μl on the first dimension origin as shown in Figure 9.11 by carefully touching the surface of the drop to the cellulose. Apply ~ 0.2–0.5 μl drops and dry between applications. Avoid touching the cellulose with the pipette tip. During each application, try to keep the wetted area to a minimum. Usually, a faint brown ring will form around the circumference of the spot.
4. Apply 0.5 μl of the marker dye to the origin at the top of the plate (X) (Figure 9.11). Begin the electrophoresis run. During electrophoresis, the green mixture will separate into yellow and blue components. The blue component provides a visual indication of the progress of the electrophoresis. The yellow component defines the position to which neutral peptides should migrate. At pH 1.9, ε-dinitrophenyl-lysine is positively charged.
5. Turn the power off when the run is complete. Disassemble the apparatus and carefully remove and air dry the plate. Do not oven dry the plates.

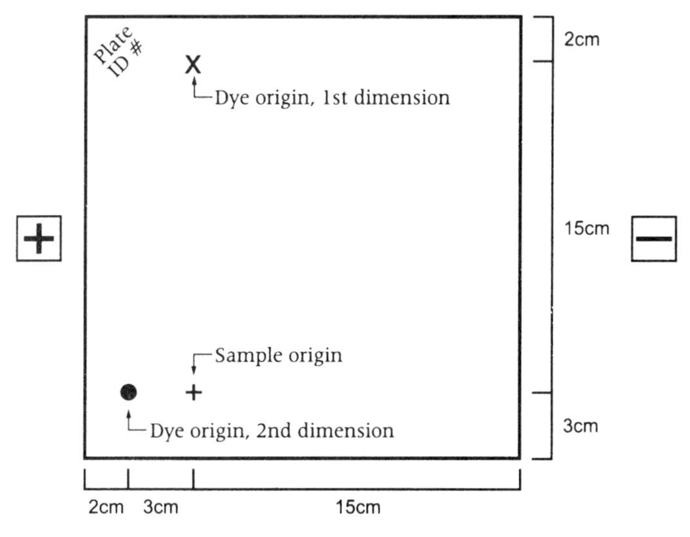

FIGURE 9.11 Locations of sample and dye origins for two-dimensional separation of phosphopeptides on 20 × 20 cm TLC plates for electrophoresis at pH 1.9 or 4.72. Marker dye for the second dimension is spotted on the left once the plate has been dried after electrophoresis. (Adapted with permission from Boyle et al, 1991.)

PROTOCOL 9.26 Second Dimension: Thin-Layer Chromatography

The second dimension is ascending thin-layer chromatography using an organic solvent system to separate peptides based on their ability to partition between the mobile (liquid) and stationary (cellulose) phases. Large glass tanks, which can accommodate the plates, are equilibrated with buffer using Whatman 3MM paper to line the tank up to the top to ensure equilibration of the vapor phase.

Materials

Green marker dye solution: see previous protocol

Chromatography tank: The tank should have a tight fitting lid sealable with silicon or vacuum grease.

Whatman 3MM filter paper

Phospho-chromatography buffer: n-butanol 750 ml, pyridine 500 ml, glacial acetic acid 150 ml, deionized water 600 ml. Buffers should be made up with reagent grade solvents and stored in glass bottles with air-tight lids.

Radioactive India ink

1. Prepare the buffer and pour it into the tank in advance of the chromatography run to allow complete equilibration. After the paper is saturated, there should be ~1 cm height of buffer remaining in the bottom of the tank.

2. Apply 0.5 μl of the green dye marker to the spot indicated in Figure 9.11 (•) for the second dimension. This will act as the marker for the second dimension and as a reference for calculating the relative mobilities of the peptides.
3. To begin chromatography, remove the lid and place the dried plate into the tank (consult Figure 9.11). Quickly replace the lid and make sure an air-tight seal is formed. Chromatography is carried out until the buffer front has reached about 3 cm from the top of the plate (~6–8 h).
4. When the run is complete, remove the plate from the tank and air dry it in a fume hood for 1 h. If peptides are to be extracted for further analysis, do not dry in a baking oven; proceed to Protocol 9.27.
5. Before autoradiography, to aid in analysis, mark the edges of the plate with radioactive India ink in an asymmetrical fashion to help orient the film.

Comments: Do not disturb a run or open the tank while a run is in progress.

PROTOCOL 9.27 Isolation of Individual Phosphopeptides from TLC Plates

This protocol describes a method for collecting the phosphopeptide identified in Protocol 9.26.

Materials

pH 1.9 buffer (Table 9.1)

1. Identify the peptide of interest from the phosphopeptide map (Protocol 9.25) by autoradiography (Protocol 7.22). Outline the spot on the cellulose plate with a blunt pencil.
2. Scrape the cellulose off the plate and collect it into a 1.5 ml microfuge tube. Add 1 ml of the pH 1.9 buffer.
3. Clear the eluate from small cellulose particles by centrifugation for 5 min at maximum speed in a microfuge.
4. Carefully remove the supernatant and lyophilize it. You are now ready to identify amino acids.

PROTOCOL 9.28 Phosphoamino Acid Analysis

Phosphoamino acid analysis is performed on the target phosphoprotein or phosphopeptide to determine which specific amino acid residue(s) are phosphorylated. The target protein or peptide, isolated from SDS-gel, PVDF membrane, or scraped off of the TLC plate is completely hydrolyzed to its amino acids (Protocol 8.11). The resulting phosphoamino acids are separated by high voltage electrophoresis in two dimensions and identified by colocalization with nonradioactive phosphoamino acid standards (Paietta et al, 1992).

Materials

TLC plates: EM Science

Phosphoamino acid standards: phosphoserine, phosphothreonine, and phosphotyrosine, each at 1 mg/ml in deionized water

Ninhydrin solution: 0.25% w/v Ninhydrin in acetone

The Hunter Thin Layer Electrophoresis apparatus HTLE 7000

Heating block

Nitrogen tank

HCl: 6 N

Speed-Vac with NaOH trap (Savant)

Fan/hair dryer

Intensifier screen

Green marker dye solution: see Protocol 9.25

1. As starting material, use either TCA-precipitated proteins that have been taken to the ethanol wash step (Protocol 9.23), proteins after performic acid oxidation and lyophilization (Protocol 9.24), or purified phosphopeptides.
2. Resuspend the dried sample in 50–100 µl of 6 N HCl, vortex mix, centrifuge briefly and incubate at 110°C for 60 min under N_2. This is a partial amino acid hydrolysis. Incubation times longer than 60 min will result in increased release of free phosphate from the phosphoamino acid residues.
3. Remove the HCl by using a Speed-Vac equipped with a NaOH trap to collect the acid. Dissolve the hydrolysates in 5–10 µl of pH 1.9 buffer (consult Table 9.1) which contains 15 parts buffer to one part of the cold phosphoamino acid standards.
4. Mark the TLC plate with a soft-leaded pencil using the diagram presented in Figure 9.12 as a guide. Four samples can be run on one TLC plate.
5. To load the samples, follow the same procedure described in Protocol 9.25. Centrifuge the sample to remove particulate material. Load 0.5 µl marker dye at the top of the plate. Load the samples in small drops of 0.5 µl. The plate is now ready for electrophoresis.
6. Using the pH 1.9 buffer, perform the electrophoresis for 20 min at 1.5 kV When the run is complete, turn off the power, remove the plate, and let it air dry using a fan or hair dryer for at least 30 min.
7. Change the buffer in the apparatus to pH 3.5 and prepare the plate for the second dimension. Place the plate in the apparatus, making sure to rotate it 90° counterclockwise (consult Figure 9.12). Electrophorese in the second dimension for 16 min at 1.3 kV. Turn off the power and remove the plate.
8. Dry the plate using a fan for 30 min or bake in a 65° oven for 10 min. To visualize the positions of phosphoserine (P-Ser), phosphothreonine (P-Thr), and phosphotyrosine (P-Tyr) stain the plate by spraying it with the freshly prepared 0.25% ninhydrin solution in acetone. Return the plate to the 65° oven for 15 min to develop the stain. Distinct purple spots will appear corresponding to the positions of P-Ser, P-Thr, and P-Tyr (see Figure 9.13). Expose TLC plate to X-ray film at –70°C with an intensifier screen. The radiolabeled phosphoamino acids will appear corresponding to the ninhydrin stained spot.

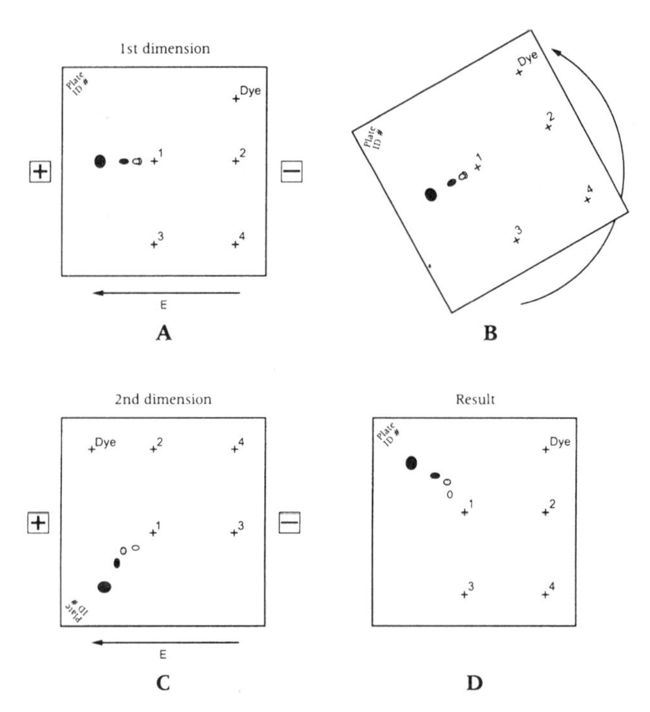

FIGURE 9.12 Preparation of the TLC plate for two dimensional phosphoamino acid analysis. The direction of the electric field E during electrophoresis is indicated by the arrow under the TLC plate (A). Prior to electrophoresis in the second dimension (C) make sure to rotate the plate 90° counterclockwise, as shown in (B). For illustrative purposes, the upper left quadrant in panel (D) represents the resolution of the phosphoamino acids after acid hydrolysis and 2-D electrophoresis. No samples were applied to origins 2, 3, and 4. The plate in (D) has been rotated clockwise 90°, back to its original position. (Adapted with permission from Boyle et al, 1991.)

PROTOCOL 9.29 Phosphoamino Acid Analysis of Phosphoproteins Isolated from PVDF Membranes

Phosphoamino acids can also be analyzed from ^{32}P-labeled phosphoproteins bound to PVDF membranes (Kamps, 1991).

Materials

Speed-Vac: equipped with a NaOH trap (Savant)

Heating block

HCl: 6 N

1. Identify the radiolabeled phosphoprotein band by autoradiography (Protocol 7.22), excise it and transfer it to a microfuge tube.
2. Wash the piece of PVDF membrane containing the protein in a large volume of water to remove the glycine and SDS.
3. Hydrolyze the protein by adding 200 μl of 6 N HCl and heating for 1 h at 110°C.

4. Transfer the acid hydrolysate to a new microfuge tube and dry the material using a Speed-Vac. Continue from Protocol 9.28, step 3.

Comments: Nitrocellulose membranes cannot be treated with concentrated acid as this will dissolve the membrane.

PROTOCOL 9.30 Identification of Phosphohistidine Residues Following Heat Treatment

Phosphohistidines, produced during the formation of phosphorylated intermediates, have been identified in several proteins (Van Etten and Hickey, 1977) including nucleoside diphosphokinase (Gilles et al, 1991). The following three methods have been used to identify the phosphoramidate, high-energy linkage (Wieland et al, 1993) suggestive of histidine phosphorylation.

In contrast to phosphotyrosine, phosphoserine, and phosphothreonine, phosphohistidine is heat sensitive. This reaction can be performed on the total cell lysate, subcellular fractions, or immunoprecipitated protein.

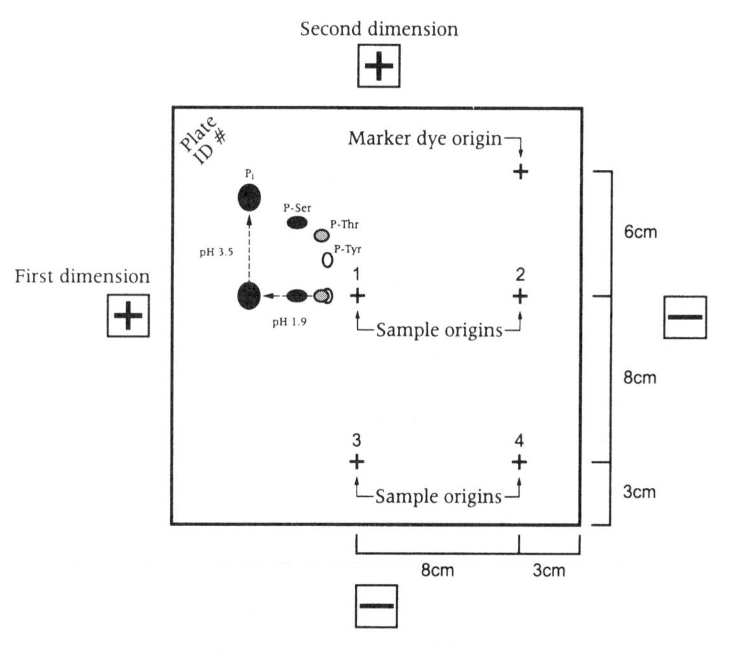

FIGURE 9.13 Diagram for marking the 20 × 20 cm TLC plate for phosphoamino acid analysis. Electrophoresis in the first dimension using pH 1.9 buffer is shown in the horizontal plane. The second dimension electrophoresis using pH 3.5 buffer is shown in the vertical plane. Four sample origins are shown. The marker dye origin is shown in the top right corner. For illustrative purposes, the upper left quadrant (sample 1) represents the resolution of the individual products after acid hydrolysis and 2D electrophoresis as described in the text. No samples were applied to origins 2, 3, and 4. (Adapted with permission of Oxford University Press from van der Geer et al, 1993.)

1. Divide the sample in half. Following the phosphorylation using [γ^{32}P-GTP] (see Protocol 9.16), add 1/4 of total reaction volume of 4 × SDS sample buffer.
2. Heat one fraction for 5 min at 95°C. Keep the control fraction at 30°C.
3. Run both samples on SDS-PAGE.

Comments: If the target phosphoprotein is histidine phosphorylated, the heat treated sample should show reduced radioactive incorporation into the target protein relative to the sample that was kept at 30°C, consistent with a phosphoramidate linkage.

PROTOCOL 9.31 Treatment with Diethyl Pyrocarbonate

Membranes that have been previously isolated from HL-60 cells are used as a source of enzyme and substrate. Diethyl pyrocarbonate is a histidine-modifying agent. At a concentration of 10 mM this reagent completely blocked the G protein β-subunit phosphorylation in HL-60 cell membranes when [γ^{32}P]GTP was used as the phosphate donor (Wieland et al, 1993). Similar concentrations of diethyl pyrocarbonate have been used to identify involvement of histidyl residues in ligand binding (Wendland et al, 1991).

Materials

Diethyl pyrocarbonate: Sigma

Triethanolamine HCl: 10 mM, pH 7.4

EDTA: 5 mM

Membrane preparation from HL-60 cells (Wieland et al, 1993)

1. Incubate isolated cell membrane preparations, 4 samples, 100 µg protein/tube, in a total volume of 200 µl with 10 mM triethanolamine HCl, pH 7.4, containing 0, 0.1, 1.0, and 10 mM diethyl pyrocarbonate for 10 min at room temperature.
2. Add 800 µl of ice-cold 10 mM triethanolamine HCl, pH 7.4, containing 5 mM EDTA.
3. Pellet the membranes for 10 min at 30,000 × g and resuspend in 10 mM triethanolamine HCl, pH 7.4.
4. Divide the treated membrane into fractions of 20 µg protein and incubate them for 3 min at 30°C. Analyze by SDS-PAGE.

PROTOCOL 9.32 Treatment of Phosphorylated Membranes with HCl and NaOH

Phosphohistidines are labile when treated with acid and are stable when treated with NaOH. If you suspect that the target protein is phosphorylated on a histidine residue, HCl treatment should lead to a loss in incorporated radioactivity. In contrast to acid treatment, the phosphohistidine signal should not decrease following treatment with NaOH.

Materials

EDTA: 250 mM

NaOH: 10 M

HCl: 10 M

1. Membrane proteins, 60 µg in 120 µl, are phosphorylated as described in Protocol 9.16 for 3 min at 30°C.
2. Divide the reaction into three equal fractions. Add 40 µl (~20 µg protein) of the phosphorylated membrane reaction mixture to 40 µl of 250 mM EDTA plus either 10 µl of H_2O, 10 µl 10 M HCl, or 10 µl of 10 M NaOH and incubate for 30 min at 30°C.
3. To neutralize, add 10 µl of either 10 M HCl or 10 M NaOH to the appropriate sample and 10 µl of H_2O to the H_2O treated control.
4. Add 30 µl of 4 × SDS sample buffer to the samples and analyze 65 µl of each fraction by SDS-PAGE followed by autoradiography.

PROTOCOL 9.33 Sulfation

Tyrosine residues can undergo a diverse range of chemical modifications in addition to phosphorylation. Other modifications to tyrosine that will not be covered are nucleotidylation to form adenyltyrosine: and halogenation to form a variety of chloro-, bromo-, and iodotyrosines (Wold, 1981), the last of which is important in the hormone thyroxin.

Tyrosine O-sulfate is a widespread derivative of tyrosine. This covalent modification has been detected in a variety of proteins and peptides ranging from extracellular matrix proteins, blood clotting enzymes, to neuropeptides. Tyrosine O-sulfation takes place in the trans-Golgi, catalyzed by a tyrosyl protein sulfotransferase that uses 3′-phosphoadenosine 5′-phosphosulfate as a sulfate donor.

It has been estimated that about 1% of proteins are Tyr sulfated in multicellular organisms, but apparently not in unicellular organisms. Sulfation introduces a permanent negative charge on the protein bound Tyr-sulfate. Although the function of sulfate is still unclear, one proposal is that the role of sulfate is in stabilizing or destabilizing intermolecular interactions (Niehrs et al, 1994).

Sequences surrounding a number of tyrosine sulfation sites have been identified. The general features suggest that the presence of three acidic residues within ten amino acids of the sulfation site is an essential recognition feature (Huttner, 1987). The absence of S-S bridges or N-linked glycans in the vicinity is also an important feature because glycosylation is an event that occurs prior to sulfation, and steric hindrance due to either of these modifications may be an adverse factor.

Analysis of protein tyrosine sulfation is performed using methodology adapted from phosphorylation studies.

Materials

[^{35}S]sulfate

TLC plates: EM Science

Barium hydroxide: 0.4M
Nitrogen tank
Tyr-SO_4
Solvent system: 1-butanol, acetic acid, water (3:1:1)

1. Metabolically label (Protocol 3.1) cells with [^{35}S]sulfate (10–50μCi/ml).
2. Prepare total cell lysates and perform immunoprecipitation as described in Protocol 3.11.
3. Identify Tyrosine-O-sulfated proteins by autoradiography (Protocol 7.22). Excise the sulfated protein band from the dried gel and hydrolyze with 0.4M barium hydroxide in the presence of nonradiolabelled carrier Tyr-SO_4 under nitrogen at 110°C for 24h. Use enough barium solution to cover the gel slice.
4. Sulfotyrosine is identified by TLC using a solvent system composed of 1-butanol acetic acid, water (3:1:1) (Karp, 1983). Alternatively, electrophoresis at pH 3.5 using TLC plates can also be used (Huttner, 1984).

C. Lipid Modification of Proteins

Lipid modifications of proteins have been recognized for many years (Schlesinger, 1981) but only recently have the structural and functional diversity of these modifications been appreciated. Covalent attachment of long-chain acyl groups to an amino acid residue results in the addition of a hydrophobic moiety to the protein. These molecules fit the definition of a proteolipid, a protein that contains a lipid moiety as part of its primary structure. Proteolipids defined in this way are analogous to glycoproteins, and phosphoproteins (Schlesinger, 1981). This modification may facilitate the interaction of the protein with hydrophobic membrane domains, promote hydrophobic protein-protein interactions or target proteins to specific cellular locations.

Fatty acids can be linked to an amino acid residue directly or indirectly as a component of a phosphatidylinositol moiety attached to a COOH-terminal amino acid through an intervening glycan structure (Low, 1989). Direct linkage of fatty acids to proteins can occur both cotranslationally and posttranslationally. Cotranslational modification involves an NH_2-terminal amino acid to which a fatty acid (usually myristate) is bound through an amide bond. Posttranslational addition of fatty acids to proteins is an acylation reaction where the fatty acids (usually palmitate) are bound to the protein by a thioester linkage.

Fatty acid acylation through thioester linkages has a more relaxed fatty acid specificity than cotranslational fatty acid acylation. Myristate, stearate, and oleate could replace palmitate as the thioester-linked fatty acid. In addition to incorporation of saturated fatty acids, polyunsaturated fatty acids can also be covalently bound to proteins (Muszbek and Laposata, 1993). The hydrophobic nature of this type of posttranslational modification lends itself to isolation of the modified peptides by HPLC.

Palmitoylation and N-Myristoylation of Proteins

The covalent attachment of long chain fatty acids to membrane proteins was first described in detail for the vesicular stomatitis virus G and Sindbis virus E1 and E2 membrane glycoproteins (Schmidt and Schlesinger, 1979; Schmidt et al, 1979). Subsequently, this posttranslational modification has been shown to be common to a wide range of prokaryotic and eukaryotic organisms. The most common fatty acids found covalently attached to proteins are myristate (C14:0), palmitate (C16:0), and stearate (C18:0). Most studies have focused on myristoylation and palmitoylation. In addition, the covalent attachment of palmitoleic acid, (C16:1), a long chain unsaturated fatty acid, to proteins in yeast (Casey et al, 1994) provided evidence for a third class of acylated proteins.

Palmitoylation is a posttranslational, reversible process (in contrast to myristoylation which is cotranslational and irreversible although some exceptions exist). Palmitic acid, a 16-carbon saturated fatty acid is attached to proteins on serine or threonine residues via an oxyester. However, in most cases, palmitate is linked to a cysteinyl residue as an acyl thioester. As the majority of palmitoylated proteins are membrane associated, it is likely that palmitate is functioning as a membrane-tethering device. Palmitate has been found attached to the structural proteins of viruses, to insulin and transferrin receptors, to $p21^{ras}$, other members of the Ras superfamily, and to CD4 (Paige et al, 1993).

The palmitoylation of certain viral glycoproteins occurs soon after translation but prior to the acquisition of resistance to Endo H, suggesting a late endoplasmic reticulum event or possibly, a cis-Golgi occurrence.

Protein N-myristoylation refers to the cotranslational linkage of myristic acid (C14:0) via an amide bond to the penultimate NH_2-terminal Gly residues of a variety of eukaryotic cellular and viral proteins (Towler et al, 1988). This modification requires prior removal of the initiating methionyl residue. The reaction is catalyzed by myristoyl-CoA:protein N-myristoyltransferase (NMT). N-Myristoyl proteins have diverse biological functions and intracellular destinations. Some examples of NMT substrates include protein kinases such as the catalytic (C) subunit of cAMP-dependent protein kinase (PK-A) and p60src, phosphatases such as calcineurin B, proteins involved in transmembrane signaling such as several guanine nucleotide-binding a subunits of heterotrimeric G proteins and the gag polyprotein precursors of a number of retroviruses.

Deletion or substitution of the Gly^2 residue of N-myristoylproteins by site-directed mutagenesis prohibits their acylation, allowing the properties of the mutant, nonmyristoylated and wild-type, N-myristoylated species to be compared and contrasted (Gordon et al, 1991).

Analysis of Bound Fatty Acids

PROTOCOL 9.34 Identification of Palmitoylated and Myristoylated Proteins

This method is used to identify the lipid species attached to cellular proteins. Only a small fraction (<0.01%) of the [^{3}H]palmitate and [^{3}H]myristate which is incorporated into the tissue culture cells is covalently attached to proteins. The remainder is incorporated into cellular lipids. Before beginning the protocol described below, prepare delipidated and dialyzed fetal calf serum by extraction with butanol: diisopropyl ether (40:60) as described by Cham and Knowles (1976).

Materials

Delipidated and dialyzed fetal calf serum
[^{3}H]palmitic acid 30–60Ci/mmol, (New England Nuclear)
[^{3}H]myristic acid, 20–40Ci/mmol, (New England Nuclear)
Lysis buffer: 10mM Na phosphate, pH 7.2, 0.14M NaCl, 0.5% Triton X-100, 0.05% SDS, 1mM $CaCl_2$, 1mM $MgCl_2$, 0.5mM iodoacetamide, 1mM PMSF, 10μg/ml aprotinin
Chloroform:Methanol: 2:1 v/v
Nitrogen tank
Fix solution: 25% isopropanol, 10% acetic acid
1 × SDS sample buffer: see Appendix C
Hydroxylamine: 1M pH 10
Tris-HCl: 1M pH 7.0,1M pH 10.0

1. Label confluent cells (10cm dish) for 4h in 4ml of Dulbecco's minimal essential medium containing 10% delipidated and dialyzed fetal calf serum with [^{3}H]palmitate (200μCi/ml) or [^{3}H]myristate (200μCi/ml).
2. At the end of the labeling period, prepare whole cell lysates as described in Protocol 3.9 or 3.10 by adding 1ml of lysis buffer per 10ml dish.
3. To remove the lipids that are not covalently attached to the proteins, treat 0.5ml of cell lysate with 30ml of chloroform: methanol solution. Separate the insoluble protein from the solvent by centrifugation at 5000 × g for 10min. Wash the precipitate five times with the same volume of chloroform: methanol. After the fifth wash the solvent should no longer be radioactive indicating that all lipids have been removed.
4. Dry the final precipitate under a stream of N_2.
5. Resuspend the proteins in 50μl of 1 × sample buffer. Load equal amounts of radiolabeled proteins onto duplicate gels and separate the proteins by SDS-PAGE.
6. Submerge the gel in fix solution for 15min then wash with H_2O to remove the acetic acid.
7. If all the samples are loaded on the same gel, cut the gel into two panels. Soak one panel for 18h (or overnight) in a solution of freshly prepared 1.0M hydroxylamine. As controls, soak the other panel in 1.0M Tris at pH 7.0 or 10.0. Prepare the gels for fluorography then dry and expose them to X-ray film.

Comments: Palmitate attached to proteins can be identified by the susceptibility of the oxy- or thioester linkage to cleavage by basic or neutral hydroxylamine. Under these conditions the amide linkage of N-myristoylated proteins is stable.

If a protein is palmitoylated, hydroxylamine incubation will remove the radioactive palmitate. Compared to the control, untreated lane, the palmitoylated target protein will no longer carry the radioactive signal. Conversely, if a protein is myristoylated the radioactivity will resist hydroxylamine treatment.

During metabolic labeling, there can be some interconversion of the radiolabeled fatty acids due to the action of enzymes involved in fatty acid elongation and β-oxidation. The label may also be incorporated into amino acids (Olson et al, 1985). Therefore, it is imperative that the chemical nature of the protein-associated radioactivity be assessed.

Isoprenylation

Isoprenoids are ubiquitous in nature and play diverse roles. The basic building block is the 5-carbon compound, isopentenyl pyrophosphate which is formed from mevalonate. The condensation of two isopentenyl units produces the 10-carbon geranyl-pyrophosphate and successive; additions of isopentenyl groups generates the intermediate farnesyl pyrophosphate and geranylgeranyl pyrophosphate. These later isoprene units have been identified as protein modifying groups. Although the farnesyl (C_{15}) and geranylgeranyl (C_{20}) groups are attached through a thioether linkage, it is included as a lipid modification (Qin et al, 1992).

The C-A-A-X tetrapeptide in which C is a Cys, A is an aliphatic residue and X is any residue, is the substrate for a series of posttranslational reactions shown in Figure 9.14. The cysteine is isoprenylated via a thioether bond and the final three amino acids are proteolytically removed. The resulting COOH-terminal cysteine residue is carboxy methylated. The hydrophobicity of the C-terminus is enhanced by the addition of a hydrophobic tail which may promote anchoring of the protein to the membrane (Maltese, 1990; Schafer and Rine, 1992).

The isoprenoids transferred by specific prenyltransferases are the C_{15} farnesyl moiety and the more abundant C_{20} geranylgeranyl moiety. The specificity of isoprenylation is influenced by the last residue of the CAAX sequence. When X = Leu the protein is usually geranylgeranylated. When X = Ser, Met or Phe, the protein is farnesylated. Proteins that are known to be farnesylated are Ras proteins, nuclear lamins A and B (Glomset et al, 1990), skeletal muscle phosphorylase kinase and at least three retinal proteins. These modifications have been shown to be important in cycling of growth-regulatory proteins between membrane compartments. Examples of geranylgeranylated proteins include the gamma subunits of G proteins and the ras related rab proteins. However, not all proteins with the CAAX motif are isoprenylated (e.g.,

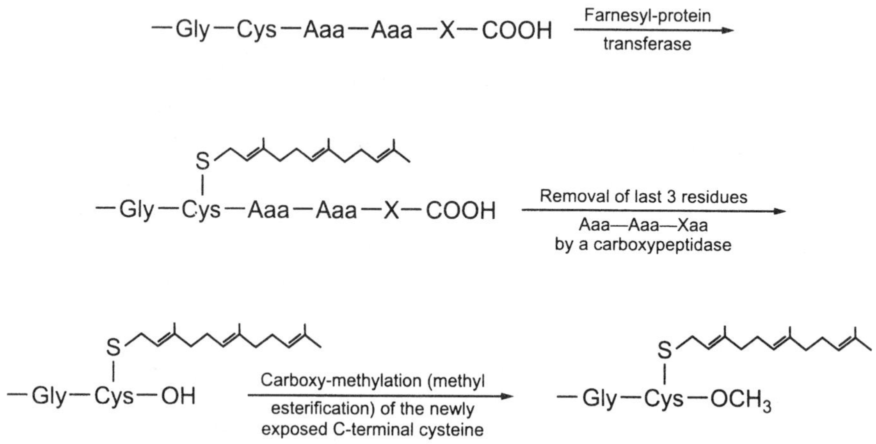

—Gly—Cys—Aaa—Aaa—X—COOH → (Farnesyl-protein transferase)

S (farnesyl)
—Gly—Cys—Aaa—Aaa—X—COOH → (Removal of last 3 residues Aaa—Aaa—Xaa by a carboxypeptidase)

S (farnesyl)
—Gly—Cys—OH → (Carboxy-methylation (methyl esterification) of the newly exposed C-terminal cysteine) S (farnesyl) —Gly—Cys—OCH_3

FIGURE 9.14 The events producing the farnesyl-carboxymethylcysteine modification. The farnesyl group is linked by a thioether bond to a cysteine at the fourth position from the carboxy terminus. The three terminal residues, Aaa-Aaa-Xaa, are removed. The newly exposed C-terminal cysteine is carboxy-methylated rendering it hydrophobic.

the $G_i\alpha$ subunit has the putative CAAX sequence -CGLF but is not modified) (Gibbs, 1991).

PROTOCOL 9.35 Metabolic Labeling with [^{3}H]Mevalonic Acid Derivatives

Cells can be metabolically labeled with [^{3}H]mevalonic acid, the biosynthetic precursor of isoprene derivatives. Alternatively, [^{3}H]mevalonolactone (Dupont-New England Nuclear) at a final concentration of 100–200Ci/ml can be used. The lactone derivative is converted *in vivo* to [^{3}H]mevalonic acid (Cox et al, 1993). The target protein is then identified by immunoprecipitation with the aid of a specific antibody.

An example of the identification of a protein that is isoprenylated *in vitro* is shown in Figure 9.15 (Firmbach-Kraft and Stick, 1993). In this case mRNA encoding lamin B3 was translated in a reticulocyte lysate either in the presence of [^{35}S]methionine or [^{3}H]mevalonic acid. The translation products were immunoprecipitated with a lamin B3 specific antibody and analyzed by SDS-PAGE followed by fluorography. Lamin B3 translated in the presence of [^{35}S]methionine migrated as two closely spaced bands (Figure 9.15, lane 1). Only the faster migrating band incorporated [^{3}H]mevalonic acid (compare Figure 9.15 lanes 1 and 2). The modified form of B3 can be distinguished from the unmodified form by SDS gel electrophoresis.

Treatment of cells with lovastatin inhibits the production of mevalonic acid, the precursor of farnesyl and geranylgeranyl groups. This effectively prevents the isotopic dilution of the tagged mevalonic acid. Extracts from lovastatin treated cells can use exogenously added [^{3}H]mevalonic acid to synthesize labeled farnesyl and geranylgeranyl pyrophosphates and prenylate existing proteins. This forms the basis for the following protocol to assay for prenylated proteins in yeast extracts (Li et al, 1993).

Materials

4 × SDS sample buffer: see Appendix C

Lovastatin

Labeling buffer: 50 mM potassium phosphate, pH 7.5, 50 mM KCl, 0.1 mM EDTA, 0.1 mM EGTA, 1 mM DTT and 1 mM PMSF

6 × Reaction buffer: 30 mM $MgCl_2$, 0.12 mg/ml $ZnSO_4$, 30 mM ATP, 72 mM phosphoenolpyruvate (Sigma), and 1500 µCi [^{3}H]mevalonate (12.7 mCi/µmol)

Pyruvate kinase (Sigma)

Spheroplasts (Deshaies and Schekman, 1989)

1. Add Lovastatin (0.1 mg/ml) to 100 ml of exponentially growing cultures of yeast $OD_{600} = 0.3$.
2. Continue to grow the yeast for 3 h at 23°C. Prepare spheroplasts and lyse in 200 µl of labeling buffer.
3. Clear the lysate by a 15 min centrifugation at maximum speed in a microfuge. Collect the supernatant and use this material for *in vitro* prenylation.
4. To 50 µl of extract add 10 µl of 6 × reaction buffer to yield the following final concentrations: 5 mM $MgCl_2$, 0.02 mg/ml $ZnSO_4$, 5 mM ATP, 12.5 mM phosphoenolpyruvate, and 250 µCi [^{3}H]mevalonate.
5. Preincubate for 30 min at 23°C then add 20 units pyruvate kinase to initiate labeling and incubate for an additional 30 min at 23°C.
6. Terminate the reaction by adding SDS sample buffer and analyze the products by electrophoresis followed by fluorography.

Comments: Approximately 40 polypeptides have been shown to be labeled *in vivo* with [^{3}H]mevalonate (Glomset et a1, 1990).

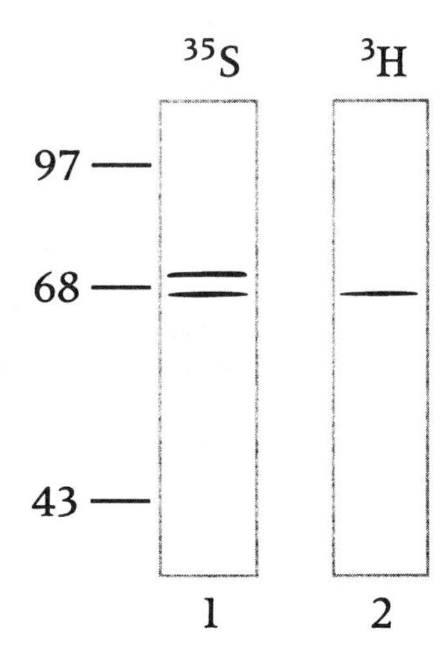

FIGURE 9.15 *In vitro* isoprenylation of Lamin B3. Synthetic lamin B3 mRNA was *in vitro* translated in the presence of [^{35}S]methionine (lanel) or [^{3}H]mevalonic acid (lane 2). Lamin B3 was immunoprecipitated with a specific monoclonal antibody and analyzed by SDS-PAGE. Proteins were visualized by fluorography and exposed to X-ray film for 1 (lane 1) or 7 d (lane 2). Comparison of the doublet in lane 1 to the single band in lane 2 indicates that about half of the B3 protein present is processed into the faster migrating isoprenylated form. (Adapted with permission from Firmbach-Kraft and Stick, 1993.)

PROTOCOL 9.36 Enzymatic Prenylation of Recombinant Proteins

Posttranslational isoprenylation of recombinant proteins can be assessed *in vitro* (Nuoffer et al, 1994). In this example, the prenyl transferases are exogenously provided in the form of rat liver homogenates. Unlike biosynthesis with [^{3}H]mevalonate (described in the previous Protocol), the labeled isoprene is provided in the form of [^{3}H]geranylgeranyl pyrophosphate (GGPP).

Materials

4 × SDS sample buffer: see Appendix C

[^{3}H]geranylgeranyl pyrophosphate (GGPP): 19.3Ci/mmol, New England Nuclear Corp.

Speed-Vac

Recombinant protein

Rat liver cytosol: 5–15μg/ml protein in 25mM HEPES-KOH, pH 7.2, 125mM KOAc prepared from rat liver homogenates as described (Davidson et al, 1992)

ATP-regenerating system: 1mM ATP, 5mM creatine phosphate, and 0.2IU rabbit muscle creatinine phosphokinase (final concentration)

1. Dry 26pmol [^{3}H]geranylgeranyl pyrophosphate (GGPP) in a Speed-Vac or lyophilizer.
2. Resuspend the GGPP in 50μl containing 1μg (~40pmol) recombinant protein, 25μl rat liver cytosol, 10mM $MgCl_2$, 1mM MES-Na, and an ATP-regenerating system.
3. Incubate for 1h at 32°C. Add 25μl of 4 × sample buffer and boil. Analyze by SDS-PAGE. Enhance the fluorographic signal by using a commercial enhancer and expose to X-ray film at –70°C.

Comments: The efficiency of the prenylation reaction can be estimated by phase separation using Triton X-114 (Bordier, 1981) as described in Protocol 9.38. When GGPP is added, the prenylated product will be recovered in the detergent phase. The expected yield will be 5–10%.

Glypiation

In eukaryotic cells, many proteins are anchored to the external surface of the plasma membrane by covalently attached glycolipids containing inositol. These anchors are referred to as glycosylphosphatidylinositols or GPIs. Nascent proteins destined to be processed to a glycosylphosphatidylinositol (GPI)-anchored form contain two hydrophobic signal peptides, one at the NH_2 terminus and another at the COOH terminus. NH_2-terminal signal peptides direct proteins into the ER (endoplasmic reticulum) and the COOH-terminal signal peptides target the partially processed protein to a site where the COOH signal peptide is posttranslationally replaced by the complete, preformed GPI anchor structure by a transamidase reaction (Udenfriend et al, 1991). This is actually an event of anchor exchange, since the primary translation product includes a relatively orthodox putative membrane spanning segment near its carboxyl terminus which is removed before or concomitant with anchor addition. Reaction of the NH_2 group on the ethanolamine of the GPI precursor with the activated carbonyl group forms an amide linkage.

GPI anchored proteins (frequently referred to as PI-G tailed proteins), are preferentially enriched at the apical plasma membrane, apparently because the GPI unit is an apical sorting signal. There have been reports of at least 100 proteins with diverse functions that have been found to contain a GPI linkage. A schematic representation of the GPI linkage and structural elements that are important for analytical methods is presented in Figure 9.16.

GPI anchored proteins, which are oriented toward the extracellular space, contain a large extracellular protein domain linked to the membrane via glycosyl-phosphatidylinositol. They share a remarkably conserved core structure consisting of ethanolamine phosphate 6 mannose α1–2 mannose α1–6 mannose α1–4 glucosamine α1–6 inositol (Tartakoff and Singh, 1992). This core glycan is precisely conserved between species, down to the level of individual glycosidic linkages. Additional groups may modify the core, introducing variation dependent on species, cell type, or protein. The amphipathic glycolipid anchor is attached covalently to the new carboxyl-terminal residue of the protein via ethanolamine to an oligosaccharide chain, which in turn is linked to the inositol ring of phosphatidylinositol as shown in Figure 9.16. This unit is inserted into the membrane noncovalently by means of the inositol phospholipid portion of the anchor.

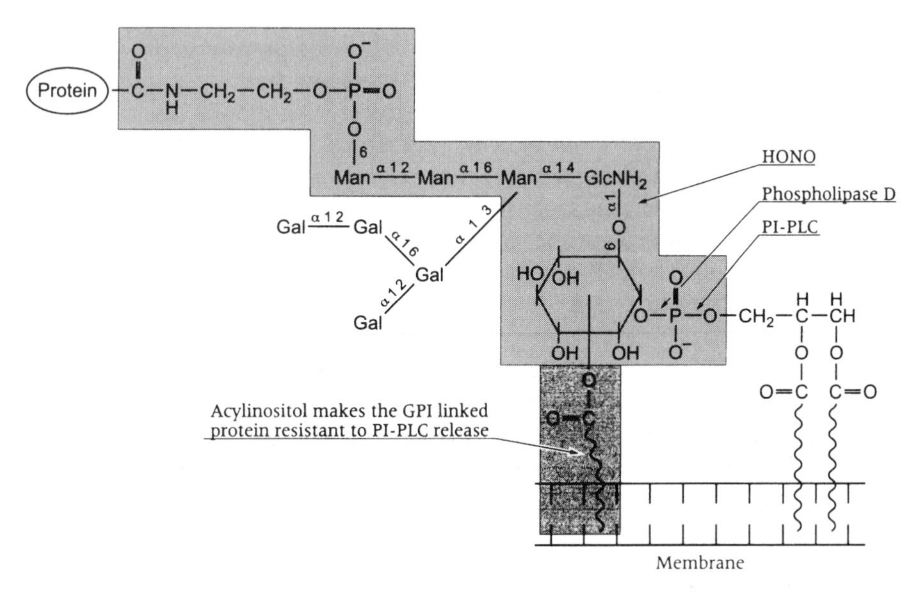

FIGURE 9.16 Schematic representation of a GPI anchored protein. The conserved core structure is shaded. Sites of cleavage for analysis are indicated. HONO, nitrous acid; PI-PLC, PI-specific phospholipase C, GPI-PLD, GPI-specific phospholipase D.

Some anchor structures such as those associated with human erythrocyte acetylcholinesterase (Roberts et al, 1988) and *T. brucii* PARP (Field et al, 1991) contain an extra fatty acid esterified to inositol which is often palmitate. The site of linkage to inositol was roughly equal to the 2- and 3-hydroxyls (see Figure 9.16). The presence of a fatty acid esterified to inositol usually confers resistance to bacterial PI-PLC. However, these acetylated anchors are susceptible to cleavage by the GPI-specific phospholipase D (Englund, 1993).

Is the Target Protein Glycosyl Phosphatidylinositol Anchored?

There are several experimental approaches that are used to determine if a protein is GPI anchored. GPI-linked proteins are released from the cell surface by phosphatidylinositol-specific enzymes such as PI specific phospholipase C (PI-PLC). Another approach used to identify GPI-anchored proteins involves detergent partitioning with Triton X-114 (Bordier, 1981). If cell extracts are subjected to Triton X-114 phase separation, GPI-anchored proteins, along with most membrane proteins, will partition into the detergent phase. Treatment with PI- or GPI-specific phospholipase C will remove the hydrophobic component of the GPI anchor, with the exception of inositol-acylated GPI anchors, which are not susceptible to PI-PLC. If the sample is again subjected to Triton X-114 detergent phase separation, the resulting hydrophilic proteins will partition into the aqueous phase. The detergent and aqueous phases are separately analyzed by SDS-PAGE to assess the cleavage and to identify the susceptible polypeptides. These experiments are typically performed with cells metabolically labeled with radioactive amino acids.

Use of Triton X-114

Triton X-114, a member of the Triton X series of detergents (Bordier, 1981), exhibits a cloud point at 20°C. At this temperature, a microscopic phase separation most likely occurs due to secondary association of small micelles into large micelle aggregates. With increased temperature, phase separation proceeds until two phases, a detergent depleted and detergent enriched phase are formed. Triton X-114 can be used to solubilize membranes and whole cells. In this fashion, integral membrane proteins which partition into the detergent phase can be separated from hydrophilic proteins.

PROTOCOL 9.37 Preparation of Triton X-114

Materials

Triton X-114: Fluka
PBS: see Appendix C
Graduated cylinder: 100 ml
Parafilm

1. Add 2 ml of Triton X-114 to 100 ml of cold PBS in a graduated cylinder.
2. Seal with parafilm and mix until the suspension is homogeneous.
3. Let stand at 37°C overnight.
4. The lower layer ~10–20 ml is enriched in detergent, and the upper layer is depleted of detergent. Aspirate off the upper layer and discard.
5. Adjust the lower layer to 100 ml with PBS and mix well. Incubate at 37°C overnight. Aspirate off the upper layer and save the lower layer, which is the stock solution of Triton X-114, 11.4%.

PROTOCOL 9.38 Fractionation of Integral Membrane Proteins with Triton X-114

Triton X-114 forms a clear micellar solution at low temperatures, which separates into two phases when warmed to 37°C. When cells are extracted with Triton-X-114, hydrophilic proteins are found in the upper, aqueous phase, and molecules with hydrophobic properties are recovered in the lower, detergent phase (Bordier, 1981).

Materials

Triton X-114 lysis buffer: 10 mM Tris-HCl pH 7.4, 0.15 M NaCl, 1 mM EDTA (TBS) containing 1% (v/v) Triton X-114

Sucrose cushion buffer: 6% sucrose, 0.06% Triton X-114 in TBS

1. Prepare a lysate from ~5×10^6 cells by adding 1 ml of Triton X-114 lysis buffer to the cells. Incubate for 1 h at 4°C with mild agitation.
2. Clarify the cell lysates by centrifugation at maximum speed for 10 min in a microfuge at 4°C.
3. Load the supernatant directly onto 100 µl of sucrose cushion buffer.
4. Place the tube in a 37°C water bath for 3 min. The contents of the tube will become cloudy. Centrifuge at $400 \times g$ at 37 °C (time and speed are not critical). Two phases will form, aqueous on top, and detergent at the bottom.
5. Transfer the aqueous phase to a fresh tube and incubate on ice. The detergent rich oily drop in the bottom of the tube should be ~50–100 µl.
6. Resuspend the detergent phase in 500 µl of cold TBS and incubate on ice
7. Repeat steps 3–6.
8. Pool the second detergent depleted aqueous phase with the first and keep on ice.
9. Resuspend the detergent rich fraction, which contains the integral membrane proteins, in 2 ml of cold PBS. Incubate on ice.
10. Re-extract the pool of detergent depleted aqueous phase by adding 50 µl of 11.4% Triton X-114 stock solution (Protocol 9.37). Mix and incubate in a 37°C water bath for 3 min. Centrifuge at $400 \times g$ for 5 min at 37°C. The aqueous phase will contain the water soluble proteins. The oily pellet from this step can be discarded.

Comments: PBS can be used in place of TBS. All solutions used for cell extraction, phase separation, and PI-PLC treatment should be ice-cold and contain protease inhibitors.

Different detergent extractability of GPI-anchored proteins has been reported in whole cells (Hooper and Turner, 1988), octylglucoside > CHAPS > DOC. The insolubility of GPI-linked proteins may reflect an association with non-GPI linked molecular components of Triton insoluble complexes. Non-GPI-linked insoluble elements may represent lipids as these complexes have been effectively solubilized only by detergents that resemble glycolipids (octylglucoside) and cholesterol (CHAPS and DOC) (Sargiacomo et al, 1993).

PROTOCOL 9.39 Digestion with Phosphatidylinositol Specific Phospholipase C (PI-PLC)

PI-PLC digestion is a simple, specific test for GPI anchorage. Following enzyme treatment and phase separation, fractions are analyzed by SDS-PAGE (Chapter 4), and the protein of interest is followed by Western blotting (Chapter 7) or identified by immunoprecipitation (Chapter 3).

GPI-linked proteins partition with the detergent phase unless they are treated with PI-specific phospholipase C (PI-PLC) (6–8U/ml from *Bacillus thuringiensis;* Sigma) which releases them from the detergent phase into the aqueous phase.

Materials

PI-PLC: from *Bacillus thuringiensis* (Sigma)

TBS: see Appendix C

Reaction buffer: 100mM Tris-HCl, pH 7.4, 50mM NaCl, 1mM EDTA

Triton X-114: 2% in TBS

Sodium deoxycholate: 1% in TBS

Trichloroacetic acid: 100% (w/v)

Triton X-114 lysis buffer: See Protocol 9.37

1 × SDS sample buffer: see Appendix C

Ammonium hydroxide

1. Perform a Triton X-114 phase separation on your cell lysate (Protocol 9.38).
2. Dilute the detergent phase (–100μl which is enriched with membrane forms of GPI anchored proteins) with reaction buffer to a final volume of 0.5ml.
3. Divide the diluted detergent phase into two fractions. Incubate one fraction with PI-PLC (6–8U/ml) with continuous mixing for 1h at 37°C. Incubate the other fraction (mock digest) in reaction buffer without enzyme.
4. Immediately after PI-PLC treatment, dilute the samples to a final volume of 1ml with TBS containing 2% Triton X-114 and separate the phases as described in Protocol 9.38. Detergent phases can be kept and analyzed. The aqueous phase contains the soluble forms of the GPI anchored protein resulting from the PI-PLC treatment.
5. The aqueous phase is in a large volume. To quantitatively precipitate the proteins add sodium deoxycholate (125μg/ml, final concentration) and

trichloroacetic acid (6% (w/v) final concentration) (Bensadoun and Weinstein, 1976) and incubate at 4°C for 30–60 min.
6. Collect the precipitated proteins by centrifugation at maximum speed at 4°C for 10 min in a microfuge.
7. Add 50 μl of 1 × SDS sample buffer to the resulting pellet and neutralize by adding microliter amounts of NH_4OH prior to boiling for 3 min. The sample should turn from yellow to blue.

Comments: A background level caused by the partitioning of membrane proteins into the aqueous phase is unavoidable because this phase contains ~0.7 mM Triton X-114, which is three times above its critical micelle concentration (CMC) and is sufficient to solubilize some plasma membrane proteins of intermediate hydrophobicity (Lisanti et al, 1988).

It is not unusual for eukaryotes to form a GPI anchor structure which includes acylinositol causing them to be resistant to PI-PLC (Walter et al, 1990), however they can be cleaved by the GPI-specific phospholipase D. This explains why some proteins that have GPI anchors are resistant to PI-PLC hydrolysis and will remain in the detergent phase when Triton X-114 phase separation is performed following enzyme treatment.

Another useful method for anchor cleavage is nitrous acid deamination. All GPIs contain a nonacetylated glucosamine. This allows nitrous acid to cleave the core glycan between glucosamine and inositol, releasing PI and a glycan that terminates in anhydromannose.

Metabolic Labeling with Precursors of the GPI Structure

The ability to radiolabel a protein with GPI precursors that are found exclusively as part of the GPI complex such as [^{3}H]ethanolamine and, to a lesser extent, [^{3}H]mannose and [^{3}H]myristic acid provides a useful means of tagging the GPI-anchored protein. Because [^{3}H]ethanolamine provides a specific tag for the carboxy terminus of a GPI-anchored protein, GPI-linked carboxy terminal peptides can be readily identified following proteolysis of a purified [^{3}H]ethanolamine-labeled GPI-anchored protein and reversed phase HPLC analysis of the proteolytic fragments can be performed (Menon, 1994).

Methods for metabolic labeling with radiolabeled monosaccharides, inositol, ethanolamine and fatty acids are described in Protocol 3.11. Things to keep in mind are that uptake of [^{3}H]mannose and [^{3}H]glucosamine is inhibited by glucose, and labeling must be performed in low glucose medium. Inositol free medium must be used for [^{3}H]inositol labeling. In all cases, use only dialyzed calf serum in the labeling medium (Menon, 1994).

Use of Anti-CRD

Digestion of GPIs with PI-PLC releases diacylglycerol and reveals a cryptic epitope referred to as the cross reacting determinant (CRD). Polyclonal antibodies to CRD have been prepared and analyzed.

The major moiety contributing to the CRD is the inositol 1,2-(cyclic)monophosphate formed during PI-PLC cleavage, although other structural features may also contribute. Reactivity of a protein with anti-CRD antibody after cleavage by PI-PLC strongly indicates the presence of a GPI anchor. True to its name, the CRD has been found across species and is a general characteristic of PI-G tail proteins. CRD positive proteins are routinely detected by Western blotting, probing with the anti-CRD antibody.

Each of the above techniques has exceptions. If one performs more than one analytical procedure, the chances of finding an artifact decrease. Absolute proof is obtained by complex structural analysis.

D. Selected Modifications

Transamidation

Transamidation of available glutamine residues is catalyzed by transglutaminases (TGs) through a calcium-dependent acyl transfer reaction. Proteins are posttranslationally modified by an acyl transfer reaction between the γ-carboxamide group of glutamine residues (see Figure 9.17) and the ε-amino group of peptide-bound lysine residues or the primary amino group of polyamines to form either ε-(γ-glutamyl)lysine or (γ-glutamyl)polyamine bonds between proteins, releasing ammonia. The resulting bonds are covalent, stable, and resistant to proteolysis. Transglutaminase-modified proteins are evident throughout the body in the fibrin network of blood clots, extracellular matrices, and the cornified features of the epidermis, callus, hair, and nail (Greenberg et al, 1991).

Acetylation

N-acetylation is the major eukaryotic co-translational modification of protein α-amino groups. During the reaction, an amide bond is formed which blocks the α-amino group to Edman degradation. This reaction is carried out by an N-terminal acetyltransferase.

Methylation

Proteins such as histones, actin, rhodopsin, and cytochrome c have been found to be methylated at the α-amino group or on the side chains

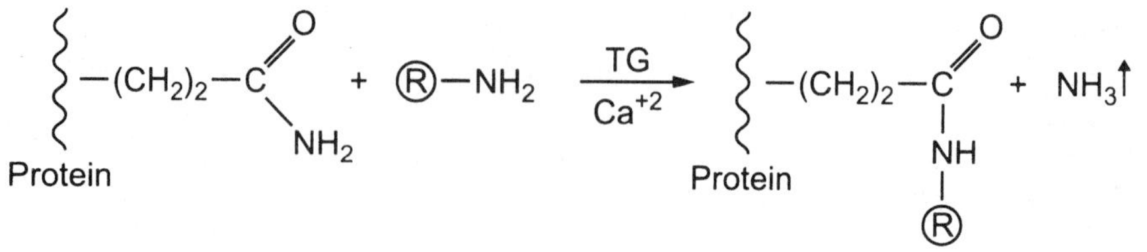

$$\text{Protein}-(CH_2)_2-C(=O)NH_2 + Ⓡ-NH_2 \xrightarrow[Ca^{+2}]{TG} \text{Protein}-(CH_2)_2-C(=O)-NH-Ⓡ + NH_3\uparrow$$

FIGURE 9.17 Transglutaminase catalyzed acyl transfer reaction. Transglutaminase catalyzes the acyl transfer reaction between the γ-carboxamide group of a peptide-bound glutamine residue with a primary amino group, designated R-NH_2. The product of the reaction is either an isopeptide bond or a (γ-glutamyl)polyamine bond and the release of ammonia.

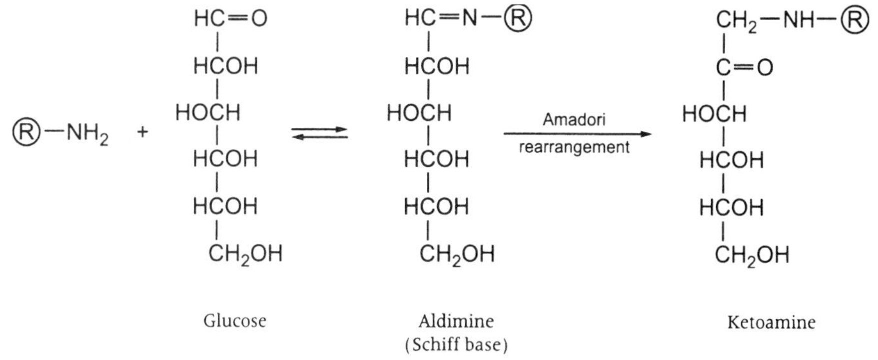

Amadori rearrangement

Glucose

Aldimine (Schiff base)

Ketoamine

FIGURE 9.18 Degradation pathway via the Amadori rearrangement.

of Lys, Arg, and His (Cantoni, 1975). The enzymes that catalyze these methylation reactions generally use S-adenosylmethionine as methyl donor. Neither the time nor the cellular site of methylation is known. Once incorporated, the methyl groups do not appear to be removed.

Hydroxylation of Proline and Lysine

An important step in the maturation and secretion of collagen is the hydroxylation of certain proline and lysine residues (Bornstein, 1974). During biosynthesis, when they appear in the sequences -Pro-Gly- and -Lys-Gly-, hydroxyl groups are enzymatically placed on the γ-carbon of proline and on the δ-carbon of lysine. Hydroxylation appears to be vital for the folding and secretion of collagen, and the δ-OH-Lys modification is necessary for the formation of intermolecular cross-links and for the attachment of glycosyl groups.

Degradation

Proteins are not immortal. They succumb to old age as their sulfur atoms become oxidized and their Gln and Asn residues become deamidated. One particular reaction associated with aging is the nonenzymatic reaction between protein amino groups and reducing sugars like glucose to form a Schiff base which then undergoes an Amadori-type rearrangement to the more stable ketoamine as depicted in Figure 9.18. The product then undergoes a variety of dehydrations to form yellowish-brown fluorescent products and cross-links between protein molecules (Monnier and Cerami, 1981). These reactions have been studied in eye lens proteins, where they produce decreased solubility with deleterious effects on vision resulting in the formation of cataracts.

Ubiquitination

Ubiquitin is a neutral 8-kDa protein. One or multiple ubiquitin molecules can be added to target molecules, in most cases by the formation of isopeptide bonds through the σ amino groups of lysine residues. Ubiquitination targets abnormal proteins for degradation and regulates the amount of several rapidly metabolized proteins. Ubiquitinated proteins have been associated with the cytoskeleton, and there is evidence that ubiquitination participates in a number of other cellular processes, covalently modifying proteins.

To establish whether a candidate protein has been modified by the addition of ubiquitin, an extract is immunoblotted with a mixture of monoclonal antibodies to ubiquitin. The blot can then be stripped of the antibodies to ubiquitin (see Chapter 7 for stripping and reprobing blots) and reprobed with antibody to the candidate protein. If the same proteins are recognized by both antibodies, one can conclude with relative certainty that they are ubiquitinated (Finley and Chau, 1991; Hochstrasser et al, 1991).

Proteolytic Processing

Proteolytic cleavage is a common covalent modification of proteins. The universal initiation codon for protein synthesis, AUG, specifies methionine as the initiator amino acid in eukaryotic proteins. Because only very few proteins are isolated with the starting amino acid still attached, it is clear that some trimming of the amino terminal Met has taken place during and/or after the assembly of the full length protein.

In addition to N-terminal trimming, extensive internal peptide bond cleavage takes place during the maturation of many proteins. These can be assigned to one of two major classes of proteolytic processing; that associated with transport and that with activation. During protein processing, specific signal peptidases remove signal peptides from the N-terminal end of the preprotein chain. Activation of proteins by exquisitely specific proteases are seen with the digestive enzymes where zymogens are transformed to active enzymes, in blood clotting cascade, complement activation and the conversion of prohormones to active hormones to mention a few examples (Krishna and Wold, 1998).

Many proteins are processed by proteolytic cleavage of the polypeptide chain at one or more peptide bonds. The polypeptide chain, before this processing reaction, is referred to as the prepro protein. The proteinase responsible must be highly specific to produce the correct mature protein. Proteolytic processing may be a mechanism for inducing conformational change to produce the biologically active molecule. The classic example is insulin shown in Figure 9.19. The preproinsulin molecule undergoes two specific hydrolysis reactions to produce first the proinsulin form then the mature insulin molecule. The mature form consists of two polypeptide chains, which are the product of a single gene, held together by interchain disulfide bonds.

Many peptide hormones are produced by proteolytic processing of much larger protein precursors. The single polypeptide chain of proopiomelanocortin is the inactive precursor for the opioid peptides corticotropin (ACTH), lipotropin, melanocyte stimulating hormone and endorphin (Nakanishi et al, 1979). The precursor molecule is split at different sites in different tissues, generating multiple combinations of bioactive peptides from a single gene product.

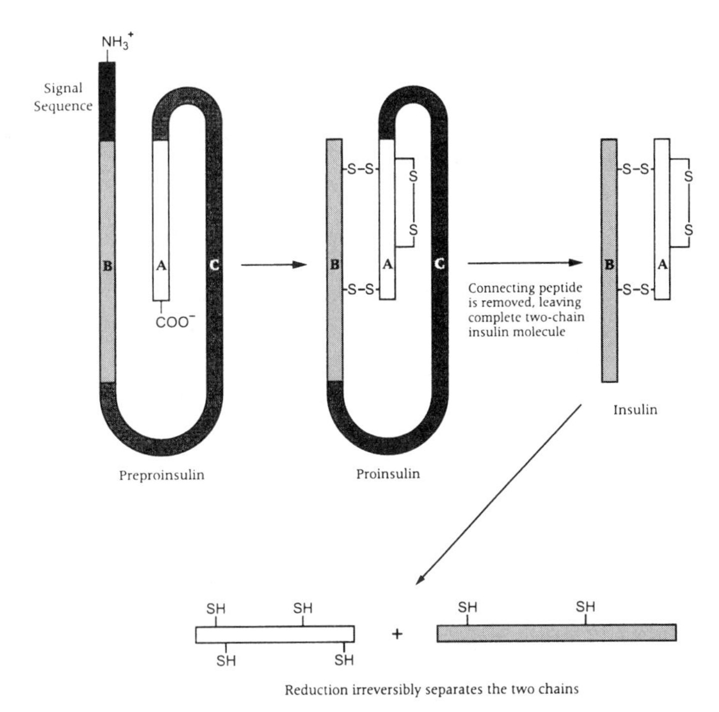

FIGURE 9.19 Proteolytic processing of insulin. Insulin is synthesized as the 110-residue preproinsulin. The 23-residue PRE peptide is removed from the amino end. The resulting proinsulin is further processed by the removal of the 35-residue C peptide. The final two-chain mature insulin molecule consists of only 51 of the original 110 residues. Adapted from BIOCHEMISTRY 2/E by Stryer. Copyright © 1981 by Lubert Stryer. Used with permission of W.H. Freeman and Company.

Proteins destined for secretion may be processed by cleavage of a leader sequence by a specific proteinase. This signal peptidase is responsible for the cleavage of the signal peptide. Cleavage occurs cotranslationally and is found both in prokaryotes and eukaryotes. Although not well characterized, the size of the signal peptide is between 16–30 amino acids. The general pattern in the leader sequences is that there are basic residues at the N-terminus followed by a hydrophobic region in the middle of 8–12 amino acids.

Processing of the target protein can be monitored by pulse-chase analysis followed by immunoprecipitation (see Chapter 3). The immature, heavier molecule will have a slower mobility than the mature target molecule. One should also be aware of the possibility of other posttranslational modifications, i.e., glycosylation that may be taking place, which could confuse the interpretation of the results.

An intriguing posttranslational modification is that in which a peptide sequence is proteolytically cleaved and then reassembled in an order that is different from the original one, or in the same order but after eliminating an internal sequence. The process is analogous to RNA splicing, where the intervening sequence is eliminated. This phenomenon was first reported by Bowles and Pappin (1988) for the plant lectin concanavalin A and reviewed by Wallace (1993). The commercial relevance of this reaction will be discussed in Chapter 11 where inteins are presented.

References

Affolter M, Watts JD, Krebs DL, Aebersold R (1994): Evaluation of 2-dimentional phosphopeptide maps by electrospray-ionization mass spectrometry of recovered peptides. *Anal Biochem* 223: 74–81

Akiyama T, Ishida J, Nakagawa S, Ogawara H, Wantanabe S, Itoh N, Shibuya M, Fukami Y (1987): Genistein, a specific inhibitor of tyrosine-specific protein kinases. *J Biol Chem* 262:5592–5595

Annan RS, Huddleston MJ, Verma R, Deshaies RJ, Carr SA (2001): A multidimensional electrospray MS-based approach to phosphopeptide mapping. *Anal Chem* 73:393–404

Ashwell G, Harford J (1982): Carbohydrate specific receptors of the liver. *Ann Rev Biochem* 51:531–554

Baenziger J, Kornfeld S (1974): Structure of the carbohydrate units of IgA_1 immunoglobulin. *J Biol Chem* 249:7270–7280

Barnea G, Grumet M, Milev P, Silvennoinen O, Levy JB, Sap J, Schlessinger J (1994): Receptor tyrosine phosphatase b is expressed in the form of proteoglycan and binds to the extracellular matrix protein tenascin. *J Biol Chem* 269:14349–14352

Bensadoun A, Weinstein D (1976): Assay of proteins in the presence of interfering materials. *Anal Biochem* 70:241–251

Blix G, Svennerholm L, Werner 1 (1952): The isolation of chondrosamine from gangliosides and from submaxillary mucin. *Acta Chem Scand* 6: 358–365

Bordier C (1981): Phase separation of integral membrane proteins in Triton X-114 solution. *J Biol Chem* 256:1604–1607

Bornstein P (1974): The biosynthesis of collagen. *Ann Rev Biochem* 43:567–603

Bourdon MA, Krusius T, Campbell S, Schwartz NB, Ruoslahti E (1987): Identification and synthesis of a recognition signal for the attachment of glycosaminoglycans to proteins. *Proc Natl Acad Sci USA* 84:3194–3198

Boyle WJ, van der Geer P, Hunter T (1991): Phosphopeptide mapping and phosphoamino acid analysis by two-dimensional separation on thin-layer cellulose plates. *Methods Enzymol* 201:110–149

Burgess AJ, Norman RI (1988): The large glycoprotein subunit of the skeletal muscle voltage-sensitive calcium channel. *Eur J Biochem* 178:527–533

Bowles DJ, Pappin DJ (1988): Traffic and assembly of concanavalin A. *Trends Biochem Sci* 13:60–64.

Cantoni GL (1975): Biological methylation: selected aspects. *Ann Rev Biochem* 44:435–451

Carmo-Fonseca M, Ferreira J, Lamond A (1993): Assembly of snRNP-containing coiled bodies is regulated in interphase and mitosis-evidence that the coiled body is a kinetic nuclear structure. *J Cell Biol* 120:841–852

Casey WM, Gibson KJ, Parks LW (1994): Covalent attachment of palmitoleic acid (C16:1) to proteins in *Saccharomyces cerevisiae*. *J Biol Chem* 269:2082–2085

Cham BE, Knowles BR (1976): A solvent system for delipidation of plasma or serum without protein precipitation. *J Lipid Res* 17:176–181

Davidson HW, McGowan CH, Balch WE (1992): Evidence for the regulation of exocytic transport by protein phosphorylation. *J Cell Biol* 116:1343–1355

Dell A, McDowell RA, Rogers ME (1992): Glycan structures in glycoproteins and their analysis by fast atom bombardment mass spectrometry. In:

Post-translational Modifications of Proteins, Harding JJ, Crabbe MJCM, eds. Boca Raton, FL: CRC Press

Deshaies RJ, Schekman R (1989): SEC62 Encodes a putative membrane protein required for protein translocation into the yeast endoplasmic reticulum. *J Cell Biol* 109:2653–2664

Dunphy WG, Rothman JE (1983): Compartmentation of asparagine-linked oligosaccharide processing in the golgi apparatus. *J Cell Biol* 97: 270–275

Elbein AD (1981): The tunicamycins—useful tools for studies on glycoproteins. *Trends Biochem Sci* 6:219–221

Elbein AD (1991): Glycosidase inhibitors: inhibitors of N-linked oligosaccharide processing. *FASEB J* 5:3055–3063

Elder JH, Alexander S (1982): Endo-β-N-Acetylglucosaminidase F: endoglycosidase from *Flavobacterium meningosepticum* that cleaves both high-mannose and complex glycoproteins. *Proc Natl Acad Sci USA* 79:4540–4544

Englund PT (1993): The structure and biosynthesis of glycosyl phosphatidylinositol protein anchors. *Annu Rev Biochem* 62:121–138

Ervasti JM, Campbell KP (1991): Membrane organization of the dystrophin-glycoprotein complex. *Cell* 66:1121–1131

Fayos BE, Bartles JR (1994): Regulation of hepatocytic glycoprotein sialylation and sialyltransferases by peroxisome proliferators. *J Biol Chem* 269:2151–2157

Ferrell JE, Martin GS (1991): Assessing activities of blotted protein kinases. *Methods Enzymol* 200:430–435

Ficarro SB, McCleland ML, Stukenberg PT, Burke DJ, Ross MM, Shabonowitz J, Hunt DF, White FM (2002): Phosphoproteome analysis by mass spectrometry and its application to *Saccharomyces cerevisiae. Nat Biotechnol* 20:301–305

Field MC, Menon AK, Cross GAM (1991): A glycosylphosphatidylinositol protein anchor from procyclic stage *Trypanosoma brucci*: lipid structure and biosynthesis. *EMBOJ* 10:2731–2739

Finley D, Chau V (1991): Ubiquitination. *Ann Rev Cell Biol* 7:25–69

Firmbach-Kraft I, Stick R (1993): The role of CaaX-dependent modifications in membrane association of Xenopus nuclear lamin during meiosis and the fate of B3 in transfected mitotic cells. *J Cell Biol* 123:1661–1670

Florman HM, Wassarman PM (1985): *O*-linked oligosaccharides of mouse egg ZP3 account for its sperm receptor activity. *Cell* 41:313–324

Frackelton AR, Posner M, Kannan B, Mermelstein F (1991): Generation of monoclonal antibodies against phosphotyrosine and their use for affinity purification of phosphotyrosine-containing proteins. *Methods Enzymol* 201:79–91

Gibbs JB (1991): Ras C-terminal processing enzymes—new drug targets? *Cell* 65:1–4

Gilles A-M, Presecan E, Vonica A, Lascu I (1991): Nucleoside diphosphate kinase from human erythrocytes. *J Biol Chem* 266:8784–8789

Glomset JA, Gelb MH, Farnsworth CC (1990): Prenyl proteins in eukarotic cells: a new type of membrane anchor. *Trends Biochem Sci* 15:139–142

Gooley AA, Packer NH (1997): The importance of protein co- and posttranslational modifications in proteome projects. In: *Proteome Research: New Frontiers in Functional Genomics*, Wilkins MR, Williams KL, Appel RD, Hochstrasser DF, eds. Berlin: Springer-Verlag, Inc.

Gordon JI, Duronio RJ, Rudnick DA, Adams SP, Gokel GW (1991): Protein *N*-Myristoylation. *J Biol Chem* 266:8647–8650

Goshe MB, Conrads TP, Panisko EA, Angell NH, Veenstra TD, Smith RD (2001): Phosphoprotein isotope-coded affinity tags approach for isolating and quantifying phosphopeptides in proteome-wide analyses. *Anal Chem* 73:2578–2586

Greenberg CS, Birckbichler PJ, Rice RH (1991): Transglutaminases: mulifunctional cross-linking enzymes that stabilize tissues. *FASEB J* 5: 3071–3077

Grinstein S, Furuya W Butler JR, Tseng J (1993): Receptor-mediated activation of multiple serine/threonine kinases in human leukocytes. *J Biol Chem* 268:20223–20231

Gudepu RG, Wold F (1996): Posttranslational modifications. In *Proteins: Analysis and Design*. Angeletti RH, ed. San Diego, CA: Academic Press, pp. 121–207

Harper JDI, Sanders MA, Salisbury JL (1993): Phosphorylation of nuclear and flagellar basal apparatus proteins during flagellar regeneration in *Chlamydomonas reinhardtii*. *J Cell Biol* 122:877–886

Hochstrasser M, Ellison MJ, Chau V Varshavsky A (1991): The short-lived MAT alpha 2 transcriptional regulator is ubiquitinated *in vivo*. *Proc Natl Acad Sci USA* 88:4606–4610

Hooper NM, Turner AJ (1988): Ectoenzymes of the kidney microvillar membrane: differential solubilization by detergents can predict a GPI membrane anchor. *Biochem J* 250:865–869

Hunter T (1991): Protein kinase classification. *Methods Enzymol* 200:3–9

Huttner WB (1984): Determination and occurrence of tyrosine *O*-sulfate in proteins. *Methods Enzymol* 107:200–223

Huttner WB (1987): Protein tyrosine sulfation. *Trends Biochem Sci* 12:361–363

Kameshita O, Fujisawa H (1989): A sensitive method for detection of calmodulin-dependent protein kinase II activity in sodium dodecyl sulfate-polyacrylamide gel. *Anal Biochem* 183:139–143

Kamps (1991): Determination of phosphoamino acid composition by acid hydrolysis of protein blotted to Immobilon. *Methods Enzymol* 201:21–28

Karlsson H, Carlstedt I, Hanson GC (1987): Rapid characterization of mucin oligosaccharides from rat small intestine with gas chromatography-mass spectrometry. *FEBS Lett* 226:23–27

Karp DR (1983): Post-translational modification of the fourth component of complement. Sulfation of the α-chain. *J Biol Chem* 258:12745–12748

Kiessling LL, Cairo CW (2002): Hitting the sweet spot. *Nat Biotechnol* 20:234–235

Kornfeld R, Kornfeld S (1985): Assembly of asparagine-linked oligosaccharides. *Ann Rev Biochem* 54:631–664

Krebs EG (1994): The growth of research on protein phosphorylation. *Trends Biochem Sci* 19:439

Krishna RG, Wold F (1998): Posttranslational modifications. In: *Proteins: Analysis and Design*. Angeletti RH, ed. San Diego, CA: Academic Press, pp. 121–206

Lasky LA (1992): Selections: interpreters of cell-specific carbohydrate information during inflammation. *Science* 258:964–969

Lee EU, Roth J, Paulson JC (1989): Alteration of terminal glycosylation sequences on N-linked oligosaccharides of Chinese hamster ovary cells by expression of β-galactoside α2,6-sialyltransferase. *J Biol Chem* 264: 13848–13855

Li R, Havel C, Watson JA, Murray AW (1993): The mitotic feedback control gene MAD2 encodes the α-subunit of a prenyltransferase. *Nature* 366:82–84

Lis H, Sharon N (1993): Protein glycosylation: structural and functional aspects. *Eur J Biochem* 218:1–27

Lisanti MP, Sargiacomo M, Graeve L, Saltiel AR, Rodriguez-Boulan E (1988): Polarized apical distribution of glycosyl-phosphatidylinositol-anchored proteins in a renal epithelial cell line. *Proc Natl Acad Sci USA* 85:9557–9561

Low MG (1989): The glycosyl-phosphatidylinositol anchor of membrane proteins. *Biochem Biophys Acta* 988:427–454

Low SH, Tang BL, Wong SH, Hong W (1994): Golgi retardation in Madin-Darby canine kidney and Chinese hamster ovary cells of a transmembrane chimera of two surface proteins. *J Biol Chem* 269:1985–1994
Maley F, Trimble RB, Tarentino AL, Plummer TH (1989): Characterization of glycoproteins and their associated oligosaccharides through the use of endoglycosidases. *Anal Biochem* 180:195–204
Maltese WA (1990): Posttranslational modification of proteins by isoprenoids in mammalian cells. *FASEB J* 4:3319–3328
McLachlin DT, Chait BT (2001): Analysis of phosphorylated proteins and peptides by mass spectrometry. *Curr Opin Chem Biol* 5:591–602
Menon A (1994): Structural analysis of glycosylphosphatiidylinositol anchors. *Methods Enzymol* 230:418–442
Mitsui K, Brady M, Palfrey HC, Nairn AC. (1993): Purification and characterization of calmodulin-dependent protein kinase III from rabbit reticulocytes and rat pancreas. *J Biol Chem* 268:13422–13433
Mizutani A, Tokumitsu H, Kobayashi R, Hidaka H (1993): Phosphorylation of Annexin XI (CAP-50) in SR-3Y1 Cells. *J Biol Chem* 268:15517–15522
Monnier VM, Cerami A (1981): Non-enzymatic browning in vivo: possible process for aging of long-lived proteins. *Science* 211:491–493
Muszbek L, Laposata M (1993): Covalent modification of proteins by arachidonate and eicosapentaenoate in platelets. *J Biol Chem* 268:18243–18248
Nakanishi S, Inoue A, Kita T, Nakamura M, Chang ACY, Cohen SN, Numa S (1979): Nucleotide sequence of cloned cDNA for bovine corticotropin-β-lipotropin precursor. *Nature* 278:423–427
Niehrs C, Beisswanger R, Huttner WB (1994): Protein tyrosine sulfation, 1993—an update. *Chem Biol Interact* 92:257–271
Nuoffer C, Davidson HW, Matteson J, Meinkoth J, Balch WE (1994): A GDP-bound form of Rabl inhibits protein export from the endoplasmic reticulum and transport between golgi compartments. *J Cell Biol* 125:225–237
Olson, EN, Towler, DA, Glaser, L (1985): Specificity of fatty acid acylation of cellular proteins. *J Biol Chem* 260:3784–3790
Paietta E, Stockert RJ, Raevskis J (1992): Alternatively spliced variants of the human hepatic asialoglycoprotein receptor, H2, differ in cellular trafficking and regulation of phosphorylation. *J Biol Chem* 267:11078–11084
Paige LA, Nadler MJS, Harrison ML, Cassady JM, Geahlen RL (1993): Reversible palmitoylation of the protein-tyrosine kinase p56lck. *J Biol Chem* 268:8669–8674
Paulson JC (1989): Glycoproteins: what are the sugar chains for? *Trends Biochem Sci* 14:272–275
Ponnambalam S, Rabouille C, Luzio PJ, Nilsson T, Warren G (1994): The TGN38 glycoprotein contains two non-overlapping signals that mediate localization to the Trans-Golgi network. *J Cell Biol* 125:253–268
Powell KA, Valova VA, Malladi CS, Jensen ON, Larsen MR, Robinson PJ (2000): Phosphorylation of dynamin I on Ser-795 by protein kinase C blocks its association with phospholipids. *J Biol Chem* 275:11610–11617
Qin N, Pittler SJ, Baehr W (1992): *In vitro* isoprenylation and membrane association of mouse rod photoreceptor cGMP phosphodiesterase α and β subunits expressed in bacteria. *J Biol Chem 267*:8458–8463
Roberts WL, Myher JJ, Kuksis A, Low MG, Rosenberry TL (1988): Lipid analysis of the glycoinositol phospholipid membrane anchor of human erythrocyte acetylcholinesterase. *J Biol Chem* 263:18766–18775
Robinson PJ, Sontag J-M, Liu J-P, Fykse EM, Slaughter C, McMahon H, Sudhof TC (1993): Dynamin GTPase regulated by protein kinase C phosphorylation in nerve terminals. *Nature* 365:163–166

Roden L, Smith R (1966): The structure of the neutral trisaccharide of the chondroitin sulfate-protein linkage region. *J Biol Chem* 241:5949–5954

Roquemore EP, Chou T-Y, Hart GW (1994): Detection of *O*-linked *N*-Acetylglucosamine (*O*-GlcNAc) on cytoplasmic and nuclear proteins. *Methods Enzymol* 230:443–460

Rudd PM, Morgan BP, Wormald MR, Harvey DJ, van den Berg CW, Davis SJ, Ferguson MAJ, Dwek RA (1997): The glycosylation of the complement regulatory protein CD59, derived from human erythrocytes and human platelets. *J Biol Chem* 272:7229–7244

Ruoslahti E, Yamaguchi Y (1991): Proteoglycans as modulators of growth factor activities. *Cell* 64:867–869

Rutishauser U, Acheson A, Hall AK, Mann DM, Sunshine J (1988): The neural cell adhesion molecule (NCAM) as a regulator of cell-cell interactions. *Science* 240:53–57

Sanders DA, Gillece-Castro BL, Stock AM, Burlingame AL, Koshland DE (1989): Identification of the site of phosphorylation of the chemotaxis response regulator protein, CheY. *J Biol Chem* 264:21770–21778

Sargiacomo M, Sudol M, Tang Z, Lisanti M (1993): Signal transducing molecules and glycosyl-phosphatidylinositol-linked proteins form a caveolin-rich insoluble complex in MDCK cells. *J Cell Biol* 122:789–807

Schafer WR, Rine J (1992): Protein prenylation: genes, enzymes, targets, and functions. *Ann Rev Genet* 30:209–234

Schlesinger MJ (1981): Proteolipids. *Ann Rev Biochem* 50:193–206

Schlessinger J (2002): A solid base for assaying protein kinase activity. *Nat Biotechnol* 20:232–3

Schlessinger J (2000): Cell signaling by receptor tyrosine kinases. *Cell* 103:211–225

Schmidt MFG, Schlesinger MJ (1979): Fatty acid binding to vesicular stomatitis virus glycoprotein: a new type of post-translational modification of the viral glycoprotein. *Cell* 17:813–819

Schmidt MFG, Bracha M, Schlesinger MJ (1979): Evidence for covalent attachment of fatty acids to Sindbis virus glycoproteins. *Proc Natl Acad Sci USA* 76:1687–1691

Sefton BM, Hunter T (1991): Preface to protein phosphorylation Part A. *Methods Enzymol* 200:xvii–xviii

Shen L, Ginsburg V (1967): Sugar analysis of cells in culture. Determination of the sugars of HeLa cells by isotope dilution. *Arch Biochem Biophys* 122: 474–480

Sickmann A, Meyer HE (2001): Phosphoamino acid analysis. Proteomics 1:200–206

Stockell Hartree A, Renwick AGC (1992): Molecular structures of glycoprotein hormones and functions of their carbohydrate components. *Biochem J* 287:665–679

Tartakoff AM, Singh N (1992): How to make a glycoinositol phospholipid anchor. *Trends Biol Sci* 17:470–473

Thomas JE, Soriano P, Brugge JS (1991): Phosphorylation of c-Src on tyrosine 527 by another protein tyrosine kinase. *Science* 254:568–571

Towler DA, Gordon JI, Adams SP, Glaser L (1988): The biology and enzymology of eukaryotic protein acylation. *Ann Rev Biochem* 57:69–99

Udenfriend S, Micanovic R, Kodukula K (1991): Structural requirements of a nascent protein for processing to a PI-G anchored form: studies in intact cells and cell-free systems. *Cell Biol Int Rep* 15:739–759

van der Geer P, Luo K, Sefton BM, Hunter T (1993): Phosphopeptide mapping and phosphoamino acid analysis on cellulose thin-layer plates. In: *Protein Phosphorylation: A Practical Approach,* Hardie DG, ed. New York: IRL Press

Van Etten RL, Hickey ME (1977): Phosphohistidine as a stoichiometric intermediate in reactions catalyzed by isoenzymes of wheat germ acid phosphatase. *Arch Biochem Biophys* 183:250–259

Wallace CJA (1993): The curious case of protein splicing: Mechanistic insights suggested by protein semisynthesis. *Protein Sci* 2:697–705

Walter EI, Roberts WL, Rosenberry TL, Ratnoff WD, Medof ME (1990): Structural basis for variations in the sensitivity of human decay accelerating factor to phosphatidylinositol-specific phospholipase c cleavage. *J Immunol* 144: 1030–1036

Weinstein J, Lee EU, McEntee K, Lai P-H, Paulson JC (1987): Primary structure of β-galactoside α2,6-sialyltransferase. *J Biol Chem* 262.:17735–17743

Wen DX, Livingston BD, Medzihradszky KF, Kelm S, Burlingame AL, Paulson JC (1992): Primary structure of Galβ1,3 (4)GlcNAc α2,3-sialyltransferase determined by mass spectrometry sequence analysis and molecular cloning. *J Biol Chem* 267:21011–21019

Wendland M, Waheed A, von Figura K, Pohlmann R (1991): *Mr* 46,000 mannose 6-phosphate receptor. *J Biol Chem* 266:2917–2923

Whiteheart SW, Passaniti A, Reichner JS, Holt GD, Haltiwanger RS, Hart GW (1989): Glycosyltransferase probes. *Methods Enzymol* 179:82–94

Wieland T, Nurnberg B, Ulibarri I, Kaldenberg-Stasch S., Schltz G, Jakobs KH (1993): Guanine nucleotide-specific phosphate transfer by guanine nucleotide-binding regulatory protein B-subunits. *J Biol Chem* 268: 18111–18118

Wold F (1981): In vivo chemical modification of proteins. *Ann Rev Biochem* 50:783–814

Yan CB, Grinnell BW, Wold F (1989): Post-translational modifications of proteins. Some problems left to solve. *Trends Biochem Sci* 14:264–268

Yurchenco PD, Ceccarini C, Atkinson PH (1978): Labeling complex carbohydrates of animal cells with monosaccharides. *Methods Enzymol* 50:175–204

Zhou H, Watts JD, Aebersold R (2001): A systematic approach to the analysis of protein phosphorylation. *Nat Biotechnol* 19:375–378

General References

Ausubel FM, Brent RG, Kingston DD, Moore JG, Seidman JA, Smith JA, Struhl K (1992): *Current Protocols in Molecular Biology.* New York: John Wiley

Burlingame AL, Boyd RK, Gaskell SJ (1998): Mass spectrometry. *Anal. Chem* 70:R647–R716

Hounsell EF, ed. (1993): *Glycoprotein Analysis in Biomedicine.* New Jersey: Humana Press

Lennarz WJ, Hart GW eds. (1994): *Guide to Techniques in Glycobiology. Methods in Enzymology, Vol. 230.* San Diego, CA: Academic Press

Schlesinger MJ (1993) *Lipid Modifications of Proteins.* Boca Raton, FL: CRC Press

Stryer L (1981): *Biochemistry.* San Francisco, CA: W .H. Freeman and Co.

10

Chromatography

Introduction

The term chromatography applies to a wide variety of separation techniques, based on the partitioning of a sample between a moving phase and a stationary phase. The invention of chromatography is generally credited to Tswett, who, in 1906 used a chalk column to separate pigments from green leaves. He referred to the process as chromatography because the term seemed to describe the colored zones moving through the column. Today, chromatography is recognized as the most powerful separation method with regard to resolution and versatility, having superior resolving power to centrifugation and ultrafiltration and capable of isolating larger quantities of protein than electrophoresis. Many different types of chromatographic methodologies have evolved, including paper, thin layer, and liquid chromatography.

Chromatography involves the isolation of components in a mixture using a medium through which a flow of liquid or gas is passed causing differential migration of the individual components. The flow is usually driven by pressure or gravity. The solutes are separated according to their interactions with the column matrix, resulting in different mobilities down a column of solid particles in the presence of a mobile phase. Chromatographic techniques are flexible in that they allow separations to be modified by changes in packing chemistry and elution buffers. Improvements in column chromatography technology have resulted from new packings that afford better resolution, higher capacity and enhanced physical stability.

Chromatography is suitable for almost any separation problem in biotechnology, ranging from the assay of picomoles of a target protein to the purification of kilogram quantities of a precious protein. Practically speaking, chromatography can be divided into partition and adsorption chromatography depending on the type of interaction that occurs between the medium and the components. In partition chromatography, the medium may be another solvent or liquid or an immobilized solid such as paper or a gel matrix like Sephadex. Specific interactions between the components to be separated and the solvent or resin are minimal. In adsorption chromatography, the resin is designed to interact with some or all of the components of the mixture. Solvents are selected to specifically increase or decrease these interactions.

There are basically four standard chromatographic techniques: gel filtration, ion exchange, hydrophobic interaction, and affinity chromatography. They all lend themselves to automation and can be run classically or in the high performance mode. The main role of chromatographic equipment is to properly initiate, monitor and control the separation process, bringing the most out from the actual separation that takes place only in the column. High performance—also known as high pressure or high speed—liquid chromatography (HPLC) is distinguished from conventional liquid chromatography by the use of sophisticated instrumentation and high efficiency columns. The advantages of HPLC over conventional liquid chromatography are increased speed, higher resolution, higher sensitivity, usually better sample recovery, and better reproducibility.

Historically, progress of a given chromatogaphic technique has been due to advances in column technology. In protein chromatography, the introduction of cellulosic ion exchangers in the early fifties, cross-linked dextrans in the late fifties, then agarose based column materials and recently rigid gigaporous packings were major steps forward.

Perhaps the most important issue for the researcher is the choice of a specific chromatography mode and when to use it. Some techniques logically follow others due to volume and buffer constraints. Ideally, every step used in a purification protocol should result in an increase in the specific activity of the target protein with minimal loss of the total activity. The specific activity of the target protein is commonly referred to in units of activity per milligram protein. A unit of activity, if not previously defined, can be set by the investigator. It is usually based on enzyme activity or a bioactivity. For example, one can arbitrarily define a unit of activity to be the amount of protein needed to give 50% stimulation in a particular bioassay. The purer the material, the more activity (units) present per mg of total protein because there is more of the biologically active target protein contributing to the total weight.

A. Important Terminology Used in Chromatography

When describing column characteristics (e.g. dimensions or volume) the term column refers to the column of gel, not the actual vessel in which the gel is held. V_t is the total volume of buffer inside and outside of the beads. The unhydrated gel matrix frequently occupies as little as 1% of the total volume of the column so the volume of the gel bed can be taken as a good estimate of V_t. The volume of gel is referred to as the bed volume or column volume, which is composed of the space occupied by the gel beads V_x, and the space occupied by the solvent surrounding the gel beads, which is also referred to as the void volume, V_o. Therefore,

$$V_t = V_o + V_x$$

The void volume is roughly estimated to be one-third of the bed volume. A molecule that is too large to enter the pores of the column matrix is said to be excluded and is eluted from the column in the void volume. The exclusion limit of a particular gel is the molecular weight of the smallest molecule that cannot enter the gel. The useful range for resolving proteins is about 55% of V_t. The actual resolution depends on diffusion and nonideal behavior in the column. V_e, the elution volume of a particular solute, is expressed as the amount of effluent solution exiting from the column between the time that the solute first penetrates the gel bed to the time that it appears in the effluent. V_e, is usually measured from the maximal peak height. The Distribution Coefficient K_D

$$K_D = (V_e - V_o)/(V_t - V_o)$$

is a useful value for the calculation of the molecular weight of a protein as demonstrated later in this chapter.

The Stokes radius (R_s) is used when discussing the apparent molecular weight of a protein as determined by gel filtration. The Stokes radius value for a protein is obtained by interpolating from a plot of Stokes radii values of standard proteins versus Erf^{-1} $(1 - K_D)$ where Erf^{-1} is the inverse error function and K_D is the distribution coefficient of a protein (Ackers, 1967).

The partition coefficient, α, is the ratio of the amounts of a substance divided or partitioned between two phases, in most cases solid particles in the presence of a flowing liquid. A solute, having a partition coefficient in a specific system of 0.8 (defined as that fraction of solute adsorbed to the solid at any time), will move at 20% of the speed of the liquid flow. At any time, 80% is adsorbed and 20% is free in solution. Individual molecules equilibrate rapidly between adsorption and solution so that each molecule travels at the same average speed over a period of time. The mobility of the solute relative to the moving liquid is $1-\alpha$. This value is referred to as the R_f.

B. Gel Filtration Chromatography

Gel filtration chromatography is also known as molecular sieving, gel permeation and size exclusion chromatography (SEC). The primary objective of gel filtration is to achieve rapid separation of molecules based on size. The gel, which is in bead form for easy column packing, consists of an open, cross-linked three-dimensional molecular network of pores into which molecules of less than maximum pore size may penetrate. The pores within the beads are of such sizes that some are not accessible to large molecules, but smaller molecules can penetrate all pores. The extent to which a molecule enters the pores depends on its shape as well as molecular weight. Therefore, separation depends on the difference in the ability of various molecules to enter these pores. Proteins above the molecular weight exclusion limit cannot enter the pores. Consequently, they come through the column as a sharp peak in the void volume. Any specific grade of a gel matrix has a certain proportion of pores that are accessible to each size of macromolecule ranging from 0–100%. Currently, it is not technically possible to make a material with just one pore size that would give a very sharp cutoff.

As the solvent moves through the column, solutes present in the original sample can do one of three things: they can run with the solvent front ($R_f = 1$) and be washed out quickly, they can be totally adsorbed to the gel matrix and remain at the starting position ($R_f = 0$), or in the most useful case, the solute is partially retarded but is eventually eluted ($0 < R_f < 1$). Very large molecules, which are too large to enter the pores, move through the chromatographic bed the fastest. Smaller molecules, which enter the gel pores, move more slowly since they go in and out of the beads as they travel down the column. Molecules that have R_f values of zero in a certain system are not amenable to gel filtration. Consequently, molecules are eluted in order of their decreasing size. The degree of retardation is a function of the specific pore size of the gel and the size of the molecule.

Separation of the proteins is achieved in the column. The results are monitored as the liquid exits from the column, usually by monitoring the UV absorbance at 280 nm. For protein purification the exact position of elution is not critical provided that the protein of interest is well separated from unwanted proteins. It is desirable to arrange for the protein to elute in the first two-thirds of the separating range where the resolution is the greatest.

A gel matrix that has been used for over 25 years is Sephadex (Pharmacia), a cross-linked dextran. Also widely used are cross-linked polyacrylamide beads (Biogels) (Bio-Rad). For very large molecules and aggregates, agarose gels have been used with much success. The major problem with the Sephadexes and polyacrylamide gels is their softness. Pressures encountered during routine operation can cause distortion and irregular packing which eventually lead to poor flow characteristics. Manufacturers have addressed this problem and have come up with extra cross-linking of dextran with acrylamide, producing more rigid beads that can be operated under higher pressures, generating faster flow rates.

It is not possible to describe all available column matrices, separation conditions, and manufacturers due to the plentiful number of alternatives. With the large variety of gels that are commercially available a few words of advice about what to purchase are in order. Unless it is known that the target protein has a molecular weight <15,000 or higher than 10^6, two or three materials will be enough to get started. Cross-linked dextran or agarose are the best matrices available. Sephacryl S-200 and S-300 or Ultrogels AcA22 and 34 are excellent choices. For small proteins Sephadex G-75 (fine grade), Ultrogel AcA54 or Biogel P-60 are equally suited. For large proteins Sepharose

TABLE 10.1 Gel Filtration Materials for Separation in the Range of 100 Da to Subcellular Particles

Company	Gel	Type	Useful Working Range (MW)
Pharmacia	Sephadex G-10	Dextran	up to 700
	G-15		up to 1,500
	G-25		100–5,000
	G-50		1,500–30,000
	G-75		3,000–70,000
	G-100		4,000–150,000
	G-150		5,000–300,000
	G-200		10,000–600,000
Bio-Rad	Biogel P-2	Polyacrylamide	100–1,800
	P-4		800–4,000
	P-6		1,000–6,000
	P-10		1,500–20,000
	P-30		2,500–40,000
	P-60		3,000–60,000
	P-100		5,000–100,000
	P-150		15,000–150,000
LKB	AcA202	Agarose	1,000–22,000
Ultrogels	AcA54	Agarose/Polyacrylamide	3,000–90,000
	AcA44		10,000–200,000
	AcA34		20,000–350,000
Pharmacia	CL-2B	Cross-linked agarose	70,000–4×10^7
Superdex	75		3,000–70,000
	200		10,000–600,000
Sepharose	CL-4B		60,000–2×10^7
	CL-6B		10,000–4×10^6
Sephacryl	S-100 HR	Cross-linked dextran/bisacrylamide	1,000–1×10^5
	S-200 HR		5,000–2.5×10^5
	S-300 HR		10,000–1.5×10^5
	S-400 HR		20,000–8×10^6
	S-500 HR		40,000–2×10^7
	S-1000 SF		5×10^5–10^8
Bio-Rad	A-0.5m	Agarose	10,000–5×10^5
Bio-Gel	A-1.5m		10,000–1.5×10^5
	A-5m		10,000–5×10^6
	A-15m		40,000–1.5×10^7
LKB	A2	Agarose	up to 4×10^6
Ultrogels	A4		up to 2×10^7
	A6		up to 5×10^7

Sephacryl S-1000 SF is used to separate large macromolecules, viruses, plasmids and subcellular organelles. Covalent bonding of dextran to highly cross-linked agarose beads allows Superdex to be used with high flow rates, generating short separation times, and permits aggressive cleaning with 1 M NaOH and 0.1 M HCl.

CL-4B, Biogel A-5.0m or Ultrogels A4 or A6 are recommended. Avoid gels that claim to cover a wide range because their resolving power will not be as good as products intended for specific ranges. For a list of gel matrixes and their useful fractionation ranges consult Table 10.1.

In choosing a matrix for size exclusion chromatography, an important consideration will be the grade of material chosen for a specific fractionation. Some gel matrices are available in grades from course to superfine. Fine particles pack better than coarse particles and give better resolution. Therefore, when resolution is the primary concern, superfine or fine grade particles should be used. For preparative separations and manipulations like desalting, where resolution is not difficult to achieve and high flow rates are desirable, medium and coarse grades should be used.

The choice of fractionation range should be made with consideration to the sizes of the main contaminants in addition to the protein of interest. Inexperienced investigators should look at the following explanations as guidelines for setting up and running a successful molecular separation using gel filtration. However, it should be emphasized that nothing is etched in stone. Alteration and manipulation of all parameters are encouraged to fit the needs of specific experiments. An important consideration is to have the component of interest on the column for the shortest time possible. With this in mind, a gel should be chosen such that the sample's expected molecular weight falls on the linear part of the selectivity curve and closer to the void volume than the bed volume. Consult Table 10.1.

Choice of Buffer

The most important consideration when choosing a buffer for gel filtration is to make sure it is compatible with the target molecule. Therefore, be aware of dissociating agents and detergents that could induce conformational changes, inactivate a biologically active molecule or dissociate proteins into subunits.

Since proteins are large molecules relative to the components in buffer systems, the proteins applied to a gel filtration column will elute from the column in the buffer in which the column has been equilibrated. Therefore, the composition of the loading buffer is not critical. This is useful if one object is to change the buffer during the size exclusion step.

Invariably there are interactions between the packing material and the biological molecules. These are often dependent on buffer composition. Partial adsorption of a protein under specific buffer conditions results in the protein eluting from the column later than would be expected for its molecular size. Most interactions of an ionic nature can be eliminated by increasing the ionic strength of the running buffer. Typically, an ionic strength of 20–50 mM is sufficient to avoid ionic interactions between the materials of interest and the gel matrix. Sodium chloride is commonly added to the buffer for this purpose. Common buffers like Tris-HCl and sodium phosphate are routinely used to control the pH.

Choice of Column Size

In gel filtration, the resolution of two separated materials increases as the square root of the column length. The ideal column dimensions allow baseline resolution of the solute molecules without significant sample dilution. The ratio of the column length to diameter will fall between 5 and 10. The choice of the column's dimensions should be based on the requirements of the experiment. Long columns should be used to obtain the best resolution. Effective bed height can be increased by using columns that are coupled in series. The length of the column depends on the required resolution, and the diameter depends on the amount of sample to be loaded. Columns narrower than 1.0 cm internal diameter may produce wall effects where materials are slightly more retarded along the walls. On the other hand, if the column is too wide, the materials may be diluted. Another disadvantage to using a wide column is that distortions at the bed surface are amplified. Dead space at the outlet of the column should be kept to a minimum to prevent dilution and remixing of separated zones.

A guide for choosing column size is the intended sample volume (and vice versa). Optimal resolution is attained with a sample size of ~1–2% of V_t. For a column with a given diameter, the bed height should be adjusted so that V_t is 50 to 100 times the sample volume. Theoretically, resolution increases with column length. However, when the length is increased to the point at which the sample volume <1% of the V_t, diffusion becomes a dominating factor because the sample must spend too much time on the column. Also, wall effects are likely to be amplified in a column that is too long.

PROTOCOL 10.1 Preparation of the Gel: Hydrating and Degassing

Gels that are supplied in powder form must be hydrated prior to use. Gels that are supplied as slurries are ready to use.

Materials

Erhlenmeyer flask with side-arm (preferably thick walled)

Beakers

Aspirator

1. Gradually add the powdered gel to an excess amount of buffer in a beaker with gentle swirling. If the gel is supplied as a slurry, go to step three.
2. Allow the gel to swell according to the manufacturer's instructions. After the gel has been hydrated, fines, which are small particles of gel that were not removed during the manufacturing process, or were generated during hydration, should be removed. Fines will decrease the column flow rate.
3. Suspend the gel in a 2–4-fold excess of buffer and allow 95% of the gel to settle out.
4. Aspirate off the remaining gel and supernatant by suction. Repeat this step.
5. Equilibrate the gel to its working temperature. Immediately prior to pouring the column, degas the gel, which will prevent the formation of bubbles that may form in the column, creating channels and irregular flow.
6. To degas the gel, place the slurry in a thick-walled Ehrlenmeyer flask with a side-arm. Seal the top with a large rubber stopper which fits over the mouth of the flask. Attach vacuum tubing to the sidearm and connect it

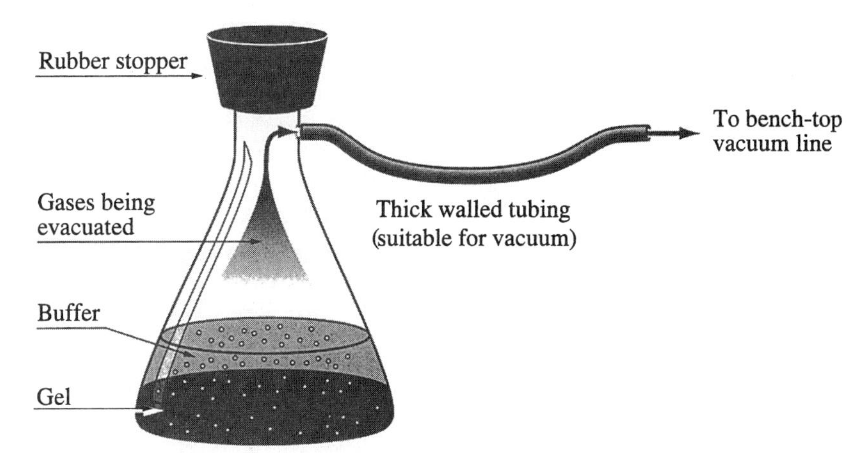

FIGURE 10.1 Degassing the gel.

to a vacuum line as shown schematically in Figure 10.1. Swirl the flask occasionally during 15 min. The gel is degassed when bubbles no longer appear.

PROTOCOL 10.2 Packing the Column

Packed columns offer high capacity per unit volume. A poorly packed column will give poor resolution. Therefore, packing the column is a crucial step.

Materials

Carpenter's level
Column and tubing
Long, thin glass rod

1. Prepare a ~50% slurry of degassed gel.
2. Equilibrate the gel to the working temperature.
3. Mount the column in a vertical position using either a carpenter's level or a home-made plumb line. Avoid direct sunlight and high traffic areas.
4. Remove air bubbles trapped in the column with the glass rod. To ensure that there is no air in the dead space at the bottom of the column, attach a buffer-filled syringe to the effluent tubing and force buffer up through the bed support.
5. Fill the column to 1/3 of its final height with buffer and then close the column outlet. Pour the gel into the column taking care not to produce bubbles. A long solid glass rod will make this step easier. Pour the gel down the glass rod that is in contact with the internal wall of the column. Whenever possible, the gel should be poured in one continuous step. This can be accomplished by using a column extension reservoir, which is commercially available, or you can improvise by using a funnel that fits flush into the column.
6. Let the gel settle for 5–10 min and then open the column outlet. This will ensure an even sedimentation.

TABLE 10.2 Approximate Maximum Flow Rates and Pressures*

Gel Type	Maximum Operating Pressure (cm H_2O)	Approximate Maximum Flow Rate (ml/min)
Sepharose CL-6B	>200	2.5
Sepharose CL-4B	120	2.2
Sepharose CL-2B	50	1.25
Sepharose 6B	200	1.2
Sepharose 4B	80	0.96
Sepharose 2B	40	0.83
Sephadex G-75	160	6.4
Sephadex G-75 Superfine	160	1.5
Sephadex G-100	96	4.2
Sephadex G-100 Superfine	96	1.0
Sephadex G-150	36	1.9
Sephadex G-150 Superfine	36	0.5
Sephadex G-200	16	1.0
Sephadex G-200 Superfine	16	0.25

With the more rigid gels like Sephadex G-10–G-50, it is not necessary to worry about operating pressure. It is also not necessary to be concerned about the operating pressure when using Sephacryl S-100 HR–Sephacryl S-500 HR.

*Data obtained from columns with 2.5 cm. diameter and a bed height of 30 cm.

(Data from *Gel Filtration Principles and Methods, 5th Edition*, page 66. Uppsala: Pharmacia Inc.)

Comments: Pouring the column from too thin a suspension or packing the column in stages will often result in a poorly packed column. If additional gel is needed, it should be added before the gel is fully settled. Gel added to a previously established bed produces visible zones or boundaries.

Do not exceed the maximum operating pressure for a particular gel (see Table 10.2). To stabilize and equilibrate the gel with the desired column running buffer, pass two or three column volumes of buffer through the column.

Flow Rate

Tall, thin columns will produce slower linear flows than short large diameter columns. Finer support materials will also produce slower flow rates than coarser materials. As a general rule, resolution decreases with increasing flow rates. Maximum resolution is obtained with a long column and a low flow rate. Highest resolution is obtained when the flow rate is maintained in the range of 2–10 cm/h where cm/h = ml/h/cm^2 (cross sectional area of the column). Increases in column diameter (cross sectional area) dramatically increase the flow rate (ml/h). However, you will probably sacrifice some resolution in favor of a shorter running time. If the flow rate is too slow, diffusion will decrease resolution.

Linear velocity (cm/hr) = volumetric flow rate (ml/min)/column cross sectional area (cm^2) × 60 min/hr.

From linear flow rate (cm/h) to volumetric flow rate (ml/min)

$$\text{Volumetric flow rate (ml/min)}$$
$$= \frac{\text{Linear flow rate (cm/h)}}{60} \times \text{column cross sectional area (cm}^2\text{)}$$
$$= \frac{Y}{60} \times \frac{\pi \times d^2}{4}$$

where
Y = linear flow rate in cm/h
d = column inner diameter in cm

From volumetric flow rate (ml/min) to linear flow rate (cm/h)

$$\text{linear flow rate (cm/h)} = \frac{\text{volumetric flow rate (ml/min)} \times 60}{\text{column cross sectional area (cm}^2\text{)}}$$
$$= Z \times 60 \times \frac{4}{\pi \times d^2}$$

where
Z = volumetric flow rate in ml/min
d = column inner diameter in cm

If peaks are well separated using a long column and a low flow rate, you could possibly increase the flow rate or load more sample onto the column. Alternatively, you could go to a shorter, wider column increasing the amount of sample that can be loaded.

Hydrostatic Pressure

The hydrostatic pressure or hydrostatic head is measured as the vertical difference between levels at the top and bottom of a column of liquid. This is also referred to as the operating pressure or buffer drop from the buffer reservoir (usually positioned above the column as depicted in Figure 10.2) to the bottom of the outlet tube below the column. The higher the buffer reservoir is placed above the column, the greater the head pressure at the top of the gel bed, resulting in a greater flow rate. In other words, the top and bottom levels are the two points where the column of liquid is in direct contact with the atmosphere. The hydrostatic pressure may be increased by lowering the outlet or raising the level of the effluent solution vessel. Conversely, pressure may be reduced by the opposite manipulation.

Hydrostatic pressures should not exceed those recommended by the manufacturers during packing and chromatography. Excess pressure will compress soft gels and reduce flow rates.

When packing the column, allow the gel to settle for 5–10 min with the column outlet closed. Then open the outlet and raise it to the height needed to establish a hydrostatic pressure below the maximum limit listed in Table 10.2. As the gel bed settles, the hydrostatic pressure can be gradually increased to the maximum value indicated in Table 10.2.

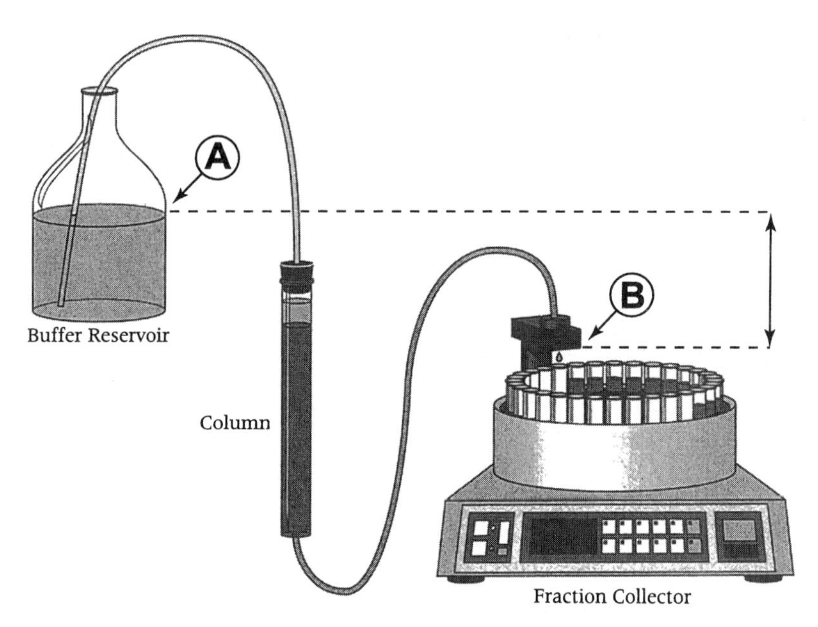

FIGURE 10.2 Operating pressure. The operating pressure is measured as the distance between the free surface in the reservoir (A) and the end of the outlet tubing (B).

Sample Application

PROTOCOL 10.3 Loading Sample onto a Drained Bed

Care must be taken to avoid disturbing the gel bed surface. An uneven bed surface will produce uneven separation of the bands and poor resolution.

1. Stop the flow of buffer to the column by attaching a pinch clamp to the incoming buffer line and open the top of the column. Remove most of the buffer above the gel surface by letting it flow through the column or by suction from above the gel with a Pasteur pipet. Never allow the bed to run dry.
2. All samples should be clarified by centrifugation or filtration prior to application to the column. This will prevent the column being clogged by precipitated or aggregated sample components.
3. Close the column outlet. Using a Pasteur pipet, gently and without disturbing the gel bed, apply the sample in a swirling motion on the inner wall of the column.
4. After the sample is fully applied, open the column outlet and allow the sample to enter the gel bed. At this point, begin to collect fractions.
5. Permit the sample to completely enter the gel then stop the flow. Gently apply a small volume of buffer to the column and allow it to enter the gel bed. Repeat this step.
6. Refill the column above the gel bed with buffer and connect the tubing to the flow system. Make sure that buffer is entering the column at the desired flow rate.

Comments: Sample application can be simplified if a glass frit or other device is used at the top of the gel bed. This provides a flat surface for sample application and prevents bed disturbances. Such devices are available commercially and can also be home-made.

Loading Sample Under the Eluent

To load the sample under the eluent it must have a higher density than the eluent. This can be accomplished by preparing the sample in a 10% glycerol buffer. The relative viscosity of the sample should not be greater than twice that of the eluent.

Using a Pasteur pipet, introduce the sample onto the top of the gel bed with the column outlet closed. As usual, care should be taken not to disturb the gel bed and to prevent mixing of the sample with the eluent. After the sample has been applied, to prevent mixing, draw a small amount of eluent into the Pasteur pipet that was used to apply the sample. The eluent flow can be restarted, and no washing is required.

Making Sure the Column Does Not Run Dry

One of the most frustrating lab experiences is to spend days carefully pouring and calibrating a column, running an overnight buffer equilibration and coming to the lab the following day and finding that the column has run dry! This problem can be easily avoided if you take advantage of the siphon effect as illustrated in Figure 10.3.

Molecular Weight Determination

Although gel filtration can be useful for protein purifications, it is also useful as an analytical tool. Traditionally, size exclusion chromatography has been used to obtain the apparent molecular weight of the target protein in its native state. In addition, as a function of migration through a precalibrated column, one can determine if aggregates have formed, or if a protein is part of a multimolecular complex. Some gels that have been repeatedly used for this method are Sephadex G-100 and G-200, Sepharose 4B, and Superose 12 in the HPLC mode although in theory, any gel can be used. The first step after pouring the column is to generate a molecular weight calibration curve by running a series of proteins with known molecular weights. Table 10.3 contains a short list of proteins with their molecular weights and Stokes radii.

The V_e of the proteins is measured from the peak of the OD_{280} reading. Any other assay can be used to locate the proteins. Molecular weights are plotted versus retention times or elution volume on a semilog scale. Once the molecular weight calibration curve for a size exclusion column has been established, it can be used to estimate the molecular weight of the target protein. Figure 10.4 demonstrates this technique.

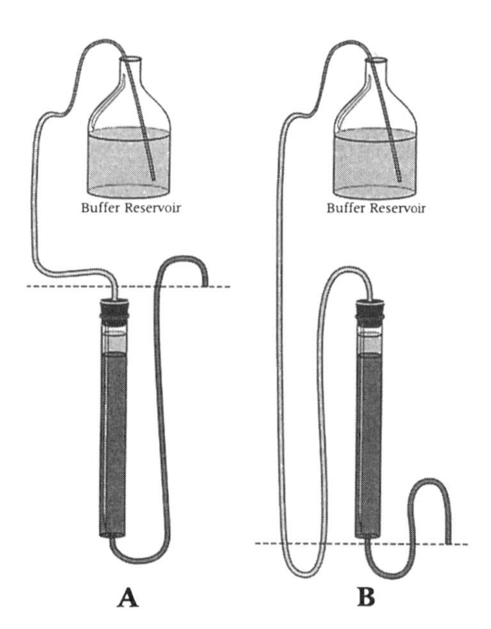

FIGURE 10.3 Safety loop arrangements that prevent a column from running dry. (A) The safety loop is placed after the column. The end of the outlet tubing is above the column. The flow will stop when the eluent in the inlet tubing reaches the level of the outlet tubing. (B) The safety loop is placed before the column. The column outlet is in any position above the lower loop of the inlet tubing. The flow stops when the eluent in the inlet tubing reaches the level of the outlet tubing. (Redrawn with permission from Cooper, 1977.)

Spin Columns Used in Gel Filtration

Spin columns are used for exchanging the buffer and removing excess salt or other undesirable small molecules from a small volume of protein solution. They can be quickly assembled from 1 ml plastic disposable syringes. For small volumes, spin-column chromatography has the advantages over conventional gravity-flow size exclusion chromatography of speed and minimal sample dilution.

TABLE 10.3 Stokes Radii and Molecular Weights

Protein	Stokes Radius (Å)	Molecular Weight (Da)
RNase A	16.4	13,700
Myoglobin	18.9	17,000
Chymotrypsinogen A	20.9	25,000
Ovalbumin	30.5	43,000
Bovine serum albumin	35.5	67,000
Lactate dehydrogenase	42.0	150,000
Aldolase	48.1	158,000
Catalase	52.2	232,000
Ferritin	61.0	440,000
Thyroglobulin		669,000

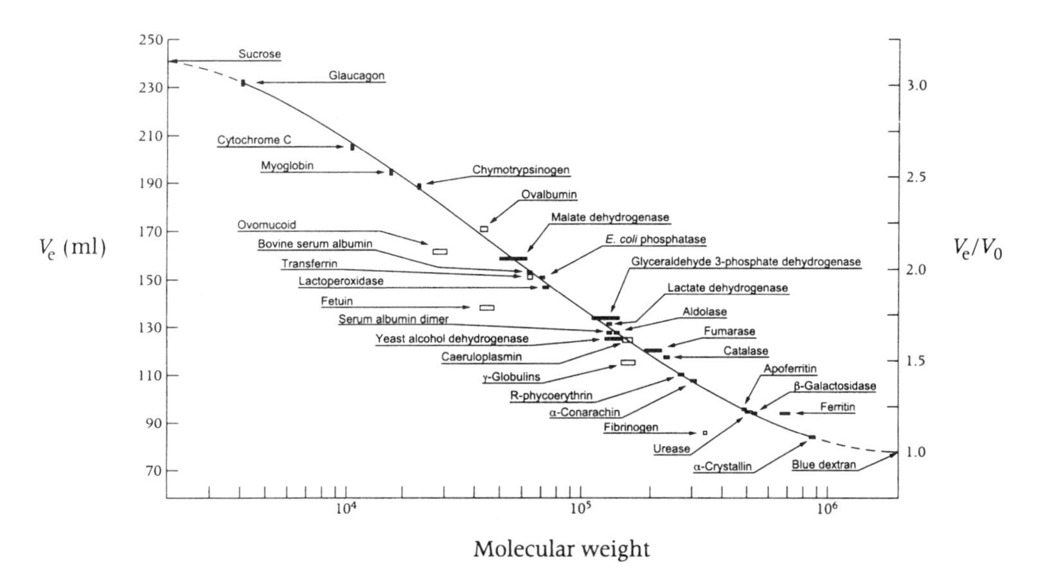

FIGURE 10.4 Molecular weight determination. Plot of the log (mol wt) of proteins versus the elution volume, V_e on a Sephadex G-200 column. The lengths of bars indicate uncertainties in molecular weights and the widths represent uncertainties in V_e (Reprinted with permission from Andrews, 1965.)

PROTOCOL 10.4 Spin Columns

Materials

Buffer
Syringe: 1 ml
Glass wool
Sephadex G-50
13 × 100 mm test tube

1. Remove the needle from a 1 ml syringe and tamp a small plug of glass wool into the bottom of the barrel with the plunger.
2. Load the syringe with Sephadex G-50 (fine grade). Add gel until the bed volume reaches 1 ml. This is your spin column.
3. Equilibrate the column with the desired buffer by passing a minimum of 2 ml of buffer through the column.
4. Place the spin column in a 13 × 100 mm glass test tube and centrifuge at 1000 rpm for 1 min. Examine the eluate. If gel beads are present, the column needs to be repoured with special care given to the plugging of the column with glass wool.
5. Apply the sample (200 µl) to the top of the gel bed. Centrifuge at 1000 rpm for 1 min. The eluted proteins are now in the same buffer used to equilibrate the column in step 3. Routinely, yields are >90%.

Comments: To obtain optimal performance, it is important to centrifuge the columns at the same relative centrifugal force for each centrifugation step. Gels other than Sephadex G-50 can be used. Spin columns are commercially available but probably not worth the expense.

PROTOCOL 10.5 Testing Fractions to Locate Protein: Bradford Spot Test

This assay is useful for quickly scanning column fractions to locate protein peaks (Harlow and Lane, 1988, adapted from Bradford, 1976).

Materials

Parafilm

Bradford reagent concentrate (Bio-Rad)

1. On a strip of parafilm, spot 8µl of each column fraction to be tested. Include a sample of elution buffer to give the background reading.
2. Add 2µl of Bradford dye concentrate (commercially available from Bio-Rad) and mix by pipetting up and down.
3. Samples that contain protein will turn blue. The sensitivity of the assay is ~10µg/ml. Detergents at concentrations over 0.2% will interfere with the reaction which is not affected by KCl, NaCl, $MgCl_2$, EDTA, and only slightly by Tris.

C. Introduction to HPLC

HPLC is usually defined as high performance, high pressure or high priced liquid chromatography. The basic elements of an HPLC system are shown in Figure 10.5. HPLC uses columns of very homogeneous, small particles. These packings provide improved chemical and physical stability, better reproducibility, and dramatically faster separations than the traditional soft gels. Columns containing high resolution packings (8, 15 or 40 micron) produce more concentrated fractions. The small particles create a high resistance to liquid flow so that the equipment has been designed to operate at relatively high pressures. The solvent is delivered to the column by a pump that provides constant, pulse-free flow at increased back pressure. The columns must be capable of withstanding the increased pressure. The detector should have a fast response time since protein peaks may pass through in a matter of seconds. The major advantages that HPLC has over open face, classical column chromatography are superior resolution and much shorter run times. The potential disadvantages are that HPLC systems are expensive and demand dedicated upkeep to insure usability and reproducibility. Usually each lab has one very disciplined individual who is in charge of the apparatus.

Take some extra time to get to know the apparatus. Most HPLC systems contain the following components; two reciprocating pumps (usually part of one unit), HPLC grade solvents, a controller, a dynamic gradient mixer, an injection unit, column guards, columns, a detector, a fraction collector, and a chart recorder and/or computer for data acquisition. The controller is a computerized unit of varying sophistication (depending on price) that is used to control flow rates, injection, gradient conditions, and collection conditions. The controller should be

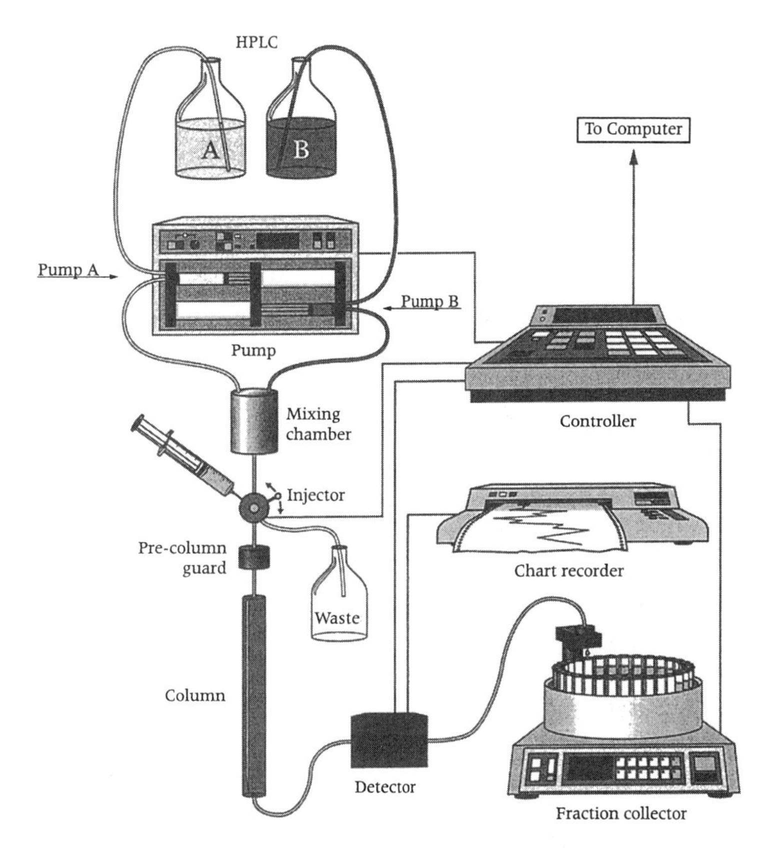

FIGURE 10.5 HPLC components. Buffers A and B are driven by pumps A and B respectively. The program is entered into the controller, which can be connected to a computer. When the material to be chromatographed is injected, the program is activated. The injected material interacts with the packing material in the column. After exiting, the material passes the UV detector and the result is recorded on the chart recorder and the data are saved in the computer for further analysis. The eluate is collected on the fraction collector in volumes that were preprogrammed into the controller panel.

interactive and user-friendly. An important component is the injection system; make sure not to inject an air bubble. It is useful to know the dead volume that exists between the detector and the end of the tubing. Excessive volume in the connecting tubing and detector causes peak broadening.

Packing Materials

Different types of packing material (also referred to as the stationary phase) are used in HPLC columns. Many of the stationary phases used in HPLC are referred to as "bonded" stationary phases. These particles are a composite consisting of a support with a sufficient mechanical stability and a surface "bonded" to the support. The surface chemistry provides the chromatographic interface for selective interaction with the substance to be separated. With non-porous materials, all binding occurs on the outer particle surface. Non-porous particles are very small (1–3 μm) allowing much higher flow rates than conventional porous materials. There is great variety as far as the chemical modification of the support surface is concerned. For size-exclusion chromatography, an inert, strongly hydrophilic coating is used. These packing materials are normally made of silica or polymers such as polystyrene and polymethacrylate. Both silica and polymer packings are available in a wide range of particle sizes and pore sizes.

Column Designs

Initially, analytical HPLC columns were made of stainless steel tubing, usually 50 or 100 cm × 2.0 mm for porous-layer beads and 25 or 30 cm × 4.0 or 4.6 mm for microparticulates. After columns were exhausted, users threw them away or, to save money, cleaned and repacked them.

Microbore columns are versions of analytical HPLC columns with smaller inner diameters (0.5–2.0 mm). The main reasons to use microbore columns are increased solvent savings and greater sensitivity for limited-sample situations. As column diameters become smaller, flow rates decrease due to the inverse square of the column radius ratio. Consequently, the system dead volume plays a bigger role in maintaining resolution.

Column tubing has evolved from stainless steel to Teflon. If an HPLC apparatus with stainless steel components is used with high salt solvents and/or extreme pH values, corrosion may occur, leading to metal ion release in the solvent. After long-term use, the reliability of the instrument itself will be impaired (Herold et al, 1991).

Biochromatographers prefer the inertness of polyetheretherketone (PEEK), which resists most of the popular HPLC solvents. In addition to being used for connective tubing, several companies have made entire columns from this material.

Column Guards

Column guards are inline filters, or short versions of analytical columns filled with the same packing material, placed between the injector and column. They collect debris such as pump-seal fragments and sample particulates, extending the life of the column. Since column guards are considerably less expensive than the analytical columns they protect, users should replace them on an as-need basis.

Detectors

UV detectors are routinely used in protein separations at a fixed wavelength of 210–220 nm where the peptide bond absorbs. Detection at 280 nm is useful for peptides or proteins containing aromatic residues. Some detectors allow the analysis of two wavelengths concurrently or have spectrum analysis capabilities, but this level of sophistication (and cost) is not required for routine protein purification.

If the peak height is greater than the width of the chart paper, the sensitivity scale of the chart recorder should be reduced in order for the top of the peak to be visualized. In this way one can determine the number of components in a chromatographic region. Alternatively, if data is fed into a computer, it can be retrieved and analyzed when the run has been completed.

Choosing the Right Conditions—Some Helpful Tips

The following short list should be considered prior to starting HPLC.

- Before deciding on which conditions will be best for a particular separation, consider all of the separation requirements. A compromise will have to be made between resolution, run time, and pressure. You can maximize any two of the factors, but in so doing, you will sacrifice the third factor (Dolan, 1994).
- To prevent gas bubble formation, the mobile, solvent phase should be degassed and filtered using a 0.2 μm filter. Degas the buffers by placing the buffer in a thick-walled Erlenmeyer flask attached to a vacuum pump or suction. When the solution ceases to bubble, it is considered degassed. Make sure that all solutions are brought to working temperature before they are used. Avoid taking a buffer that has been stored at 4°C and immediately using it at 20°C. Low temperatures promote gas solubility. Therefore, degas buffers that have been stored in the cold. Any solution containing nonvolatile buffer salts should be filtered through a 0.2 μm filter prior to degassing. Filter buffer stock solutions (~1 M) and then add the concentrate to degassed water to prepare the working solution. Volatile buffers should be added after degassing.
- Ensure that no air is trapped in the pump heads or in the precolumn tubing by shunting the column inlet to waste and running the pumps for a few minutes. This is referred to as priming the pumps.
- Connect the solvent flow to the column and start the pump. For getting started, use a flow rate of 1 ml/min, which is reached gradually in 0.1 ml/min increments over ~1 min. Equilibrate the column by pumping solvent through the column until the detector output is stable.
- To reduce backpressure increase due to particle fouling, remove precipitated material from the sample by either centrifugation or filtration. Be sure that components of the sample do not precipitate when they come in contact with the mobile phase. The use of a column guard or precolumn filter unit is recommended.
- The sample is usually introduced onto the column in a small volume (10–100 μl) by loading it into a sample loop and is then flushed onto the column with solvent. When loading the sample loop, do not use a pointed needle as this could damage the injection valve. When loading the injection loop take care not to introduce air bubbles.
- Prior to injecting another sample, reequilibrate the column to the initial running conditions.
- Exercise caution when using halides in the mobile phases especially with metal equipment. Halides have a corrosive effect on metal. Wash the equipment thoroughly after use.
- Silica based packing should not be routinely used above pH 8. Also avoid the use of chelating agents with silicas.

HPLC: Size Exclusion

The chromatographic separation of a protein mixture by size exclusion, (SE)-HPLC, is dependent upon the molecular sizes of the components. The principles are the same as regular size-exclusion chromatography. Proteins larger than the pore size will be excluded (i.e., not retained or retarded) and will be eluted in the void volume of the column.

When components are retarded by interacting with the column matrix by electrostatic or hydrophobic interactions, the elution volumes will no longer accurately reflect their molecular sizes. In these cases the apparent molecular size will be smaller than it actually is. Conversely, when using silica-based supports and low ionic strength buffers, early elution resulting from ionic repulsion may occur. This would be interpreted as an overestimation of the molecular weight. To minimize these types of interactions, the ionic strength of the buffer is increased or an organic solvent (i.e., methanol or acetonitrile) is added to the buffer.

Separations can be optimized by altering the flow rate. Generally, reducing the flow rate results in increased resolution. The sample volume should not exceed 1% of the column volume. As the sample size increases, peaks broaden and resolution deteriorates. The selection of a column will depend on the molecular weight of the target protein. As a first approximation, use a column from the TOSOHAAS SW series. For the separation of globular proteins <30,000 use a G2000SW column; for molecular weights from 30,000 to 500,000 use a G3000SW; and for molecular weights >500,000 a G4000SW.

D. Ion Exchange Chromatography: Separation on the Basis of Charge

Ion exchange is the most commonly practiced chromatographic method of protein purification due to its ease of use and scale-up capabilities. This technique exploits the amphoteric character of a protein (net positive in low pH buffer and negative in a high pH buffer). In ion exchange separations, the distribution and net charge on the protein's surface determines the interaction of the protein with the charged groups on the surface of the packing material. The support medium, which has covalently attached positive functional groups, is referred to as an anion exchanger if mobile negatively charged anions will be the exchanged species. Conversely, an exchanger with covalently bound negative groups is a cation exchanger, the mobile counter ions being cations. The charges on the protein and the packing material must be opposite for the exchange interaction to occur. Proteins that interact weakly with the ion exchanger will be weakly retained on the column resulting in short retention times. Conversely, proteins that strongly interact with the ion exchanger will be retained and have longer retention times. The nature of the matrix determines its physical properties such as its mechanical strength and flow characteristics, its behavior towards biological substances and, to a certain extent, its capacity. Basically, three major groups of materials are used in the construction of ion exchangers: polystyrene, cellulose and polymers of acrylamide and dextran. The acrylamide and dextran polymers have the advantage of having molecular sieving properties, being able to separate on the basis of size and charge.

One way to elute proteins is by increasing the ionic strength of the buffer. The competing salt ions (counterions) in the buffer displace the adsorbed sample molecules because their high concentration forces the equilibrium in favor of their adsorption in place of protein molecules. In a complex mixture, each component has a distinct net charge and requires a specific ionic strength to elute it from the column. When a salt gradient of increasing concentration is used, proteins are eluted at specific salt concentrations, effectively fractionating the mixture. Alternatively, the proteins can be eluted by altering the pH. As the pH approaches the pI of a protein, it loses its charge and elutes from the exchanger. When using pH gradients with a cation exchanger, the pH gradient increases; with an anion exchanger, it decreases. The variables that can be manipulated, such as buffer pH, buffer composition, the gradient slope, and the gradient forming salt offer a wide range of options for optimizing separations.

Simplified Theory of Ion Exchange

Proteins are bound to ion exchange matrices by reversible, electrostatic interactions. A separation is obtained because the diverse arrays of molecular species have different affinities for the exchanger. The adsorbed proteins are eluted in order of least to most strongly bound molecules, usually by increasing the ionic strength of the elution buffer. Ion exchange can tolerate the application of a large sample volume, often much greater than the volume of the column itself. This property is extremely useful, enabling the application of a dilute solution from the previous purification step. There are three possible fates for a particular protein on an exchange column: it can adsorb totally and immovably (with the partition coefficient $\alpha = 1$); it does not adsorb at all ($\alpha = 0$); or it adsorbs partially ($0 < \alpha < 1$), and moves down the column, emerging when $1/(1 - \alpha)$ column volumes of liquid have passed through. If the protein of interest has an α value of zero or slightly greater than zero, it will not be retained and pass through the column with the starting buffer. This could be useful if many other proteins in the extract are retained. Even if the degree of purification is not great, the proteins that are removed by adsorption might otherwise prove to be difficult to remove at a later stage.

The interacting charged groups on proteins are mainly carboxyl ($—COOH \leftrightarrow —COO^- + H^+$) and amino or tertiary amino ($—NR_2 + H^+ \leftrightarrow —NR_2H^+$). An anion exchanger binds proteins through their unprotonated carboxyl groups and repels protonated amino groups and vice versa for a cation exchanger. Thus the affinity of a molecule for a specific matrix can be experimentally manipulated by varying conditions such as ionic strength and pH. The result is that the net charge on a molecule can be made more strongly negative or positive by altering the solution parameters. If an anion exchanger is being used, a nega-

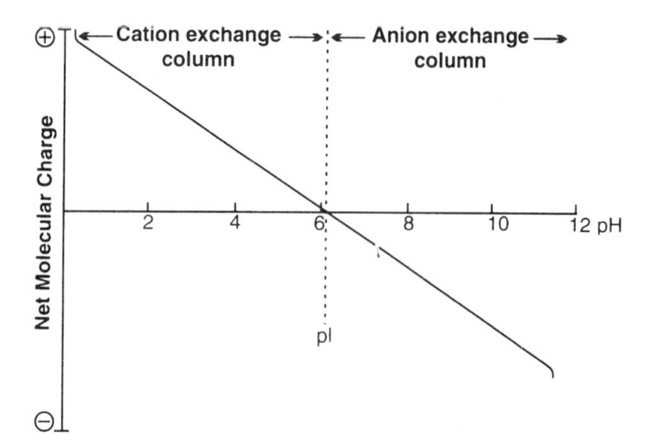

FIGURE 10.6 Guide to choosing a cation or anion exchange column based on the pI of the target protein. (Figure kindly provided by Bio-Rad laboratories, Life Science Group, Hercules, CA).

tively charged molecule will bind more avidly if the pH is raised. If the pH is lowered, the molecule is made more positive and will then bind more strongly to a cation exchanger.

Knowing the isoelectric point of your protein is a great advantage. The isoelectric point (pI) is the pH at which a molecule is electrically neutral. At a pH above its pI, the protein carries a negative charge and anion exchange columns should be used; below the pI, it carries a positive charge and a cation exchanger is recommended. This principle is illustrated schematically in Figure 10.6. When a molecule with a specific net charge is applied to an ion exchange column of the opposite charge, it adsorbs by electrostatic forces. Neutral molecules and molecules with the same charge as the exchanger do not bind and are eluted in the void volume (V_o) of the column.

When using ion exchange chromatography, the investigator has the option of choosing whether to bind the target molecule or let it pass through the column. Generally, it is more advantageous to adsorb the protein of interest, which usually results in a greater degree of purification.

The pI value of a protein is a measure of the theoretical total charge distribution on the target protein. However, the tertiary structure of a protein is such that not all of the charged residues are on the surface of the molecule, able to interact with an exchanger. The interaction can be localized at a particular site where there is a high density of like charges. Due to microenvironmental charge densities it is possible for a protein to be retained by both types of exchangers when in an environment close to the isoelectric point.

Functional Groups on Exchange Columns

The presence of a charged group is a fundamental property of a particular ion exchanger. The type of group determines the strength of the exchanger and their total number determines the capacity.

The most common functional groups for anion exchange are quaternary amines, diethylaminoethyl (DEAE) and polyethylenimine (PEI). For cation exchange the most common functional groups are sulfopropyl (S or SP) and carboxyl (C or CM). (See Tables 10.4A and 10.4B.)

Sulphonic and quaternary amino groups are used to form strong ion exchangers. The other groups form weak ion exchangers. Strong and weak refer to the extent of the variation of ionization with pH. Strong ion exchangers are completely ionized over a wide pH range. Weak ion exchangers start to lose their charge at pH values below six for cation exchangers or above nine for anion exchangers. Weak ion exchangers with DEAE or CM groups are preferred when separating very labile components.

DEAE tertiary ammonium exchangers continue to be the most widely used for protein separations in the pH range of 7–9. The DEAE group maintains a constant positive charge that is neutralized by a counter ion, usually Cl^-. Other anions are capable of competing for the positive DEAE group, including proteins that have a net negative charge because of their amino acid composition.

The capacity of an ion exchanger is a quantitative measure of its ability to take up exchangeable counter ions. Total capacity is the amount of charged and potentially charged groups per gram of dry ion exchanger. However, available or real capacity depends on the specific experimental conditions. Factors that can influence the capacity are accessibility of the functional groups, ionic strength of the eluent, the nature of the counter ions, pH and temperature.

Choice of Exchanger Matrix

When deciding what type of exchanger to use, the first consideration should be the pH stability of the protein. If the activity of the object protein is more stable below its pI then a cation exchanger would be a good choice. If the activity is more stable above the pI, an anion

TABLE 10.4A Anion Exchangers and Functional Groups

Anion Exchanger	pH Range	Functional Group
Aminoethyl (AE-) (intermediate)	2–10	$—OCH_2CH_2NH^+_3$
Diethylaminoethyl (DEAE-) (weak)	2–9	$—OCH_2CH_2N^+H(CH_2CH_3)_2$
Quaternary aminoethyl (QAE-) (strong)	2–10	$—OCH_2CH_2N^+(C_2H_5)_2 — CH_2CH(OH)CH_3$

TABLE 10.4B Cation Exchangers and Functional Groups

Cation Exchanger	pH Range	Functional Group
Carboxymethyl (CM-) (weak)	3–10	$—OCH_2COO^-$
Phospho (intermediate)	2–10	$—PO_4H_{2-}$
Sulphopropyl (SP-)	2–12	$—CH_2CH_2CH_2SO_{3-}$

exchanger is recommended. If the material is stable over a wide pH range, either type of exchanger can be used.

The pH of the starting buffer is chosen so that the material of interest will be bound to the exchanger in bioactive form. If the isoelectric point of the protein is known, then the pH of the buffer can be adjusted to be at least 1 pH unit above or below the pI of the substance to facilitate adequate binding. Charge distribution affects the strength with which a protein binds to an exchanger. Multiple positive charges in close proximity will enhance binding to a cation exchanger over that expected based on total surface charge. If anionic groups are within positively charged patches, repulsive forces will affect the pK's of the side chains and interaction with the exchanger.

Preparation of the Exchanger

The exchanger matrix is swelled (precycled) to expose a larger percentage of the exchanger's charged groups. Precycling is important when the exchanger is based on a cellulose matrix. Swelling is performed by suspending the exchanger in a solution of either strong acid or strong base. For DEAE-cellulose, treatment with HCl converts all of the groups to the charged species ($C_2H_4N^+H(C_2H_5)_2$). When all of the functional groups are positively charged, there is mutual electrical repulsion that results in the maximal amount of swelling. For a cation exchanger like CM-cellulose, all the carboxylic acid groups are treated with strong base resulting in a uniform creation of carboxylate ions. Once the matrix is fully swelled, the acid or base is washed away, and the resin is treated with base in the case of the anion exchanger or acid in the case of the cation exchanger. This converts the functional groups of the DEAE and CM to their free base and free acid forms, which are the forms that are most easily equilibrated with the desired counterions. Since strong acid and base tend to decompose the exchanger, the treatments should be limited to periods less than 1 h with concentrations not exceeding 0.5 N.

Fines, small particles generated during manufacturing, or as the result of excessively vigorous mixing, are removed after the matrix has been fully swelled. Fines are easily removed by repeatedly suspending the exchanger in a large volume of water, allowing 95% of the resin to settle and slowly decanting off the slower sedimenting material.

The resin is now ready to be equilibrated with the appropriate counter ion. This is accomplished by passing a large volume of the desired counterion in high concentration over the resin. Excess counter ions are removed by washing the exchanger with a large volume of water or dilute buffer.

Choice of Buffer

The pH should be within the operational range of the ion exchanger and also be compatible with the protein. A working ionic strength for the application buffer is 0.01–0.05 M. For anion exchange common starting buffers are Tris or phosphate in a pH range of 7–9. The eluting buffer is usually the starting buffer, containing in addition 0.05–1.0 M

TABLE 10.5 Suitable Buffer Compositions for Ion Exchange Chromatography

pH	Buffer
For Anion Exchangers*	
8.5–9.2	Diethanolamine, or 2-amino-2-methyl-1,3-propanediol
7.8–8.9	Tris
7.4–8.4	Triethanolamine
6.6–7.7	Imidazole†
6.2–7.1	Bis-Tris
5.6–6.6	Histidine†
For Cation Exchangers**	
3.6–4.3	Lactate
4.3–5.2	Acetate
5.2–5.8	Picolinic acid
5.8–6.8	MES
6.3–7.1	ADA
6.8–7.8	MOPS
7.2–8.2	TES
7.8–8.9	Tricine

*Anion exchange buffers are adjusted to the indicated pH with HCl or the base form of the indicated buffer.

†Histidine and imidazole complex with divalent cations.

**Cation exchange buffers are adjusted to the indicated pH with KOH or the acid form of the buffer.

Data from Scopes (1987), page 117. Reprinted with permission of Springer-Verlag.

NaCl. Common starting buffers for cation exchange are bis-Tris, MES and acetate in a pH range of 5–7. A more detailed list of buffers is presented in Table 10.5.

Batch Adsorption

In batch adsorption, the gel matrix is placed in a test tube or beaker. The solutions are added to the matrix and mixed. The gel is allowed to settle and the supernatant is removed and replaced with another solution. Elution is performed in a step-by-step manner. Contaminants are removed in an elution step prior to eluting the protein of interest. The primary requirement for successful batch elution is that the partition coefficient, α, of the target protein be very close to one and that a large amount of the proteins in the extract have lower α values. Batch methods offer the advantages of speed and the possibility of treating large volumes of material. However, unless α is >0.98, some losses will be inevitable. Columns are slower and may clog, especially when dealing with crude extracts.

Defining the exact conditions cannot be stressed enough. It is important to note salt concentration and pH of buffers, bed volume of exchanger used and where in the protocol buffer changes were performed.

PROTOCOL 10.6 Selecting the Starting pH

A preliminary small-scale experiment using batch adsorption should provide the information necessary for choosing the starting pH and also give an accurate estimation of the salt concentration needed to elute the target protein. For anion exchangers, test between pH 5–9; for cation exchangers between pH 4–8. The pH intervals should be 0.5–1.0 pH unit. The pH of the buffer to be used in an experiment should allow the target protein to bind, but be as close to the point of release as possible. If too low or too high a pH is chosen, elution becomes difficult, and extremely high salt concentrations may have to be used.

Materials

Starting buffer solutions, varied pH, identical ionic strength
Sepharose ion exchanger (DEAE or CM)

1. Equilibrate 0.2–0.5 ml of Sepharose ion exchanger with the starting buffers by washing the gel five times with 10 ml of buffer.
2. Add known amounts of protein solution to the pre-equilibrated exchanger (step 1).
3. Gently mix the sample and gel for 5–10 min and allow the gel to settle.
4. Remove the supernatant and save.
5. Wash the ion-exchanger test fractions twice with 1 ml of starting buffer. Keep the washes.
6. Elute the target protein in a step-wise manner by adding 0.5 ml increments of starting buffer containing increasing concentrations of NaCl from, 0.05 M, 0.1 M, 0.2 M, 0.4 M, 1.0 M. Collect each increment.
7. Assay all fractions for the target protein.

Comments: For a large-scale purification, choose a starting buffer that binds the target protein completely and permits recovery with 0.1–0.4 M NaCl.

PROTOCOL 10.7 Packing an Ion Exchange Column

Packing an ion exchange column is similar and perhaps easier than packing a gel filtration column. When pouring a column, keep dead spaces to a minimum to prevent remixing of separated zones.

1. Equilibrate the chosen ion exchanger to the working temperature before packing the column.
2. Degas the ion exchanger (Protocol 10.1).
3. Fill the column ~1/3 full with buffer and close the outlet valve.
4. Using starting buffer, prepare a 50–75% slurry of the chosen exchanger and pour it into the column, making sure not to trap air bubbles. Whenever possible, pour the exchanger all at once by adding a column extension reservoir if necessary.
5. Open the outlet valve and allow the gel to pack.

6. Adjust the column outlet so that the operating pressure is about 1 cm H_2O/cm bed height. Rigid gels like DEAE-Sephacel and Sepharose can be packed at higher flow rates. Pack the column using a slightly higher flow rate than that to be used during the experiment.
7. Pass at least two bed volumes of buffer through the column to stabilize the bed and equlibrate the exchanger. Measure the pH of the effluent. It should be the same as the equilibration buffer. Examine the column to make sure that the gel surface is level and that there are no trapped bubbles.

Experimental Tips

- The amount of sample to be loaded is dependent on the capacity of the exchanger. Do not overload the exchanger.
- One great advantage of ion-exchange chromatography is that large volumes can be applied to the column. When the experiment is designed so that the target protein binds to the exchanger, a large volume of dilute solution can be applied. In this manner, ion exchange serves as a means of concentrating a sample in addition to purification.
- The ionic composition of the sample should be the same as the starting buffer. This can be attained by either dialysis or gel filtration, or by simply diluting the sample.
- To reduce the chances of clogging the column, always either filter or centrifuge the material prior to applying it to the column.
- The diameter of the column can be adjusted on the basis of the capacity required for a specific separating job.
- Since peaks are broadened by both slow flow rates and by nonequilibrium conditions, a flow rate that minimizes peak width should be chosen. One ml/min is frequently a good compromise.

Elution-Step or Linear Gradient?

Elution of the target protein bound to the exchanger can be achieved by varying the pH, increasing the ionic strength or both. Proteins are routinely eluted from anion exchangers either by decreasing the pH of the eluting buffer or increasing the ionic strength. For elution from a cation exchanger, the pH of the eluting buffer is increased or the ionic strength is increased. Changing the pH of a substance towards its isoelectric point causes it to lose its net charge and be eluted from the exchanger. In practice, pH elution results in fairly broad peaks compared to salt elution. Occasionally, conditions can be chosen such that sample components are separated by elution with starting buffer. The time of the run can be long and some materials may be irreversibly bound. If all retarded substances are eluted, column regeneration is not required.

Step-by-step ionic strength elution is accomplished by changing the ionic strength of a buffer in premixed, defined increments. Step-by-step elution is technically easier to perform than gradient elution because the gradient making apparatus is not required. Compact peaks containing the protein of interest can be obtained by the incremental application of a buffer with a higher ionic strength. However, unless the step increase of salt is quite large, expect a tailing effect where the protein of interest is also found in the next, higher ionic strength fraction. The second peak is usually not a different form of the protein, and it would therefore be erroneous to deduce the presence of isoforms. With a large increase in salt concentration, many other proteins in addition to the protein of interest will be eluted. Stepwise elution sacrifices resolution for simplicity.

Ionic strength gradients are easily prepared by gradually mixing two buffers of differing ionic strength. The buffer can flow to the column by gravity or with the aid of a pump. The ionic strength will increase linearly. Two examples of gradient makers are shown in Figure 10.7. The gradient maker shown in Figure 10.7B can be made in the lab with bottles, rubber stoppers and glass tubing.

Gradient elution produces symmetrical peaks and better resolution because during the elution, the zones tend to sharpen. As it encounters an area of low ionic strength, the leading edge is retarded while the trailing edge tends to move at a higher speed due to the increased ionic strength.

Generally, step-wise elution is used to determine the conditions necessary for the elution of the target molecule. Subsequently, the fine-tuning that will optimize the purification will be done using gradient elution based on the properties learned from the step-by-step elution. Following the run, the ionic strength of the fractions should be measured with a conductivity meter. This parameter will define the ionic strength necessary to elute the target protein. Knowing this information, the experimental parameters of subsequent runs can be adjusted to improve the elution of the target molecule. For example, if the target molecule is eluting with 0.15M NaCl in the buffer, the binding and washing can be performed with 0.05M NaCl and the gradient from 0.05 to 0.25M NaCl.

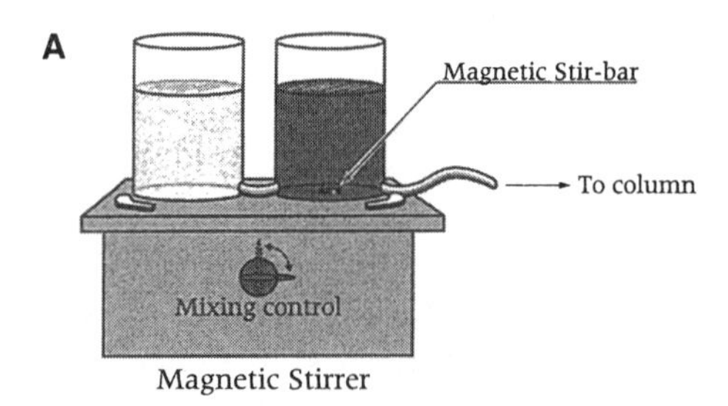

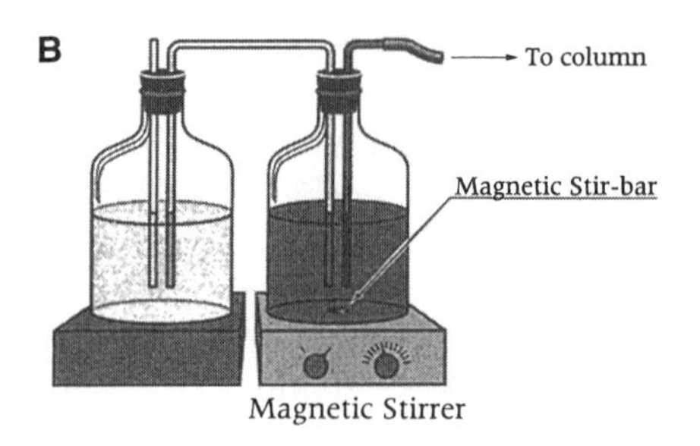

FIGURE 10.7 Gradient makers.

PROTOCOL 10.8 Regeneration of Sephadex Ion Exchangers

For repeated runs of a specific separation, it is recommended to reuse the same ion exchanger. Ion exchangers can be rejuvenated without having to repack the column.

Materials

pH paper
NaOH: 0.1M

1. Wash the exchanger with salt solutions containing the desired counter ion, increasing the ionic strength until an ionic strength of about two has been reached.
2. Wash with 0.1M NaOH followed by distilled water.
3. Equilibrate the column with starting buffer. Check the pH of the eluent to be sure you are at the pH of the starting buffer. To remove lipids the column can be washed with alcohol and nonionic detergents. If in doubt, consult the product literature or call the company's tech service department.

PROTOCOL 10.9 Regeneration of Sepharose Ion Exchangers

Sepharose ion exchangers are routinely regenerated by passing a high ionic strength buffer through the column and then equilibrating the exchanger with running buffer. A more rigorous regeneration protocol is presented below.

Materials

Sodium acetate buffer: 1M sodium acetate-HCl, pH 3.0
NaOH: 0.5M

1. Wash the column with 1 bed volume of acetate buffer.
2. Wash the column with 1.5 bed volumes 0.5M NaOH. Leave this solution on the column overnight.
3. Wash the column with 1.5 bed volumes of sodium acetate buffer.
4. Reequilibrate the column with starting buffer.

PROTOCOL 10.10 Chromatofocusing

Chromatofocusing is a unique column chromatographic method with broad applications developed by Sluyterman and Elgersma (1978) and Sluyterman and Wijdenes (1981). It is a special type of IEC where the pH of the eluent is continuously changing. It combines speed with high resolution to separate molecular species on the basis of their isoelectric points (pI). Any lab with equipment for routine gel filtration can use the same pumps and columns for chromatofocusing. It is possible to chromatofocus in the presence of nonionic detergents such as Triton X-100 and NP-40. Other forms of detergents interfere with separation and resolution. Crude protein mixtures are applied to a column that is equilibrated at a pH slightly above the assumed pI of the target protein. Proteins that carry a charge identical to that of the column are excluded, and like conventional ion-exchange chromatography, those of opposite net charge are retained.

A self-generating pH gradient is formed using ampholyte type buffers that combine high buffering power with low ionic strength. The exchanger, polyethyleneimine-agarose, is similar to DEAE-adsorbents. The pH gradient is generated within the column by the continuous addition of an acidic form of the ampholyte, Polybuffer (Pharmacia Publications, 1981), which is adjusted to a pH lower than the pI value of the target protein. As a result, a steady pH gradient emerges from the column, and proteins are eluted in sharp, well separated bands at pHs close to their isoelectric points. Since the linear flow rate of eluent through the column is faster than the development of the pH gradient, proteins are desorbed repeatedly throughout the length of the column, contributing to the focusing effect. Chromatofocusing can separate molecules of almost identical pI, resolving differences of less than 0.05 pH units.

For best results choose a long, narrow column. As a working approximation, 100mg of protein can be loaded per 10ml of gel bed. Choose a pH interval such that the isoelectric point of the protein of interest falls in the middle of the pH gradient.

Materials

Column exchanger resin

Polybuffer: Commercial polybuffers are available from Pharmacia: Polybuffer 74 (pH range 4 to 7), Polybuffer 96 pH 6 to 9, Pierce (Buffalyte 3–10 and 4–8) and Polysciences (PolySep 47).

Starting buffer

1. Choose and equilibrate the appropriate exchanger resin, either PBE 118 (used for a gradient between pH 11–8) or PBE 94 (used between pH 9–4) with start buffer. The pH of the start buffer should be slightly above (~0.4 pH unit) the upper limit of the desired pH gradient.
2. Choose the appropriate Polybuffer stock and prepare the working Polybuffer solution according to your specific experimental needs by adjusting the pH of the Polybuffer solution to the pH value selected for the lower limit of the pH gradient and adjusting the volume. Usually ten column volumes are used over a pH interval of three units. The less the Polybuffer is diluted the steeper will be the gradient, eluting the proteins closer together. Conversely, diluted Polybuffer and large volumes will result in long elution times, broad peaks and low sample concentration.
3. Prior to loading the sample, apply 5ml of Polybuffer to the column, which will protect the sample from pH extremes.
4. Equilibrate the sample in start buffer or Polybuffer, whichever buffer pH is most suitable for the stability of the sample, and apply it to the column.

The volume of sample should be such that the total sample volume is applied before the target protein is eluted.

5. Elute the column with Polybuffer. The pH gradient is self-generating during the run. About 1.5–2.5 bed volumes of buffer pass through the column before the gradient begins. Therefore, the total volume of Polybuffer required is ~12.5 bed volumes.
6. Monitor the eluate at 280nm. In addition, measure the pH of each fraction soon after collection to insure that the gradient was generated satisfactorily.

Comments: There have been reports in which chromatofocusing was the sole chromatographic method employed. In a purification protocol, chromatofocusing is often used as the final step. It can concentrate large volumes of dilute material into tight bands of much smaller volume while taking to virtual homogeneity already enriched fractions. However, this should not be taken to mean that chromatofocusing should be used exclusively as a final step.

Removing the Polybuffer

Polybuffer can be an unwanted byproduct of chromatofocusing. Although it does not interfere with the Bradford protein determination, Polybuffer will complex with copper ions interfering with the Lowry protein determination assay. More importantly, it may alter the biological behavior of the purified protein. Three methods are suggested for removing Polybuffer from the purified proteins.

- The simplest method to remove the Polybuffer is precipitation of the protein with ammonium sulfate. Solid ammonium sulfate is added to a final concentration of 80–100% saturation. The sample is allowed to stand at 4°C 1–2h until the protein precipitates and is then collected by centrifugation and washed once with saturated ammonium sulfate. The Polybuffer will stay in solution.
- If the protein of interest is greater than 25kDa, the Polybuffer can be removed by gel filtration using Sephadex G-75 or G-100. Choose a gel that excludes the target protein.
- The target protein can be run through a specific affinity matrix that will bind the protein and let the Polybuffer run through unretarded.

HPLC-Ion Exchange Chromatography

The same principles that apply for classical ion exchange chromatography are applicable for high performance ion exchange chromatography. The column packing is designed to facilitate an intimate contact between the mobile phase and the stationary phase. The chromatographic surface should selectively retain the components to be separated with favorable desorption kinetics. For interactive chromatographic methods like ion-exchange and hydrophobic interaction chromatography, the surface coating embodies the respective fixed ionogenic functions and mildly hydrophobic groups that have the appropriate chemical structure and functional concentration.

If the pI of the target protein is known and it is <7, then anion-exchange is recommended. If the pI is >7, then cation exchange should be used. In anion exchange, proteins containing a higher ratio of acidic to basic charges will be retained by the positively charged amines on the support surface. Conversely, in cation exchange, proteins with an excess of basic charges will be attracted to the negatively charged carboxyl, or other acidic groups. (See Table 10.6.)

After use, the ion-exchange column should be washed with water to remove any of the buffer salts and then stored in 50% methanol. As with almost all chromatography, never allow the column to dry out.

Membrane Adsorbers

Increasing flow rates across column beads cause reduced contact between solutes and available adsorption sites. This results in a reduction in capacity and band spreading during elution. These detrimental effects have been lessened by a new technology in the form of Membrane Adsorbers (MA), which combine ion exchange properties with membrane technology to provide a rapid, low pressure ion exchange liquid chromatography in an integral, cross-linked regenerated cellulosic matrix (Sartorius Corp). Membrane Adsorbers utilize functional groups covalently bonded to the surface of a cross-linked, non-collapsible cellulose membrane. A high surface area entirely accessible to the convection flow through the membrane permits high flow rates (up to 100 ml/min with 15 cm^2 units) without channeling and high binding and elution. Traditionally packed columns offer flow rates up to 10 ml/min. The membrane matrix is resistant to a wide spectrum of chemicals, does not swell or distort in buffer solutions and is compatible with pH values 2–13.

MA units are operated in an adsorption/desorption mode, capturing biomolecules in low ionic strength, dilute solutions and eluting with increased salt concentrations. Some advantages of the MA are that close contact of charged functional groups in the adsorptive matrix with the biomolecules permits efficient separations. In addition, the membrane provides low non-specific adsorption with binding capacities up to 2 mg/cm^2 at flow rates up to 50 ml/min. Protein recoveries with MA units are >90% with retention of bioactivity typically >90%. Properties are presented in Tables 10.7 and 10.8.

Membrane adsorber units can be operated with a syringe using luer lock connections, a centrifuge or connected to a peristaltic pump.

TABLE 10.6 Buffers and Buffer Components for Ion-Exchange HPLC of Proteins

Mode	Buffers 10–50 mM	Sodium Salts	Denaturants/ Detergents	Organic Solvents
Anion Exchange	Tris-Cl	Chloride	Urea (4–7 M)	Methanol (40%)
	Bis-Tris	Acetate	CHAPS (0.05%)	Acetonitrile (30%)
	Phosphate	Phosphate		Propanol (20%)
Cation Exchange	$NaPO_4$	Chloride	Urea (4–7 M)	Methanol (40%)
		Acetate	CHAPS (0.05%)	Acetonitrile (30%)
		Phosphate	SDS (0.02%)	Propanol (20%)

TABLE 10.7 Technical Specifications of Sartobind Membrane Adsorbers

Membrane:	Cross-linked, regenerated cellulose in 3 or 5 layers
Adsorption areas:	$15\,cm^2$ or $100\,cm^2$
Bed heights:	0.75 mm and 1.25 mm
Bed diameters:	26 mm and 50 mm
Bed volumes:	Bed volumes 0.3 ml and 2.0 ml
Flow rates:	1–100 ml/min/15 cm^2 unit 1–200 ml/min/200 cm^2 unit
Protein binding capacities:	10–30 mg/15 cm^2 unit 70–200 mg/100 cm^2 unit
Max. operating pressures:	105 psi (7 bar)/15 cm^2 unit 90 psi (6 bar)/200 cm^2 unit
pH compatibility:	2–13
Regeneration:	1 N NaOH for 1 hr 1 N HCl or 70% EtOH, backflushing is possible

(From *Genetic Engineering News*, May 1, 1995, page 22. Reprinted with permission.)

Because the binding capacity increases linearly with adsorption area, systems can be scaled up simply by using multiple smaller units connected inline. MA units can be cleaned and regenerated with 1 M NaOH or 1 N HCl.

Applications include concentration and purification of proteins, screening for optimal purification conditions for uncharacterized proteins, and protein purification and characterization. MA units should prove especially useful for the purification of proteins that lose activity during lengthly separation protocols (Weiss and Henricksen, 1995).

Perfusion Chromatography

In Perfusion Chromatography, developed by Perseptive Biosystems, transport into the particles occurs by a combination of convection and diffusion. The innovation lies in the pore morphology of the polymer-

TABLE 10.8 Rapid Adsorption/Desorption for Concentration or Purification of Proteins

MA Unit	†S15	S100	‡Q15
Adsorption area	$15\,cm^2$	$100\,cm^2$	$15\,cm^2$
Protein	rat/mouse lgG1*	mouse/mouse lgG1*	bovine gamma-globulin**
Initial concentration	0.02 mg/ml	0.04 mg/ml	0.1 mg/ml
End concentration	0.4 mg/ml	12 mg/ml	5 mg/ml
Loading buffer (A)	25 mM MES, pH 5.9	25 mM MES, pH 5.9	25 mM Tris-HCl, pH 9.0
Elution buffer		A + 250 mM KCl	A + 1 M KCl
Flow rate	50 ml/min	100 ml/min	100 ml/min
Starting volume	100 ml	0 ml	4 ml
Concentration factor 20X	300X	50X	
Purity (%)	>90	>90	—
Processing Time	3–4 min	35 min	3–4 min

*A supermatant of serum-free tissue culture with contaminants BSA, human transferrin and bovine insulin.
**Pure, diluted protein was used.
†Functional groups sulfuric acid strongly acidic cation exchange.
‡Functional groups are strongly basic anion exchanger.
(From *Genetic Engineering News*, May 1, 1995, page 22. Reprinted with permission.)

FIGURE 10.8 Structure of the POROS bead. The filled circles designate functional groups (absorptive sites) and arrows indicate mobile phase motion. (Reprinted from Belenkii and Malt'sev (1995), by permission of Eaton Publishing Co., Natick, MA).

ized column beads as shown in Figure 10.8. The throughpore structure creates a larger surface area thus increasing diffusion through the column. This makes the capacity of Poros packing much higher at higher flow rates as compared to conventional packings. This innovation can be applied to ion exchange, reversed phased and affinity chromatography of proteins and peptides. (Afeyan et al, 1990)

E. Hydrophobic Interaction Chromatography (HIC)

In many cases, the target protein and impurities have a similar Stokes radius or isoelectric point. When such difficulties are encountered, the overall effectiveness of the purification scheme may be significantly enhanced by adding a fourth chromatographic dimension. Hydrophobic interaction chromatography (HIC) is a technique in which proteins are separated on the basis of the differing strengths of their hydrophobic interactions with a gel matrix that contains uncharged hydrophobic groups. Since this method makes use of the hydrophobic sites on the surface of the protein, it is a generally applicable separation technique. Proteins and polypeptides interact with the hydrophobic surface by adsorption rather than partition. Aqueous buffer systems are used in HIC. The protein is eluted with a gradient of decreasing ionic strength. It is ideal for use immediately after ammonium sulfate precipitation where the initial high salt conditions enhance the hydrophobic interactions. Because the target protein is eluted in a low salt concentration, it is recommended to perform HIC prior to ion exchange thus avoiding a dialysis step.

Simplified Theory of HIC

The HIC matrix is designed to be the most hydrophobic and/or abundant binding site in the system. The presence of salting-out ions decreases the availability of water molecules in solution, which serves to enhance hydrophobic interactions. As a result of these factors, proteins will preferentially bind to the HIC support. The bound material can then be selectively desorbed by decreasing the salt concentration in the mobile phase. Changes in temperature, pH and ionic strength will influence the separation and should be closely monitored.

The strength of the hydrophobic interaction is increased when the concentration of neutral salt is increased. The proteins are commonly adsorbed to the matrix at high ionic strength (e.g. 4M NaCl or 1M ammonium sulfate). Ammonium sulfate is the most frequently used salt for HIC mobile phases because it does not alter the activity of most proteins, which presumably retain their native configuration. Some salts are more effective than others in promoting hydrophobic interactions. In Figure 10.9, both anions and cations can be arranged in order of increasing hydrophobic interactions (the Hofmeister series) (Arakawa and Timasheff, 1984).

The ionic species are arranged in descending order of chaotropic effect, which leads to decreases in the strength of the hydrophobic interaction between the protein and the matrix.

Adsorbed substances are eluted by lowering the ionic strength or changing the buffer composition to contain an ion with a higher chaotropic effect (see above). After the sample has been applied to the column in high salt buffer, the adsorbed proteins are eluted by applying a gradient of decreasing ionic strength. Lowering the salt concentration results in a reduction of hydrophobic interactions. Proteins are eluted under nondenaturing conditions. Other techniques that reduce adsorption include polarity-lowering agents such as ethanol and ethylene glycol, and detergents. Harder to elute adsorbed proteins may be eluted with a combination of a decreasing salt gradient together with an increasing ethylene glycol gradient.

PROTOCOL 10.11 Protein Fractionation by HIC

Octyl and phenyl-Sepharose (Pharmacia) are derivatives of Sepharose CL-4B that contain hydrophobic n-octyl and phenyl groups respectively. Phenyl Sepharose is less hydrophobic than octyl Sepharose. Extremely hydrophobic proteins may adsorb too strongly to octyl Sepharose and require strong eluting conditions, in which case the phenyl Sepharose would be a better choice. Although the binding capacities for individual proteins may vary, the manufacturer reports

Increasing salting-out effect

←

$PO_4^{-3} > SO_4^{-2} > CH_3COO^- > Cl^- > Br^- > NO_3^- > ClO_4^- > SCN^-$

Cations that have an increasing hydrophobic effect are:

$NH_4^+ > Rb^+ > K^+ > Na^+ > Cs^+Li^+ > Mg^{+2} > Ca^{+2} > Ba^{+2}$

→

Increasing salting-in effect

FIGURE 10.9 Hofmeister series.

that both phenyl and octyl-Sepharose can bind roughly 15–20 mg human serum albumin or 3–5 mg β-lactoglobulin per ml of gel.

Prior to HIC, the sample ionic strength should be adjusted with salt and buffered near the pI of the target protein (if that information is known). The column should be equilibrated with the same buffer.

In the protocol described below, proteins are fractionated using sequential elution with reduced ionic strength followed by a gradient of increasing concentration of ethylene glycol.

Materials

Equilibration buffer: 0.8 M $(NH_4)_2SO_4$, 0.025 M potassium phosphate ~pH 7
Gradient maker

1. Equilibrate the gel and protein sample with equilibration buffer and pour the column (see Protocol 10.2).
2. Equilibrate the column with at least three column volumes of equilibration buffer.
3. Apply the sample to the column.
4. Wash the column with at least two column volumes of equilibration buffer.
5. Elute with three column volumes of a decreasing linear gradient of $(NH_4)_2SO_4$ from 0.8–0.0 M in 0.025 M potassium phosphate buffer ~pH 7.
6. If the target protein did not elute during the decreasing salt gradient used in step 5, elute the column with an increasing linear gradient of 0–80% (v/v) ethylene glycol in 0.025 M potassium phosphate buffer pH 7.

Comments: Elution conditions necessary for optimum protein separation are derived empirically. Failure to elute the target protein with a decreasing ionic-strength gradient should be overcome by using stronger elution conditions or switching to a less hydrophobic matrix, if available.

PROTOCOL 10.12 Solid Phase Extraction Cartridges

Prior to performing reversed phase (RP) HPLC (described below), it is recommended to perform a pre-run with a solid phase extraction (SPE) cartridge. SPE cartridges are prepacked mini-columns that are smaller, less efficient versions of regular HPLC columns (Majors, 1994). They can be mounted on syringes, enabling convenient, rapid separations. Passage through an inexpensive, disposable SPE cartridge will clean samples before passing them through an expensive HPLC column. Using SPE cartridges, the hydrophobicity of the target protein and conditions necessary to elute the target protein can be approximated. In addition, the SPE cartridge provides a purification of its own, removing contaminants with hydrophobicities unlike those of the target molecule but which are still present in the semi-purified extract at this stage. The following protocol is used for Sep-Pak C_{18} (Waters) SPE cartridge.

Materials

Sep-Pak Cis (Waters)
Methanol
Trifluoroacetic acid (TFA): 0.1% TFA in water
10 ml syringe

1. Wash the Sep-Pak cartridge with five column volumes of 100% methanol.
2. Equilibrate the cartridge with trifluoroacetic acid (TFA) solution. Do not allow the Sep-Pak cartridge to run dry.
3. Using a 10 ml syringe, apply the protein solution to the Sep-Pak. Collect the flow through and reapply it.
4. Wash the cartridge with five column volumes of TFA solution. This should elute polar molecules.
5. Perform a step-by-step elution with five column volume steps of increasing amounts of methanol 10%, 30%, 70%, 100% all containing 0.1% TFA. Collect the fractions and assay them for the presence of the target molecule. If methanol or TFA interferes with the detection assay, take aliquots and remove the methanol and TFA by evaporation with a Speed-Vac prior to the assay.

Comments: Cartridges may be conditioned at flow rates up to 25 ml/min (steps 1 and 2). Load the sample and elute at flow rates below 10 ml/min. The reversed-phase cartridges may have a capacity up to 100 mg of strongly retained substances.

Reversed Phase HPLC

In reversed-phase chromatography (RPC), the chromatographic surface is strongly hydrophobic due to fixed nonpolar groups at the support surface. The adsorbent surface can carry alkyl chains of differing lengths, C_1-C_{18}, or can be composed of an organic polymer providing an apolar surface. RPC is carried out in an aqueous solution containing a water miscible organic solvent such as acetonitrile or methanol of varying concentration. RPC relies on hydrophobic interactions between a protein's nonpolar amino acid side chains and the packing surface. The hydrophobic patches of the protein interact with the apolar surface of the adsorbent surface causing the proteins to be retained. The separation depends mainly on two factors: (1) the strength of the hydrophobic interactions of each component of the complex solution with the hydrophobic surface of the column matrix; and (2) the elution strength of the organic solvent in the mobile phase. Proteins adsorb to the hydrophobic surface and remain there until the organic content of the eluent rises high enough to weaken the interactions between the proteins and the column matrix. As the concentration of the organic solvent increases, the adsorbed polypeptides are eluted in order of least (most polar species) to most strongly bound molecules (least polar species). Once the critical concentration of organic solvent is reached, the polypeptide quickly elutes, accounting for the sharp peaks and high resolution obtained when using RP-HPLC. RP-HPLC separations are rapid and are very reproducible.

Polypeptides adsorb to hydrophobic surfaces upon entering the column and remain adsorbed until the organic modifier reaches the desorbing concentration. Once molecules have desorbed, they interact only slightly. RP-HPLC separates polypeptides based on subtle differences in well defined hydrophobic moieties based on small differences in amino acid sequence and on conformation. The concentration of organic modifier molecules needed to desorb a polypeptide is precise and takes place within a narrow window. The sudden desorption when the critical concentration is reached produces sharp peaks.

Reversed phase is not generally used for the purification and recovery of biological activity of molecules >30 kDa. However, RP-HPLC is the method of choice for separating peptides derived from chemical cleavage or enzymatic digestion of a purified protein, purity tests and other applications where denatured proteins are acceptable. The matrix is made up of silica gel supports modified by alkylation where the chain length of the alkyl substituent varies in length between C_{18} (octadecyl) and methyl. The most popular sizes are C_4, C_8, and C_{18}. C_{18} packings bind proteins more tightly and are more likely to cause denaturation than C_8 or C_4 packings. Dilute trifluoroacetic acid (TFA 0.1% v/v) is often used in the buffer because it is a good solvent for proteins and does not absorb in the far UV (205–220 nm) where the peptide bond absorbs. The incorporation of 0.1% TFA in the buffer system acidifies the mobile phase and also acts as an ion-pair type reagent to enhance the hydrophobic nature of peptides by neutralizing charges. Another advantage to using TFA is that it is volatile and easy to remove.

Elution is usually carried out using a gradient of increasing concentrations of organic solvent, exploiting a competitive interaction between the protein, the matrix, and the solvent. The organic component must possess sufficient polarity to keep proteins in solution and sufficient apolarity to elute the protein from the matrix. The most commonly used solvents may be arranged in an eleutropic series (see below). The use of stronger solvents such as acetonitrile in the mobile phase will decrease the retention time. The optimum solvent composition will be derived by trial and error.

Eleutropic Series of Solvents for Use in Reverse Phase Chromatography

Water
Methanol
Acetonitrile
Ethanol
Tetrahydrofuran
n-Propanol

The solvents are listed by decreasing polarity. The more nonpolar the solvent, the stronger is its eluting power for proteins on RP-HPLC. HPLC grade low-UV absorbance solvents should always be used.

The most commonly used organic modifier is acetonitrile because of its low viscosity, high volatility and UV transparency at low wavelengths.

One technique for developing an RP-HPLC separation procedure is to inject the extract containing the target protein and run a complete gradient from 0% to 100% acetonitrile. The acetonitrile concentrations corresponding to elution times of the first and last peaks can then be assessed. The next step is to run a test separation using these estimated concentrations for the initial and final gradient values.

A peak may be composed of more than one component. True purification requires an additional step that will resolve the peptides that compose the peak. This could be accomplished by changing conditions, i.e., running at pH 7 using either 20mM ammonium acetate or 20mM triethylammonium acetate. An alternate approach would be to perform an isocratic run at a fixed, constant percentage of acetonitrile slightly lower than the percentage of the apparent elution of the peptide of interest. For example, if the target protein elutes at 45% acetonitrile, then an isocratic separation at 40% may separate the individual components of a peak.

Before using a column, remove the organic solvent that was used for storage with degassed HPLC-grade water by starting the flow with 100% organic solvent and making a linear gradient to 100% water over 15–20min at a flow rate of 1ml/min.

When not in use for more than 2 days, store the column in 100% methanol.

Reversed Phase HPLC for the Isolation of Peptides

RP-HPLC is often used at late stages of a purification process. It is frequently used as a polishing step, useful for endotoxin removal, DNA removal and virus removal. Endotoxins, which are highly hydrophobic lipopolysaccharides strongly adsorb to RP columns. They can be removed by using acetic acid/ethanol washes. RP-HPLC also removes salts.

The standard method for separating peptide mixtures is RP-HPLC with gradients of increasing concentrations of acetonitrile in the presence of trifluoroacetic acid (TFA) as illustrated in Figure 10.10 (Stone et al, 1989). A selected list of fifty different polypeptides and proteins that have been purified by reversed phase HPLC has been compiled by Hearn and Aguilar (1988). These authors also include the type of column used, and elution conditions.

When the concentration of the peptides is 200pmol or larger the digestion products are chromatographed on standard size 4.6mm × 15–25cm columns with flow rates of 0.4–1.0ml/min. With small amounts of product, <200pmol, try narrow bore 2.1mm × 25cm columns with a flow rate of 0.15ml/min. Flow rates can be decreased to increase resolution.

The eluate is monitored in the 210–220nm absorption range. Fractions of not more than 0.3ml are collected in microfuge tubes, which are immediately capped to prevent evaporation. Once the system has

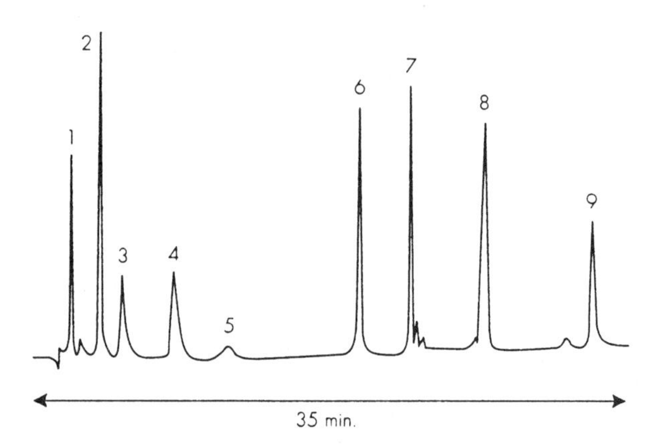

FIGURE 10.10 Elution profile of peptides and proteins from a C_{18} RP-HPLC column. (Reprinted from *The Waters Chromatography Handbook, 1993–1994,* with permission of the Waters Corporation).

been mastered, the HPLC can be programmed to collect only the peaks and in smaller volumes.

An introductory program is presented below.

Buffer A: 0.060% TFA

Buffer B: 0.056% TFA in 80% Acetonitrile

Program the HPLC apparatus as follows:

0–63 min	2–37% B
63–95 min	37–75% B
95–105 min	75–98% B
105–110 min	98% B
110–115 min	98%–0% B
115–130 min	0% B

Before beginning the product analysis wash the system extensively with the gradient program listed above to remove any UV absorbing material.

Identify peaks due to contaminants by running samples of reagents alone. This is accomplished by a dummy run, injecting the solvent used to dissolve the digest and using the identical program that will be used with the digest.

Difficult separations are addressed by modifying the slope and composition of the gradient.

Multidimensional Liquid Chromatography

Multidimensional separation techniques have the potential to provide greater sensitivity than a 1-D approach. Various 2-D liquid chromatography systems including coupled liquid chromatography and capillary electrophoresis techniques have been reviewed (Liu et al, 2002).

A high-throughput method for protein identification from complex mixtures has been developed using a biphasic column (a column containing two different packing materials) called multidimensional protein identification technology (MudPIT) (Link et al, 1999). MudPIT represents an alternative to the 2-D SDS-PAGE and MS approach to analysis of proteins. The method couples two dimensions of liquid chromatography, strong cation exchanger (SCX) and RP, to tandem mass spectrometry (MS/MS).

The biphasic MudPIT column has three major advantages. The original column used SCX resin and RP C_{18} in a fused silica capillary column. By using a column with an i.d. of 100 μm the flow rate can be very low thereby increasing the sensitivity of MudPIT. In addition, because SCX separates by charge and RP separates by hydrophobicity the combination is non-overlapping. Having both packing materials in the same column eliminates valves and dead volume, commonly associated with columns run in tandem.

Complex peptide mixtures are loaded onto a biphasic microcapillary column packed with a SCX and reverse phase materials. Peptides are first displaced from the SCX to the RP by increasing concentrations of salt in the buffer and eluted off the RP into the MS/MS. The micro-column is then reequilibrated and an additional salt step of higher concentration displaces peptides from the SCX matrix to the RP where they are eluted directly into the MS/MS (Washburn, 2001).

MudPIT technology was extended to a triphasic column consisting of Aqua C_{18}. Partisphere strong cation exchanger and a hydrophilic interaction chromatography material (Wu et al, 2003).

F. Affinity Chromatography

The development of affinity chromatography introduced an element of rational design into an otherwise empirical area of research. This method takes advantage of a property that is unique to the protein of interest by isolating it on the basis of its biospecific interaction with an immobilized ligand. The power of the approach is that **under ideal circumstances the target protein can be purified in a single step**.

Immunoaffinity chromatography uses immobilized antibodies to selectively absorb a protein with exquisite specificity at high affinity. By applying a protein mixture to a suitable antibody immobilized on a resin, washing off unbound or weakly bound material and eluting the specific protein with the appropriate elution agents, both a significant purification and simultaneous concentration can be obtained.

Beaded agarose gels are most frequently used as the immobilizing matrix because they provide high porosity, which allows free access to macromolecules with good mechanical strength and flow properties. Most importantly, they can be readily derivatized using well-defined chemical procedures.

A comprehensive list of antibodies and proteins that are commercially available, along with the vendor's name and address, is presented in the Linscott Directory of Immunological and Biological Reagents. It should be noted that the term affinity chromatography is used with a wide latitude in this section. Affinity is used in the general sense to mean attraction, so that adsorbents such as dyes, immobilized metals, and mixed-function ligands, which do not necessarily interact at biospecific binding sites will also be mentioned.

Immunoaffinity Purification

The availability of a specific antibody that recognizes the protein of interest, be it a polyclonal or monoclonal, enables the investigator to use immunoaffinity methodology. Polyclonal antibodies are relatively easy to produce by immunizing goats or rabbits, but are heterogeneous with respect to epitope specificity and binding properties. Monoclonal antibodies (mAbs) are difficult and expensive to produce but have several advantages in immunoaffinity chromatography. Once a hybridoma cell line is established, it can be used to produce a virtually unlimited supply of antibody with inherent reproducible properties. Most importantly, the antibody binds to a single binding site or epitope and has homogeneous binding and elution properties. Because of the unique properties of the antigen-antibody interaction, no other type of chromatographic technique is likely to yield a greater purification (1000- to 10,000-fold) in a single step.

Several factors contribute to the success of an affinity chromatography purification step. Three of the most critical are the starting purity of the antigen, the affinity of the antibody for the antigen and the ease with which the antibody-antigen bond can be broken.

Immunoaffinity purification is a three step process:

- The specific antibody is coupled to the matrix. Either monoclonal or affinity-purified polyclonal antibodies are covalently attached to a solid-phase matrix by methodology described below.
- The target antigen is bound to the antibody-matrix and contaminating molecules are removed by washing.
- The antibody-antigen interaction is broken, and the antigen is released and collected in the eluate.

The simplest immunoaffinity procedure, from the point of view of preparation, makes use of commercially available matrices that contain reagents that specifically bind antibodies. The most commonly used of these matrices is protein A Speharose. Protein A contains four high affinity binding sites that specifically bind to the F_c domain of immunoglobulins making the antigen binding site maximally accessible. After the antibody is bound, the interaction may be stabilized by cross-linking (Schneider et al, 1982). Antibodies with a low affinity for protein A are coupled to the protein A beads through an intermediate anti-immunoglobulin antibody which itself has a high affinity for protein A. The most commonly used intermediate antibodies are prepared in rabbits. Alternatively, antibodies can be coupled directly to activated Sepharose beads.

For some antibodies, in particular mouse IgG_{1a}, the affinity for protein A can be increased by changing the binding conditions, using high salt concentrations that favor the hydrophobic bond found in protein A-F_c binding.

An alternative to protein A is protein G (Bjorck and Kronvall, 1984), which has a somewhat different spectrum of binding affinities for antibodies from different species and classes than protein A (see Table 3.3). Protein G beads are commercially available. The techniques described below can be used for both protein A and protein G beads.

When using protein A or protein G beads, antibodies that are present in the antigen preparation may themselves be bound to protein A or protein G if these sites are still available.

PROTOCOL 10.13 Direct Antibody Coupling to Protein A Beads

Protein A binds to the F_c portion of the antibody molecule. Therefore, the antigenic binding sites are oriented outward, allowing for maximum antigen binding efficiency, making this the method of choice when the antibody binds with high affinity to protein A or protein G (Schneider et al, 1982) (see Table 3.3). Chemical cross-linking with dimethylpimelimidate (DMP) further stabilizes the complex, overcoming the problem of antibody leakage. The reaction is shown in Figure 10.11.

Materials

Protein A Sepharose (Pharmacia)
Borate buffer: 0.2M sodium borate, pH 9.0
Dimethylpimelimidate (DMP, Pierce)
Ethanolamine: 0.2M, pH 8.0
PBS: see Appendix C
Sodium azide: 20% (w/v)

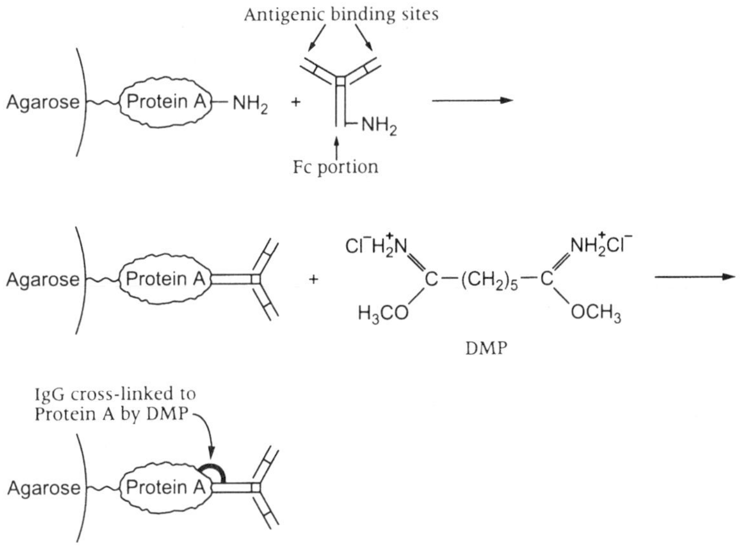

FIGURE 10.11 Cross-linking antibody to protein A agarose with DMP.

1. Mix the antibodies with the protein A beads and incubate at room temperature for 1h with gentle rocking. Antibody can be from serum, tissue culture supernatant, ascites or purified solutions. Try to bind 2mg of antibody per ml of wet beads.
2. Wash the beads twice with 10ml of borate buffer per ml of beads. Collect the beads by gentle centrifugation.
3. Following the second wash, resuspend the beads in 10 volumes of borate buffer. Add enough solid dimethylpimelimidate to make a final concentration of 20mM. For example, to 1ml of wet beads add 10ml of borate buffer and 57mg of DMP. Crosslinking with DMP must be performed above pH 8.3.
4. Mix for 30–45min at room temperature on a rocker or shaker.
5. To stop the reaction, collect the beads by gentle centrifugation then wash the beads in 0.2M ethanolamine, pH 8.0. Incubate the beads for 2h at room temperature in 0.2M ethanolamine with gentle mixing.
6. After the final wash, resuspend the beads in PBS with an anti-bacterial agent such as 0.02% sodium azide or 0.01% merthiolate. When stored at 4°C, the cross-linked beads are stable for about a year.

Comments: To check the efficiency of the cross-linking, remove the equivalent of 10µl of beads, boil them in 1% SDS, and run on SDS-PAGE and stain with Coomassie blue. Good coupling has been attained if there is no heavy-chain band (55kDa) visualized. If there is a small amount of heavy chain, wash the coupled beads with 100mM glycine, pH 3, to remove any antibodies that are bound noncovalently to the protein A molecules.

This method can be used with protein G in place of protein A.

Coupling reactions can be performed directly in a disposable plastic minicolumn (Pierce), providing a convenient chamber for performing various coupling reactions. It has a top and a bottom cap and can be easily sealed. Mixing can be done simply by rocking the column to keep the gel suspended. A blood rocker (Aliquot Mixer, HemaTek, Miles) is ideal for this purpose. In this way, the gel can be mixed, reacted, washed, and used for affinity separations without being removed from the column.

PROTOCOL 10.14 Indirect Antibody Coupling to Protein A Beads

This technique allows antibodies that do not bind avidly to protein A to be coupled to protein A through a bridging immunoglobulin.

Materials

Protein A (Sepharose or agarose)

Rabbit anti-Ig (Species specific toward the antibody to be immobilized)

Borate buffer: 0.2M sodium borate, pH 9.0

Dimethylpimelimidate (DMP, Pierce)

1. Bind the anti-immunoglobulin antibodies (usually prepared in rabbits) to the protein A beads by gentle rocking at room temperature for 1h. If

possible, use a concentration that will theoretically saturate the protein A sites ~2 mg/ml of purified antibodies per 1 ml of wet beads.

2. Collect the beads by gentle centrifugation and add the antibody to be used for immunoaffinity chromatography. The antibodies that are conjugated can be from any source. Incubate for 1 h with gentle rocking. It is desirable that the amount of immunoaffinity antibody added is saturating (~2 mg/ml of beads).
3. Collect the beads by gentle centrifugation and wash them twice with 10 volumes of borate buffer per ml of wet beads.
4. After the second wash, resuspend the beads in 10 volumes of borate buffer and add enough solid DMP to bring the final concentration to 20 mM. The remainder of the protocol is identical to Protocol 10.14 for direct coupling of antibody to protein A.

PROTOCOL 10.15 Preparation of Affinity Columns

Small scale separations (less than 5 ml of gel) can be performed easily using the plastic disposable minicolumns shown in Figure 10.12, which are commercially available from Pierce and BioRad. A convenience of minicolumn use is the stop-flow characteristic of the top porous disk. When column buffer is applied, it will flow through until it reaches the level of the top of the disk. At that point, capillary force in the porous

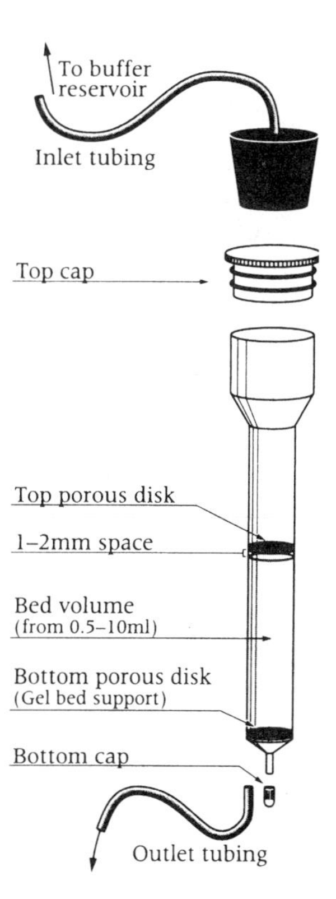

FIGURE 10.12 Components of the mini-column.

disk effectively stops flow and prevents the gel bed from drying out. This is an extremely useful feature, allowing the accurate collection of fraction volumes, since the amount of buffer applied will be the same as that collected up to the point at which the flow stops.

As a word of caution, the disks may be extremely hydrophobic and difficult to wet completely with aqueous buffers. In addition, their hydrophobicity may cause nonspecific protein binding. This potential problem is addressed in the first step of the protocol.

Materials

Tween 20: 0.1% (v/v)
Minicolumns and accessories
Binding buffer
Gel
0.02% sodium azide (if storing)

1. Before use, treat the disks with 0.1% Tween 20 blocking nonspecific sites and wetting the inner pores. This process also eliminates trapped air in the disk, which could impair the flow rate.
2. With water in the empty column, push the disk to the bottom. You are now ready to pour the column.
3. Prepare a 50% (v/v) slurry of the affinity gel in binding buffer.
4. Place the bottom cap on the end of the column.
5. Empty the column of water and add the desired amount of gel slurry. Let the gel settle with the column in the vertical position.
6. Place a porous disk on top of the liquid in the column and carefully depress it so that it is 1–2 mm above the level of the settled gel.
7. For storage, keep ~2 ml of water or buffer containing 0.02% sodium azide above the top disk to prevent the gel from drying out. Store in an upright position with both top and bottom caps in place.
8. To use the packed minicolumn, first remove the top cap then remove the bottom cap. This prevents air from being drawn into the gel from the bottom.
9. Allow the storage solution to drain through or remove it from the top with a Pasteur pipet. You are now ready to equilibrate the column.

Flow Rate

The rate (measured in ml/min) at which eluent flows through the column is called the flow rate. If the flow rate is too fast, the mobile phase will pass the beads before the target molecule can diffuse into the inner structure of the porous beads resulting in inefficient binding. If the flow rate is too slow, run times are painfully long and can lead to artifactual results (overlapping peaks). The flow rate must therefore be optimized to balance these tendencies. Flow rate optimization is largely a matter of trial and error and experience.

If the flow rate is agonizingly slow, the affinity support may have become clogged. Unpack the column and thoroughly wash the gel in batch using a Büchner or fritted glass filter funnel. Repour the column. Affinity matrices can often be restored to their original chromatographic potential by periodic cleanups.

Most affinity chromatography steps can be performed using gravity flow (i.e., without auxiliary pumping systems). The actual linear flow (in cm/min through the gel bed) will depend on the physical dimensions of the column.

Binding Antigens to Immunoaffinity Matrices

An important feature of immunoaffinity purification is that the antigen of interest be removed quantitatively from the crude solution. This can be easily evaluated by examining the first wash for the presence of the desired antigen. If after a 1–2h incubation there is still antigen in the wash, indicating less than quantitative binding of the target molecule to the affinity matrix, then conditions should be modified. Occasionally, an affinity interaction will require more time for optimal binding than the normal column flow rate will allow. More efficient binding of the target molecule can often be accomplished by incubating the sample on the column for periods as long as overnight by stopping the flow after the sample has been applied. Alternatively, if the volume of sample is greater than the column bed volume, the sample can be recycled through the column repeatedly. The simplest modifications to try are to increase the quantity of beads and increase the binding time. The amount of antibody on the bead can also be increased, although this would entail an additional coupling. High-affinity antibodies are more efficient at removing the antigen of interest from solution than low-affinity antibodies. High-affinity antibodies can be used at lower concentrations with shorter incubation times, conditions that lead to lower backgrounds when analyzing the results.

Nonspecific Interactions

Nonspecific binding is the adsorption of components in the crude or semi-purified extract that do not have targeted sites of interaction with the immobilized ligand. Nonspecific interactions may be ionic, hydrophobic, or be due to cross-reactivity inherent in the immobilized ligand.

Nonspecific interactions originating in the matrix are often due to the hydrophobic or ionic character of the support material. If the ligand does not sufficiently saturate the active sites on the matrix, or the activation and coupling chemistry does not modify this property, significant adsorption of unwanted molecules may occur. Optimization of the binding buffer composition usually will eliminate nonspecific binding.

Crossreactivity due to the coupled ligand is the result of substances that have broad specificities. For example, immobilized dyes will have some binding affinity for more than one component of a complex proteinaceous sample. Similarly, immobilized metal chelators may have affinity for more than one metal binding protein in a mixture. This type of nonspecificity may be overcome by a more careful selection of potential affinity ligands or by developing elution conditions that will separate the target molecule from unwanted contaminants.

Another technique for removing contaminants, especially if the starting material is sera, cell extracts, or conditioned media is to pre-clear the extract by passing it through a control column consisting of beads coupled to a ligand that is structurally similar to the affinity ligand but with different binding properties. This procedure will remove molecules from the crude mixture that would adhere nonspecifically to either the insoluble support (the beads) or the ligand.

Affinity matrices may sometimes require a blocking step much like that performed on Western blots (see Chapter 7). In this case, nonspecificity is again caused by surface characteristics of the matrix, usually consisting of surface areas of hydrophobicity. To prevent this type of interaction, protein buffer additives that play no active role in the affinity recognition are used to noncovalently block and eliminate nonspecific sites of adsorption.

PROTOCOL 10.16 Blocking the Affinity Matrix

This protocol should only be used if preliminary results indicate nonspecific binding.

1. To effectively coat and block all sites, incubate the affinity matrix in the column for 1 h at 37°C, several hours at room temperature, or overnight at 4°C with any of the following blocking additives added to the binding buffer (Hermanson et al, 1992):
   - 5% nonfat dry milk
   - 1–3% BSA
   - 1% casein
   - 0.25% gelatin
   - 1–3% horse serum
   - 0.1% Tween 20
2. Wash the column extensively with binding buffer. You are now ready to load the column.

Elution of Antigens from Immunoaffinity Matrices

Elution of the target protein from the affinity matrix is potentially the most frustrating step in affinity chromatography. Because of the strength of the antigen-antibody interaction, it is often difficult to elute the antigen from an immunoaffinity column. At this stage you know that the target protein is bound to the affinity matrix; it is just a matter of recovering it in highly purified, active form. Elution may require harsh elution conditions, such as extreme pH, denaturing agents, or chaotropic salts that disrupt protein structure. The optimal elution buffer will liberate the bound target molecule in a minimum volume and maintain activity or the structural integrity of the purified molecule.

Antigens and antibodies are bound to each other by a combination of ionic bonding, hydrogen bonding, van der Waals forces, and hydrophobic interactions (Frost et al, 1981). Eluting the antigen from the antibody is accomplished by breaking the bonds between the antibody and the antigen. Elution conditions are designed to create a shift in the composition of the mobile phase so that the optimal binding environment created by the binding buffer is destroyed. The most direct approach involves displacement of the bound material with a counterligand. The conditions for elution are determined empirically and often will depend on the desired state of the antigen. There is a logical strategy, or sequence of eluents to consider when approaching a new immunoaffinity application.

Specific elution with excess antigen should be considered first. However, this strategy is often impractical due to the cost and lack of availability of the reagent. Sometimes the affinity interaction is pH dependent so the binding constant is very low at pH values that are not optimal. In these cases, the target protein can be eluted with a buffer that has a pH that differs from the binding buffer. The most commonly used desorption method is acid elution with 0.1 M glycine-HCl, pH 2.5, or sodium citrate, pH 2.5. If these fail, try introducing a surface tension-reducing agent (to lessen hydrophobic and van der Waals interactions). Add up to 1% (v/v) of a nonionic detergent like Triton X-100, 1 M propionic acid or 10–50% ethylene glycol to the acid eluent to disrupt hydrophobic interactions between antigen and antibody.

Base elution is used less frequently than acid elution, but in some cases it has proven to be more effective. Elution with 50 mM diethylamine, pH 11.5 or 1 M NH_4OH, has been shown to be effective with membrane glycoproteins and with certain antigens that precipitate in acid but are stable in base (Izuta and Saneyoshi, 1988).

Chaotropic agents disrupt the tertiary structure of proteins and have been used to dissociate antigen-antibody complexes by disrupting ionic interactions, hydrogen bonding, and sometimes hydrophobic interactions. Chaotropic anions are effective in the order: $SCN^- > I^- > ClO_4^- > Br^- > Cl^-$. Chaotropic cations are effective in the order: guanidine > $Me^+ > K^+ > Na^+$. Reagents such as 8 M urea, 6 M guanidine hydrochloide and 6 M NaSCN disrupt most protein-protein interactions. Unfortunately they may also destroy the activity of the target protein. Selection of the mildest chaotropic or denaturing agent at the lowest possible concentration that will just effect elution is the best strategy for insuring rapid elution with the best chance of retaining activity.

Many proteins depend on complexed metals to form the correct binding site orientation for interaction with a ligand or substrate. The presence of a metal such as calcium, magnesium or zinc in the binding buffer is usually required for effective binding. Therefore, to elute the target protein, include a chelating agent in the elution buffer. EDTA or the calcium specific chelator EGTA at concentrations in the range of 1–10 mM is usually sufficient to effect elution.

Studies were undertaken to find a method for identifying mAbs that bind tightly but will release under gentle elution conditions. A type of mAb, called a polyol-responsive mAb (PR-mAb) has ideal properties for use in immunoaffinity chromatography (Thompson and Burgess, 2001). PR-mAbs bind very tightly to the specific antigen under standard conditions, but release antigen when eluted under mild, nondenaturing conditions in aqueous solution at neutral pH supplemented with a polyol, such as ethylene glycol or propylene glycol, and a non-chaotropic salt such as NaCl or ammonium sulfate (Burgess and Thompson, 2002).

PROTOCOL 10.17 Eluting the Antigen

Finding the right elution conditions for your target protein is an empirical process of rationally designed trial and error. Begin with acid conditions, then basic elution buffers. If these do not elute the antigen, try other additives. As a guideline try: acid, base, 3–5 M $MgCl_2$, 5–10 M LiCl, water, 25–50% ethylene glycol, 5–20% dioxane, 1–5 M thiocyanate, 2–5 M guanidine, 2–8 M urea, 0.5–2% SDS. Use combinations of eluting agents. Increasing salt concentrations increase hydrophobic forces. One disadvantage is that the increased salt concentration will also increase nonspecific interactions.

1. Wash the column extensively with ten column volumes of a buffer which will allow a quick change to the desired elution conditions.
2. To test potential elution buffers, pass 3 bed volumes of buffer through the column and collect each individually. If high or low pH steps are used, the collection tubes should have the appropriate neutralizing buffer.
3. Evaluate the fractions with an assay specific for the target protein.
4. If the first elution buffer fails, try something else. Rinse the column between elution attempts. Do not discard these rinses. Test them also for target protein.

Comments: Buffers that do not elute the target protein can be used for washing the column and removing potentially sticky contaminants.

Some noneluting buffers, especially high salt, 0.5–1 M NaCl, may stabilize the antigen-antibody complex by increasing hydrophobic interactions. These conditions could be employed in the binding step when low affinity antibodies are used.

Elutions tend to be more effective when performed on beads packed in a column than in batch. In a column, once the antigen-antibody interaction is broken, the antigen will be washed away and will have less opportunity to rebind to the matrix.

Avoid using dithiothreitol, mercaptoethanol and other reducing agents as they will break the disulfide linkages between the heavy and light chains of the antibody molecule.

The adsorbents can be reused many times provided that they are cleaned immediately after use with solutions of urea or SDS, and stored in the presence of a bacteriocidal agent such as azide or merthiolate.

If everything you have tried has failed, remove ~50 µl of resin from the affinity column and boil it in SDS sample buffer. Run the sample on an SDS gel. Assay for the presence of the target protein. If you have data on the migration of the target protein or have an antibody that will recognize it on a Western blot, you will at least be assured that the target protein is present.

Ligand Affinity Chromatography

Choice of Ligand and Immobilization Chemistry

Once the decision has been made to use affinity chromatography as a purification technique, you must choose which ligand to couple. There is an ever-expanding list of immobilized ligands that are commercially available (see the Sigma catalog). Obviously, the simplest way to obtain the ligand of interest immobilized on either Sepharose or agarose is to purchase it ready made. Alternatively, you must choose to either purchase the coupling matrix preactivated or to perform the activation in the lab. Activated adsorbents enable successful, convenient immobilization of ligands without the need for complex chemical syntheses or special equipment. A wide range of coupling chemistries has been developed for immobilization of the ligand through a specific functional group. Activated matrices are commercially available, and one should consider one of these before activating a matrix in-house. The original method for activating Sepharose using CNBr involves the reaction of hydroxyl groups on the agarose matrix with CNBr (Cuatrecasas et al, 1968). Although CNBr activation is still used, it has been gradually replaced by other chemistries such as N-hydroxysuccinimide ester derivatives of agarose.

Immobilization chemistries for coupling ligands to solid supports are well defined for compounds having functional groups that are suitable to facilitate the attachment. The types of functional groups used for attachment include primary amines, sulfhydryls, aldehydes, carboxylic acids, and hydroxyls. Usually, the solid phase matrix is first activated with a compound that is reactive toward one or more functional group. The ligand is then added and a covalent linkage between the ligand and the support matrix results in ligand immobilization. Some examples of useful activation chemistries include CNBr, carbonyldiimidazole (CDI), carbodiimide, epoxy, divinyl sulfone, toluene sulfonyl chloride, and N-hydroxysuccinimide ester (NHS).

PROTOCOL 10.18 Immobilization of Proteins on *N*-Hydroxysuccinimide Ester Derivatives of Agarose

N-Hydroxysuccinimide (NHS) esters of agarose react with amine-containing ligands, forming stable amide bonds between ligand and matrix. Affi-Gel 10 and Affi-Gel 15 (Bio-Rad), affinity supports with *N*-Hydroxysuccinimide esters of a derivatized crosslinked agarose, are depicted schematically in Figures 10.13A and 10.13B. Both matrices couple to amine-containing ligands spontaneously in aqueous or non-aqueous solutions as shown in Figure 10.13C. Affi-Gel 10 contains a neutral 10-atom spacer arm while Affi-Gel 15 contains a cationic charge in its 15-atom spacer arm, which enhances coupling efficiency for acidic proteins at physiological pH. These supports offer the following advantages:

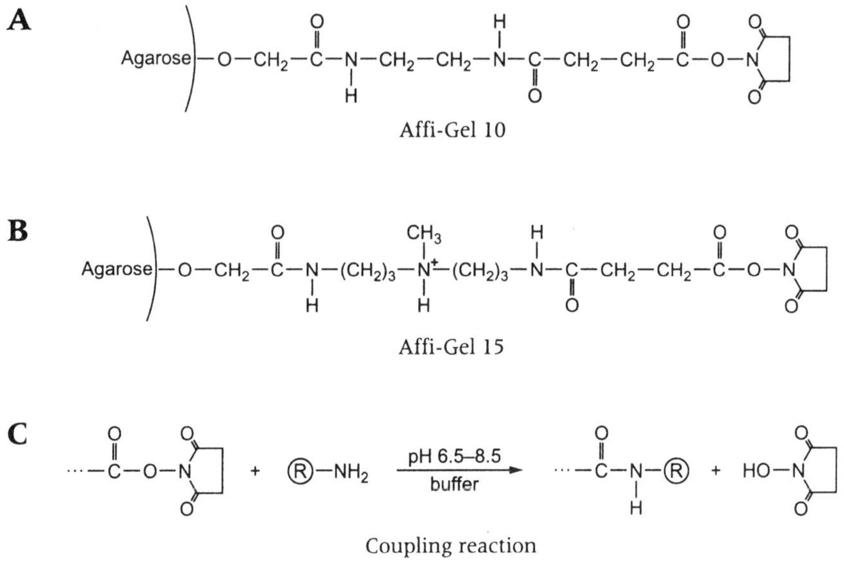

A

Affi-Gel 10

B

Affi-Gel 15

C

pH 6.5–8.5 buffer

Coupling reaction

FIGURE 10.13 (A) AM-Gel 10; (B) AM-Gel 15; (C) Coupling reaction.

- High stablility in chaotropic agents and from pH 2–11
- Rapid, gentle coupling within 4h
- High capacity (up to 35mg protein per ml)
- Stable ligand attachment

Materials

Affi-Gel 10 or Affi-Gel 15 (Bio-Rad)
Buchner funnel
Ethanolamine: 1M, pH 8.0

1. Dissolve the protein in a suitable buffer i.e., MES, MOPS, HEPES, acetate, or bicarbonate. **Do not use Tris or glycine!** They contain primary amino groups that will couple to the gel. The protein concentration should be in the range of up to 25mg protein per ml of gel (if possible) and stored until use at 4°C.
2. Transfer the desired quantity of gel slurry to a Büchner funnel or glass fritted funnel. Wash the gel with at least three bed volumes of cold deionized H_2O by applying a gentle suction. Do not let the gel dry. Preparation time should be as short as possible, not to exceed 15min.
3. Transfer the moist gel cake to a test tube and add the protein solution at 4°C. Add at least 0.5ml of ligand solution per ml of gel and mix to make a uniform suspension. Mix the gel slurry gently using a rocker, shaker or wheel for 1h at room temperature or for 4h at 4°C.
4. Block any remaining active ester groups by adding 0.1ml of 1M ethanolamine-HCl, pH 8.0, per ml of gel.
5. Check the coupling efficiency by comparing the OD_{280} of the original ligand solution with the OD_{280} of the solution at the end of the coupling reaction.

Comments: When coupling at neutral pH (6.5–7.5), Affi-Gel 10 is recommended for proteins with isoelectric points of 6.5 to 11 (neutral or basic proteins). Affi-Gel 15 is recommended for proteins with isoelectric points below 6.5 (acidic proteins).

Unbound sample should be recovered and can be reused without further treatment.

The stability of attachment of amine-containing ligands using active ester gels provides optimum recycling of the gel and freedom from cross-contamination of affinity-purified material due to leakage of ligand.

Toluene Sulfonyl Chloride (Tosyl Chloride)

Tosyl activated agarose is used to couple primary amines. Hydroxyl groups on the surface of the cross-linked agarose bead are reacted with tosyl chloride to produce a sulfonated support as shown in Figure 10.14. The sulfonates couple to primary amines or thiols to yield a stable affinity support (Nilsson and Mosbach, 1980). Therefore, avoid reducing agents and primary amines in the buffer because these reagents will compete in the coupling reaction. The tosyl group introduced on activation reacts rapidly to give a very stable secondary amine from the primary amine ligand. Tosyl activated agarose couples to primary amines at a pH optimal range of 9–10.5. Recommended buffers are HEPES, phosphate, bicarbonate and borate. Reducing agents and primary amines may be used to inactivate any unreacted sulfonyl groups after the ligand has been immobilized. At neutral pH the secondary amine group is charged so that it behaves similarly to CNBr activated product; however, the linkage is much more stable (Nilsson and Mosbach, 1984).

Pseudo-Affinity Adsorbents

Pseudo-affinity ligands are synthetic small molecules that have affinities for certain proteins. Table 10.9 lists a wide range of ever-expanding, group specific, ready to use matrices that have an affinity

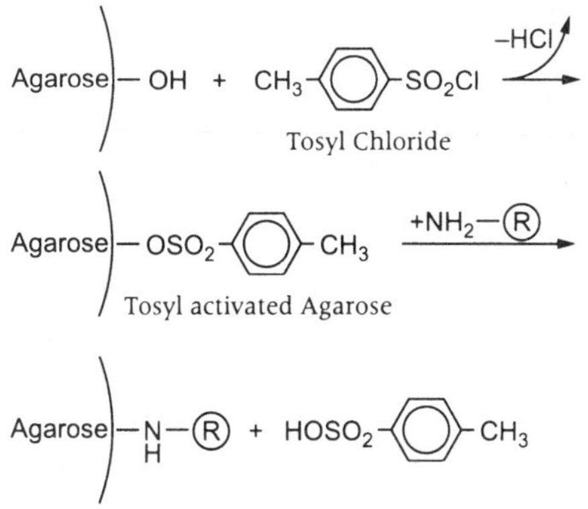

Agarose—OH + CH_3–C$_6$H$_4$–SO_2Cl $\xrightarrow{-HCl}$
Tosyl Chloride

Agarose—OSO_2–C$_6$H$_4$–CH_3 $\xrightarrow{+NH_2-Ⓡ}$
Tosyl activated Agarose

Agarose—NH—Ⓡ + $HOSO_2$–C$_6$H$_4$–CH_3

FIGURE 10.14 Activation of Tosyl-activated agarose and protein coupling.

TABLE 10.9 Group Specific Matricies

Product	Applications
Affi-prep polymyxin (Bio-Rad)	endotoxin removal
Affi-gel 501 (Bio-Rad)	adsorption of sulfhydryls
Affi-gel Heparin (Bio-Rad)	Heparin binding proteins, coagulation proteins, lipoproteins, growth factors
Affi-gel blue (Bio-Rad)	Enzymes requiring adenyl-containing cofactors (NAD^+ and $NADP^+$), albumin, coagulation factors, interferon
Arginine Sepharose (Pharmacia)	Serine proteases
Benzamidine Sepharose (Pharmacia)	Serine proteases
Calmodulin Sepharose 4B (Pharmacia)	ATPases, protein kinases, phosphodiesterases
Gelatin Sepharose 4B (Pharmacia)	fibronectin
Glutathione Sepharose 4B (Pharmacia)	S-transferases, glutathionc-dependent proteins
IgG Sepharose 6FF (Pharmacia)	Protein A fusion conjugates
Red Sepharose CL-6B (Pharmacia)	NADP + dependent enzymes, carboxypeptdase G
Avidin Agarose (Pierce)	Retrieving biotin labeled proteins
Biotin Agarose (Pierce)	Retrieve avidin-labeled proteins, purify biotin binding proteins
N-Acetyl-D-Glucosamine Agarose (Pierce)	Affinity adsorbents for lectins (WGA) and glycosidases
D-Galactose Agarose (Pierce)	Purifying carbohydrate binding proteins with D-galactose specificity
p-Chloromercuribenzoate (Pierce)	Purifying sulfhydryl-containing proteins

for an expanded range of related substances rather than for a single molecular species.

Dye adsorbents such as Cibacron Blue F3GA (Figure 10.15A) qualify as pseudo affinity adsorbents because the dye bears no obvious resemblance to the true biological ligand, leading Scopes (1987) to refer to them as "pseudo-affinity ligands". Cibacron blue was discovered serendipitously to bind to protein when it was used coupled to high molecular weight dextran to determine the void volume of size exclusion chromatography. Cibacron Blue, Procion Red H-E3B, and other reactive triazinyl dyes, have been used successfully in purifying dehydrogenases both NAD- and NADP specific, as well as other proteins (Ashton and Polya, 1978; Clonis and Lowe, 1980). Dye affinity separations have some of the following advantages:

- Protein capacity can be high
- Ligand leaching is minimal
- High resolution can be achieved
- Leached ligand is easy to remove from the final product

Immobilized heparin (Figure 10.15B) is useful for the isolation of blood proteins such as platelet-secreted anti-heparin proteins (Paul et al, 1980), human lipoprotein lipase (Becht et al, 1980) and heparin-binding growth factors (Lobb et al, 1986).

Immobilized *p*-chloromercuribenzoate (Figure 10.15C) is effective for purifying sulfhydryl-containing proteins. It has been used in a protocol to purify L-pyrolidonecarboxylate peptidase from *Bacillus amyloliquifaciens* (Tsuru et al, 1978).

Lectin Affinity Chromatography

Lectins are proteins that bind selective carbohydrate moieties. It is this specificity that has made them extremely effective for the purification of glycoproteins. For lectins to be useful in a purification protocol they must be coupled to a solid phase that will allow specific binding of the glycoprotein to the carbohydrate binding site of the immobilized lectin. A large number of lectins are now commercially available immobilized on agarose. (See the catalogs of EY, Pierce, Pharmacia, and Vector Labs, to name a few.)

Binding of glycoproteins to lectin columns is pH dependent; therefore samples should be buffered with either phosphate or Tris at a pH in the physiological range (6.8–7.5).

Lectin chromatography can be run in the presence of relatively high salt concentrations because binding between the sugar residue and the

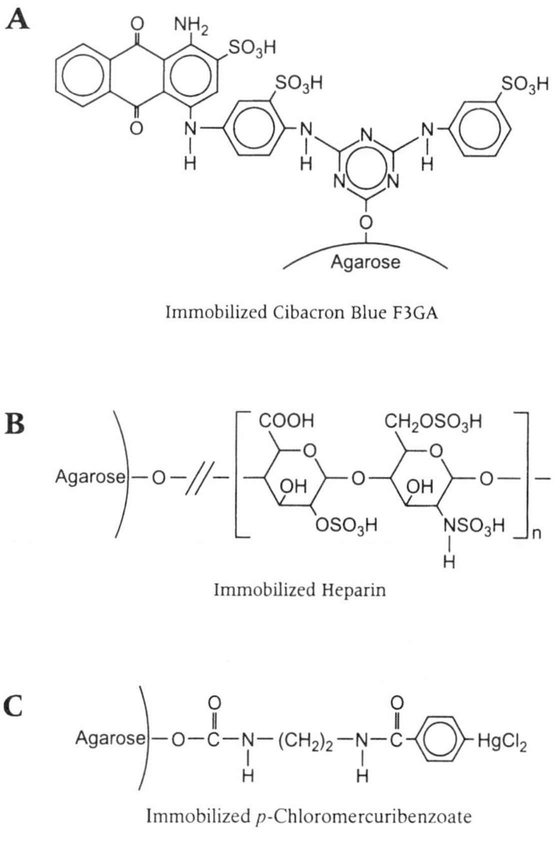

A

Immobilized Cibacron Blue F3GA

B

Immobilized Heparin

C

Immobilized *p*-Chloromercuribenzoate

FIGURE 10.15 Pseudo-affinity ligands. (A) Cibacron Blue F3GA; (B) Heparin; (C) *p*-Chloromercuribenzoate.

lectin is not due to ionic interaction. The presence of high salt concentrations in the buffer also prevents nonspecific binding of proteins to the matrix. Lectin affinity chromatography can be performed in the presence of non-ionic detergents.

Some lectins require divalent cations for glycoprotein binding (consult Table 7.1) which should be added to the sample prior to chromatography and also are present in the buffer at a final concentration of 1 mM.

Many glycoproteins of biological interest are membrane bound and must be solubilized. There are a number of methods available for disrupting protein-lipid and protein-protein interactions in the membrane. By far the most amenable to preserving biological activity and permitting lectin chromatography is by using nonionic detergents. (See Chapter 6.)

Once the buffer composition is determined, the crude protein mix is applied to the lectin column and unbound material is removed by washing with a vast excess of column buffer. The specifically bound proteins are eluted by adding the competing sugar to the buffer at concentrations of 0.1–0.5 M.

PROTOCOL 10.19 Glycoprotein Purification Using Wheat Germ Agglutinin (WGA)

Elution of glycoproteins from a lectin column is illustrated with WGA. The same protocol can be used with other lectins with minor modifications, specifically adding the appropriate divalent cation to the buffer, and using the appropriate sugar.

Materials

WGA agarose

Lysis buffer: 50 mM Tris-HCl, pH 7.4, 100 mM NaCl, 1% Triton X-100 and proteolytic enzyme inhibitors

Wash buffer: 50 mM Tris-HCl, pH 7.4, 100 mM NaCl, 0.1% Triton X-100 and proteolytic enzyme inhibitors

Elution buffer: 50 mM Tris-HCl, pH 7.4, 100 mM NaCl, 0.1% Triton X-100, 0.3 M N-acetyl-D-glucosamine and proteolytic enzyme inhibitors

1. Lyse the cells with lysis buffer at 4°C and centrifuge the lysate for 15 min at 10,000 × *g* to remove insoluble material. Discard the pellet.
2. Incubate the supernatant at 4°C with WGA agarose. This step can be performed in bulk in a tube using a rocker or directly in a mini-column. The incubation can vary from 30 min to overnight.
3. Collect and save the unbound supernatant and wash the WGA beads three times with ice-cold wash buffer that contains 0.1% Triton X-100 (in place of the 1% Triton X-100 of the lysis buffer).
4. To specifically elute the glycoproteins that have affinity for the WGA, incubate the WGA beads with elution buffer in bulk. Alternatively, the WGA beads can be transferred to a column and the elution buffer applied and fractions collected and analyzed.

Comments: Prior to lysis, the cells may be metabolically labeled. The concentration of the eluting sugar can vary between 0.1–0.5M.

PROTOCOL 10.20 Immobilized Metal Chelate Chromatography (IMAC)

Immobilized metal affinity chromatography is a rapid, highly effective method for enriching or purifying a desired protein (Porath and Olin, 1983). This technique separates proteins according to their surface density of thiol, histidine, tryptophan and phenylalanine residues that act as electron donors when they are on the exposed surface of proteins. They can interact with electron acceptors, such as the transition metal ions Cu^{+2} or Ni^{+2}, and Zn^{+2} that are immobilized on a carrier matrix by using a ligand with strong chelating properties such as iminodiacetic acid (IDA) or nitrilotriacetic acid (NTA). Other metals such as cobalt, manganese, and magnesium can also form complexes which are much weaker than those formed with zinc, nickel and copper. The most appropriate metal ion for a particular application must generally be determined empirically. The transition metals copper and zinc form complexes with histidine, imidazole and the thiol group of cysteine. In order to bind to an immobilized metal, these amino acid residues must be situated on the surface of the protein. This interaction is pH dependent with neutral pH providing the strongest and most selective level of adsorption. The amino acids responsible for metal binding at neutral pH are ranked in the following order of bond strength:

Cys > His > > Asp; Glu > Trp, Tyr, Lys, Met, Asn, Gln, Arg

Separation of proteins in IMAC is caused by the interaction of the electron donor groups on the surface of the protein with the electron acceptor groups immobilized on the stationary phase. The protein-metal ion coordination complex is allowed to form at an intermediate pH that will assure that the interacting amino acids on the protein are in the deprotonated state. Desorption is performed by competition using a strong electron donor like imidazole or decreasing the pH to pH 4.0 to protonate the interacting amino acids.

Materials

Column matrix: Iminodiacetate substituted agarose

Divalent cation solutions: 1mg/ml

Equilibration buffer: 0.02M phosphate, pH 7.5, 0.5M NaCl

Elution buffer: 0.1M sodium acetate, pH 4–6, 0.5M NaCl

Chelating buffer: 50mM EDTA in 50mM sodium phosphate buffer pH 7, 0.5M NaCl

1. To determine the best metal and the optimal concentration, construct a series of small test columns each pre-equilibrated with a different metal ion at concentrations from 1–5mg/ml. Charge the column matrix with metal ions in aqueous solution ($ZnCl_2$ or $CuSO_4{\cdot}5H_2O$ at a concentration of 1mg/ml).

2. Equilibrate the column and protein sample with equilibration buffer.
3. Apply the protein sample to the column.
4. Wash the column extensively with equilibration buffer.
5. Elute the bound protein using a reduced pH elution buffer. The lower the pH, the more effective the elution. All elution buffers should contain salt. A displacement elution can also be used with imidazole and histidine in the elution buffer.
6. Elute more strongly bound proteins using chelating buffer.
7. To regenerate the column, wash with 0.05M EDTA which regenerates the column by stripping off bound metal ions.

Comments: An elution buffer alternative is: 0.02M phosphate, pH 7.5, 0.5M NaCl, 50mM imidazole.

The high association constants when purifying a poly-His tagged protein allows the use of detergents, high salt, (e.g. 2M KCl), 6M guanidine, or 8M urea to desorb contaminants.

PROTOCOL 10.21 Purifying a Histidine Tagged Recombinant Fusion Protein

A popular use for the zinc chelate columns is for purifying recombinant proteins that have been genetically engineered to contain a histidine tag (Chapter 11). The following protocol is an example of the use of a Chelating Sepharose Fast Flow® (Pharmacia) column to purify a target protein from Baculovirus Sf9 conditioned media (Drake and Barnett, 1992).

Materials

Thimerisol

NaCl

Sodium phosphate: 0.5M, pH 7.5

Chelating Sepharose Fast Flow: Pharmacia

Equilibration buffer: 10mM HEPES, pH 7.3, 0.5M NaCl, 0.1% Tween 20 and 0.01% thimerosal

Column prep buffer: 50mM sodium phosphate, pH 7.5, 0.5M NaCl, 0.1% Tween 20, 0.01% thimerosal

1. Harvest the Sf9 cell products 5 days after infection (cells should be almost totally lysed and their contents shed into the media) by centrifuging at 5,000 × *g* for 15min.
2. Add thimerosal to the supernatant to a final concentration of 0.01%.
3. Adjust the cell free lysate to contain 0.5M NaCl, 50mM sodium phosphate, pH 7.5, 0.1% Tween 20. You will have 5ml of lysate containing ~35mg of total protein.
4. Perform the following procedures at room temperature. Equilibrate a 1.5ml column of Chelating Sepharose Fast Flow® with equilibration buffer.
5. Wash the column with 10ml equilibration buffer containing 10mM $ZnCl_2$.

6. Wash the column with 30 ml of equilibration buffer and then 30 ml of column buffer.
7. Load the cell free lysate onto the column and wash with 30 ml of column buffer.
8. Elute the target protein with 5 ml increments of 20, 30, 40, 50, and 100 mM imidazole in column buffer.
9. Analyze the fractions for the presence of the recombinant histidine tagged target protein by SDS-PAGE followed by Western blotting.

PROTOCOL 10.22 Hydroxylapatite Chromatography

Hydroxylapatite (HA) $(Ca_5(PO_4)_3OH)_2$, is a crystalized form of calcium phosphate that has been found useful for the fractionation and purification of proteins, nucleic acids and viruses. HA separates macromolecules by differential surface binding which is ionic in nature, but also has specific affinity properties. Its selectivity is not primarily dependent on molecular weight, charge density or isoelectric point, which makes HA chromatography a valuable complement to other separation techniques.

HA has negligible adsorptive capacity for low molecular weight substances such as nucleotides, salts and amino acids. Interaction between negatively charged groups on the protein surface and Ca^{+2} groups on HA play an important role in protein adsorption.

Proteins are usually adsorbed to HA in a low ionic strength low concentration phosphate buffer at neutral pH and are eluted by increasing the phosphate concentration. The interaction of protein with HA is not affected by high content of NaCl or $(NH_4)_2SO_4$ which makes HA chromatography suitable for protein purification in the presence of these ions. Phosphate buffers are used for elution with a step or linear gradient increase in their concentration while the pH is held constant. The higher the phosphate concentration, the less strongly the proteins are adsorbed.

Conditions for binding and eluting specific proteins are reached empirically. Initial experiments can be conveniently performed by adsorbing the protein in a buffer that is 10 mM phosphate and eluting with incremental twofold concentration increases of phosphate. Use two column volumes for each step to obtain complete elution of each peak. A 0.5 M phosphate concentration is usually sufficient to remove all adsorbed protein.

One technical difficulty encountered when using HA is low flow rates that is often due to fines. The following protocol removes the fine particles.

Materials

Hydroxylapatite

Packing buffer: 3 mM sodium phosphate, pH 6.8

1. Make up 10 liters of packing buffer. Warm the buffer to 55°C and degas.
2. Suspend about 50 g of HA powder in ~700 ml of warm packing buffer and stir gently with a glass rod. The slurry should be handled with great care

since rough treatment creates fines by fracturing the crystals. For this reason, avoid magnetic stirring. To reduce contact with CO_2, try to keep the slurry between 45–50°C.

3. Allow the gel to settle for ~5 min. Remove fines by aspirating off the cloudy liquid above the settled gel.
4. Repeat step 2 about eight times until a relatively sharp zone of settled particles is obtained.

Comments: Carbon dioxide binds to HA which might form a crust at the top of the column bed. This may lead to pressure incline and bed compression. Therefore, after several runs, remove the top centimeter of gel and replace it. To reduce crust formation use degassed buffers.

References

Ackers GK (1967): A new calibration procedure for gel filtration columns. *J Biol Chem* 242:3237–3238

Afeyan NB, Fulton SP, Gordon NF, Mazaroff I, Varady L, Regnier FE (1990): *Bio/technology 8:203–206*

Andrews P (1965): The gel-filtration behavior of proteins related to their molecular weights over a wide range. *Biochem J* 96:595–606

Arakawa T, Timasheff SN (1984): Mechanism of protein salting in and salting out by divalent cation salts: balance between hydration and salt binding. *Biochemistry* 23:5912–5923

Ashton A, Polya GM (1978): The specific interaction of Cibacron and related dyes with cyclic nucleotide phosphodiesterase and lactate dehydrogenase. *Biochem J* 175:501–506

Becht I, Schrecker O, Klose G, Greten H (1980): Purification of human plasma lipoprotein lipase. *Biochem Biophys Acta* 620:583–591

Belenkii BG, Malt'sev VG (1995): High-performance membrane chromatopraphy. *BioTechniques* 18:228–231

Bjorck L, Kronvall G (1984): Purification and some properties of streptococcal protein G, a novel IgG-binding reagent. *J Immunol* 133:969–974

Bradford MM (1976): A rapid and sensitive method for quantitation of microgram quantities of protein utilizing the principle of protein-dye binding. *Anal Biochem* 72:248–254

Burgess RR, Thompson NE (2002): Advances in gentle immunoaffinity chromatography. *Curr Opin Biotechnol* 13:304–308

Clonis YD, Lowe CR (1980): Triazine dyes, a new class of affinity labels for nucleotide-dependent enzymes. *Biochem J* 191:247–251

Cooper TC (1977): *The Tools of Biochemistry.* New York: John Wiley

Cuatrecasas P, Wilchek M, Anfinsen CD (1968): Selective enzyme purification by affinity chromatography. *Proc Natl Acad Sci USA* 61:636–643

Dolan JW (1994): LC troubleshooting-obtaining separations, part III: adjusting column conditions. *LC·GC* 12:520–524

Drake L, Barnett T (1992): A useful modification of cDNA that enhances purification of recombinant protein. *BioTechniques* 12:645–650

Frost RG, Monthony JF, Engelhorn SC, Siebert CJ (1981): Covalent immobilization of proteins to N-hydroxysuccinimide ester derivatives of agarose. *Biochim Biophys Acta* 670:163–169

Hancock WS, ed. (1984): *Handbook for the Separation of Amino Acids, Peptides, and Proteins*, Vol. 1,2. Boca Raton, FL: CRC Press

Harlow E, Lane D (1988): *Antibodies: A Laboratory Manual.* Cold Spring Harbor, NY: Cold Spring Harbor Laboratory

Hermanson GT, Mallia AK, Smith PK (1992): *Immobilized Affinity Ligand Techniques.* San Diego: Academic Press

Herold M, Rozing GP, Curtis JL (1991): Recovery of biologically active enzymes after HPLC separation. *BioTechniques* 10:656–662

Izuta S, Saneyoshi M (1988): AraUTP-Affi-Gel 10: a novel affinity adsorbent for the specific purification of DNA polymerase a-primase. *Anal Biochem* 174:318–324

Link AJ, Eng J, Schielz DM, Carmack E, Mize GJ, Morris DR, Garvik BM, Yates JR III (1999): Direct analysis of protein complexes using mass spectrometry. *Nat Biotechnol* 17:676–682

Linscott WD (1994): *Linscott's Directory of Immunological and Biological Reagents.* Santa Rosa, CA: Linscott Co.

Liu H, Lin D, Yates JR III (2002): Multidimensional separations for protein/peptide analysis in the post-genomic era. *BioTechniques* 32:898–911

Lobb RR, Harper JW Fett JW (1986): Purification of heparin-binding growth factors. *Anal Biochem* 154:1–14

Majors RE (1994): Twenty-five years of HPLC column development—a commercial perspective. *LC·GC* 12:508–518

Nilsson K, Mosbach K (1980): p-Toluenesulfonyl chloride as an activating agent of agarose for the preparation of immobilized affinity ligands and proteins. *Eur J Biochem* 112:397–02

Nilsson K, Mosbach K (1984): Immobilization of ligands with organic sulfonyl chlorides. *Methods Enzymol* 104:56–69

Paul D, Niewiarowski S, Varma KG, Rucinski B, Rucker S, Lange E (1980): Human platelet basic protein associated with antiheparin and mitogenic activities: purification and partial characterization. *Proc Natl Acad Sci USA* 77:5914–5918

Pharmacia Fine Chemicals AB Publications (1981): *Chromatofocusing with Polybuffer™ and PBE™.* Uppsala: Pharmacia, Inc.

Porath J, Olin B (1983): Immobilized metal ion affinity adsorption and immobilized metal ion affinity chromatography of biomaterials. Serum protein affinities for gel-immobilized iron and nickel ions. *Biochemistry* 22: 1621–1630

Schneider C, Newman RA, Sutherland DR, Asser U, Greaves MF (1982): A one-step purification of membrane proteins using a high efficiency immunomatrix. *J Biol Chem* 257:10766–10769

Scopes RK (1987): *Protein Purification Principles and Practice, Second Edition.* New York: Springer-Verlag

Sluyterman LAA, Elgersma O (1978a): Chromatofocusing: isoelectric focusing on ionexchange columns. I. General principles. *J Chromatogr* 150:17–30

Sluyterman LAA, Elgersma O (1978b): Chromatofocusing: isoelectric focusing on ionexchange columns. II. Experimental verification. *J Chromatogr* 150:31–44

Sluyterman LAA, Wijdenes J (1981a): Chromatofocusing III. The properties of a DEAE-agarose anion exchanger and its suitability for protein separations. *J Chromatogr* 206:429–440

Sluyterman LAA, Wijdenes J (1981b): Chromatofocusing IV Properties of an agarose polyethyleneimine ion exchanger and its suitability for protein separations. *J Chromatogr* 206:441–447

Thompson NE, Burgess RR (2001): Identification and use of polyol responsive monoclonal antibodies for use in immunoaffinity chromatography. In: *Current Protocols in Molecular Biology.* New York: John Wiley & Sons, 11.18.1–11.18.9

Tsuru D, Fujiwara K, Kado K (1978): Purification and characterization of L-pyrrolidonecarboxylate peptidase from *Bacillus amyloliquefaciens*. *J Biochem* 84:467–476

Tswett MS (1906): Adsorption analysis and chromatographic method: application to the chemistry of chlorophyll. *Berichte der Deutschen Botanischen Gesellschaft* 24:385

Washburn MP, Wolters D, Yates JR III (2001): Large-scale analysis of the yeast proteome by multidimensional protein identification technology. *Nat Biotechnol* 19:242–247

Weiss AR, Henricksen G (1995): Membrane Adsorbers for rapid and scaleable protein separations. *Gen Engineer News*, May 1, 1995, p22

Wu CC, MacCoss MJ, Howell KE, Yates JR (2003): A method for the comprehensive proteomic analysis of membrane proteins. *Nat Biotechnol* 21:532–538

Zechmeister L (1951) Early History of Chromotagraphy. *Nature* 167:405–406

General Reference

Kenney A, Fowell S, eds. (1992): *Practical Protein Chromatography*. Totowa, NJ: Humana Press

11

Recombinant Protein Techniques

The time is approaching when a new protein will be designed on the drawing board, using predictive algorithms, and its subsequent synthesis, via cloning or peptide coupling, will offer new and interesting challenges for biochemists and molecular biologists.

—*Gerald D. Fasman, 1989*

Introduction

The isolation and characterization of proteins parallels the major technical advances in protein chemistry. The first proteins that were characterized were those that could be isolated in large amounts. With the availability of more sensitive techniques, smaller and smaller quantities were required to produce detailed structural information. Newly evolving technologies, in conjunction with expression systems, have allowed the production of large amounts of any proteinacious factor. The isolation and structural characterization of proteins now depends only on the ingenuity of the investigator in devising an appropriate assay.

The mechanisms by which primary structural information is translated into a biologically relevant molecule with a precisely defined three-dimensional structure remain a mystery. Recombinant technology has infused new opportunities and new ways of looking at protein chemistry. There are now novel ways of identifying new proteins and new choices for expression vector and host, assay system, purification protocol and characterization of the recombinant product. In principle, any protein, however rare in nature, can be generated in limitless quantities. Now that there are so many systems available, the challenge becomes choosing the one that is most applicable for the expression of your recombinant protein. The first step is identifying the host organism that best fits your needs.

Recombinant methods fill a void for expressing rare messages in heterologous cell lines that can be grown in culture. Purification of a recombinant protein has several advantages over the isolation of the native counterpart. Perhaps the most significant is that the recombinant protein is relatively abundant. Another important advantage is that expression systems can be manipulated, and tailor made systems elegantly developed.

Expression, the directed synthesis of a foreign gene product, is the next logical step for investigators who have isolated a gene and want to study the protein that it encodes. The ideal expression system would provide for the synthesis of a sufficient quantity of recombinant protein and permit its isolation in highly purified form. A gene whose product is to be expressed is introduced into a plasmid or other expression vector and is then introduced into living cells by transformation or transfection. Typically, expression vectors contain the necessary promotors and other genetic machinery to direct the synthesis of large amounts of mRNA corresponding to the introduced gene. They may include sequences that allow for autonomous replication of the plasmid within the host, traits that allow cells containing the vectors to be preferentially selected, and sequences that increase the efficiency with which the mRNA is translated.

The number of expression systems that are currently available has grown since the early days of *E. coli*, *S. cerevisiae* and Chinese Hamster ovary cells. Today, producer organisms range from *E. coli* to whole animals that can be bled or milked to produce industrial quantities of the target protein (Hodgson, 1993). Insect cells have increased in popularity as laboratory-scale expression systems. Fungi and plant systems have also been developed.

When choosing a system, one should consider the intended application of the expressed protein because this will dictate which expression system to use. For proteins with a simple structure and no secondary modifications, prokaryotic expression systems may be faster and more straightforward than other more elaborate expression systems. However, many proteins of interest are complex in structure, requiring extensive secondary modifications, like glycosylation, amination or phosphorylation in order to assume a structure that is functionally and immunologically comparable to the native protein.

The wide-spread use of protein engineering has led to the development of procedures such as site-specific mutagenesis. The protein chemist can now specifically replace labile amino acid residues. For instance, recombinant-derived interleukin-1α has been stabilized against a pH-dependent covalent reaction, deamidation, by replacing a particularly labile asparagine residue with serine (Wingfield et al, 1987).

This chapter will not cover basic techniques in molecular biology. The reader is referred to the following excellent manuals: *Current Protocols in Molecular Biology* (Ausubel et al, 1989), K *Guide to Molecular Cloning. Methods in Enzymology Series, Vol 152* (Kimmel and Berger, 1987) and *Molecular Cloning* (Maniatis et al, 1982).

Recombinant Protein for Antibody Production

When the aim of the project is to use the expressed protein as an antigen for antibody production, two basic approaches are used to reliably make the protein of interest and to rapidly purify it. Fusion proteins with specific tags can be synthesized then purified by affinity chromatography. Alternatively, the protein can be expressed under conditions that cause it to precipitate in inclusion bodies and can then be purified. The protein is extracted from the inclusion bodies, run on SDS gels and the specific band identified, isolated, and used as the immunogen.

Protein for Biochemical or Cell Biological Studies

The generation of fusion proteins by recombinant genetic techniques has become routine in many labs. Some advantages in using a recombinant protein over isolation from its natural source include: (1) ready availability of raw material; (2) high expression levels, (particularly useful for proteins of low abundance) and (3) creation of modified fusion proteins. An affinity tag is usually added to the target protein to create a recombinant protein that is recognized by an affinity matrix, thereby greatly simplifying the purification of the genetically engineered protein. Ideally, the newly introduced tag sequence can be removed by specific proteolysis leaving the recombinant target protein intact.

A word of caution concerning recombinant proteins: do not assume that the recombinant protein will behave identically to the native protein. Test both molecules in parallel in every assay system.

The requirements for efficient translation of a heterologous protein have been found to be complex. Fusion protein technology avoids many problems encountered when attempting to express a foreign gene. In this approach, the cloned gene is generally introduced into an expression vector 3′ to a sequence coding for the amino terminus of a highly expressed protein which will be the carrier protein. The carrier sequence, which can be as short as one amino acid and as large as a complete protein, provides the necessary signals for good expression. The expressed fusion protein will have its N-terminal region encoded by the carrier derived protein domains, which can often be exploited in purifying the fusion protein with preexisting or commercially available antibody directed against the carrier.

There are three potential problems one should be aware of before deciding to make fusion proteins. These are: (1) the solubility of the expressed protein; (2) the stability of the expressed protein; and (3) the presence of the carrier protein.

When biological activity of a recombinant protein is an important consideration, the choice of an appropriate host cell type for expression becomes important. Heterologous gene expression systems may utilize either prokaryotic (e.g., *Escherichia coli*) or eukaryotic (e.g., yeast, insect or mammalian cell) hosts.

Hydrophobic membrane proteins appear to be toxic for *E. coli* and are expressed in this system in very low levels if at all. However, several membrane proteins have been successfully expressed in baker's yeast, *Saccharomyces cerevisiae*, and were found to incorporate into cellular membranes.

A. *In vitro* Transcription and Translation

In vitro translation systems have been developed to provide a eukaryotic cell free environment for translating foreign mRNA into proteins. *In vitro* synthesized proteins are extremely useful for a variety of purposes. This technique, coupled with mutagenesis, can be used to probe the primary structure of a protein for functional properties. The production of a series of truncated proteins can be useful for determining subunit structure and for detecting epitopes that are important for protein-protein interactions, including susceptibility to proteases, immunogenic epitopes, and enzyme active sites.

When appropriate, translation products can be used to extend the analysis of the target molecule by cotranslational processing and core glycosylation made possible by the addition of canine pancreatic microsomal membranes to a standard translation reaction. Processing events are detected as shifts in the apparent molecular weight of the translated molecule.

In vitro transcription and translation can be broken down into three steps. (1) The protein-coding sequence is cloned into a vector containing a promotor for SP6 or T7 RNA polymerase. (2) The specific mRNA is then transcribed from the DNA template. (3) The desired protein is synthesized (with [^{35}S]methionine). The translation product is then identified and analyzed by immunoprecipitation and SDS-PAGE.

Translation reactions are generally performed with commercially available kits. Up-dated, newer models are being continuously introduced and although they are relatively expensive they are highly recommended. The two most popular systems are the wheat germ extract and the rabbit reticulocyte lysate.

In vitro translation reactions can be directed by mRNA or by DNA templates that are transcribed *in vitro*. When using mRNA synthesized *in vitro*, the presence of a 5′ cap structure greatly enhances translational activity.

Once you have a high quality DNA template, it is possible in a single day to cleave the DNA, transcribe the template into mRNA, synthesize the target protein, and analyze it by SDS-PAGE and autoradiography. More typically, these procedures are carried out over two days.

PROTOCOL 11.1 Preparation of the DNA Template

This method requires knowledge and skills in molecular biology techniques. The protein coding sequences must be contiguous, not interrupted by introns, and must also be in frame in the correct orientation, downstream of the promotor. The initiation codon should be the first AUG in the RNA to be synthesized. For best results, the initiation codon should be within 100 bases of the 5′-end of the RNA.

1. Subclone the protein-coding DNA sequence of interest into a plasmid vector that contains a promotor for SP6 or T7 RNA polymerase at a site downstream of the promotor. Suitable vectors are commercially available from Stratagene and Clonetech.

2. Prepare plasmid DNA by CsCl gradient centrifugation.
3. Digest ~10μg DNA with a restriction endonuclease that has a site just downstream of the termination codon and does not cut within the protein coding sequence.
4. Run the digestion products on an agarose gel. Identify the band of interest, and isolate it by any one of the methods that are being used routinely in the lab.
5. Purify the cut DNA by phenol extraction and ethanol precipitation. Resuspend in 50μl of TE buffer. This should be enough for ten *in vitro* transcription and translation reactions.

PROTOCOL 11.2 *In vitro* Transcription—Preparation of the mRNA

This method is based on the cloning of protein-coding sequences into a vector containing a promoter for SP6 or T7 RNA polymerase. Messenger RNA is then transcribed from the DNA template using SP6 or T7 RNA polymerase (Ausubel et al, 1989).

Materials

Tris buffered phenol
Isobutanol
Ammonium acetate: 10M
Ethanol: 100%; 70%
TE buffer: 10mM Tris, pH 7.5, 1mM EDTA

1. Prepare the following reaction mixture at room temperature:
   8μl H_2O
   5μl DNA (~1μg)
   5μl 5 × ribonucleoside triphosphate mix (5mM each ATP, UTP, CTP, 5mM diguanosine triphosphate (G-5′ppp5′-G)TP, 0.5mM GTP)
   2.5μl 10 × SP6/T7 RNA polymerase buffer (0.4M Tris-HCl, pH 7.5, 0.1M $MgCl_2$, 50mM DTT, 0.5mg/ml BSA)
   2.5μl 10mM spermidine (for SP6 RNA polymerase only)
   1μl RNasin (30 to 60U)
   1μl SP6 or T7 RNA polymerase (5 to 20U)
   Adjust the final volume to 25μl.
2. Incubate the mixture at 40°C for 60min. When using the T7 RNA polymerase, substitute 2.5μl of H_2O for the spermidine.
3. Extract the RNA by adding 25μ1 of Tris buffered phenol, vortex and transfer the aqueous phase (top) to a new tube. Extract twice with 25μl isobutanol.
4. Add 6μl of 10M ammonium acetate and 70μl ethanol (100%) and precipitate on dry ice for 20min. Centrifuge at 4°C at maximum speed in a microfuge, and wash the precipitated RNA pellet once with 70% ethanol.
5. Resuspend the RNA in 25μl TE, add 6μl of 10M ammonium acetate and 70μl ethanol (100%), reprecipitate and wash once with 70% ethanol. Resuspend the RNA in 10μl TE. Use immediately or quick-freeze and store at –70°C.

Comments: For efficient *in vitro* translation, the mRNA should contain a 5′-capped structure that is normally found in eukaryotic mRNAs (Melton et al, 1984). For this reason, 90% of the GTP in the reaction mixture is replaced by diguanosine GTP making 90% of the synthesized RNA contain a 5′-capped structure.

PROTOCOL 11.3 Guanylyltransferase Catalyzed Addition of a G(5′)ppp(5′)G Cap to mRNA

Proper capping of RNA promotes correct initiation of protein synthesis. Uncapped RNA is rapidly degraded by cellular RNases. Capped RNA is also translated more efficiently in reticulocyte lysate and wheat germ *in vitro* translation systems. For efficient translation, a m7G(5′)ppp(5′)G cap structure must be added to the mRNA. Although this structure is present when mRNA is isolated from cells, when RNA is transcribed *in vitro*, the cap structure must be added synthetically.

The cap can be added during the transcription reaction as described in Protocol 11.2, or enzymatically in a separate reaction using guanylyltransferase.

Materials

Guanylyltransferase

Phenol:chloroform: 1:1

Chloroform

Reaction mixture: 50 mM Tris-HCl, pH 7.9, 1.25 mM $MgCl_2$, 6 mM KCl, 2.5 mM DTT, 100 µg/ml BSA, 1 u/µl RNasin, 50–100 µM S-adenosylmethionine, 40 µM GTP

Ethanol: 100%

1. Mix synthesized mRNA (~1 µg) with guanylyltransferase (1 u/µg RNA) in 30 µl of reaction mixture.
2. Incubate at 37°C for 45 min.
3. Extract with phenol:chloroform, then with chloroform alone and precipitate with ethanol.

Comments: While the cap is routinely added onto synthetic transcripts generated during *in vitro* transcription, it can be incorporated into the RNA in both the forward, m7G(5′)ppp(5′)G(pg) and reverse orientation G(5′)pppm7G(pN) leading to the synthesis of two isomeric populations of RNA of approximately equal proportion (Pasquinelli et al, 1995). RNA capped with reverse 5′ caps cannot be translated, resulting in a 50% reduction in expected yield of protein. One way to prevent incorporation of cap analog in the reverse orientation thereby maximizing translation efficiency is to use the recently described antireverse cap analog (ARCA) (Stepinski et al, 2001). In ARCA, the 3′ OH group closer to 7mG is replaced with —OCH_3, effectively restricting transcription initiation to the remaining —OH group, generating RNAs with cap analog incorporated only in the functional orientation. Therefore, the ARCA cap results in synthesis of capped RNAs that are 100% translatable.

PROTOCOL 11.4 *In vitro* Translation: Protein Synthesis

Proteins are synthesized in a cell free system. This protocol is usually performed with commercially available kits. You will need access to a scintillation counter.

Materials

In vitro translation kit

[^{35}S]methionine: Usually sold as translation grade, 1400 Ci/mmol (The [^{35}S]methionine should have the highest specific activity commercially available)

Scintillation vials

Scintillation fluid

Trichloroacetic acid (TCA): 25% and 5%

Acetone

1. Heat the template mRNA for 10 min at 67°C and immediately cool on ice.
2. To a 0.5 ml polypropylene microcentrifuge tube, add the RNA to the appropriate reagents of the *in vitro* translation kit and follow the manufacturer's instructions. Typically use about 5–15 µCi of [^{35}S]methionine per translation reaction. In parallel, be sure to include a negative control consisting of the translation reaction lacking RNA. Some manufacturers will include RNA to be used as a positive control to check the technique and reagents.
3. Incubate the reticulocyte translation reactions for 60 min at 30°C.
4. To determine the effectiveness of the translation, remove 2 µl of the translation product and add to 250 µl of 1 N NaOH/2% H_2O_2 solution in a 1.5 ml microcentrifuge tube. Incubate for 15 min at 37°C. The NaOH hydrolyzes aminoacyl tRNAs, preventing them from being included in the incorporation calculation. The H_2O_2 removes the red color from the reaction, preventing any quenching during scintillation counting.
5. Add 1 ml of cold 25% TCA and incubate for 15 min on ice.
6. Collect the precipitated protein by filtering under vacuum on glass fiber filters (Whatman GF/A). Wash the filter with 3 ml of 5% cold TCA and then with 3 ml of acetone. Allow the filter to dry.
7. Repeat the wash with the blank and control reactions. Put the filter in a scintillation vial, add scintillation fluid and determine incorporated [^{35}S]methionine using a scintillation counter.

Comments: If the counts from the experimental sample are at least three times greater than the counts of the blank, translation has occurred. The positive control should be approximately seven times the blank. If the positive control included with the kit incorporated radioactivity and the sample did not, the experimental RNA was either quantitatively not sufficient or may be degraded. If the positive control did not incorporate radioactivity, then the kit may be defective.

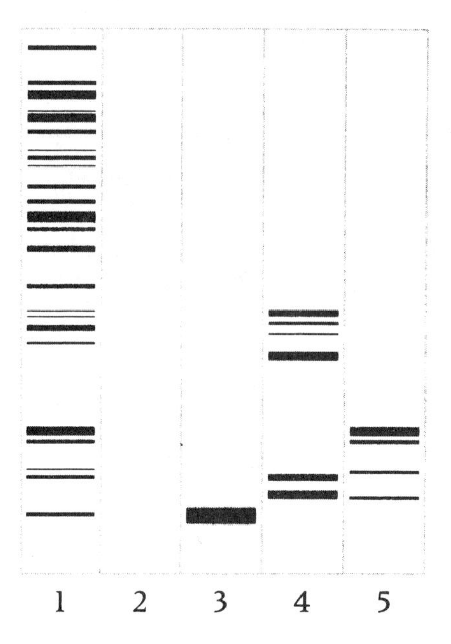

FIGURE 11.1 Translation products of various mRNAs. The indicated mRNAs were translated with the Wheat Germ Extract system (Promega). Lane 1: Brome Mosaic Virus RNA; lane 2: Wheat Germ Extract minus exogenously added RNA; lane 3: rabbit globin mRNA; lane 4: *E. coli* β-lactamase mRNA; lane 5: *S. cerevisiae* α-factor mRNA. (Adapted with permission of the Promega Corporation.).

The RNA used for *in vitro* translation can be total RNA, which should be used at ~10 μg in 2 μl or poly(A^+) RNA at 2 μg in 2 μl.

The translated products can be analyzed directly by SDS-PAGE followed by fluorography or can be immunoprecipitated.

Direct analysis should only be used when a pure RNA species is translated. In this case the protein generated should be radiochemically pure, yielding only a single radioactive band of the expected molecular weight. Other proteins will be present that were components of the translation mix, however, these will not be radioactive. In many cases, minor radioactive bands are observed which are smaller than the expected molecular weight. These are usually due to premature termination of translation or proteolysis. Translation products from a wide spectrum of mRNAs are shown in Figure 11.1.

When the RNA to be translated is either total RNA or poly(A^+), best analytical results are obtained by performing immunoprecipitation with a specific antibody. For an additional control, divide the translation product in half and perform two immunoprecipitation reactions, one with the specific antibody and the other with a non-binding antibody. Following a positive signal as determined by fluorography, autoradiography or phosphorimaging you should be confident that the target protein has been synthesized.

PROTOCOL 11.5 Cotranslational Processing Using Canine Pancreatic Microsomal Membranes

Processing events such as signal peptide cleavage, membrane insertion, translocation, and core glycosylation can be evaluated by performing the translation in the presence of canine pancreatic microsomal membranes that contain the enzymatic machinery to perform all of the above functions. Commercially available microsomal membrane preparations have been isolated free from contaminating membrane fractions, endogenous membrane bound ribosomes and mRNA. The preparations should contain signal peptidase and core glycosylation activities. Increasing the amount of membranes in the reaction increases the proportion of polypeptides translocated into vesicles but reduces the total number of polypeptides synthesized.

Materials

Reticulocyte lysate
Amino acid mixture (minus methionine)
[^{35}S]methionine, Translation grade, highest specific activity available
Microsomal membranes (Promega)
Your mRNA

1. Mix the following components on ice in the order given in a sterile microfuge tube:
   17.5 µl reticulocyte lysate
   0.5 µl 1 mM amino acid mixture (minus methionine)
   2 µl [^{35}S]methionine).
   2.2 µl H_2O
   1.8 µl microsomal membranes
   1.0 µl RNA in H_2O
2. Incubate at 30°C for 60 min.
3. Analyze the results by SDS-PAGE followed by fluorography.

Comments: Some kits provide positive controls for signal processing and glycosylation. To become familiar with the protocol, use these controls in parallel with your mRNA.

PROTOCOL 11.6 Translocated Products Are Resistant to Protease Digestion

It may be difficult to determine if a processing event has occurred by SDS-PAGE alone. Two additional assays are described to aid in the analysis.

If a newly synthesized protein is translocated into a vesicle, it should be protected from the effect of exogenously added proteases by the vesicle membrane. In this system, the microsomal membrane is provided. The assay uses protease resistance as a measure of translocation.

Materials

Triton X-100: 10% (v/v) stock
Trypsin or Proteinase K

1. To confirm that translocation has occurred, divide the translation product in half. Add Triton X-100 to one fraction to make a 0.1% final concentration.

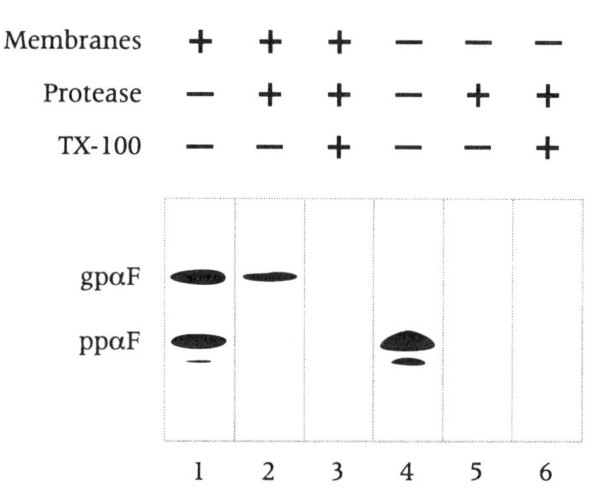

FIGURE 11.2 Protection of the translocated product from protease digestion. *In vitro* transcription and translation were performed. The presence of membranes, proteinase K, and Triton X-100 are indicated above the figure. The presence of yeast microsomal membranes resulted in the glycosylation of the propeptide (pro-α-factor) (lanes 1 and 2). Translocation is demonstrated by the protection of glycosylated pro-α-factor (gpαF) from protease digestion in the presence of intact membranes (lane 2) and its digestion when the membranes are disrupted by the addition of Triton X-100 (lane 3). Untranslocated prepro-α-factor (ppαF) was protease sensitive even in the presence of membranes (lane 2). No translocation occurred in the absence of membranes (lane 4). (Adapted with permission from Bush et al, 1991).

2. Add proteinase K or trypsin to a final concentration of 0.1 mg/ml to both fractions.
3. Incubate on ice for 30 min and then analyze by SDS-PAGE.

Comments: The detergent treated fraction should be susceptible to exogenously added protease. If the target protein has been translocated it should be protected by the membrane and resistant to proteinase digestion. The expected results of this type of experiment are illustrated in Figure 11.2 using the translocation and glycosylation of mating pheromone precursor, prepro-α-factor of *S. cerevisiae* (Bush et al, 1991).

PROTOCOL 11.7 Was the Translational Product Glycosylated? Endoglycosidase H (Endo H) Analysis

Endo H cleaves the internal N-acetylglucosamine residues of high mannose carbohydrates, resulting in a shift in apparent molecular weight on SDS gels to a position close to the unglycosylated species (see Chapter 9).

Materials

SDS: see Appendix C

2 × SDS sample buffer

Endo H
Sodium citrate buffer: 0.1M, pH 5.5

1. Divide the translation product from Protocol 11.5 in half. Treat one fraction with 0.1% SDS, in sodium citrate buffer and incubate at 37°C for 12h. Incubate the other fraction in citrate buffer without SDS.
2. Add 1.3–2mU of Endo H to both reactions.
3. In addition to *in vitro* translated products from Protocol 11.5, prepare *in vitro* translation products without post-translational modifications according to Protocol 11.4.
4. Analyze by SDS-PAGE.

Comments: The fraction incubated with Endo H in the presence of SDS should not be protected from Endo H digestion and should migrate close to the nonglycosylated species, indicating that the translation product had been glycosylated. If there is no change in mobility the translation product was not glycosylated.

Protein Transduction: A Method for Introducing Exogenous Proteins into Cells

There are many fast and efficient techniques for introducing transcriptionally competent DNA into cells, but options for delivering functionally active proteins into cells are limited. Protein transduction is a relatively new approach where functionally active peptides or proteins enter a cell by crossing its plasma membrane.

The mammalian cell membrane is refractory to most proteins and peptides. The most widely used methods for introducing membrane-impermeant molecules into cells were microinjection and electroporation. Both of these techniques are invasive and disrupt the cell membrane.

The serendipitous finding that some proteins can enter cells when added to the media led to the identification of short basic peptide stretches from these proteins that can traverse the plasma membrane and carry the rest of the protein with them (Schwarze and Dowdy, 2000). Two widely studied proteins with transduction domains are the herpes simplex virus (HSV) protein VP22 and the human immunodeficiency virus (HIV)-1 transcriptional activator Tat.

Protein transduction domains (PTD) are generally short peptides consisting of 10–16 amino acids. Structurally dissimilar, their only common feature appears to be the presence of numerous positively charged lysine and arginine residues. PTD mediated transduction does not occur through endocytosis.

Protein transduction has many advantages over DNA transfection as a method to express proteins in cells. It is well known that some mammalian cell lines are difficult to transfect. All mammalian cells tested to date are receptive to protein transduction. In a transfection experiment it takes from a few hours to one to three days to detect expression of the target protein, whereas a transduced protein can be detected in about 10 minutes. By varying the amount of protein added to the culture medium, researchers can control the final intracellular concentration (Bonetta, 2002).

B. Recombinant Gene Products in *E. coli*: Expression, Identification and Characterization

Many gene products of biological interest are naturally produced in minute quantities, making it extremely difficult to study them. One solution has been the development of recombinant vector systems that achieve expression of cloned genes in prokaryotic hosts, especially *E. coli*, which are widely used for the expression of recombinant proteins. A wide variety of coding regions can be expressed in bacteria either on their own or as fusion proteins. Indeed, *E. coli* have been used extensively for the expression of a large number of gene products at levels sufficient for biochemical analysis and even product development. As a recombinant protein expression host, *E. coli* offer many advantages. They are very easy to manipulate, grow rapidly, and have simple nutritional requirements.

The expression of foreign genes from plasmids transformed into *E. coli* is increasingly being used to investigate the functions of proteins. Two disadvantages of bacterial expression systems are: (1) eukaryotic proteins will not be processed; and (2) proteins expressed in large amounts will often precipitate. The reasons suggested for insolubility include incorrect folding and aggregation of the expressed protein, association of the expressed protein with *E. coli* host proteins, and co-aggregation of the fusion protein with bacterial membranes (Frankel et al., 1991). Overproduction may not be problematic because methods have been developed for the recovery of recombinant protein from inclusion bodies. However, if any kind of posttranslational modification of the recombinant protein is necessary for future studies, then bacteria is probably not the system of choice.

The design of *E. coli* expression systems involves inserting the coding region of the foreign gene into a multicopy vector system, which is usually a plasmid. The insert is then transcribed and translated. To be efficiently expressed, the foreign gene must be placed under the transcriptional and translational control of regulatory elements recognized by the *E. coli*. A number of vectors have been developed, incorporating many features designed for high efficiency and control.

The inserted gene is often placed adjacent to a translation initiation signal recognized by the bacterial translation machinery, or the insert is introduced in-frame to a gene or gene fragment that expresses well in *E. coli*. The vectors are designed with many convenient restriction sites adjacent to the translation initiation signal. The inserted gene is commonly fused to the N-terminal coding region of a bacterial gene resulting in the synthesis of a fusion protein. In this way the recombinant gene will be under the control of regulatory elements that will permit its expression in *E. coli*. By fusing the coding sequence of the target protein with the coding sequence of a polypeptide domain with high affinity to a ligand, any protein can be efficiently purified, allowing for the identification and characterization of proteins that would otherwise be difficult to select, much less purify, from cellular extracts.

The gene product may be produced in *E. coli* either constitutively or in an inducible manner. The production of the recombinant protein is monitored in several ways: direct visualization of the protein band following SDS-PAGE of an aliquot of bacterial lysate, antibody detection, or a functional assay. Ideally, one would like the fusion protein to accumulate to high levels and be stable.

The design and synthesis of a chimeric protein should meet the specific experimental needs. Fusion proteins are frequently planned with the development of antisera in mind. If the function of the chimeric protein is preserved, then detection and characterization should be achievable. A common approach is to fuse a heterologous gene to a well defined bacterial gene like *lac*Z (β-galactosidase). High titer antisera are commercially available that recognize the amino-terminal portion of β-galactosidase permitting immunological detection of the resulting chimeric gene product.

Expression and Purification of *lac*Z and *trp*E Fusion Proteins

E. coli fusion protein technology offers the advantage of overproducing a protein of interest with relative ease. The principal application for overproduced fusion protein in *E. coli* is for antibody production. A potential disadvantage is that it is often associated with an insoluble inclusion body. However, the initial insolubility can be utilized as a significant purification step.

Another useful feature when using a *lac*Z fusion construct is that β-galactosidase (β-gal), the product of the *lac*Z gene, can be used as a means of identifying the fusion protein. Further, antibodies to β-gal can be used to affinity purify the fusion protein and to follow the purification of the fusion protein by Western blot analysis of the various fractions.

The strategy is straightforward. The gene of interest is cloned in the correct reading frame into the polylinker located at the end of the *lac*Z gene of a pUR fusion-protein expression vector. The pUR family of plasmid vectors contains the *lac*Z gene transcribed by the UV5 promotor (Rüther and Müller-Hill, 1983). The correct transformant is selected, grown and then induced by depression of the UV5 promotor with isopropyl-β-thiogalactopyranoside (IPTG). Extracts are prepared and checked for the presence of fusion protein by SDS-PAGE. Ideally, the fusion protein should be enriched enough to be used as an antigen.

In addition to the *lac*Z system, other fusion protein constructs have been developed and used successfully. The *trp*E system is based on fusing the foreign gene of interest with the *trp*E gene in pATH2 plasmids and transforming *E. coli* (Spindler et al, 1984; Dieckmann and Tzagoloff, 1985). This method is illustrated in Protocol 11.9.

Plasmid vectors containing the gene coding for staphylococcal protein A have also been developed. These vectors allow fusion of any gene to the protein A moiety, resulting in fusion proteins that can be purified in a single step procedure by IgG affinity chromatography. The pRIT2 vector was designed for the temperature inducible expression of intracellular fusion proteins in *E. coli* (Nilsson et al, 1985).

To successfully perform Protocols 11.8 and 11.9 you will need to be familiar with basic molecular biology techniques.

PROTOCOL 11.8 *lacZ* Induction

Materials

E. coli, strain HB101

LB/ampicillin plates (Maniatis et al, 1982)

LB/amp media

Isopropyl-β-thiogalactopyranoside (IPTG): 100 mM

1. Subclone the gene of interest into a pUR vector in the correct reading frame. Transform competent *E. coli*, strain HB101, and select transformants on LB/ampicillin plates.
2. Perform minipreps from two to five isolated colonies to be sure the colony contains the insert in the correct orientation. Streak LB/amp plates from the miniprep liquid culture.
3. Inoculate 2–5 ml of LB/amp media with a single colony, validated by miniprep analysis. Grow overnight at 37°C, shaking.
4. Add 1 ml of the overnight culture to 400 ml LB/amp media in a 2-liter flask. Grow at 37°C with vigorous shaking until OD_{600} reaches 0.5.
5. Add 1.6 ml of IPTG solution (0.4 mM final). Grow the bacteria an additional 2 h.
6. Harvest the culture by centrifugation for 10 min at 4000 × *g* at 4°C and discard the supernatant.
7. Proceed with Protocol 11.10

Comments: To prepare LB, add 10 g tryptone, 5 g yeast extract and 10 g NaCl to 1 L distilled water, and autoclave. To prepare LB plates, add 15 g agar to LB and autoclave. When the media has cooled down to ~ 55°C, add the desired antibiotic from a concentrated stock.

PROTOCOL 11.9 Induction of the *trpE* Fusion Protein

The *trpE* fusion protein is produced by removing tryptophan from the media and adding β-indoleacrylic acid.

Materials

E. coli strain DH5α

M9 medium: 6 g Na_2HPO_4, 3 g KH_2PO_4, 0.5 g NaCl, 1 g NH_4Cl in 1 L distilled water. Adjust to pH 7.4 and autoclave. Add 2 ml of filter sterilized 1 M $MgSO_4$ containing 1% Casamino Acids, 20 μg/ml tryptophan, and 150 μg/ml ampicillin.

β-indoleacrylic acid

1. Use the pATH2 or pATH22 plasmid and *E. coli* strain DH5α.
2. Grow the transformed bacteria overnight in M9 medium.
3. For induction, pellet the bacteria and remove them from the tryptophan-containing medium by resuspending them in tryptophan-free M9, Casamino Acid medium, centrifuging the cells and resuspending the pellet in 100 ml of tryptophan-free medium and incubate for 1 h.
4. Add β-indoleacrylic acid to a final concentration of 20 μg/ml and incubate for an additional 4 h.
5. Harvest the culture by centrifugation for 10 min at 4000 × *g* at 4°C. Discard the supernatant and proceed with Protocol 11.10

PROTOCOL 11.10 Preparation of the Protein Extract

Materials

PBS: See Appendix C

HEM buffer: 25 mM HEPES, pH 7.6, 0.1 mM EDTA, 12.5 mM $MgCl_2$, which includes anti-protease cocktail

Sonicator

Lysozyme: 50 mg/ml

1. Vigorously resuspend the bacterial pellet in 10 ml PBS containing only antiprotease cocktail and transfer to a 15 ml conical centrifuge tube. Centrifuge for 10 min at 3000 × *g* and discard the supernatant.
2. Resuspend the bacterial pellet in 4 ml of HEM buffer.
3. Add 40 μl of the lysozyme solution (0.5 mg/ml final) and incubate on ice for 30 min.
4. Disrupt the bacteria by sonication equipped with a microtip, if available. Use a setting that will generate vigorous churning and position the probe in the bacterial suspension so that foam is not produced. Use two 15 sec bursts with a 1 min break between bursts placing the tube in ice to avoid overheating.
5. Centrifuge the sonicate for 15 min at 27,000 × *g* (15,000 rpm in a Sorvall SS-34 rotor).
6. Collect the supernatant and save. Check for the presence of fusion protein. Save the pellet. It contains the induced protein in an insoluble form.

Comments: Techniques have been described to partially purify the relatively insoluble fusion protein product from bacterial protein by digesting the cell pellet with lysozyme (3 mg/ml), sonicating three times for 15 sec each burst, on ice, and successively washing with 20 ml of ice-cold 0.5% Nonidet P-40 in 0.3 M NaCl followed by 1 M NaCl in 10 mM Tris-HCl, pH 7.5, and finally, 10 mM Tris-HCl, pH 7.5 (Holzman et al, 1990). Samples were then analyzed by SDS-PAGE.

PROTOCOL 11.11 Solubilization of the Fusion Protein

Materials

HEM buffer: 25mM HEPES, pH 7.6, 0.1mM EDTA, 12.5mM $MgCl_2$, antiprotease cocktail (Roche)

Guanidine-HCl

Ultracentrifuge and Beckman 60Ti rotor

1. To solubilize the fusion protein, resuspend the pellet (see Protocol 11.10) in 2ml of HEM buffer. The pellet may be difficult to resuspend. A few short bursts of sonication may be helpful.
2. Add 2ml of HEM buffer containing 8M guanidine-HCl. Incubate with gentle shaking for 30min at 4°C.
3. Centrifuge for 30min at 4°C at 87,000 × *g* (35,000rpm in a Beckman 60Ti rotor).
4. Transfer the supernatant to a dialysis bag and dialyze in three steps of not less than 3h per step. First, versus 500ml HEM buffer 1M in guanidine-HCl, then two additional changes with HEM without guanidine-HCl. As the guanidine is removed by dialysis, proteins will begin to precipitate.
5. Transfer the contents of the dialysis bag to a centrifuge tube and centrifuge for 5min at 12,000 × *g* (10,000rpm in an SS-34 rotor). Save the supernatant. It should contain ~ 1mg/ml protein and be colorless. The supernatant contains ~ 1% of the total amount of fusion protein. Save the pellet. It still contains the majority of the fusion protein.
6. Determine the protein concentrations of all fractions (Chapter 5) and check for the presence of the fusion protein using SDS-PAGE. Load 5–10μg of protein per lane. In addition, it is useful to compare extracts containing the fusion protein with control extracts from uninduced bacteria or from bacteria containing the expression vector but no insert.

Comments: The molecular weight of β-gal is 116kDa. The *trp*E gene product expressed from pATH2, (anthranilate synthetase component I from *Salmonella typhimurium*) is approximately 33kDa. Therefore, the molecular weight of the fusion protein can be predicted from the length of the open reading frame that was ligated into the vector.

PROTOCOL 11.12 Purification of Eukaryotic Proteins from Inclusion Bodies in *E. coli*

Expression of cloned eukaryotic genes in *E. coli* is a popular way of obtaining large amounts of protein. The overexpressed proteins accumulate in the cytoplasm in insoluble inclusion bodies. Despite their insolubility, inclusion bodies can be used as the source of protein for further purification. Because they are dense, inclusion bodies can be easily separated from the majority of the contaminants by centrifugation at the appropriate *g* force. The inclusion bodies are then washed, and the protein of interest is selectively solubilized (Lin and Cheng, 1991).

Materials

Buffer A: 20 mM Tris-HCl, pH 7.5, 20% sucrose, 1 mM EDTA

Buffer P: PBS without Ca^{2+} and Mg^{2+}, 5 mM EDTA, 10 µg/ml leupeptin, 10 µg/ml aprotinin, 2 mM PMSF

Sonicator: Sonifier Cell Disrupter (Model W185 Heat Systems Ultrasonics, Plainview, NY)

RNase T1

DNase I

Buffer W: PBS without Ca^{2+} and Mg^{2+}, 25% sucrose, 5 mM EDTA and 1% Triton X-100

Denaturation buffer: 50 mM Tris-HCl, pH 8.0, 5 M guanidinium HCl, 5 mM EDTA

Renaturation buffer: 50 mM Tris-HCl, pH 8.0,1 mM DTT, 20% glycerol solution and protease inhibitors

1. Resuspend the cell pellet from a 1 liter culture in 50 ml of buffer A and incubate on ice for 10 min.
2. Pellet the cells by centrifugation for 10 min at 6,000 × *g* and resuspend them in ice cold water and incubate them at 4°C for 10 min. This treatment results in the lysis of the outer cell wall.
3. Centrifuge the lysate at 8,000 × *g* for 15 min. Resuspend the pellet, consisting of spheroplasts, in 10 ml of buffer P.
4. Disrupt the cell membranes by sonication, three 30 sec pulses at 50 W with 30 second pauses between pulses. Add RNase µl (1.3×10^3 U/10 ml) and DNase I (400 µg/10 ml) to the sonicate and incubate for 10 min at room temperature.
5. Dilute the suspension further with 40 ml of buffer P. Pellet crude inclusion bodies by centrifugation at 13,000 × *g* for 30 min.
6. Resuspend the pelleted inclusion bodies in 40 ml of buffer W. Incubate on ice for 10 min then centrifuge for 10 min at 25,000 × *g*. Repeat this washing step twice.
7. Resuspend the inclusion bodies in 10 ml of denaturation buffer and sonicate for 5 sec at 50 W to facilitate solubilization of the aggregated proteins.
8. Incubate the suspension on ice for 1 h, then centrifuge at 12,000 × *g* for 30 min. Add the supernatant to 100 ml of renaturation buffer and stir gently overnight at 4°C to renature the proteins.
9. Clarify the supernatant by centrifugation at 13,500 × *g* for 30 min. The supernatant should contain highly purified expressed proteins at concentrations of 0.02–0.12 mg/ml which can be aliquotted and stored at –70°C without additional concentration steps.

Comments: For analyzing the expression of fusion proteins in extracts of transformed *E. coli*, it is informative to include some of the following controls: (1) vector alone, uninduced; (2) vector alone, induced; (3) fusion protein construct, uninduced; and (4) fusion protein construct, induced.

C. Affinity Tags

The use of genetically-encoded affinity tags is now a standard method of purifying proteins (reviewed in Hannig and Makrides, 1998). Genetic engineering is being applied to aid in the purification of recombinant proteins. The addition of specifically designed tags has enabled the development of novel strategies for downstream processing that can be employed for the efficient recovery of recombinant proteins (Nygren et al, 1994). Two different tags have been fused to the same protein to effect high purification and immunodetection (Lu et al, 1997).

A fusion tag, also referred to as affinity tags and epitope tags are short peptide or protein sequences that are easily recognizable by commercially available tag-specific antibodies. For purification of biochemically active proteins, affinity tags enable high affinity and selectivity for binding to specific resins to facilitate purification under conditions that retain activity (Nilsson et al, 1997). An incomplete list of popular tags is presented in Table 11.1. Affinity tag systems share the following features (Terpe, 2003):

- one-step adsorption purification
- a minimal effect on tertiary structure and biological activity
- easy and specific removal to produce the native protein
- simple and accurate assay of the recombinant protein during purification
- applicability to a number of different proteins

Epitope tags can be viewed as handles to identify and purify proteins. These systems, described below, have gradually replaced the first generation *lacZ* and *trpE* systems. Factors such as stability, solubility, the ability of the protein to be secreted, protein folding, and purification

TABLE 11.1 Epitope tags

Tag	Size	Media
HA	9 aa	Anti-HA affinity resins
FLAG	8–23 aa	Anti-FLAG affinity gels
c-myc	10 aa	Monoclonal antibody
T7·Tag	11 aa	T7·Tag antibody agarose
Peptide 38	5–14 aa	
Softag 1	8–13 aa	
Poly-Arg	5–6 aa	Cation exchange resin
His6	6–19 aa	Ni-NTA resins
S·Tag	15 aa	S-protein agarose
Strep-tag II	8 aa	StrepTactin
CCXXCC	6–17 aa	
GST	26 kDa	GST Sepharose
MBP	40 kDa	Amylose resin
Cellulose-binding domain	11 kDa	Cellulose matrix
Protein A	7–31 kDa	
ChitinBP	56 kDa	Chitin resin
Pinpoint (biotin)	13 kDa	SoftLink resin
TAP	22 kDa	
Calmodulin binding peptide (CBP)	2.96 kDa	Calmodulin resin

Table modified from Burgess and Thompson, (2002).

conditions should be taken into consideration when choosing a suitable expression strategy for a specific protein.

Genetic engineering has provided the technology to alter the properties of proteins to facilitate their purification. By the insertion of DNA sequences on either the 3′ or the 5′ end of the translated DNA, additional amino acids can be inserted, creating a fusion protein. The newly modified target protein may now be more suitable for purification by immunoaffinity, metal chelate, or ion exchange chromatography. The final step in a purification protocol should be the rapid and total removal of the added stretch of amino acids. The fusion protein should not interfere with the protein's biological activity and should not affect the protein's folding. Although none of the tags shown in Table 11.1 is perfect for all applications, the availability of kits should enable the researcher to clone the protein, express it in a suitable host, and purify it by an affinity method utilizing the tags that are genetically engineered into the expression vector to be part of the fusion protein.

Fusion proteins have evolved into highly sophisticated purification systems. Many vectors offer the option of engineering the fusion protein so that the epitope tag will be located either N-terminal or C-terminal to the target protein. Tags supporting single-step purifications have the advantage of convenience and yield. Tags supporting two sequencial affinity steps combine two different tags on the same protein, which are normally separated by an enzyme-cleavable linker sequence (Aebersold and Mann, 2003).

One should be aware that an epitope tag may interfere with the structure and function of the recombinant protein. The effect of an epitope tag cannot be predicted a priori as the folding of the fusions cannot be accurately predicted. In rare cases epitope tags may not be detected by the standard molecular detection methods due to interactions with the fusion polypeptide that block the antigenic site. The tag may not be surface-exposed. Proteins may be functionally inactive as fusion proteins. Inevitably, there will be fusion proteins that cannot be expressed, solubilized or purified. This may be due to the loss of a cofactor, inappropriate buffers, or other incompatible conditions.

Removing the Tag

The presence of affinity tags may affect important functions or properties of the target protein. Therefore, removing the tag is often desirable. Cleavage of many of the tags described below can be accomplished with a site-specific protease with no loss of protein activity. Removal of the protease after cleavage is made easier by using a recombinant protease that itself contains an affinity tag or using a biotinylated protease. Cleavage of the tag without using a protease is also possible by introducing a self-splicing intein, discussed below. The most commonly used proteases are enterokinase, thrombin and factor Xa.

Enterokinase is often used to remove tags from the N-terminus of the fusion protein. It specifically recognizes the pentapeptide (DDDDKX) and cleaves at the carboxyl site of lysine. Interestingly and conveniently, the FLAG-tag (DYKDDDK) contains an enterokinase recognition site.

Thrombin is a protease that is widely used to cleave tags. In contrast to enterokinase and factorXa, thrombin cleavage results in the retention of the two amino acids on the C-terminal side of the cleavage point of the target protein. The optimal cleavage site for α–thrombin has the structures of X_4-X_3-P-R-[K]-X_1'-X_2', where X_4 and X_3 are hydrophobic amino acids and X_1' and X_2' are non-acidic amino acids (Chang, 1985). Some frequently used recognition sites are L-V-P-R-G-S and L-V-P-R-G-F (Terpe, 2003).

Factor Xa has been used successfully to remove N-terminal affinity tags (Pryor and Leiting, 1997). Factor Xa cleaves at the carboxyl side of the four-amino acid peptide I-E[D]-G-R-X_1 where X_1 can be any amino acid except arginine and proline. The cleavage reaction is routinely performed between 4 and 25°C.

Glutathione-S-Transferase (GST) Fusion Proteins

Foreign polypeptides can be expressed as fusions to the C terminus of glutathione-S-transferase (GST). The GST protein is small, 27.5 kDa, compared to other carriers (e.g., β-gal). A series of expression vectors was developed referred to as pGEX, in which cDNAs are expressed as GST fusion proteins in a form that allows them to be purified rapidly using nondenaturing conditions (Smith and Johnson, 1988). The pGEX vectors carry protease recognition sequences located between the GST coding region and the multiple cloning site. The fusion proteins typically remain soluble within the bacteria and can be purified from lysed cells because of the affinity of the GST moiety for glutathione, which is immobilized on agarose beads. The fusion protein is adsorbed onto glutathione-agarose, washed extensively and eluted by competition with free, reduced glutathione. The fusion protein is then incubated with a site-specific protease, removing the GST. The free glutathione in the elution solution is removed by ultrafiltration, and the GST is removed by re-absorption of the GST to glutathione-agarose.

The main advantage of the GST system is that the majority of the fusion proteins remain soluble. Therefore, harsh, denaturing conditions are not needed at any stage of purification, and the expressed protein will stand a better chance of retaining its functional activity and antigenicity. In most cases the GST carrier does not influence the antigenicity or functional activity of the foreign polypeptide.

An additional advantage of the pGEX system is that transformants that express the fusion protein can be rapidly identified and quickly screened for ease of purification by affinity adsorption on glutathione-agarose beads.

A built-in feature of the pGEX vectors is the *lac* repressor (the product of the *lacI* gene) that represses the expression of the GST fusion protein. Upon induction with IPTG, the GST fusion protein is expressed. pGEX vectors are available in three forms to ensure that the foreign DNA is in frame. Each pGEX vector contains an open reading frame encoding GST, followed by unique restriction sites. The cloning sites are present in a different reading frame in each of the three vectors. First, choose a vector that will express the foreign protein in frame with GST. If the removal of GST is desirable, choose either pGEX2 or 3.

PROTOCOL 11.13 Production and Analysis of GST Fusion Protein Transformants (Small Scale)

Laboratory skills in molecular biology are necessary to successfully perform this protocol.

Materials

LB/amp plates

Isopropyl-β-thiogalactopyranoside (IPTG): 100 mM stock in distilled water

PBS: see Appendix C

Sonicator

Glutathione agarose beads: Pharmacia

2 × SDS sample buffer: see Appendix C

Sterile tooth picks

1. Subclone the DNA fragment into the appropriate pGEX vector in the correct reading frame. Transform competent *E. coli* and select transformants on LB/amp plates. As a control, include empty vector. If the insert is being cloned with ends generated by only one endonuclease, treat the vector with calf intestine phosphatase following digestion to minimize background.
2. With sterile tooth picks, pick colonies and inoculate 2 ml of LB/amp media. In addition, streak out the picked colony on an LB/amp master plate. Grow the liquid cultures shaking at 37°C until visibly turbid (3–5 h).
3. Induce the expression of the fusion protein by adding IPTG to a final concentration of 0.1 mM from the 100 mM stock. Incubate an additional 2 h.
4. Transfer the majority of the culture to a microfuge tube. Centrifuge for 30 sec at maximum speed and decant off the supernatant. Resuspend the pellet in 300 μl cold PBS. Remove a 10 μl aliquot for analysis.
5. On ice, lyse the bacteria using a sonicator with a microtip. Holding the tube on ice, sonicate until the turbid suspension of bacteria becomes translucent, which is routinely less than 15 sec at a high output. Centrifuge the sonicate in a microfuge at maximum speed at 4°C for 5 min.
6. Add 50 μl of a 50% slurry of glutathione-agarose beads to each supernatant and mix for 2–5 min at room temperature. Wash the beads three times with cold PBS.
7. Add 50 μl of 2 × SDS sample buffer to the washed beads and to the aliquot of bacteria from step 4. For comparison purposes, include bacteria that received the pGEX vector without the gene for the heterologous protein. Boil the samples and load onto a 10% SDS-polyacrylamide gel. Following the run, stain the gel with Coomassie Blue.

Comments: Transformants expressing the desired fusion protein will not have the prominent GST band at 27.5 kDa. Rather, a larger, unique species should be present which can be predicted knowing the molecular weight of the target protein. If the inserted DNA can be incorporated into the vector in two orientations, the mobilities should be enough to identify the desired construct. If not, conventional restriction endonuclease analysis will be required. If the fusion protein has adsorbed to the glutathione agarose beads, large scale purification can be initiated.

PROTOCOL 11.14 Purification of GST Fusion Proteins

Fusion proteins are often found associated with the pellet in the clarifying centrifugation step after sonication of the bacteria (i.e., step 5 of Protocol 11.13). This method describes the solubilization of the fusion protein from the pellet, which substantially increases the yields of the recombinant proteins (adapted from Grieco et al, 1992).

Materials

LB/amp media

Isopropyl-β-thiogalactopyranoside (IPTG): 100 mM stock

Sonicator

Triton X-100: 10% stock

Sarkosyl™ buffer: 1.5% N-lauroylsarcosine (Sarkosyl™), 25 mM triethanolamine, 1 mM EDTA, pH 8.0

$CaCl_2$: 1 M

Elution buffer: 50 mM Tris-HCl, pH 8.0, and 5 mM reduced glutathione

PBS: See Appendix C

Glutathione-agarose beads

1. Inoculate a colony of the pGEX transformant into 100 ml LB/amp media and grow overnight shaking at 37°C.
2. Dilute the culture 1:10 into 1 liter of fresh LB/amp and split into two 2-liter flasks and grow for 1 h at 37°C.
3. Add IPTG to a final concentration of 0.1 mM and continue incubation another 3–7 h.
4. Harvest the bacteria by centrifugation at 5000 × *g*. Discard the supernatant and resuspend the pelleted bacteria in 10–20 ml of cold PBS.
5. Place the tube on ice and sonicate to lyse the bacteria. Adjust the sonicator output so that disruption occurs within 30 seconds. Avoid excessive sonication because other undesirable proteins may be released.
6. Centrifuge the sonicate at 10,000 × *g* for 10 min at 4°C.
7. Adjust the supernatant to contain 1% in Triton X-100 and incubate on ice. Resuspend the pellet in 8 ml of Sarkosyl buffer and mix for 10 min at 4°C then centrifuge at 10,000 × *g* for 10 min at 4°C.
8. Add Triton X-100 and $CaCl_2$ to the supernatant to final concentrations of 2% and 1 mM respectively.
9. Pool the first supernatant and the supernatant from the sarkosyl-resuspended pellet and apply it to 1 ml of a 50% slurry of glutathione-agarose beads and mix gently at room temperature. (The capacity of the glutathione-agarose is ~8 mg protein/ml hydrated beads.) Wash with 50 ml of cold PBS and pellet the beads with a short, low speed centrifu-

gation in a refrigerated benchtop centrifuge. Repeat the PBS wash two more times. Collect the glutathione-agarose beads with a short, low-speed centrifugation and discard the supernatant.

10. Add 1 ml of elution buffer and mix gently for 2 min.
11. Centrifuge the beads and collect the supernatant. Repeat the elution three times. Keep each supernatant separate and analyze the fractions by SDS-PAGE.
12. Store the eluted fusion protein aliquotted at –70°C.

Comment: The addition of Sarkosyl™, aids in the solubilization and recovery of the recombinant protein that had aggregated and precipitated after cell lysis. The presence of the detergent does not affect the subsequent affinity chromatography and increases yields.

PROTOCOL 11.15 Removing the GST from the Fusion Protein

Using a computer equipped with the appropriate software, inspect the amino acid sequence of the target protein to verify that it lacks cleavage sites recognized by thrombin or factor Xa. Factor Xa recognizes and cleaves at the carboxy side of Arg in Ile-Glu-Gly-Arg. If the target protein contains sites that will be cleaved by thrombin and factor Xa, do not use this method. To decrease potential loss of product due to nonspecific proteolysis, use protease inhibitors in the buffers.

The method given takes into account the potential insolubility of the fusion protein.

Materials

Wash buffer: 1% Triton X-100 in PBS

Thrombin cleavage buffer: 50 mM Tris-HCl, pH 8.0, 0.15 M NaCl, 2.5 mM $CaCl_2$

Xa cleavage buffer: 50 mM Tris-HCl, pH 7.5, 0.15 M NaCl, 1 mM $CaCl_2$

Thrombin

Factor Xa

1. Wash the glutathione-agarose beads containing the adsorbed fusion protein twice with 20 ml of wash buffer.
2. For pGEX2T, wash the beads once with thrombin cleavage buffer. For pGEX3X, wash the beads once with factor Xa cleavage buffer. Resuspend the beads in a small volume, <1 ml of the appropriate cleavage buffer.
3. Estimate the amount of fusion protein bound to the beads by removing a small aliquot of beads and performing SDS-PAGE.
4. For pGEX2T, add thrombin (0.2–1% w/w fusion protein, or try 5 ml (4 NIH units) of thrombin) for 24 h. For pGEX3X add factor Xa (1% w/w fusion protein) and incubate for 1 h at 25°C.
5. Recover the released protein and wash the beads four additional times with one bed volume of wash buffer, saving the eluted fractions. Analyze the eluted fractions by SDS-PAGE.

Comments: Optimizing the growth conditions and altering the induction period can greatly improve the yield of the fusion protein. If the fusion protein appears to be toxic to the bacteria, try delaying the addition of the IPTG (which induces fusion protein synthesis) until the culture is denser and also decrease the time that the bacteria are growing in the presence of IPTG. Conversely, if the bacteria tolerate the fusion protein and it is stable, the yield should be improved by a longer induction time (Gearing et al, 1989).

PROTOCOL 11.16 His-Tag Purification System

A potential solution to the dilemma of purifying a recombinant protein with unknown properties may be the incorporation of a short histidine oligomer at either the COOH or NH_2 terminus of the protein (Hochuli et al, 1988).

QIA*express* (QIAGEN Inc, Chatsworth CA) is based on the selectivity of nickel-NTA (Nitrilo-Tri-Acetic acid) linked to Sepharose CL-6B for proteins with an affinity tag consisting of six consecutive histidine residues (6xHis tag). The 6xHis tag can be placed at either the N- or C-terminus of the recombinant protein allowing it to be positioned away from known active sites. The Ni-NTA resin has a binding capacity of 5–10mg of tagged protein per ml of resin. The affinity of the Ni-NTA resin is very high, enabling the contaminating proteins to be washed away. Tagged proteins can be purified to >95% purity in one step. Insert the coding sequence into one of the QIA*express* expression vectors (QIAGEN), or insert a 6xHis tag sequence into your own expression system.

Prepare a cleared lysate containing the protein of interest with the engineered 6xHis tag and mix it with the Ni-NTA resin. The protein can be present in high or low concentrations under denaturing or native conditions. To remove contaminants, the resin can be washed in either batch or column format. Elution is accomplished by adding imidazole to the buffer to displace the 6xHis tagged protein from the resin. Alternatively, lowering the pH to 5–6 will also cause the 6xHis tagged protein to dissociate from the nickel ions.

The following protocol is presented as an example to illustrate the successful use of the 6xHis tag and Ni-NTA matrix for the purification of recombinant synaptotagmin (Nonet et al, 1993).

Materials

Ni-NTA Sepharose CL-6B resin: (QIAGEN)

Isopropyl- β-thiogalactopyranoside (IPTG): 100mM stock

Sonication buffer: 6M guanidine-HCl, 100mM NaH_2PO_4, 10mM Tris-HCl, pH 8.0, 10mM 2-ME

Wash buffer A: 8M urea, 100mM NaH_2PO_4, 10mM 2-ME, 10mM Tris-HCl, pH 8.0

Wash buffer B: Same as wash buffer A except pH 6.3

Elution buffer: Same as wash buffer A except pH 4.5

Dialysis solution: 0.1% Triton X-100 in PBS

1. Grow up bacteria containing the target gene fused to the 6xHis tag. To produce the fusion protein, induce bacteria with 0.4mM IPTG for 1–4h.

Collect the cells by centrifugation and resuspend them in 10ml of sonication buffer.

2. Sonicate the suspension by four 30sec bursts using a microtip at maximal power.
3. Remove the debris from the total sonicate by centrifugation at $10,000 \times g$ for 10min.
4. Incubate the supernatant with 2ml of the Ni-NTA resin that had been preequilibrated with sonication buffer.
5. Wash the resin and bound fusion protein four times with wash buffer A by centrifugation at $500 \times g$ for 30sec followed by four washes with wash buffer B.
6. Elute the fusion protein with 3×2ml volumes of elution buffer.
7. Dialyze the eluate versus PBS, 0.1% Triton X-100.

Comments: The 6xHis tag rarely affects function of the purified proteins. It only contributes 0.72kDa to the fusion protein. It is therefore not necessary to remove it in order to recover biologically active proteins. An alternate protocol for preparing the bacterial extract is described by Nuoffer et al (1994).

Some proteins, particularly multimeric proteins formed from 6xHis-tagged monomers, may not elute using 250mM imidazole. This occurs when several 6xHis tags from each multimeric complex bind to nickel ions attached to the Ni-NTA resin. In this case, use increasing concentrations of imidazole up to 1M.

Alternatively, proteins can be eluted by adding EDTA (50–100mM) to the elution buffer. EDTA chelates and strips the nickel ions from the resin, freeing the bound proteins. Free EDTA and EDTA-nickel complexes can be removed from the eluate by a variety of techniques including dialysis and gel filtration. Since EDTA treatment strips the nickel from the resin, the Ni-NTA resin must be recharged with nickel before reuse (McGarvey, 2002).

The following reagents do not affect the binding of the 6xHis tag to Ni-NTA resin: 10mM 2-ME, 1M NaCl, 30% ethanol, 50% glycerol, 0.05% Triton X-100, 0.05% Tween 20, 0.01% Sarkosyl.

FLAG® Biosystem

The FLAG® epitope, an eight amino acid peptide (N-Asp-Tyr-Lys-Asp-Asp-Asp-Asp-Lys-C), is useful for immunoaffinity purification of fusion proteins. This system enables the fast, efficient purification of any protein fused to the FLAG peptide due to efficient expression and secretion of FLAG fusion proteins by the pFLAG-1 expression vector, and the fact that the small FLAG peptide does not affect biological activity. The FLAG system has been used for detection of proteins expressed in a variety of cell types, including bacterial (Su et al, 1992), yeast (Hopp et al, 1988) and mammalian cells (Kunz et al, 1992). FLAG tagged proteins can be detected on Western blots and used in immunoprecipitation protocols.

The FLAG sequence is hydrophilic and includes the cleavage recognition sequence (Asp-Asp-Asp-Asp-Lys) of enterokinase (Hopp et al,

1988). The FLAG peptide can be fused with virtually any protein for expression by the pFLAG-1 vector. (FLAG® and Anti-FLAG® are trademarks of Immunology Ventures.) The FLAG fusion protein can be rapidly purified with the AntiFLAG® M1 affinity column that binds the FLAG epitope in a calcium-dependent manner, allowing gentle elution by the addition of chelating agents. The FLAG can then be easily removed with enterokinase, and detected with the anti-FLAG Ml monoclonal antibody. The FLAG peptide is present on the protein surface for easy antibody recognition and subsequent removal.

A limitation of the anti-FLAG M1 MAb is that it is specific for the N-terminal FLAG fusion proteins produced following removal of a secretion signal sequence in which the first amino acid of the FLAG sequence is at the very N-terminus of the fusion protein. In an effort to extend the usefulness of the FLAG system, new expression vectors have been developed from the original pFLAG-1 expression vector. A second generation monoclonal antibody (anti-FLAG M2) has been developed for use in the affinity purification of FLAG fusion proteins (Brizzard et al, 1994). Bound N-terminal Met-FLAG and C-terminal FLAG fusion proteins can be eluted by competition with FLAG peptide.

c-myc

The human c-myc epitope (EQKLISEEDL), engineered to be expressed at the N- or C-terminus of the target protein, can be used for easy detection of the target protein from mammalian, insect, yeast, *Arabidobsis* or bacterial cell lysates. The murin anti-c-myc antibody 9E10 is commonly used in cell biology and protein engineering (Evan et al, 1985).

Strep-tag II™

The Strep-tag II (Sigma-Genosys) allows bacterial expression of foreign proteins in *E. coli* followed by a single-step purification under mild conditions. The expression vectors pASK-IBA allow expression of N- or C-terminal fusion proteins that can then be purified by immobilization on a StrepTactin column. The method is based on a short peptide sequence of eight amino acids (Strep-tag) (Schmidt and Skerra, 1993), which binds strongly to streptavidin. The Strep-tag peptide can be used for detection and quantitative purification of the recombinant proteins. Induction of expression is followed by protein secretion into the periplasmic space of the host bacteria where the signal peptide is removed. The Strep-tag containing recombinant protein, contained in the periplasmic extract, is then purified in a single step by affinity chromatography on immobilized streptavidin. Competitive elution of the fusion protein is achieved by adding small amounts of desthiobiotin, a biotin analog, to the elution buffer. Due to its small size, the Strep-tag peptide need not be removed for most applications and can be used in an ELISA or Western blot assay.

Maltose Binding Protein (MBP) Fusion Proteins

The gene of interest is inserted into a pMAL vector (New England Biolabs) downstream from the *malE* gene that encodes the maltose binding protein (MBP) (Ausubel et al, 1993, New England Biolabs catalogue). The pMAL-2 vector includes the *malE* signal sequence that directs the fusion protein through the cytoplasmic membrane to the periplasm, allowing recovery of the expressed protein from the periplasmic space. This results in the expression of large amounts of the MBP-fusion protein which is then purified by a one step affinity chromatographic procedure using an amylose resin. The fusion protein is eluted from the affinity matrix in purified form with maltose (Kellerman and Ferrenci, 1982). The purified fusion protein is cleaved with factor Xa and the target protein is separated from MBP by passing the mixture over an amylose column, retarding the MBP and letting the target protein flow through. Expression from the pMAL vectors can yield up to 100 mg fusion protein from a one liter culture.

The vectors also include a sequence coding for the recognition site of factor Xa that cleaves at the carboxy side of ARG in Ile-Glu-Gly-Arg. This modification allows the protein of interest to be cleaved from MBP after purification, eliminating any vector-derived residues from the target protein (Nagai and Thogersen, 1987). These steps are illustrated in Figure 11.3 for the purification of paramyosin.

S Tag™ System

Novagen Inc. introduced the S Tag protein tagging system based on the interaction of the 15 amino acid S Tag peptide with the 104 amino acid S-protein ($K_d = 10^{-9}$M) derived from pancreatic ribonuclease A. The S-protein-S-peptide complex is enzymatically active, whereas neither component displays activity by itself (Richards and Wyckoff, 1971).

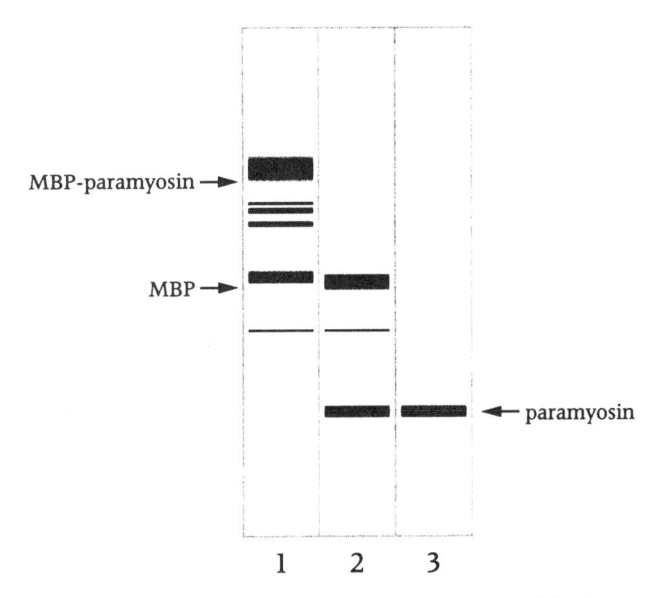

FIGURE 11.3 SDS-gel of fractions from the purification of MBP-paramyosin. Lane 1: purified fusion protein eluted from amylose column with maltose. Lane 2: purified fusion protein after factor Xa cleavage. Lane 3: paramyosin fragment eluted from the second amylose column. (Adapted with permission from the New England Biolabs Catalog Copyright © 1996/97).

These characteristics make the S Tag peptide ideal for recombinant proteins. Any fusion protein carrying the S Tag peptide can be conveniently detected and purified using the S-protein (Kim and Raines, 1993). The unique property of reconstituting enzymatic activity by the S Tag peptide-S-protein interaction allows sensitive quantitative measurement of any fusion protein by a simple assay.

PinPoint™

DNA coding for the protein of interest is cloned into a PinPoint™ vector (there are three that differ only in the reading frame) downstream of a sequence encoding a peptide that becomes biotinylated *in vivo*. The tag, a 122 amino acid subunit of the transcarboxylase complex from *Propionibacterium shermanii* adds about 13kDa to the molecular weight of the protein of interest. Specifically, amino acid residue 88 (lysine) of the tag is biotinylated. The PinPoint™ purification system is based on the interaction of the biotinylated fusion protein with monomeric avidin SoftLink™ soft release resin (Promega). A schematic diagram of the expression and purification of the target protein in the PinPoint™ system is shown in Figure 11.4.

The cell lysate is passed over an avidin affinity column, binding the biotinylated fusion protein which is then eluted after washing using mild, nondenaturing conditions (5mM biotin). A factor Xa protease cleavage site is located between the biotinylated peptide and the target

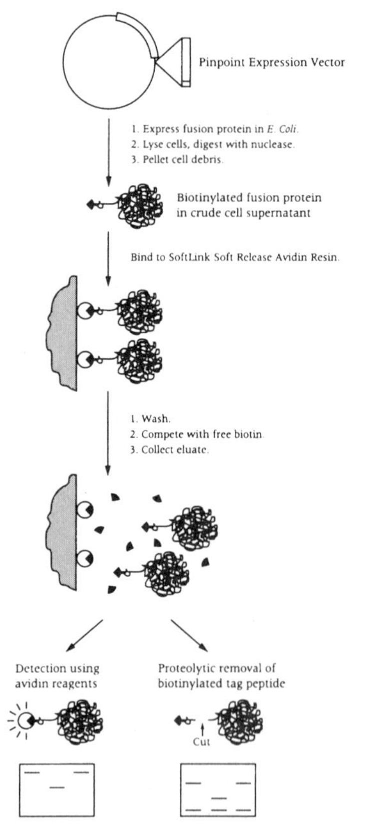

FIGURE 11.4 Recombinant protein expression and purification using the PinPoint™ purification system. (Adapted with permission of the Promega Corporation).

protein, allowing the cleavage and purification of the protein of interest.

To monitor purification, Streptavidin Alkaline Phosphatase can be used to detect the biotinylated fusion protein in a Western blot format. Following factor Xa cleavage, the biotin tagged peptide can be removed from the solution that contains the target protein using column chromatography. The system generally yields 1–5 mg of recombinant protein per liter of culture.

PinPoint fusion proteins have been expressed and biotinylated in *E. coli* cell lines JM109 and HB101. Theoretically, the biotin ligase should be conserved in all *E. coli* strains.

HA Tag

Small peptides corresponding to sequences from the hemagglutinin (HA) of the influenza virus (HAI) were used to raise antibodies that reacted with the HAI protein itself (Green et al., 1982). A short HA derived 9 amino acid peptide, Tyr-ProTyr-Asp-Val-Pro-Asp-Tyr-Ala, was shown to contain the complete antigenic determinant (Wilson et al, 1984). A vector was constructed which encoded the HAI epitope (Field et al, 1988). Monoclonal antibody directed against the HA epitope is commercially available (Berkeley Antibody Company). The HA peptide can be routinely synthesized. The fusion protein is designed to contain the HA epitope extending from the amino-terminal end of the target protein (Bennett et al, 1993) or the carboxy terminus (Wrana et al, 1994). The HA epitope can be utilized for purifying the target protein or for following the fate of the target protein.

Staphylococcal Protein A and ZZ

Staphylococcal protein A and its derivative ZZ have been used successfully in several different hosts, such as bacteria (Nilsson and Abrahmsén, 1990), yeast (Zueco and Boyd, 1992), and insect cells (Andersons et al, 1991). Elution is routinely accomplished by lowering the pH of the buffer to pH 3. Competitive elution strategies based on competitor proteins can also be developed. The human polyclonal IgG used as the ligand in this system can be replaced by recombinant F_c fragments, thus avoiding the use of human serum protein in the purification protocol.

CCXXCC

The bis-arsenical fluorescein dye FlAsH, specifically recognizes short α-helical peptides containing the sequence CCXXCC (Thorn et al, 2000). Proteins tagged with this cysteine-containing helix bind specifically to FlAsH resin and can be eluted in a fully active form. The interaction is readily reversed by incubation with small dithiols such as dithiothreitol. The introduction of the tag on the C- or N-terminal side of the target protein did not change the expression level or the solubility of the protein when expressed in *E. coli*.

FlAsH affinity chromatography is a highly specific protein purification method. The primary amine on the β-alanyl FlAsH readily reacts with N-hydroxysuccinamide (NHS) agarose beads to give a stable covalent linkage (Thorn et al, 2000).

Intein-Mediated Purification with an Affinity Chitin-Binding Tag (IMPACT^{TM})

Protein splicing involves a precise excision of an internal segment, the intein, from a protein precursor and a subsequent ligation of the flanking regions resulting in the production of two polypeptides. Since the initial discovery of protein splicing, more than 40 inteins have been identified (Perler et al, 1997). Inteins are in-frame sequences that are bounded by Cys, Ser at the N termini and Asn at the C termini and with few exceptions have His as the penultimate residue. Inteins undergo self-catalyzed cleavage from a larger protein precursor by a self-catalytic protein splicing mechanism (Xu et al, 2000). *In vivo,* the resulting spliced ends undergo subsequent fusion.

The IMPACT system utilizes the inducible self-cleavage activity of engineered protein splicing elements (inteins) to purify recombinant proteins in a single affinity step (Chong et al. 1998). The steps to protein purification using this system are presented in Figure 11.5. The gene encoding the target protein is inserted into the multiple cloning site (MCS) of a pTYB vector to create a fusion between the C-terminus of the target protein and the N-terminus of a gene encoding the intein. The DNA encoding a small 5 kDa chitin-binding domain from *Bacillus circulans* is included as part of the C-terminus of the intein to facilitate the affinity purification of the fusion protein. Crude extracts of *E. coli*

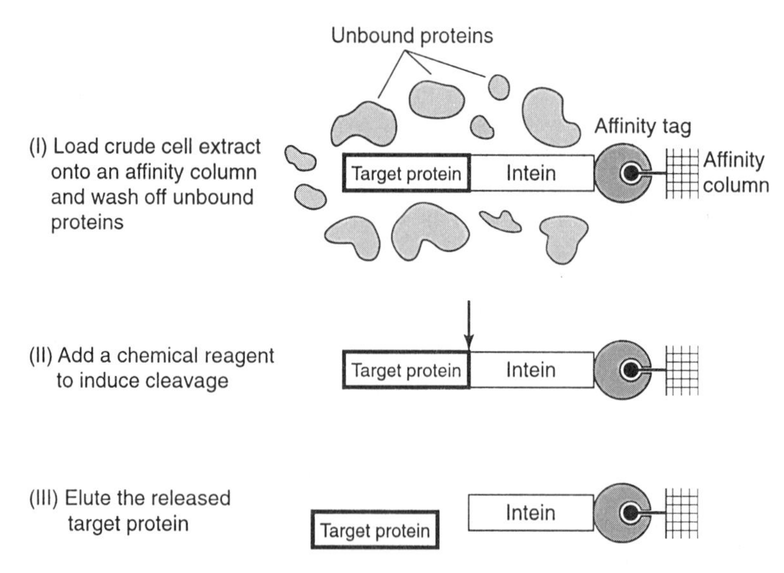

FIGURE 11.5 Principles of intein-mediated protein purification. A target protein is fused to the N-terminus of the intein which is in turn fused to an affinity tag (the chitin binding domain). The fusion protein in a crude cell extract is purified by adsorption to the chitin affinity column (I) The intein is induced to undergo on-column self-cleavage (arrow) (II). The target protein is specifically released from the column and is eluted in purified form (III). Reprinted with request from Chong et al, 1997.

induced to express the fusion protein are passed through a chitin column. The fusion protein binds to the chitin column while contaminants are washed out. The fusion protein is then induced to undergo an intein-mediated self-cleavage on the column during an overnight incubation at 4°C in the presence of either DTT or 2-mercaptoethanol. The target protein is released while the intein-chitin binding domain of the fusion protein remains bound to the column. Advantages of this system include the simultaneous purification and cleavage of the fusion protein at low temperature and the release of the fusion partner without the use of proteases. This method is sold in kit form by New England Biolabs as the IMPACT I System.

Green Fluorescent Protein (GFP)

GFP has become a valuable tool for visualizing various molecular events in the cell. Originally reported in 1994 from the jellyfish *Aequora victoria*, protein fusions with GFP are routinely used in transient transfections to localize proteins in live cells (Chalfie et al, 1994). Derivatives of GFP have become popular as protein labels, indicators of gene expression and more recently as visible reporters of pH and other conditions in living cells, thus the term "live biochemistry". Protein activities and interactions can be imaged and localized within a single cell, allowing correlation with the cell cycle, migration and morphogenesis. The GFP-tagged proteins can easily be monitored in both live and fixed cells by fluorescence microscopy. The GFP technology has advanced so that derivatives of GFP can be used to follow localization of two tagged proteins in the same cell (Ellenberg, 1999).

TAP System

The TAP system is composed of a multi-tag, tandem affinity purification (TAP) tag (Puig et al, 2001). It consists of the protein of interest, a calmodulin binding peptide, and the immunoglobulin binding domain of protein A for immobilization. The domains are separated by a tobacco etch virus (TEV) protease cleavage site that consists of a seven amino acid high specificity recognition site. Cleavage occurs between the conserved glutamine and serine in E-X-X-Y-X-Q-S, where X can be various amino acids but not all are tolerated (Parks et al, 1994). The tagged proteins are initially bound to a solid support modified by immunoglobulins, recovered by TEV proteolysis and then bound to a calmodulin column from which they can be selectively eluted by increased Ca^{2+}. TAP tags significantly reduce background noise, but probably result in the loss of some of the weaker binding partners during the purification procedure (Rigaut et al, 1999).

D. Expression of Foreign Proteins in Eukaryotic Cells

Expression and Isolation of Recombinant Proteins from Yeast

Yeast, one of the simplest eukaryotic organisms, contains all the major membrane-bound subcellular organelles found in higher eukaryotes. Overexpression of most proteins in yeast is full length, and most posttranslational modifications are similar to those of higher eukaryotes. For these reasons, *S. cerevisiae*, baker's yeast, has been referred to as the *E. coli* of eukaryotic organisms (Mak et al, 1989). Because of its similarity to higher eukaryotes, yeast is an attractive model for studying biochemical processes. However, the presence of a highly complex rigid cell wall makes this organism difficult to lyse.

The *Pichia pastoris* expression system offers economy, ease of manipulation, the ability to perform complex posttranslational modifications and high expression levels. *P. pastoris*, a methylotropic species, uses methanol as a carbon source. The system does not require tissue culture hoods or CO_2 incubators and the inducer, methanol, and other reagents are relatively inexpensive. Moreover, with *Pichia* the theoretical upward limit is grams per liter (Morrow, 1995).

P. pastoris can be easily manipulated at the molecular genetic level, it can express foreign proteins at high levels both intracellularly and extracellularly and it can perform many higher eukaryotic protein modifications including glycosylation, disulfide bond formation and proteolytic processing. The *P. pastoris* expression system uses an extremely efficient, tightly regulated promoter from the alcohol oxidase I gene (*AOX1*) that is used to drive the expression of the foreign gene (Cregg and Madden, 1988). The *AOX1* promoter is strongly repressed in cells grown on glucose and most other carbon sources, but is induced over 1000-fold when the cells are shifted to a medium containing methanol as a sole carbon source. Although expression of heterologous proteins can be done in shaker-flasks, protein levels are much higher in fermenter cultures. In fermenters, where parameters can be controlled, it is possible to achieve ultra-high densities >100 g/L dry cell weight; >400 g/L wet cell weight (Lin Cereghino et al, 2002). The move from low-density shaker flask to high-density, small-volume fermenter will require re-optimization of culture conditions. Production of foreign proteins in fermenter cultures of *P. pastoris* is a three-stage process. Procedures ranging from strain construction to purification of the recombinant protein are presented (Higgins et al, 1998).

PROTOCOL 11.17 Preparation of Protein Extracts from Yeast

Protein purification from yeast is still time-consuming and labor intensive due to the need to obtain the protein in soluble form. Three methods are commonly used for releasing proteins from yeast. The yeast cell wall can be enzymatically digested. This treatment causes the cells to become spherical in shape (spheroplasts). The conversion of yeast to spheroplasts by zymolase treatment is followed by osmotic shock lysis and Dounce homogenization. Alternatively, cells can be frozen immediately in liquid nitrogen and then lysed by grinding in a Waring-type blender in the presence of liquid nitrogen (Sorger and Pelham, 1987). The method, described in greater detail below (Ausubel et al, 1993), involves the mechanical disruption of the yeast cells with a glass bead mill (Klekamp and Weil, 1982). It is important that the beads be the correct size and the temperature be kept between 0–4°C.

Materials

Disruption buffer: 20 mM Tris-HCl, pH 7.9, 10 mM $MgCl_2$, 1 mM EDTA, 5% glycerol, 1 mM DTT, 0.3 M $(NH_4)_2SO_4$, protease inhibitors

Glass beads: Prepare glass beads (0.45 to 0.55 mm) by soaking them for 1 h in concentrated nitric acid. Rinse the beads thoroughly with water then dry the beads in a baking oven. Cool the beads to room temperature and store at 4°C until needed.

Bead Beater

1. Grow the yeast in appropriate media to an OD_{600} of between 1–5 (mid-log phase) with vigorous shaking.
2. Harvest the cells by centrifugation, 5 min at 1500 × *g* at 4°C. Preweigh the centrifuge tubes.
3. Determine the packed cell volume by converting from the wet weight of the harvested yeast pellet. Compare the weight of the harvested cells to that of the preweighed centrifuge tubes or bottles. The weight of the packed cells is approximately equal to the packed cell volume (1 g = 1 ml). One liter of yeast strain BJ926 (a diploid strain) with OD_{600} = 1.0 yields a cell volume of ~2–3 ml.
4. Resuspend the cells in 4 vol of ice-cold H_2O and centrifuge at 4°C for 5 min at 1500 × *g*. Discard the supernatant.
5. Resuspend the pellet in 1 vol of disruption buffer.
6. Mix the cell paste with 2 vol of glass bead disruption buffer, and add 4 vol of prechilled, acid-washed glass beads.
7. Transfer the suspension to an appropriately sized Bead Beater vessel and add disruption buffer to almost fill the vessel. The volume of buffer required to fill the vessel should not exceed 1 cell-suspension volume. Attach the blade and cap assembly, ensuring that all air is excluded from the vessel to prevent foaming.
8. Grind at high speed for 60 sec and then incubate on ice 1–2 min. Repeat three to five times. Monitor the amount of cell breakage microscopically.
9. Allow the glass beads to settle and decant the supernatant. Rinse the vessel with four vol of disruption buffer. Allow the beads to settle and pool the supernatants.
10. Centrifuge the pooled supernatants at 4°C for 60 min at 12,000 × *g*. Collect the supernatant, which represents the crude extract. For long term storage, aliquot the crude extract into small tubes, quick freeze in liquid nitrogen and store at –80°C.

Comments: If your lab is not equipped with a Bead Beater, a screw-capped centrifuge tube and a vortex will be an adequate substitute. Transfer the cell suspension (from step 6 above) to an appropriately sized screw-capped centrifuge tube. The suspension should not fill more than 60–70% of the capacity of the tube. Mix the suspension by vortexing at maximum speed for 30 to 60 sec at 4°C. Place the tube on ice for 1 to 2 min. Repeat the procedure three to five times. Check for cell breakage microscopically. Proceed to step 8 above.

Certain mutant strains of *S. cerevisiae* have fragile cell walls. These cells grow in media containing osmotic stabilizers like 10% sorbitol or

1.5% NaCl but lyse spontaneously on transfer to a hypotonic solution (Broker, 1994). The strain 4STLU has been found to be the most effective strain with respect to both the expression rate and also the extraction yield.

The extraction yield of soluble protein will be ~20% compared with the 100% value obtained by using a glass bead mill.

Expression of Proteins in Insect Cells Using Baculoviral Vectors

Baculoviruses have become a popular system for overproducing recombinant proteins in eukaryotic cells. The baculovirus system has some advantages compared to bacterial expression systems. The insect cell line is relatively easy to infect, transfect, and culture. The insect cells infected by the baculoviruses accomplish most eukaryotic post-translational modifications using many of the protein modification, processing and transport systems present in higher eukaryotic cells. Although insect cells cannot complete protein glycosylation, the sugar chain will end in a mannose and not be processed to contain the penultimate galactose or terminal sialic acid. It is possible to obtain large amounts of recombinant protein, which remains soluble in insect cells, in contrast to insoluble proteins produced in bacterial expression systems.

Baculoviruses (of the family Baculoviridae) are insect-specific viruses that have received much interest due to their usefulness as biopesticides. They have also found widespread use as recombinant protein expression vectors. A commonly used viral isolate is the *Autographa californica* nuclear polyhedrosis virus (AcMNPV). Insect cells that are infected with this virus accumulate high levels of the polyhedrin protein in the nucleus. The polyhedrin gene is expressed to very high levels during the normal viral life cycle, but in cell culture it is dispensable, allowing its replacement by almost any heterologous gene of interest. The gene of interest is inserted into baculovirus under the control of the polyhedron promoter. Since the foreign gene will also be controlled by the polyhedrin promotor, high levels of the foreign gene product will be produced.

The baculovirus infects cultured lepidopteran insect cells, the most commonly used laboratory line being Sf9 cells, established from the army worm *S. frugiperda*. The baculovirus system is lytic. The baculovirus multiplies in the infected cells and kills them, causing them to break open and spill their contents into the surrounding medium. The recombinant protein is therefore produced in batches, with repeat infection required for each batch.

Companies have been able to create stable cell lines using plasmids (rather than baculovirus) to deliver genes into insect cells. In this manner the production of the recombinant protein is continuous rather than batch. Although the stable system has the advantage of not having to reinfect the cells, the expression levels have been reported to be less than in the lytic system.

Recombinant fusion proteins have been reported to be expressed at levels ranging from 1–500 mg/L. Purification of the recombinant protein usually requires fewer steps compared to purification from mammalian cell culture, tissues, or bacterial cultures due to the lower abundance of proteases in insect cells and the high concentration of recombinant versus cellular protein synthesis. The baculovirus can accommodate large fragments of foreign DNA giving this system the capability of expressing a larger recombinant than possible in a bacterial system. Finally, baculoviruses are noninfectious to vertebrates and do not pose a health hazard.

Because baculoviruses infect invertebrate cells, the protein processing reactions differ from those in vertebrate cells. Some posttranslational modifications will be identical to those that occur in higher eukaryotic cells but there may be exceptions. Insect cells have been shown to be capable of myristoylation, palmitoylation and phosphorylation. Cleavage of signal sequences, removal of hormonal prosequences and polyprotein cleavages have also been reported. Protein targeting seems to be conserved between insect and vertebrate cells. Proteins can be secreted and localized to the nucleus, cytoplasm or plasma membrane.

The use of the Baculovirus system requires experience in molecular biology. A rough experimental outline and the approximate time frame is presented below.

- Establish the Sf9 cell line (a clonal isolate of *Spodoptera frugiperda* cells). This will take 3–4 weeks.
- While establishing the cell culture, select an appropriate transfer plasmid and perform the molecular biological manipulations of inserting the gene into the vector, validating fidelity, and preparing the plasmid DNA.
- Produce a recombinant virus. Cotransfect *S. frugiperda* cells with AcMNPV viral DNA and plasmid DNA. This will take ~5 days.
- Purify the recombinant virus from transfection supernatants by plaque assay. This will take 3–4 weeks.
- Propagate the virus and confirm the recombinant and determine the steady state mRNA and protein levels.

Disadvantages of the Baculovirus system are that the setup is time-consuming, laborious and expensive. For an in-depth treatment of the Baculovirus system, the interested reader is encouraged to read Summers and Smith, (1987). Excellent technical help is provided by Invitrogen Corp., which supplies a baculovirus expression system called MAXBAC™.

Expression of Foreign Proteins in Mammalian Cells

The introduction of genes into eukaryotic cells has received an increasing amount of attention for studying gene regulation, DNA replication, and cellular transformation at the molecular level. Improved gene transfer into eukaryotic cells has also increased the feasibility of gene transfer therapy. Mammalian systems offer advantages for the expression of higher eukaryotic proteins that are normally available in limited quantities. Expression in mammalian cells is practical for small and medium scale work. The advantages of mammalian cells as hosts for the expression of gene products obtained from higher eukaryotes are that the signals for processing and secretion are usually recognized. The recombinant proteins will faithfully accumulate in the correct cellular compartment. In addition, the physiologic consequences of the foreign protein can be evaluated.

Efficiency of transient and stable transfection varies widely between different cell types, cell lines and promotors used. Adequate gene expression in one cell line may not predict what will happen in another cell line.

A prime consideration for choosing a particular mammalian expression system is whether the cell line chosen will be appropriate for expression of the gene of interest. When evaluating which approach to take in expressing a gene, consider the goal of the expression project. If expression is required to demonstrate functional activity or to characterize an activity, then transient expression in COS cells is recommended. If a large quantity of protein >1 mg is required, then stable transfection is preferred.

Transfection: Expression of Recombinant Proteins in Eukaryotic Systems

Cells that are transiently transfected will express the newly acquired DNA and synthesize the encoded protein over a period of several days to several weeks. Production of the recombinant protein usually peaks from 24–72 h after transfection. Eventually, because cells containing the foreign DNA grow more slowly, they are lost from the population. Efficiency of expression from transient transfection depends upon: (1) the number of cells that take up the transfected DNA, which depends on the length of time that the cells are incubated in the presence of the foreign DNA; (2) the gene copy number; and (3) the quality of the DNA and the solutions. Transfection efficiency is referred to as the percentage of the total number of transfected cells that express the transgene.

Transient transfections can be performed on cells prior to a stable transfection to ensure the protein of interest can be expressed in the chosen cell line. Transient transfections are frequently used to verify that the transfected gene is indeed being expressed. This can be accomplished by Western blot or immunoprecipitation. Transient transfections can be used to study the effects of engineered mutations on gene activity or protein function. Transient transfections can also be used to screen and isolate genes from cDNA libraries that were put into mammalian expression vectors and transfected into cell lines. Cells are selected, isolated and cloned based on their ability to express a particular activity in the host cells.

Unfortunately, there are also drawbacks associated with transient transfections. It is difficult to scale up the production of the foreign gene product because only a portion of the cells will have taken up the foreign DNA. The high copy number produced in a transient transfection could be lethal to the host cell.

COS cells have been used successfully in transient transfection protocols. The COS cell line was derived from African green monkey kidney cells that were transformed with an origin-defective SV-40 (Gluzman, 1981). The COS cells express high levels of SV-40 large T antigen which is required to initiate viral DNA replication at the origin of SV-40. This allows the copy number of transfected plasmids, which contain the SV-40 origin of replication, to be >100,000 copies per cell. Each transfected COS cell will express several thousand to several hundred thousand copies of the protein 72h posttransfection.

In most cases the expressed protein produced in COS cells is biologically active. Although COS cells are able to carry out posttranslational modifications, they may not modify the expressed protein in exactly the same way as the cell that normally produces it.

Mammalian expression vectors will contain:

- An SV-40 origin of replication for amplification to high copy number in COS cells.
- An efficient promotor element for high-level transcription initiation.
- mRNA processing signals, cleavage and polyadenylation signals, and intervening sequences.
- Polylinker sites also referred to as a multiple cloning site (MCS).
- Selection markers.
- Plasmid backbone permitting the propagation in bacterial cells.
- Inducible expression system.

Specific promotor sequences are required for induction based on exposure to a variety of stimuli like β-interferon, heat shock, heavy metal ions, and steroids. If the effect of an expressed protein on a cellular process is being studied, one must determine: (1) that the inducing stimulus does not interfere with the cellular process; (2) the induction factor (fold increase) and the difference between the basal and induced level of protein expression; and (3) the maximal level of achievable expression.

Two frequently used transfection protocols are described below. Other methods that will not be descibed are Lipofection (Felgner et al, 1987), electroporation (Toneguzzo et al, 1988; Andreason and Evans, 1989), and viral mediated gene transfer (Muzyczka, 1989).

PROTOCOL 11.18 Transfection of DNA into Eukaryotic Cells with Calcium Phosphate

A precipitate containing $CaPO_4$ and DNA is formed by slowly mixing a HEPES-buffered saline solution with a solution containing $CaCl_2$ and DNA. The precipitate adheres to the surface of cells. ~10% of the cells will take up the DNA through an as yet undetermined mechanism.

Materials

$CaCl_2$: 2.5M

15ml conical centrifuge tubes

HEPES buffer

PBS: see Appendix C

HEPES buffered saline 2 × (HBS): 16.4g NaCl, 11.9g HEPES, 0.21g Na_2HPO_4, 800ml H_2O. Titrate to pH 7.05 with 5N NaOH, add H_2O to 1000ml and filter sterilize. **The pH of this solution is critical.** The optimal range is 7.05–7.12. Test the solution for transfection efficiency by mixing 0.5ml of the 2 × HBS solution with 0.5ml of 250mM $CaCl_2$ and vortex. A fine precipitate should form that is visible microscopically. If the precipitate does not form there is something wrong. Store aliquots frozen at –20°C in 50ml.

1. Split the cells into 10cm tissue culture plates the day before transfection. The optimal density of cells is that which will produce a near confluent monolayer when the cells are harvested or split into selective media 48h following the initiation of transfection. Feed the cells with 9ml of complete media 2–4h before beginning the protocol.
2. Ethanol precipitate the DNA to be transfected and air-dry the pellet in a tissue culture hood. Resuspend the pellet in 450µl of sterile water and add 50µl of 2.5M $CaCl_2$ The amount of DNA optimal for transfection varies from ~1.0–50µg/10cm plate.
3. Place 500µl of 2 × HBS in a 15ml conical centrifuge tube. Bubble the 2 × HBS solution and add the DNA/$CaCl_2$ solution drop by drop. Immediately vortex the solution.
4. Allow the precipitate to stand for 20min at room temperature.
5. With a Pasteur pipet, distribute the precipitate evenly over a 10cm dish and then gently agitate the plate.
6. Incubate for 4–16h under standard growth conditions. Remove the medium, wash the plate twice with PBS and refeed the plate with 10ml of complete medium. The amount of time that the precipitate is left on the cells will be determined empirically.
7. For transient transfections, harvest at desired, optimal times, usually 48–72h. For stable transfectants, allow the cells to double twice before plating them on selective medium.

PROTOCOL 11.19 Glycerol Shock

A glycerol or DMSO shock step has been reported to increase the transfection efficiency. This can be used in combination with Protocol 11.18.

Materials

Glycerol shock solution: 10% (v/v) of glycerol or DMSO in complete medium, filter sterilize

PBS

1. Incubate cells for 4–6h as in step 6, Protocol 11.18.
2. Remove the medium and add 2.0ml of glycerol shock solution (alternatively, a 10% DMSO solution can be used).

3. Let the plate stand at room temperature for 3 minutes.
4. Add 5ml of PBS to the plate, swirl gently and remove shock media. Excessive exposure to glycerol or DMSO will kill cells. Therefore the PBS dilution step is important.
5. Wash the plate twice with PBS and then feed the cells with complete medium.

Comments: The HEPES based technique has been used to analyze replication and promotor function using transient transfection protocols in which cells are harvested 48–60h after the start of transfection. Calcium phosphate transfection is more efficient than DEAE-dextran mediated transfection (Protocol 11.20) for formation of stable cell lines. Electroporation (Wong and Neumann, 1982) and liposome mediated transfection protocols (Felgner et al, 1987) can also be used to produce stable cell lines.

Two main reasons for unsuccessful transfections are that the HBS may no longer be at optimum pH, and the $CaCl_2$ may have deteriorated during prolonged storage. If transfections do not work, freshly prepare both solutions.

PROTOCOL 11.20 Transfection Using DEAE-Dextran

Although this method is simpler and more reproducible than the calcium phosphate method, it is not useful for the production of stably transfected cell lines.

Materials

Nu-serum: (Collaborative Biomedical Products, Becton Dickinson)
DEAE-Dextran: 10mg/ml in TBS
TBS: see Appendix C
PBS: see Appendix C
DMSO solution: 10% DMSO in PBS

1. Perform the transfection on cells that are 30–50% confluent after ~3 days of growth.
2. Precipitate the desired plasmid DNA (~4µg/plate) in a microfuge tube. Remove the ethanol and allow the pellet to air dry in the tissue culture hood.
3. Resuspend the pellet in 40µl of TBS. (The amount of plasmid that will produce the best result is determined experimentally. If the same DNA is used to transfect more than one plate, precipitate the DNA in one tube and then take the appropriate amount from the resuspended pellet. This avoids multiple precipitations of small quantities of identical DNA.)
4. Aspirate the medium from the plate and wash once with PBS. For a 10cm dish, add 4ml of complete DMEM made up with 10% Nu-serum. (Nu-serum has been found to be better than calf serum or fetal calf serum in allowing the cells to tolerate DEAE-dextran for an extended incubation.)
5. Add resuspended DNA to 80µl of prewarmed DEAE-dextran solution. Add the DNA slowly while shaking the tube.

6. Add the DNA/DEAE-dextran drop-by-drop to the plate. Disperse the drops uniformly over the plate and swirl gently. The final concentration of DEAE-dextran is 200 μg/ml. When the experimental design calls for a final volume of <60 μl, do not add drop-by-drop; rather, mix in a sterile tube with 4 ml of complete medium. Pipet the mixture up and down several times, then add it to the plate.
7. Aspirate the DNA/DEAE-dextran medium. The cells might look unhealthy.
8. Shock the cells by adding 5 ml of the DMSO solution. Incubate for 1 min at room temperature. Aspirate the DMSO solution and wash the plate with 5 ml of PBS. Aspirate off the PBS and add 10 ml of complete medium.
9. Incubate the plate under standard conditions and analyze at the appropriate time. Analysis could include:
   - Preparation of cell extract to assay for a reporter gene.
   - Assay the media for the presence of a secreted protein.
   - Prepare RNA
   - Perform immunohistochemistry.

Comments: For final DEAE-dextran concentrations of <50 μg/ml use a 1 mg/ml stock solution of DEAE-dextran.

Optimize the protocol for a given cell type, concentration of DNA, concentration of DEAE-dextran, duration of transfection and the time course of expression of the transfected gene.

It has been reported that chloroquine diphosphate increases transfection efficiency. If chloroquine is to be included, add it to a final concentration of 100 μM in the medium and then swirl the dish. Do not leave the chloroquine on the cells for more than 4 h.

The peak of plasmid replication in transfected COS cells occurs 48–72 h posttransfection. Protein production starts to accumulate 24 h posttransfection but peaks 72 to 96 h posttransfection. When expressing cell-surface or cytoplasmic proteins, harvest the cells 72 to 96 h posttransfection. Transfected cells continue to produce protein for up to one week posttransfection, and the conditioned culture media should be harvested a week posttransfection.

PROTOCOL 11.21 Stable Transfections

Analysis of function can be performed using mammalian cell lines that have been transfected with the gene of interest in a stable, integrated form. Since one in 10^4 cells in a transfection will stably integrate DNA, a selection marker is necessary to permit the isolation of the stable transfectant. The gene of interest is transfected into a cell line along with a gene that expresses a selectable marker. The cells are allowed to grow under selection for ten doublings before individual colonies are picked and expanded into cell lines. Popular selection markers are the neomycin phosphotransferase gene encoding resistance to G418 and the hygromycin phosphotransferase gene, encoding resistance to hygromycin.

Stable and high expression of transgenes depends on the strength of the promoter, the processing of the mRNA and the nature of the transgene. The choice of the right promoter for each cell type for optimal expression of the transgene is an important consideration. The expression of the transgene may depend more on the promoter than the method of gene transfer.

1. Make sure that the cell line to be transfected is able to grow as isolated colonies. The day before transfection, split a confluent dish of COS or CHO cells 1:15. In theory, any cell line can be used.
2. Transfect the desired gene using 5 to 10 μg of plasmid DNA per 10 cm dish. Use at least a 5:1 molar ratio of plasmid containing the gene of interest to the plasmid containing the selection marker. This ratio statistically ensures that every cell that picks up the gene of interest will also take up the selection marker. As a control, use carrier DNA instead of the plasmid containing the gene to be studied. (It is not necessary to physically link the selection marker gene to the gene of interest when a 5:1 ratio is used.)
3. Allow the cells to double twice under nonselective conditions. Split the cells 1:15 into selection medium and allow the cells to grow ~10–12 days making sure that distinct, isolated colonies retain their identity. Move the dishes as infrequently as possible to prevent formation of sibling colonies that may form if a cell from the original colony floats away and lands elsewhere on the dish.
4. To subclone the stably transfected cells, pick large healthy colonies which should contain ~500–1000 cells.

Comments: A frequently used selection marker is aminoglycoside phosphotransferase (APH, neo, G418). G418 blocks protein synthesis in mammalian cells by interfering with ribosomal function. The cells will not die immediately. G418 is routinely added to the tissue culture medium to a final concentration of 100–800 μg/ml depending upon the susceptibility of the cell line to the drug. Prepare G418 in a highly buffered solution (e.g., 100 mM HEPES pH 7.3). Addition of the drug should not alter the pH of the medium.

PROTOCOL 11.22 Picking Stable Colonies

Materials

Cloning cylinders
Trypsin for splitting cells

1. Ten days to two weeks after placing the cells in selection media, examine the dishes for the presence of colonies. Hold the plate above your head. You are looking for opaque patches. Circle the area on the plate and examine the plates microscopically.
2. Choose large healthy colonies. Coat one end of a cloning cylinder with sterile vacuum grease by touching the cylinder to grease that has been autoclaved in a glass petri dish. Place the cylinder around the colony to be picked.
3. Rinse the colony with prewarmed trypsin.

4. Add a few drops of prewarmed trypsin to the cloning cylinder. Wait 1–2min then fill the cylinder with medium. Harvest the contents of the cylinder by pipetting the contents up and down repeatedly.
5. Plate the harvested cells in wells of a 12 well plate or a 35mm tissue culture dish.
6. As the cells grow out split them as necessary. Do not let them form large colonies.

Comments: Another quick method to pick colonies is to first locate and mark them. Aspirate off the medium from the plate then rub the isolated colony with a sterile cotton tipped applicator. Immediately put the applicator tip in a small amount of trypsin to disperse the cells, then rinse the suspension with media and deposit the cells in a small culture dish or twelve well plate.

References

Aebersold R, Mann M (2003): Mass spectrometry-based proteomics. *Nature* 422:198–207

Andersons D, Engström Å, Josephson S, Hansson L, Steiner H (1991): Biologically active and amidated cecropin produced in a baculovirus expression system from a fusion construct containing the antibody-binding part of protein A. *Biochem J* 280:219–224

Andreason GL, Evans GA (1989): Optimization of electroporation for transfection of mammalian cell lines. *Anal Biochem* 180:269–275

Ausubel FM, Brent R, Kingston RE, Moore DD, Seidman JG, Smith JA, Struhl K (1993): *Current Protocols in Molecular Biology*. New York: John Wiley

Bennett MK, Garc'a-Arrarás JE, Elferink LA, Peterson K, Fleming AM, Hazuka CD, Scheller RH (1993): The syntaxin family of vesicular transport receptors. *Cell* 74:863–873

Bonetta L (2002): Getting proteins into cells. *Scientist* 16:38–40

Brizzard BL, Chubet RG, Vizard DL (1994): Immunoaffinity purification of FLAG epitope-tagged bacterial alkaline phosphatase using a novel monoclonal antibody and peptide elution. *BioTechniques* 16:730–735

Broker M (1994): Isolation of recombinant proteins from *Saccharomyces cerevisiae* by use of osmotically fragile mutant strains. *BioTechniques* 16:604–610

Burgess RR, Thompson NE (2002): Advances in gentle immunoaffinity chromatography. *Curr Opin Biotechnol* 13:304–308

Bush GL, Tassin A-M, Fridn H, Meyer DI (1991): Secretion in yeast. *J Biol Chem* 266:13811–13814

Chalfie M, Tu Y, Euskirchen G, Ward WW, Prasher DC (1994): Green fluorescent protein as a marker for gene expression. *Science* 263:802–805

Chang JY (1985): Thrombin specificity. Requirement for apolar amino acids adjacent to the thrombin cleavage site of polypeptide substrate. *Eur J Biochem* 151:217–224

Chong S, Mersha FB, Comb DG, Scott ME, Landry D, Vence LM, Perler FB, Benner J, Kucera RB, Hirvonen CA, Pelletier JJ, Paulus H, Xu M-Q (1997): Single-column purification of free recombinant proteins using a self-cleavable affinity tag derived from a protein splicing element. *Gene* 192: 271–281

Chong S, Williams KS, Wotkowicz C, Xu M-Q (1998): Modulation of protein splicing of the Saccharomyces cerevisiae vacuolar membrane ATPase intein. *J Biol Chem* 273:10567–10577

Cregg JM, Madden KR (1988): Development of the methylotrophic yeast, *Pichia pastoris*, as a host system for the production of foreign proteins. *Dev Ind Microbiol* 29:33–41

Dieckmann CL, Tzagoloff A (1985): Assembly of the mitochondrial membrane system. *J Biol Chem* 260:1513–1520

Ellenberg J, Lippincott-Schwartz J, Presley JF (1999): Dual-color imaging with GFP variants. *Trends Cell Biol* 9:52–56

Evan GI, Lewis GK, Ramsay G, Bishop JM (1985): Isolation of monoclonal antibodies specific for human c-myc proto-oncogene product. *Mol Cell Biol* 5:3610–3616

Fasman GD (1989): *Prediction of Protein Structure and the Principles of Protein Conformation.* New York: Plenum Press

Felgner PL, Gadek TR, Holm M, Roman R, Chan HW, Wenz M, Northrop JP, Rinhold GM, Danielsen M (1987): Lipofection: A highly efficient, lipid mediated DNA transfection procedure. *Proc Natl Acad Sci USA* 84:7413–7417

Field J, Nikawa J-I, Broek D, MacDonald B, Rodgers L, Wilson IA, Lerner RA, Wigler M (1988): Purification of a RAS-responsive adenylyl cyclase complex from *Saccharomyces cerevisiae* by use of an epitope addition method. *Mol Cell Biol* 8:2159–2165

Frankel SR, Sohn R, Leinwand L (1991): The use of sarkosyl in generating soluble protein after bacterial expression. *Proc Natl Acad Sci USA* 88: 1192–1196

Gearing DP, Nicola NA, Metcalf D, Foote S, Willson TA, Gough NM, Williams RL (1989): Production of leukemia factor in *E. coli* by a novel procedure and its use in maintaining embryonic stem cells in culture. *Bio/Technology* 7:1157–1161

Gluzman Y (1981): SV-40 transformed simian cells support the replication of early SV-40 mutants. *Cell* 23:175–182

Green N, Alexander H, Olson A, Alexander S, Shinnick TM, Sutcliffe JG, Lerner RA (1982): Immunogenic structure of the influenza virus hemagglutinin. *Cell* 28:477–487

Grieco F, Hay JM, Hull R (1992): An improved procedure for the purification of protein fused with glutathione S-transferase. *BioTechniques* 13: 856–857

Hannig G, Makrides SC (1998): Strategies for optimizing heterologous protein expression in *Escherichia coli. Trends Biotechnol* 16:54–60

Higgins DR, Cregg JM (1998): *Methods in Molecular Biology: Pichia Protocols.* Totowa: NJ: Humana Press

Hochuli E, Bannwarth W Döbeli H, Gentz R, Stüber D (1988): Genetic approach to facilitate purification of recombinant proteins with a novel chelate adsorbent. *Bio/Technology* 6:1321–1325

Hodgson J (1993): Expression systems: a user's guide. *Bio/Technology* 11:887–893

Holzman LB, Marks RM, Dixit VM (1990): A novel immediate-early response gene of endothelium is induced by cytokines and encodes a secreted protein. *Mol Cell Biol* 10:5830–5838

Hopp TP, Prickett KS, Price VL, Libby RT, March CJ, Cerretti DP, Urdal DL, Conlon PJ (1988): A short polypeptide marker sequence for recombinant protein identification and purification. *Bio/Technology* 6:1204–1210

Kellermann OK, Ferenci T (1982): Maltose-binding protein from *Escherichia coli. Methods Enzymol* 90:459–463

Kim J-S, Raines RT (1993): Ribonuclease S-peptide as a carrier in fusion proteins. *Protein Science* 2:348–356

Kimmel AR, Berger SL, eds. (1987): *Guide to Molecular Cloning.* Methods in Enzymology Series, Vol 152. New York: Academic Press

Klekamp MS, Weil PA (1982): Specific transcription of homologous class III genes in yeast *Saccharomyces cerevisiae* soluble cell-free extracts. J *Biol Chem* 257:8432–8441

Kunz D, Gerard NP, Gerard C (1992): The human leukocyte platelet-activating factor receptor. *J Biol Chem* 267:9101–9106

Lin K, Cheng S (1991): An efficient method to purify active eukaryotic proteins from the inclusion bodies in *Escherichia coli. BioTechniques* 11:748–753

Lin Cereghino GP, Lin Cereghino J, Ilgen C, Cregg JM (2002): Production of recombinant proteins in fermenter cultures of the yeast *Pichia pastoris. Curr Opin Biotech* 13:329–332

Lu Q, Bauer JC, Greener A (1997): Using S pombe as host for expression and purification of eukaryotic proteins. *Gene* 200:135–144

Mak P, McDonnell DP, Weigel NL, Schrader WT, O'Malley BW (1989): Expression of functional chicken oviduct progesterone receptors in yeast (*Saccharomyces cerevislae*). *J Biol Chem* 264:21613–21618

Maniatis T, Fritsch EF, Sambrook J (1982): *Molecular Cloning.* Cold Spring Harbor, NY: Cold Spring Harbor Laboratory

McGarvey D (2002) *Qiagen News* 2:30

Melton DA, Krieg PA, Rebagliati MR, Maniatis T, Zinn K, Green MR (1984): Efficient *in vitro* synthesis of biologically active RNA and RNA hybridization probes from plasmids containing a bacteriophage SPG promoter. *Nucl Acids Res* 12:7057–7070

Morrow KJ (1995): Optimizing expression systems for the enhancement of protein purification. *Gen Eng News* May 1:16–17

Muzyczka N, ed. (1989): *Eukaryotic Viral Expression Vectors: Current Topics in Microbiology and Immunology.* Berlin: Springer-Verlag

Nagai K, Thøgersen HC (1987): Synthesis and sequence-specific proteolysis of hybrid proteins produced in *Escherichia coli. Methods Enzymol* 153: 461–481

Nilsson B, Abrahmsén L (1990): Fusions to staphylococcal protein A. *Methods Enzymol* 185:144–161

Nilsson B, Abrahmsén L, Uhlén M (1985): Immobilization and purification of enzymes with staphylococcal protein A gene fusion vectors. *EMBO J* 4:1075–1080

Nilsson J, Stahl S, Lundeberg J, Uhlen M, Nygren PA (1997): Affinity fusion strategies for detection, purification, and immobilization of recombinant proteins. *Protein Exp Purif* 11:1–16

Nonet ML, Grundahl K, Meyer BJ, Rand JB (1993): Synaptic function is impared but not eliminated in *C. elegans* mutants lacking synaptotagmin. *Cell* 73:1291–1305

Nuoffer C, Davidson HW, Matteson J, Meinkoth J, Balch WE (1994): A GDP-bound form of Rabl inhibits protein export from the endoplasmic reticulum and transport between golgi compartments. *J Cell Biol* 125:225–237

Nygren PA, Ståhl S, Uhlén M (1994): Engineering proteins to facilitate bioprocessing. *Trends Biotechnol* 12:338–349

Parks TD, Leuther KK, Howard ED, Johnston SA, Dougherty WG (1994): Release of proteins and peptides from fusion proteins using a recombinant plant virus proteinase. *Anal Biochem* 216:413–417

Pasquinelli AE, Dahlberg JE, Lund E (1995): Reverse 5′ caps in RNAs made *in vitro* by phage RNA polymerases. *RNA* 1:957–967

Perler FB, Olsen GJ, Adam E (1997): *Nucleic Acids Res* 25:1087–1094

Pryor KD, Leiting B (1997): High-level expression of soluble protein in *Escherichia coli* using His_6-tag and maltose-binding-protein double-affinity systems. *Prot Expr Purif* 10:309–319

Puig O, Caspary F, Rigaut G, Rutz B, Bouveret E, Bragado-Nilsson E, Wilm M, Seraphin B (2001): The tandem affinity purification (TAP) method: a general procedure of protein complex purification. *Methods* 24:218–229

Richards FM, Wyckoff HW (1971): Bovine pancreatic ribonuclease. In: *The Enzymes, Vol 4,* Boyer PD, ed. New York: Academic Press

Rigaut G, Shevchenko A, Rutz B, Wilm M, Mann M, Séraphin B (1999): A generic protein purification method for protein complex characterization and proteome exploration. *Nat Biotechnol* 17:1030–1032

Rüther U, Müller-Hill B (1983): Easy identification of cDNA clones. *EMBO J* 2:1791–1794

Schmidt TG, Skerra A (1993): The random peptide library-assisted engineering of a C-terminal affinity peptide, useful for the detection and purification of a functional Ig Fv fragment. *Protein Eng* 6:109–122

Schwartze SR, Dowdy SF (2000): *In vivo* protein transduction: intracellular delivery of biologically active proteins, compounds and DNA. *Trends Pharmacol Sci* 21:45–48

Smith DB, Johnson KS (1988): Single-step purification of polypeptides expressed in *E. coli* as fusions with glutathione S-transferase. *Gene* 67: 31–40

Sorger PK, Pelham HRB (1987): Purification and characterization of a heat-shock element binding protein from yeast. *EMBO J* 6:3035–3041

Spindler KR, Rosser DSE, Berk AJ (1984): Analysis of Adenovirus transforming proteins from early regions 1A and 1B with antisera to inducible fusion antigens produced in *Escherichia coli. J Virol* 49:132–141

Stepinski J, Waddell C, Stolarski R, Darzynkiewicz E (2001): Synthesis and properties of mRNAs containing the novel "anti-reverse" cap analogs 7-methyl(3′-O-methyl)GpppG and 7-methyl (3′-deoxy)GpppG. *RNA* 7:1486–1495

Su X, Prestwood AK, McGraw RA (1992): Production of recombinant porcine tumor necrosis factor alpha in a novel *E. coli* expression system. *BioTechniques* 13:756–762

Summers MD, Smith GE (1987): *A Manual of Methods for Baculovirus Vectors and Insect Cell Culture Procedures.* Texas Agricultural Experiment Station Bulletin No. 1555, College Station, TX

Terpe K (2003): Overview of tag protein fusions: from molecular and biochemical fundamentals to commercial systems. *Appl Microbiol Biotechnol* 60:523–533

Thorn KS, Naber N, Matuska M, Vale RD, Cooke R (2000): A novel method of affinity-purifying proteins using a bis-arsenical fluorescein. *Prot Sci* 9:213–217

Toneguzzo F, Keating A, Flynn S, McDonald K (1988): Electric field-mediated gene transfer: characterization of DNA transfer and patterns of integration in lymphoid cells. *Nucleic Acids Res* 16:5515–5532

Wilson IA, Niman HL, Houghten RA, Cherenson AR, Connolly ML, Lerner RA (1984): The structure of an antigenic deteminant in a protein. *Cell* 37:767–778

Wingfield PT, Mattaliano RJ, MacDonald HR, Craig S, Clove GM, Gronenborn AM, Schmiesner U (1987): Recombinant derived IL-1α, replacing a labile asparagine with a serine. *Protein Eng* 1:413–27–36

Wong TK, Neumann E (1982): Electric field mediated gene transfer. *Biochem Biophys Res Commun* 107:584–587

Xu M-Q, Paulus H, Chong S (2000): Fusions to self-splicing inteins for protein purification. *Methods Enzymol* 236:376–418

General References

Goeddel DV, ed. (1990): *Gene Expression Technology.* Methods in Enzymology Series, Vol 185. New York: Academic Press

Grossman L, Wu R, eds. (1987): *Recombinant DNA,* Parts D, E, R Methods in Enzymology Series, Vols 153,154,155. New York: Academic Press

Seetharam R, Sharma SK, eds. (1991): *Purification and Analysis of Recombinant Proteins.* New York: Marcel Dekker, Inc.

Wu R, ed. (1993): *Recombinant DNA,* Parts G, H, I. Methods in Enzymology Series, Vols 216, 217, 218. New York: Academic Press

Appendixes

A

Safety Considerations

Safety is a primary concern in the laboratory. Many of the techniques that are presented can be hazardous. Wherever possible, dangerous aspects of a method will be emphasized. Here are a few things to keep in mind when handling chemicals, radioactivity, biological agents, and using heavy machinery. **Whenever possible, don't work alone. Know where the fire extinguisher is located.**

The CRC Handbook of Laboratory Safety is a comprehensive manual that provides an in depth treatment to all of the areas mentioned below, and many other additional issues and situations that may be encountered in the lab.

First Aid: Emergency Procedures

Fast responses in all cases are essential.

Eye Contact: Wear safety goggles in the lab. If an accident happens, flush eyes with large amounts of water. Get medical attention.

Skin Contact: Flush skin with large amounts of water while removing contaminated clothing and shoes. Get medical attention. Wash clothes before rewearing them.

Inhalation: Bring the victim to a source of fresh air. If breathing has stopped, perform artificial respiration. Keep the person warm and get medical attention.

Ingestion: Check the material safety data sheet (MSDS) for the specific chemical for the immediate response. Do not necessarily induce vomiting in all cases. Never give anything by mouth to an unconscious or convulsing person. Get medical attention immediately.

Chemicals: Most chemicals used in biochemical procedures are potentially dangerous. Many solvents such as chloroform, butanol, ether, and mercaptoethanol should be handled in a fume hood with protection for clothing, skin and respiration. Know where to store hazardous chemicals like strong acids and bases. Know what to do in the event of a chemical spill. **Always wear a labcoat.** Reagents can be carcinogens or mutagens like diaminobenzidine and ethidium bromide. **Never pipette by mouth.** Find a suitable pipetting device that you are comfortable using. It can be a bulb, a pump, manual or battery powered.

Radioactivity: Make sure that all institutional rules for the safe handling of radioactivity are strictly followed. Take precautions when disposing of radioactive waste. Familiarize yourself with your institution's guidelines for the safe handling and use of radioactive substances. Know how to deal with spills should they occur. Make sure that the proper shielding, storage, and disposal materials are available. Keep track of the location and use of all radioactive material.

Here are some guidelines to keep in mind when working with radioactive materials:

1. Always work in your lab's designated areas when using isotopes.
2. Do not allow contaminated material to accumulate in the working area.
3. Use appropriate shielding.
4. Dress appropriately, film badge, labcoat, safety glasses, double gloves, and lead apron with certain isotopes.
5. Do not add acids to radioiodine solutions. The volatility of ^{125}I and ^{131}I is enhanced at low pH.
6. Never pipette by mouth!
7. Whenever possible use disposable equipment.
8. If skin or hands become contaminated, wash the area with commercially available decontaminating solutions.
9. In the event of a spill, decontaminate the equipment immediately. If the spilled isotope is iodine, wipe the contaminated surfaces with 0.1 M NaI, 0.1 M NaOH, and 0.1 M $Na_2S_2O_3$.
10. Do not eat or store food in the lab.
11. Behave responsibly, do not ignore spills. Your lab might lose its license to work with isotopes.
12. When in doubt, contact your radiation safety office. They are there to help you, not to shut down your lab.

Equipment Hazards: Some of the pieces of equipment used in these techniques are potentially dangerous if not handled properly. Consult the literature accompanying the equipment for safety information. Exercise care when lifting heavy objects.

It is your responsibility to be aware of all potential hazards and know how to deal with them safely. Adhere to the instructions provided by suppliers of reagents, radioactivity and equipment. In addition, the Occupational Safety and Health Administration (OSHA) and your institution have information regarding potential hazards and how to deal with them safely.

B

Antibody Preparation

Production of Polyclonal Antisera in Rabbits

When an antigen such as a protein is injected into a laboratory animal, the different parts of the protein molecule, referred to as epitopes or determinants, act as targets for the production of antibodies. Polyclonal antibodies are produced when several different plasma cells secrete antibody molecules that are each specific for a different epitope. The result is a mixture that may contain several different types of antibody directed against various determinants of the antigen, called polyclonal antisera. Arguably the most important factor in producing a highly specific polyclonal antiserum is the purity of the antigen used for the immunization. Any contaminating proteins can potentially induce a strong immune response when injected in the presence of adjuvant. After a primary immunization, naive B cells are stimulated to differentiate into antibody producing plasma cells. Specific antibody usually begins to appear in the serum 5 to 7 days after the injection. The antibody titer continues to rise and peaks around day 12, after which it decreases. In addition to differentiating into antibody-producing cells, the antigen-stimulated B cells will also form a population of "memory" B cells which become activated after a booster injection. A booster injection results in a higher titer which is sustained for a longer period. As a consequence of the existence of the memory B cells, a smaller amount of antigen is required to stimulate the strong, long lasting secondary response.

The use of adjuvants greatly enhances the antibody titer. The antigen forms an emulsion with the adjuvant which results in a slow, long-lasting release of antigen and components of the adjuvant, which causes activation of the T cell population and maximally stimulates antibody production. **Complete Freund's adjuvant is usually used for the primary immunization and incomplete Freund's adjuvant is adequate for the boosts.** New adjuvants have been developed which minimize trauma to the animal and still give excellent antibody production.

Before performing the primary immunization it is extremely useful to bleed the animal. This serum, also referred to as the "prebleed", will be used as a control and for comparison with the results obtained when using the specific antibody as it becomes available. Another alternative is to collect blood from a rabbit immunized with an irrelevant antigen.

Prior to immunization a choice must be made (if a choice is possible) whether to immunize with the antigen in the native or denatured state. Antibodies specific for denatured forms of the antigen are useful for Western blotting and screening of cDNA or genomic libraries. However, for functional studies it is important that the antibody interact with the native conformation of the target protein.

In the presence of an adjuvant, the protein antigen is injected either intramuscularly, intradermally, or subcutaneously into a rabbit. Four to eight weeks after the priming immunization, the rabbit is given booster injections at intervals of 2-3 weeks. Seven to 10 days after each boost the rabbit is bled and serum is prepared.

Antibody production is often performed by a core facility or on a contract basis by a commercial lab. These facilities will do everything (for a fee) from the synthesis and conjugation of the antigenic peptide to a carrier, to injecting, bleeding, and even assaying the serum for the specific immune response.

PROTOCOL B.1 Preparation of Antigen-Adjuvant Emulsion

Materials

Complete Freund's adjuvant

Glass syringes and double-ended locking hub connector (both available from Luer-Lok, Becton Dickson)

1. Add 2ml of complete Freund's adjuvant to a 2ml solution of purified antigen (1-2mg/ml in PBS). This takes into account loss of emulsion during the procedure.
2. Prepare the emulsion by adding the antigen and adjuvant to two 3ml glass syringes fitted with locking hubs with a double-ended locking hub connector. Make sure that the connections are tight. Form the emulsion by pushing the contents back and forth vigorously.
3. Test the emulsion by allowing a drop to fall onto the surface of water in a beaker. The emulsion is ready to use if the drop remains tight on the surface of the water.
4. Transfer the emulsion to a 3ml syringe. Attach a 22 gauge needle and remove any air bubbles. You are now ready to immunize. Work quickly, as the emulsion can separate out into phases.

PROTOCOL B.2 Intramuscular Immunization (IM)

1. Place the rabbit on a flat surface.
2. Clean the area to be injected with 70% ethanol.

3. Keeping the rabbit well restrained, insert the needle ~1 cm into the thigh muscle of each hind leg. Inject 0.5 ml of antigen emulsified in complete Freund's adjuvant (as described above) into each thigh muscle.
4. Boost schedule: Four weeks after the primary immunization, boost the rabbit with 1 mg of antigen emulsified in incomplete Freund's adjuvant prepared as described above. Two weeks after the first boost, repeat the boost immunization.

PROTOCOL B.3 Intradermal Immunization

Intradermal and subcutaneous injections are usually attempted when the antigen is suspected of being poorly immunogenic.

1. Prepare the antigen emulsion as previously described in Protocol B.1.
2. Shave the back of the rabbit and swab the area liberally with disinfectant. Hold the animal firmly against a flat surface.
3. Stretch the area of the skin to be immunized between thumb and index finger. Using a glass syringe with a 24 gauge needle, insert the needle at a 30° angle just under the outer skin layer, bevel facing up. Lower the syringe so that it is almost lying on the rabbit's back and inject 100 µl between the dermal layers. Repeat this procedure at four other sites. Do not inject more than 100 µl at each site. Small lumps should appear at the injection sites. The injection sites may become ulcerated about two weeks following immunization.
4. Boost the animal intramuscularly using complete Freund's adjuvant as described above. If a high titer antibody is not attained following two IM boosts, immunize the rabbit intradermally, but only at two sites.

PROTOCOL B.4 Subcutaneous Immunization

1. Prepare the antigen emulsion in complete Freund's adjuvant as previously described.
2. In a pinching movement, pull the skin away from the underlying muscle. Insert a 24 gauge needle through the skin, being careful not to penetrate the muscle. Inject 0.5 ml at four different sites.
3. Boost the animal IM using imcomplete Freund's adjuvant as described.

PROTOCOL B.5 Bleeding the Rabbit and Serum Preparation

Ten days after the second booster injection the rabbit is bled from the marginal vein of the ear as shown in Figure B.1. Although the central artery is more prominent, do not attempt to draw blood from this vessel.

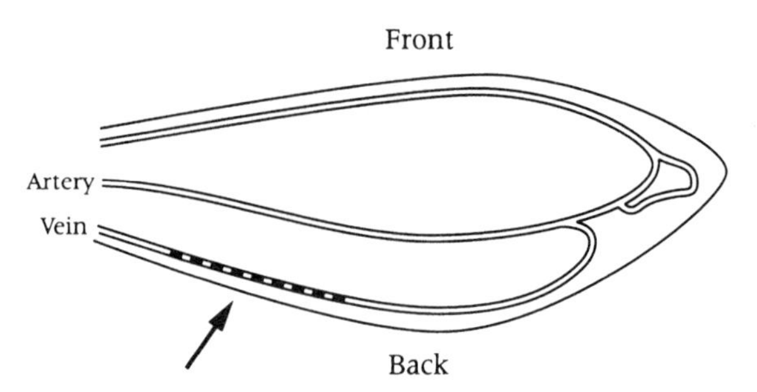

FIGURE B.1 The circulatory system of the rabbit ear. The area of the vein, to be longitudinally cut, is marked by a dashed line.

1. Place the rabbit in a restrainer with its ear extended. Make sure that the animal cannot move freely.
2. Shave the area of interest on the ear and swab the area with 70% ethanol. Using a heating lamp, warm the animal. Tap the vein lightly.
3. Using a new scalpel, make a small cut, ~0.5 cm long perpendicular to the vein. Make sure that the cut has gone cleanly through the vein and is not just a superficial scratch. Hold the ear so that the blood drips into a 50 ml plastic centrifuge tube allowing the blood to flow down the side of the tube to reduce hemolysis.
4. When enough blood has been collected (not to exceed 50 ml), stop the bleeding by wiping the area of the cut with petroleum jelly and apply pressure to the cut with a piece of gauze.
5. Allow the blood to stand for two hours at room temperature before placing it in the refrigerator overnight. The next day, a clot should have formed. Loosen it from the sides of the tube with a wooden applicator. Do not break up the clot, remove it with the applicator stick.
6. Transfer the serum to an appropriate centrifuge tube and pellet any remaining red cells and debris by a 10 min centrifugation at 5000 × *g*. Collect the serum and store appropriately.

Comment: Alternate ears between bleeds to allow the wound to heal completely.

PROTOCOL B.6 Precipitation of IgG with Saturated Ammonium Sulfate

The total IgG fraction in serum can be precipitated at a final concentration of 33% $(NH_4)_2SO_4$. At this concentration most proteins remain in solution.

1. Prepare the saturated ammonium sulfate (SAS): 450 g ammonium sulfate to 500 ml H_20. Heat the solution until the $(NH_4)_2SO_4$ dissolves completely. Filter it while it is still warm. Allow it to cool. Crystals will form. Do not remove them. Adjust the pH of the cooled solution to pH 7.5 with ammonium hydroxide. For the 33% SAS solution needed for washing the pellet, mix 33 ml of SAS with 67 ml of PBS.

2. Add 1 vol. SAS dropwise to 2 vol. of serum, ascites or tissue culture supernatant with constant mixing at 4°C.
3. Allow the precipitate to form for at least 2h with constant mixing. Centrifuge the suspension at 12,000 × g for 20min.
4. Decant the supernatant and wash the pellet by resuspending it in a volume of cold 33% SAS equivalent to the original volume of serum. Pellet the precipitate as above and resuspend in 5–10% of the original volume.
5. Dialyze the IgG solution for ~48h at 4°C against the desired buffer to fully remove the ammonium sulfate.

Purification of Antibody Using Protein A Affinity Columns

Antibody can be purified from serum, culture supernatants, or ascites fluid by using affinity columns of immobilized Protein A. Not all murine and rat Ig classes and subclasses bind to Protein A. The binding avidity of murine IgGs is in the following order: $IgG_{2b} > IgG_{2a} = IgG_3 >> IgG_1$ while IgM ad IgE fail to bind. For rat IgGs, the only subclasses that bind are IgG_{2c} and IgG_1 to a lower degree. The binding of murine IgGs is pH dependent, IgG_1 binds a pH 8.0 and not a pH 7.2.

PROTOCOL B.7 Purifying Total Ig

1. Swell the Protein A Sepharose in 0.1M phosphate, pH 8.0, for 30min. if it is in powdered form. (1.5g of dry Protein A Sepharose yields about 5ml of hydrated gel and has a total capacity of ~20mg/ml of mouse immunoglobins but a lower capacity, 5–6mg for IgG_1. See Table 3.3)
2. Prepare a small column using either a commercially available mini-column or a Pasteur pipette by plugging it with either glass wool or a glass bead and then adding about 0.5ml of Protein A Sepharose.
3. Add one volume of 0.5M phosphate buffer to 4 volumes of serum.
4. Slowly apply the diluted serum to the Protein A Sepharose column. Reapply the effluent a total of 3 times.
5. Wash the column with phosphate buffer until the OD_{280} is less than 0.02 which takes between 30–50ml of buffer.
6. For elution of total IgG, apply 0.1M acetic acid and collect 1ml fractions. Immediately neutralize the eluted IgG by collecting the effluent in a tube containing 50µl of 1M Tris pH 8.0. Alternatively, and/or in addition, the effluent can be immediately dialyzed against PBS.

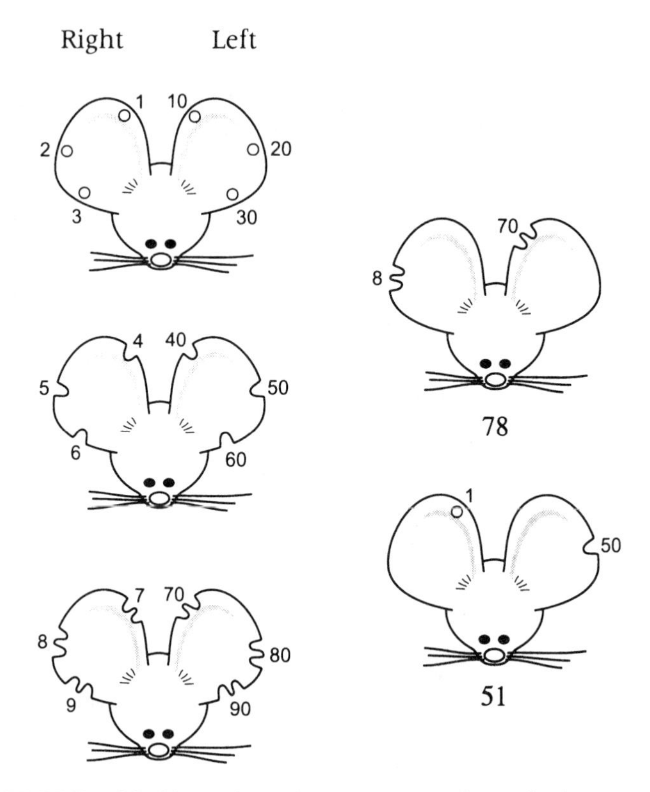

FIGURE B.2 Marking mice using an ear punch numbering system.

Numbering Mice

When mice are kept in a transgenic colony, for hybridoma production, or any other system, it is desirable to have an unambiguous numbering system. This can be done with a metal ear tag or with the ear punch system presented in Figure B.2. If the cage is numbered it should not be necessary to go higher than number twenty, for a large cage.

C

Solutions

TABLE C.1 Commercial Strengths of Common Laboratory Chemicals

Compound	Molecular Weight	Molarity	% (by weight)	Specific Gravity	Formula
Acetic acid, glacial	60.05	17.4	99.5	1.05	CH_3COOH
Hydrochloric acid	36.5	11.6	36	1.18	HCl
Nitric acid	63.02	16	71	1.42	HNO_3
Sulfuric acid*	98.1	18	85	1.84	H_2SO_4
Phosphoric acid	80	18.1	85	1.70	H_3PO_4
Ammonium hydroxide	35	14.8	28	0.90	NH_4OH
Lactic acid	90.1	11.3	85	1.2	$CH_3CH(OH)COOH$
Potassium hydroxide	56.1	13.5	50	1.52	KOH
Sodium hydroxide	40	19.1	50	1.53	$NaOH$
Sodium carbonate	106	1.04	10	1.10	$NaCO_3$
2-Mercaptoethanol	78	14.4	100		$HSCH_2CH_2OH$
Hydrogen peroxide	34		30 or 35	1.46	H_2O_2

*When making a solution of H_2SO_4 always add concentrated H_2SO_4 to H_2O. Never add H_2O to the acid.

Water

The preparation of buffers and biological solutions requires copious amounts of water. Preparing laboratory or reagent grade water entails the removal of contaminating ions (sodium, chloride, and calcium), dissolved ionized gases (carbon dioxide), organic chemicals, and microorganisms or other particles that may have found their way into the water supply.

Water purity is defined in terms of specific resistance (megohm-cm), specific conductance (microhms per cm), total silica, total organic carbon (TOC), pyrogen levels and bacterial count among other attributes.

The two most popular methods of preparing purified water are filtration and distillation. Distilled water is produced when water is heated to its gaseous state and then allowed to recondense. Impurities are left behind as the water vapor condenses. Many labs are equipped with stills that produce large amounts of distilled water. However, distillation does not always remove dissolved gases, the most common of which is carbon dioxide. For HPLC, trace amounts of carbon dioxide can lead to errant results.

Reagent grade water can also be prepared by deionization, using a charged resin to which oppositely charged compounds stick as water is passed through. With a system of cartridges containing resins or filters, the Milli-Q water system (Millipore Corp.) combines several methods into one system.

Ultrapure, reagent grade water, also known as Type I water has a specific resistance of at least 18 megohms-cm at 25°C and meets other criteria specified by United States Pharmacopeia (USP) and the International Organization of Standardization (ISO).

For the preparation of buffers and solutions, use the purest water available.

Molarity

A mole of a particular substance is the number of grams equal to the atomic or molecular weight of the substance. The term gram molecular weight is often used as a definition of a mole. One mole of any substance will contain approximately 6.02×10^{23} particles (Avogadro's number). Molarity is a number that expresses the number of moles of substance in 1 L of solution. A 1.0 molar solution (1 M) contains 1 mole of solute per liter of solution.

$$\text{Molecular weight} \times \text{Molarity} = \text{grams/L}.$$

A millimole is 1/1000 mole and 1 mol equals 1000 mmol.

The following easy conversion table will save time and greatly reduce errors when going from moles to mM.

$$\begin{aligned} 1\,\text{M} &= 1\,\text{mol/L} \\ &= 1\,\text{mmol/ml} \\ 1\,\text{mM} &= 1\,\text{mmol/L} \\ &= 1\,\mu\text{mol/ml} \\ &= 1\,\text{nmol}/\mu\text{l} \\ 1\,\mu\text{M} &= 1\,\mu\text{mol/L} \\ &= 1\,\text{nmol/ml} \\ &= 1\,\text{pmol}/\mu\text{l} \end{aligned}$$

Choosing and Preparing Buffers

A buffer is only as good as the pH meter that is used for its preparation (Table C.2). Therefore, the pH meter should be calibrated regularly using commercial buffer standards. Buffers are solutions used to maintain a constant pH by absorbing and releasing protons. They consist of an acid and its conjugate base (e.g. acetic acid and acetate) and are formed by mixing the appropriate amount of each to achieve the required pH. Buffer pH is temperature sensitive. Therefore, check the pH at the temperature at which the buffer is intended to be used. pK

TABLE C.2 Preparation of Commonly Used Buffers*

A. Glycine-HCl Buffer[1]

Stock Solutions

A: 0.2M solution of glycine (15.01 g. in 1000 ml.).

B: 0.2M HCl.

50 ml. of A + x ml. of B, diluted to a total of 200 ml.

x	pH	x	pH
5.0	3.6	16.8	2.8
6.4	3.4	24.2	2.6
8.2	3.2	32.4	2.4
11.4	3.0	44.0	2.2

B. Citrate Buffers[2]

Stock Solutions

A: 0.1M solution of citric acid (21.01 g. in 1000 ml.).

B: 0.1M solution of sodium citrate (29.41 g. $C_6H_5O_7Na_3 \cdot 2H_2O$ in 1000 ml.; the use of the salt with $5^1/_2$ H_2O is not recommended).

x ml. of A + y ml. of B, diluted to a total of 100 ml.

x	y	pH
46.5	3.5	3.0
43.7	6.3	3.2
40.0	10.0	3.4
37.0	13.0	3.6
35.0	15.0	3.8
33.0	17.0	4.0
31.5	18.5	4.2
28.0	22.0	4.4
25.5	24.5	4.6
23.0	27.0	4.8
20.5	29.5	5.0
18.0	32.0	5.2
16.0	34.0	5.4
13.7	36.3	5.6
11.8	38.2	5.8
9.5	41.5	6.0
7.2	42.8	6.2

C. Acetate Buffer[3]

Stock Solutions

A: 0.2M solution of acetic acid (11.55 ml. in 1000 ml.).

B: 0.2M solution of sodium acetate (16.4 g. of $C_2H_3O_2Na$ or 27.2 g. of $C_2H_3O_2Na \cdot 3H_2O$ in 1000 ml.).

x ml. of A + y ml. of B, diluted to a total of 100 ml.

x	y	pH
46.3	3.7	3.6
44.0	6.0	3.8
41.0	9.0	4.0
36.8	13.2	4.2
30.5	19.5	4.4
25.5	24.5	4.6
20.0	30.0	4.8
14.8	35.2	5.0
10.5	39.5	5.2
8.8	41.2	5.4
4.8	45.2	5.6

Continued

TABLE C.2 *Continued*

D. Phosphate Buffer[1]

Stock Solutions

A: 0.2*M* solution of monobasic sodium phosphate (27.8g. in 1000ml.).

B: 0.2*M* solution of dibasic sodium phosphate (53.65g. of $Na_2HPO_4 \cdot 7H_2O$ or 71.7g. of $Na_2HPO_4 \cdot 12H_2O$ in 1000ml.).

x ml. of A + y ml. of B, diluted to a total of 200ml.

x	y	pH	x	y	pH
93.5	6.5	5.7	45.0	55.0	6.9
92.0	8.0	5.8	39.0	61.0	7.0
90.0	10.0	5.9	33.0	67.0	7.1
87.7	12.3	6.0	28.0	72.0	7.2
85.0	15.0	6.1	23.0	77.0	7.3
81.5	18.5	6.2	19.0	81.0	7.4
77.5	22.5	6.3	16.0	84.0	7.5
73.5	26.5	6.4	13.0	87.0	7.6
68.5	31.5	6.5	10.5	90.5	7.7
62.5	37.5	6.6	8.5	91.5	7.8
56.5	43.5	6.7	7.0	93.0	7.9
51.0	49.0	6.8	5.3	94.7	8.0

E. Tris(hydroxymethyl)aminomethane-maletate (Tris-maleate) Buffer[4]

Stock Solutions

A: 0.2*M* solution of Tris acid maleate (24.2g. of tris(hydroxymethyl)aminomethane + 23.2g. of maleic acid or 19.6g. of maleic anhydride in 1000ml.).

B: 0.2*M* NaOH.

50ml. of A + x ml. of B, diluted to a total of 200ml.

x	pH	x	pH
7.0	5.2	48.0	7.0
10.8	5.4	51.0	7.2
15.5	5.6	54.0	7.4
20.5	5.8	58.0	7.6
26.0	6.0	63.5	7.8
31.5	6.2	69.0	8.0
37.0	6.4	75.0	8.2
42.5	6.6	81.0	8.4
45.0	6.8	86.5	8.6

F. Barbital Buffer[5]

Stock Solutions

A: 0.2*M* solution of sodium barbital (veronal) (41.2g. in 1000ml.).

B: 0.2*M* HCl.

50ml. of A + x ml. of B, diluted to a total of 200ml.

x	pH
1.5	9.2
2.5	9.0
4.0	8.8
6.0	8.6
9.0	8.4
12.7	8.2
17.5	8.0
22.5	7.8
27.5	7.6
32.5	7.4
39.0	7.2
43.0	7.0
45.0	6.8

TABLE C.2 *Continued*

Solutions more concentrated than 0.05*M* may crystallize on standing, especially in the cold.

G. Tris(hydroxymethyl)aminomethane (Tris) Buffer

Stock Solutions

A: 0.2*M* solution of tris(hydroxymethyl)aminomethane (24.2g. in 1000ml.).

B: 0.2*M* HCl.

50ml. of A + *x* ml. of B, diluted to a total of 200ml.

x	pH
5.0	9.0
8.1	8.8
12.2	8.6
16.5	8.4
21.9	8.2
26.8	8.0
32.5	7.8
38.4	7.6
41.4	7.4
44.2	7.2

H. Boric Acid-Borax Buffer[6]

Stock Solutions

A: 0.2*M* solution of boric acid (12.4g. in 1000ml.).

B: 0.05*M* solution of borax (19.05g. in 1000ml.; 0.2*M* in terms of sodium borate).

50ml. of A + *x* ml. of B, diluted to a total of 200ml.

x	pH	*x*	pH
2.0	7.6	22.5	8.7
3.1	7.8	30.0	8.8
4.9	8.0	42.5	8.9
7.3	8.2	59.0	9.0
11.5	8.4	83.0	9.1
17.5	8.6	115.0	9.2

I. Glycine-NaOH Buffer[1]

Stock Solutions

A: 0.2*M* solution of glycine (15.01g. in 1000ml.).

B: 0.2*M* NaOH.

50ml. of A + *x* ml. of B, diluted to a total of 200ml.

x	pH	*x*	pH
4.0	8.6	22.4	9.6
6.0	8.8	27.2	9.8
8.8	9.0	32.0	10.0
12.0	9.2	38.6	10.4
16.8	9.4	45.5	10.6

J. Carbonate-Bicarbonate Buffer[7]

Stock Solutions

A: 0.2*M* solution of anhydrous sodium carbonate (21.2g. in 1000ml.).

B: 0.2*M* solution of sodium bicarbonate (16.8g. in 1000ml.).

x ml. of A + *y* ml. of B, diluted to a total of 200ml.

x	*y*	pH
4.0	46.0	9.2
7.5	42.5	9.3
9.5	40.5	9.4
13.0	37.0	9.5
16.0	34.0	9.6

Continued

TABLE C.2 *Continued*

19.5	30.5	9.7
22.0	28.0	9.8
25.0	25.0	9.9
27.5	22.5	10.0
30.0	20.0	10.1
33.0	17.0	10.2
35.5	14.5	10.3
38.5	11.5	10.4
40.5	9.5	10.5
42.5	7.5	10.6
45.0	5.0	10.7

[1]S.P.L. Sørensen, *Biochem. Z.* **21,** 131 (1909); **22,** 352 (1909).

[2]R.D. Lillie, "Histopathologic Technique," Blakiston, Philadelphia and Toronto, 1948.

[3]G.S. Walpole, *J. Chem. Soc.* **105,** 2501 (1914).

[4]G. Gomori, *Proc. Soc. Exptl. Biol. Med.* **68,** 354 (1948).

[5]L. Michaelis, *J. Biol. Chem.* **87,** 33 (1930).

[6]W. Holmes, *Anat. Record* **86,** 163 (1943).

[7]G.E. Delory and E.J. King, *Biochem. J.* **39,** 245 (1945).

and pK_a are frequently used interchangeably in the scientific literature although pK_a is more correct. Buffers stabilize pH most effectively at their pK value, but can generally be used within 1 pH unit on either side of the pK. Generally, the pK_a of a weak acid or base indicates the pH of the center of the buffering region. Typically, buffers are used in the range of 10–50 mM, with Tris, phosphate and acetate buffers being the most commonly used.

Most biological happenings occur in the pH range of 6 to 8. However, many commonly used buffers suffer from serious drawbacks in this range that would make their use prohibitive. Phosphate is a weak buffer above pH 7.5. It also precipitates or binds to many polyvalent cations, and is frequently a participating or inhibiting metabolite. Tris is a poor buffer below pH 7.5. It is also a potentially reactive primary amine. For these reasons, try to choose a buffer system that fulfills the following criteria:

- High solubility in aqueous systems. This feature permits its use from concentrated buffer stocks.
- Chemical stability.
- Availability in pure form.
- Does not interfere with the function of the protein.
- Buffers should not absorb light in the visible or ultraviolet regions of the spectrum. Absorption at wavelengths greater than 230–240 nm may interfere with spectrophotometric assays.

There is often the need to prepare buffers without mineral cations. For these situations tetramethylammonium hydroxide can be used. The basicity of this quaternary amine is equivalent to that of sodium or potassium hydroxide.

"Good" Buffers

Hydrogen ion buffers covering the range $pK_a = 6.15–8.35$ that are alternatives to the conventional buffers have been in use since 1966 (Good et al, 1966). Table C.3 lists some of these buffers and their useful pH range. The pH of these materials in solution will not be near the pKa and they will not become buffers until the pH is adjusted. If the material is a free acid, adjust to the working pH with sodium hydroxide, potassium hydroxide, tetramethylammonium hydroxide or other appropriate base. Buffer materials that are free bases are adjusted by the addition of a suitable acid.

Common Laboratory Solutions

Phosphate Buffered Saline (PBS)

Dissolve 8g NaCl, 0.2g KCl, 1.44g Na_2HPO_4, and 0.24g KH_2PO_4 in 800ml bidistilled H_2O. Adjust the pH to 7.4 with HCl. Add H_2O to 1L.

Phosphate Buffered Saline with Calcium and Magnesium (PBS^{++})

Add $CaCl_2 \cdot 2\ H_2O$ (130mg/L), $MgCl_2 \cdot 6\ H_2O$ (100mg/L) to PBS.

Phosphate Buffered Saline + Glucose (PBS+G) pH 7.6

10ml 1.0M Phosphate buffer pH 7.6, 8.2g NaCl, 10g glucose, add water to 1 liter

TABLE C.3

Product	Useful pH Range
MES	5.5–6.7
BIS-TRIS	5.8–7.2
ADA	6.0–7.2
PIPES	6.1–7.5
ACES	6.1–7.5
MOPSO	6.2–7.6
BES	6.4–7.8
MOPS	6.5–7.9
TES	6.8–8.2
HEPES	6.8–8.2
TAPSO	7.0–8.2
POPSO	7.2–8.5
EPPS	7.3–8.7
TRICINE	7.4–8.8
BICINE	7.6–9.0
TAPS	7.7–9.1
CHES	8.6–10.0
CAPS	9.7–11.1

(Data from Good et al, 1966; Good and Izawa 1968).

Ethylenediamine Tetraacetic Acid, Disodium Salt (EDTA) 0.5M, pH 8.0

Add 186.1g EDTA to 800ml of H_2O. Stir vigorously. Adjust the pH to 8.0 with NaOH (~20g of NaOH pellets). The disodium salt of EDTA will not dissolve until the pH is adjusted to ~pH 8.0.

Trichloroacetic Acid (TCA) 100% Solution

Add 227ml of H_2O to a bottle containing 500g TCA. The solution will be 100% (w/v) TCA.

Sodium Dodecyl Sulfate (SDS) 10%

10g SDS in 100 H_2O ml. Store at room temperature. At low temperatures the SDS will precipitate. Heat gently to redissolve. SDS is also called sodium lauryl sulfate. Wear a mask when weighing SDS.

4× SDS Sample Buffer (Reducing)

Sucrose, 4.0g, 2-Mercaptoethanol, 2.0ml, SDS, 0.8g, 1M Tris-HCl pH 6.8, 2.5ml, Bromophenol blue, 0.001% w/v. Bring to 10ml with water. Aliquot and store frozen until needed. To prepare 1× or 2× sample buffer, dilute with water. Note that if a given protocol calls for non-reducing sample buffer, substitute water for the reducing agent 2-mercaptoethanol.

Phenylmethyl Sulfonyl Fluoride (PMSF)

Use caution. Highly toxic. Prepare as a 200mM stock solution in isopropanol. Mol wt 174. Dissolve 384mgs in 10ml of isopropanol. Store at room temperature, and away from light. Add immediately prior to use as it decomposes rapidly in aqueous solutions.

Ammonium Persulfate (APS)

10% w/v. Dissolve 100mg in 1.0ml water. Store at 4°C. Make fresh weekly.

Tris Buffered Saline (TBS)

Working solution: 10mM Tris, pH 7.6, 150mM NaCl. This solution can be prepared as a 10× stock. For 1L of 10 × stock: 12.2g Tris, 87.75g NaCl. Dissolve and bring to the desired pH with concentrated HCl then add H_2O to 1 liter.

Tris Buffered Saline with Tween 20 (TTBS)

Prepare 1 liter of 1× TBS from the 10× stock and add 0.5ml of Tween 20 (Sigma) to a final concentration of 0.05%.

RIPA Buffer

10mM Tris-HCl, pH 8.0, 140mM NaCl, 1% Triton X-100, 0.5% sodium deoxycholate, 0.1% SDS.

N,N,N′,N′-Tetramethylene-Diamine (TEMED)

Dithiothreitol (DTT)

Also known as Clelands' reagent. Mol wt 154. Dissolve 154 mg/ml of water or buffer and dilute to the desired concentration. Use freshly made or store aliquots at −20°C.

Protein A Sepharose Suspension

3% w/v in PBS, 0.02% sodium azide. Store refrigerated.

Western Transfer Buffer for Tris-Glycine and Tris-Tricine Gels

Working solution: 3.03 g Tris, 14.42 g glycine, 10% methanol and add H_2O to 1 liter. This solution can be prepared as a 10× stock. The methanol is optional and should be replaced with water.

TBE

Working solution: 0.089 M Tris-borate, 0.089 M boric acid, 0.002 M EDTA. To prepare 1 liter of 10× concentrated stock: 108 g Tris, 55 g boric acid, 40 ml 0.5 M EDTA (pH 8.0), add H_2O to 1 liter.

SSC

Working solution: 0.15 mM NaCl, 0.015 M sodium citrate. To prepare 1 liter of 20× concentrated stock: 175.4 g NaCl, 88.2 g sodium citrate, add H_2O to 1 liter.

Extinction Coefficients

Beer's law is stated as:

$$T = 10^{-abc}$$

T is % Transmittance, a is an absorptivity constant, specific for each absorbing material, termed the absorption coefficient, b is the light path length (which is usually 1 cm), and c is the concentration of absorbing material. Although transmittance was initially used as the predominant measure of absorption, it has been replaced by absorbance or optical density, denoted as A and O.D. respectively. Absorptivity, a, has been largely replaced by the use of the molar extinction coefficient, E, which is defined as the absorbance of a 1 M solution of a pure compound under standard conditions of solvent, temperature and wavelength. Therefore, in its most used form, Beer's law is $A = Ebc$. Molar extinction coefficients for many substances can be found in the literature.

Example: If E (the extinction coefficient) is known, the concentration can be calculated from the absorbance as follows:

$$c = A/Eb$$

b, the light path of the cuvette is usually 1.0 centimeter.

$$c = A/E$$

E for NADH at 340 nm = 6.22×10^3 and the absorbance at 340 nm = 0.311.

$$\text{Therefore,}\quad c = \frac{0.311}{6.22 \times 10^3} = 0.05 \times 10^{-3}\ \text{mol/L}.$$

Usually, before the extinction coefficients can be of practical value, the concentration must be adjusted to conform to the working limits of the spectrophotometer. This is done by diluting the sample and then making a mathematical correction of the results. To illustrate this point, a 1M solution of bilirubin should have an absorbance of 60,700 at 453 nm in chloroform at 25°C, when measured in a cuvette with a 1 cm light path. Since spectrophotometers are most accurate between 0.1 and 1.5 absorbance units, a dilution should be made so that the absorbance falls within these limits. If a 1M solution had an absorbance of 60,700 a 1/60,700M solution would have an absorbance of 1.0. Therefore, a 1/121,400M solution would have an absorbance of 0.5 which falls within the desired range.

When working in the UV range below 340 nm always either use quartz cuvettes or the new plastic disposable cuvettes especially made for reading in the UV region.

The molar extinction coefficients for many known proteins are available (Kirshenbaum, 1976).

D

Nucleic Acids

1 μg of 1,000 bp DNA = 1.52 pmole
1 pmole of 1,000 bp DNA = 0.66 μg

Spectrophotometric Conversions

1 OD_{260} unit of double-stranded DNA = 50 μg/ml = 0.15 mM (in nucleotides)
1 OD_{260} unit of single-stranded DNA = 33 μg/ml = 0.10 mM (in nucleotides)
1 OD_{260} unit of single stranded RNA = 40 μg/ml = 0.11 mM (in nucleotides)

DNA/Protein Conversions

1 kb of DNA codes 333 amino acids ≈ 3.7×10^4 MW protein
10,000 MW protein = 270 bp DNA
30,000 MW protein = 810 bp DNA
50,000 MW protein = 1.35 kb DNA
100,000 MW protein = 2.7 kb DNA

Oligonucleotide Concentrations

pmol/ml = $OD_{260}/(0.01 \times N)$

N = the number of bases in the oligonucleotide
OD_{260} value is after the dilution factor

PROTOCOL D.1 Fluorometric Estimation of DNA Concentrations

The determination of DNA concentrations in solution is valuable information. The outcome of various types of experiments depends on the accurate determination of DNA concentrations. PCR reactions can be inhibited by excessive DNA concentrations. No amplification product will be synthesized if DNA is below the amplification threshold. It is important to know the DNA concentration when performing ligation reactions and transfections. DNA concentrations have been estimated in solution either from its absorbance at 260 nm or fluorescently with the use of specific dyes.

The method presented below, was designed to be performed using an optical thermal cycler. Volumes can be scaled-up and determinations can also be performed using a fluorometer (Gonzalez and Saiz-Jimenez, 2003).

Materials

Fluorometer or optical thermal cycler

SYBR® Green I (Molecular Probes): 1:10,000

TE buffer, pH 8.0

1. Prepare dilutions of the test DNA solution ranging from 10^{-8} to 0.5 μg/μl.
2. To either tubes or plates, mix 1 μl DNA solution, 44 μl TE buffer and 5 μl SYBR Green I stock. The final concentration of SYBR Green is 1:100,000.
3. In parallel, construct a standard curve using known concentrations of dsDNA.
4. Measurements were performed using a program at constant temperature for 1 min at 25°C, four cycles with two steps for 12 sec each at 25°C, obtaining fluorescence measurements during the second steps, and a final 1 min at 25°C.
5. These steps result in three estimates which are averaged to obtain the final estimate of dsDNA in the solution.

Comments: Linear relationships between fluorescence and DNA concentration occur in a five-fold range between 0.1 and 10^{-6} μg/μl.

If a fluorometer is not available, an alternative is to use the DNA DipStick™ (Invitrogen). This method gives a result within 10 min and does not require specialized instrumentation, ethidium bromide, radioactivity, or agarose gels.

RNA Precipitation

Add 1/10 volume of 3M sodium acetate, pH 5.2 and 2.2 volumes of 100% ethanol. Keep the suspension at –20°C for at least 1 hour. Collect the RNA by centrifugation.

TABLE D.1 The Codon Dictionary. The third nucleotide of each codon is less specific than the first two. The codons read in the 5′ → 3′ direction. For example, UUA = pUpUpA = leucine. The three termination [End] codons are UAA, UAG, and UGA. The initiation codon is AUG

The Genetic Code.

1st Position	2nd Position: U		C		A		G		3rd Position
U	UUU	Phe	UCU	Ser	UAU	Tyr	UGU	Cys	U
	UUC	Phe	UCC	Ser	UAC	Tyr	UGC	Cys	C
	UUA	Leu	UCA	Ser	**UAA**	Stop	**UGA**	Stop	A
	UUG	Leu	UCG	Ser	**UAG**	Stop	UGG	Trp	G
C	CUU	Leu	CCU	Pro	CAU	His	CGU	Arg	U
	CUC	Leu	CCC	Pro	CAC	His	CGC	Arg	C
	CUA	Leu	CCA	Pro	CAA	Gln	CGA	Arg	A
	CUG	Leu	CCG	Pro	CAG	Gln	CGG	Arg	G
A	AUU	Ile	ACU	Thr	AAU	Asn	AGU	Ser	U
	AUC	Ile	ACC	Thr	AAC	Asn	AGC	Ser	C
	AUA	Ile	ACA	Thr	AAA	Lys	AGA	Arg	A
	AUG	Met	ACG	Thr	AAG	Lys	AGG	Arg	G
G	GUU	Val	GCU	Ala	GAU	Asp	GGU	Gly	U
	GUC	Val	GCC	Ala	GAC	Asp	GGC	Gly	C
	GUA	Val	GCA	Ala	GAA	Glu	GGA	Gly	A
	GUG	Val	GCG	Ala	GAG	Glu	GGG	Gly	G

The codons read in the 5′ → 3′ direction.
Termination codons are in bold.

TABLE D.2 Amino Acid Code Degeneracy

Amino acid	Number of codons
Ala	4
Arg	6
Asn	2
Asp	2
Cys	2
Gln	2
Glu	2
Gly	4
His	2
Ile	3
Leu	6
Lys	2
Met	1
Phe	2
Pro	4
Ser	6
Thr	4
Trp	1
Tyr	2
Val	4

TABLE D.3 Amino Acid and Codon Usage. Reprinted with permission of Cold Springs Harbor Laboratory Press, form Harlow and Lane, 1988

	Amino acid usage[a]			Condon usage[b]	
Amino Acid	*E. coli*[c]	human[d]	Codon	*E. coli*[c]	human[d]
Alanine	11.1	7.0	GCU	0.33	0.31
			GCC	0.18	0.40
			GCA	0.28	0.17
			GCG	0.21	0.12
Leucine	7.9	10.4	UUA	0.11	0.05
			UUG	0.11	0.09
			CUU	0.12	0.11
			CUC	0.12	0.22
			CUA	0.03	0.07
			CUG	0.72	0.46
Valine	7.5	6.2	GUU	0.37	0.13
			GUC	0.12	0.27
			GUA	0.28	0.09
			GUG	0.23	0.50
Glycine	7.2	5.7	GGU	0.46	0.15
			GGC	0.40	0.44
			GGA	0.06	0.17
			GGG	0.08	0.24
Arginine	6.5	5.0	CGU	0.46	0.09
			CGC	0.32	0.19
			CGA	0.05	0.10
			CGG	0.06	0.15
			AGA	0.08	0.24
			AGG	0.03	0.23
Lysine	6.4	7.0	AAA	0.72	0.45
			AAG	0.28	0.55
Serine	6.3	8.1	UCU	0.27	0.17
			UCC	0.21	0.26
			UCA	0.13	0.11
			UCG	0.14	0.07
			AGU	0.11	0.11
			AGC	0.14	0.29
Isoleucine	6.1	2.9	AUU	0.39	0.23
			AUC	0.52	0.64
			AUA	0.08	0.13
Threonine	5.8	5.6	ACU	0.36	0.20
			ACC	0.38	0.47
			ACA	0.09	0.21
			ACG	0.17	0.12
Glutamic Acid	5.5	7.3	GAA	0.67	0.40
			GAG	0.33	0.60
Aspartic Acid	5.2	4.9	GAU	0.48	0.38
			GAC	0.52	0.62
Glutamine	4.2	4.5	CAA	0.31	0.26
			CAG	0.69	0.74
Asparagine	3.5	3.5	AAU	0.29	0.34
			AAC	0.71	0.66

TABLE D.3 *Continued*

Amino Acid	Amino acid usage[a] *E. coli*[c]	human[d]	Codon	Condon usage[b] *E. coli*[c]	human[d]
Proline	3.5	4.9	CCU	0.14	0.24
			CCC	0.11	0.41
			CCA	0.20	0.24
			CCG	0.54	0.11
Methionine	2.2	1.8	AUG	1.00	1.00
Phenylalanine	3.6	4.5	UUU	0.50	0.35
			UUC	0.50	0.65
Tyrosine	2.6	3.6	UAU	0.54	0.47
			UAC	0.46	0.53
Histidine	2.5	2.5	CAU	0.64	0.42
			CAC	0.36	0.58
Tryptophan	1.2	1.3	UGG	1.00	1.00
Cysteine	1.1	3.4	UGU	0.45	0.30
			UGC	0.55	0.70

[a]Amino acid usage given in percentage of the total amino acid composition.
[b]Fractional use of each codon for a particular amino acid.
[c]Grantham et al. (1981).
[d]Lathe (1985).

E

Modifications and Motifs

Nomenclature

Proteins are often referred to in shorthand according to the predicted or actual molecular weight of the protein ($\times 10^{-3}$) preceded by "p" for protein or "pp" for phosphoprotein, and followed by the three-letter italicized and superscripted acronym for the gene (e.g., $pp60^{v\text{-}src}$).

Viral oncogene and cellular protooncogene products are distinguished by v- and c-prefixes respectively.

Cloned protein kinase genes, whose functions are not known, have generally been given three letter acronyms.

Human protein kinase genes are styled with three upper case italicized letters (e.g., *TRK*). The protein encoded by *TRK* is referred to as TRK. For protein kinases from other species, the protein products are referred to as Trk.

Sequence Requirements, Motifs and Signals for Processing, Modification, and Translocation

Many proteins are molecular mosaics composed of a wide variety of conserved sequence motifs which comprise structurally distinct domains. One extremely effective method for the characterization of a new protein involves the comparison of its amino acid sequence with the sequences of previously characterized proteins. The rapid increase in the accumulation of sequence data, owing to recombinant DNA technology, has made database searching routine if not mandatory. Sequence comparison methods allow the identity of proteins to be established. The search for functional motifs in proteins and for sites of covalent modification will become increasingly important as a "first approximational" aid to the study of a purified protein of unknown structure and function.

A protein superfamily is a group of proteins that are believed to be, on the basis of their sequence similarity, evolutionarily related. In general, sequence comparison alone does not prove that proteins are homologous (evolved from the same ancestor) but can only indicate that a similarity exists. This likeness usually translates to functional similarity, irrespective of their evolutionary relationships.

It is clearly not sufficient to determine the primary structure of a protein or deduce it from the DNA sequence and expect that this will explain all the properties of a protein. Elucidation of the complete covalent structure of a protein requires the knowledge of the amino acid sequence, as described above, and the chemical nature and positions of all the modifications to the protein that are necessary for its correct function, regulation, and antigenicity.

When an investigator has succeeded in sequencing the target protein or translating a cDNA sequence, a careful analysis of the amino acid sequence may yield a great deal of structural and functional information about the protein. Consensus sequences have been defined for many of the known posttranslational modifications and functional domains. This information may suggest a function for a previously uncharacterized protein.

The term "consensus sequence" refers to those sequence elements including and surrounding the site to be modified. These sequence elements are considered essential for its recognition. It generally takes the form of a short linear sequence of amino acids indicating the identity of the minimum set of amino acids comprising such a site and their position relative to the residue to be modified (Kennelly and Krebs, 1991).

The sequences of many proteins contain short, conserved motifs that are involved in recognition and targeting. These sequences are often separate from other functional properties of the molecule in which they are found. These motifs appear in the primary, linear structure of the protein. It does not include elements from different polypeptide chains or from widely scattered portions of a single polypeptide chain. Therefore, they are not the result of distant segments being brought together as the protein assumes its native conformation. The conservation of these motifs varies: some are highly conserved while others permit substitutions that retain only a certain pattern or charge across the motif.

The usefulness of consensus sequence analysis is based on its simplicity. Summarizing the complexities of the substrate recognition sequence as sets of short sequences has facilitated the evaluation of a large body of observations. A word of caution: the existence of an apparent consensus sequence does not assure that a protein is modified. Consensus sequence information functions best as a guide whose implications must be confirmed or refuted experimentally.

The amino acid which is modified is denoted by an asterisk (*). Where two amino acids function interchangeably, both are listed with a slash (/) separating them. Sequence positions judged to be recognition neutral are denoted by an X.

Many of the modifications mentioned below are discussed in greater detail in Chapter 9.

Signal Peptide Characteristics

Proteins destined for the endoplasmic reticulum, Golgi, lysosome, plasma membrane and the exterior of the cell are initially transported across the membrane of the endoplasmic reticulum (Rapoport et al, 1996). Although there is little amino acid homology among the numerous signal sequences examined, several common characteristics exist: these proteins contain an N-terminal signal peptide (SP) of 15–25 residues. SPs consist of a short, positively charged N-terminal region (n-region), a 7–15 residue hydrophobic core (h-region), and a more polar 3–7 residue C-terminal region (c-region) leading up to the signal peptidase cleavage site between the SP and the mature protein (von Heijne, 1990). The characteristics of eukaryotic SPs are essentially shared by the SPs of prokaryotic and archael proteins (Claros et al, 1997). The study of signal sequence structure and function has been approached from biophysical, biochemical, and biological perspectives. Comparison of all known signal sequences reveals no regions of strict homology. The cleavage site shows the strongest conservation, as expected, since it must be recognized by the signal protease. The length of the hydrophobic core distinguishes it from membrane spanning sequences (24 + 2 residues long) and from hydrophobic sequences of globular proteins (6–8 amino acids long). When examining the presumptive amino acid sequence of a protein, if the first amino acids are predominantly hydrophobic they probably represent a signal or leader sequence. This stretch of amino acids is removed during processing by specific proteases and will not be present on the mature protein.

Signal Peptidase Recognition Site

—A-X-A—
(Gierasch, 1989)

Transmembrane Domain

A continuous region of ~24 amino acids, usually close to the C terminal of the protein is characteristic of a transmembrane domain. Proteins can span the membrane many times. If your cDNA contains a sequence indicating a transmembrane domain the chances are good that you are dealing with an integral membrane protein.

Protein Modification Sequences

Protein N-Myristoylation

Protein N-myristoylation refers to the cotranslational linkage of myristic acid (C14:0) via an amide bond to the NH_2-terminal Gly residues of a variety of eukaryotic cellular and viral proteins (Towler et al, 1988). The reaction is catalyzed by myristoyl-CoA:protein N-myristoyltransferase (NMT).

N-Myristoylproteins have diverse biological functions and diverse intracellular destinations. Some examples of NMT substrates include protein kinases such as the catalytic (C) subunit of cAMP-dependent protein kinase (PK-A) and p60src, phosphatases such as calcineurin B, proteins involved in transmembrane signaling such as several guanine nucleotide-binding a subunits of heterotrimeric G proteins and the gag polyprotein precursors of a number of retroviruses.

Deletion or substitution of the Gly^2 residue of N-myristoylproteins by site-directed mutagenesis prohibits their acylation, allowing the properties of the mutant, nonmyristoylated and wild-type, N-myristoylated species to be compared and contrasted (Gordon et al, 1991).

Consensus sequence:

```
  Q
  N
MGAT/SXSX
  D      DP
  F
  Y
```

X is any amino acid. Residues above the line are tolerated, residues below the line are inhibitory (Magee and Hanley, 1988).

Isoprenylation

C-terminal C-A-A-X

Where C is a Cys, A an aliphatic residue and X any residue.

The CAAX motif signals posttranslational isoprenylation of Cys via a thioether linkage followed by a proteolytic cleavage, C/AAX and carboxymethylation of Cys-COOH (Qin et al, 1992). The hydrophobicity of the C terminus is enhanced by the addition of a hydrophobic tail which may promote anchoring of the protein to the membrane. The isoprenoids transferred by specific prenyltransferases are the C_{15} farnesyl moiety and the C_{20} geranylgeranyl moiety. The specificity of isoprenylation is influenced by the last residue of the CAAX sequence. When X = Leu the protein is usually geranylgeranylated. When X = Ser, Met or Phe the protein is farnesylated. For more about isoprenylation see Chapter 9.

N-Linked Glycoprotein Recognition Signal

— N-X-S/T-

Carbohydrate is added to the asparagine residue.

X can be any amino acid.

Recognition Signal for the Attachment of Glycosaminoglycans to Proteins

A pair or triplet of acidic amino acids closely followed by **S-G X-G**

X can be any amino acid (Bourdon et al, 1987).

Tyrosine Sulfation

-Neg-*Neg*-Tyr-Neg-Neg-

(Aitken, 1990)

Neg is an acidic, negatively charged, amino acid, aspartate, glutamate or tyrosine sulfate. The italicized acidic residue at position-1 N-terminal to the sulfated tyrosine is the strongest determinant.

RNA Binding Motif

K-G-(Y/F)-G-(Y/F)-V-X-(Y/F)
(Dreyfuss et al, 1988)

Protein Kinase Recognition Sequence Motifs

Protein kinases play important roles in the regulation of many cellular processes, altering the functions of their target proteins by phosphorylating specific serine, threonine, and tyrosine residues. Identification of phosphorylation specific site sequences has provided valuable information as to how these enzymes recognize their substrate proteins. Using this data it is now possible to identify potential phosphorylation sites in newly sequenced proteins (Kemp and Pearson, 1990).

The progress in recognizing specific phosphorylation site motifs for many protein kinases has led to the expectation of being able to scan protein sequences and identify potential phosphorylation sites for given protein kinases. One must proceed cautiously however and do not accept the phosphorylation site motifs as "canons" of recognition.

The phosphoacceptor residue will be considered to be at the zero position. The adjacent N-terminal amino acids will be designated –3, –2, –1, 0 and the -COOH terminal +1, +2, +3.

cAMP Dependent Protein Kinases

R-R/K-X-S*/T* > R-X-S*/T*
(Feramisco et al, 1980)

Phosphorylation Sites for Protein Kinase C

R/K-X-X-S*/T*-X-R/K
(Kemp and Pearson, 1990)

Substrates for p34^{cdc2} Protein Kinase

S*/T*-P-X-Z
X being a polar amino acid; Z is generally a basic amino acid (Moreno and Nurse, 1990).

Calmodulin Dependent Protein Kinases

R-X-X-S*/T
(Czernik et al, 1987)
Casein Kinase I
S(P)-X_2-S*/T*
S(P) denotes that this serine is already phosphorylated (Kennelly and Krebs, 1991).

Casein Kinase II (CKII)

S*/T*-(D/E/S(P)$_{1\text{-}3}$, $X_{2\text{-}0}$)
CK II requires D, E or serine phosphate, (S(P)) immediately C-terminal (+1 to +3) to the phosphoacceptor S*/T* (Kennelly and Krebs, 1991).

Extracellular Signal Regulated Kinases (ERKs)

(also known as microtubule-associated protein type 2 kinases)
X-S*/ T*-Pro-X
The phosphorylated site is flanked on its C terminus by a proline residue and X can be any amino acid (Joseph et al, 1993). The cdc2 serine / threonine kinases also phosphorylate substrates that contain this recognition sequence and are also members of this proline-directed protein kinase family.

ATP—Binding Site Motif

G-X-G-X-X-G

GTP—Binding Consensus Sequence

N-K-X-D
(Low et al, 1992)

Subcellular Localization Motifs

The signals and mechanisms that govern the selective localization of proteins to specific intracellular compartments or distinct surface domains are being extensively investigated (Hong and Tang, 1993).

Endoplasmic Reticulum Retention Signal

-K-D-E-L In animal cells, carboxy-terminal
-H-D-E-L In yeast
(Munro and Pelham, 1987)

Endoplasmic Reticulum Localization Signal of Type I Membrane Proteins

-K-X-K-X-X carboxy-terminal
(Shin et al, 1991)

Trans-Golgi Network Localization Signal

-S-X-Y-Q-R-L
The critical tyrosine residue is located at the end of a tight turn within the cytoplasmic domains of both type I and type II membrane proteins. The tetrapeptide sequence YQRL is a necessary and sufficient cytoplasmic domain signal to target integral membrane proteins to the trans Golgi network (TGN) (Wong and Hong, 1993).

Lysosomal Targeting Signal on Cytosolic Proteins

K-F-E-R-Q
(Dice, 1990)

Coated Pit Localization Signal

F-D-N-P-V-Y
(Yokode et al, 1992)

Endocytosis Recognition Motif

Y-X-R-F
(Collawn et al, 1990)

Basolateral Sorting Sequence

R-N-X-D-X-X-S/T-X-X-S
Position number 1-10
(Yokode et al, 1992)

Nuclear Localization Signal, Consensus Sequence

K-R / K-X-R / K
X = any amino acid
The **bipartite motif** (Dingwall and Laskey, 1991) consists of a cluster of two adjacent basic amino acids separated by any ten amino acids from a second cluster, in which three of the next five amino acids are also basic (Chelsky et al, 1989).

The "Classical" Cell Binding Domain

R-G-D
(Pierschbacher and Ruoslahti, 1984)

Hematopoietin Receptor Superfamily

W-S-X-W S
(Cosman et al, 1990)

Retrovirus Aspartic Protease Family

D-T/S-G and G-R-N/D
(Weber et al, 1989)

IgA-Protease Cleavage Site

P-P-T-P
(Pohlner et a1, 1992)

Paired Basic Amino Acid Residue-Processing Motif

K/R-X-X-R
The dibasic site is present in many mammalian proproteins and prohormones (Barr, 1991). The most common processing sites found in mammals are simple pairs of basic residues, K-R and R-R, together with the more prevalent K/R-X-K/R-R. The endoproteases that are involved belong to the subtilisin family of serine proteases.

Protein Databases

Computerized sequence comparison has become a basic laboratory skill in the biological sciences. The number, size and complexity of the databases used by biological scientists have increased dramatically. Gene sequences now accumulated in the data banks ensure that the identification of a new gene sequence can in many cases immediately lead to an assignment of function by analogy. Comprehensive, up-to-the-minute lists of amino acid and nucleotide sequence data are available and easily accessible through computer networks. The majority of these databases are available on the Internet, which links most major academic and industrial research centers. Internet is an international network that links computers world-wide. A good printed introduction to Internet services is available (Kehoe, 1992). There are many information services for the biosciences available through the Internet. These services are usually free and up-to-date. Various servers and tools exist to help retrieve relevant information. These databases, like GenBank™/ EMBL Data Bank, SWISS-PROT, and NBRK should be routinely screened when new sequence is obtained.

There is no task more crucially dependent on up-to-date databases than searching a new sequence for similarity to an existing sequence. Sequence comparisons are performed by self-comparison using the BLAST algorithm (Altschul et al, 1990) which identifies many, but not all, biologically significant sequence similarities. The BLAST algorithm is conveniently amenable to statistical analysis. The BLAST server searches databases that are updated on a daily basis (Henikoff, 1993). By screening databases, you can immediately discover if the newly obtained amino acid or nucleotide sequence is unique, similar or even identical to a previously described protein. This information will influence the direction that the project will take. Perhaps the protein that you are attempting to purify has previously been cloned or an antibody that recognizes the target protein is commercially available.

The World Wide Web (WWW) has simplified the search for finding the "right" piece of information. WWW, which originated at CERN in Geneva, is a global information retrieval service which merges the power of worldwide networks, hypertext and multimedia (Berners-Lee et al, 1992). ExPASy is a WWW server with an emphasis on data relevant to proteins. Through ExPASy one can get access to the SWISS-PROT database of annotated protein sequences (Bairoch and Boeekmann, 1993) and the SWISS-2DPAGE database of two-dimensional gel electrophoresis images (Appel et al, 1993). SWISS-PROT can be searched by protein description, entry name, accession number, or referenced author name. From any SWISS-PROT entry, the user can access corresponding entries in other databases. The ExPASy WWW server has links from SWISS-PROT to EMBL (nucleotide sequences), PDB (three-dimensional structures, Medline (bibliographic references), PROSITE (protein sites and patterns) (Bairoch 1993), OMIM (human genes and genetic diseases), FlyBase (Drosophila genomic data), REBASE (restriction enzymes), SWISS-3DIMAGE (images of three-dimensional structures of proteins), and SWISS-2DPAGE (Appel et al, 1994).

SWISS-2DPAGE contains data on proteins identified on 2D-PAGE. The database contains images of reference maps showing the location of identified proteins. Maps are available for human liver, plasma, red blood cells, HepG2 cells, HepG2-secreted proteins, lymphoma, CSF, macrophage-like cell line, and erythroleukemia cells. Remarkably, if the protein of interest has not yet been identified on a 2D-PAGE map, the theoretical location may be displayed! One can select a reference 2D-PAGE map, and by clicking on a spot, retrieve the corresponding database entry, http://www.expasy.ch/ch2d/. Information retrieval has never been so versatile.

F

Centrifugation

Centrifugation is a technique used to separate particles from a solution. In the biological sciences the particles are usually cells, subcellular organelles, large molecules, or aggregates. There are two types of centrifugation procedures: preparative, used to isolate specific particles; and analytical, which involves measuring physical properties of a sedimenting particle.

As a rotor spins in a centrifuge, each particle in the suspension sediments at a rate which is proportional to the centrifugal force applied. In addition, the viscosity of the sample solution and the physical properties of the particle affect the sedimentation rate. At a fixed centrifugal force and liquid viscosity, the sedimentation rate of the particle is proportional to its size (molecular weight) and the difference between its density and the density of the solution. Proteins in solution tend to sediment at high centrifugal fields, thereby overcoming the opposing tendency of diffusion. Centrifugal methods are routinely used to separate mixtures of proteins, other macromolecules and organelles. The charts and tables that are presented in this section should provide most investigators with the information necessary to choose a suitable rotor and reach the required speed. Some key terms will also be reviewed.

The relative centrifugal field, **RCF**, is the ratio of the centrifugal acceleration at a specified radius and speed, $\mathbf{r\omega^2}$, to the standard acceleration of gravity, g, according to the following formula:

$$RCF = r\omega^2/g$$

The angular velocity $\boldsymbol{\omega}$, or rate of rotation is measured in radians per second ($\omega = 2\pi\text{rpm}/60$, or $\omega = 0.10472 \times \text{rpm}$).

r = radius in mm measured from the center of the spindle to the bottom of the rotor bucket, and g is the standard acceleration of gravity, $9807\,\text{mm/S}^2$.

After substitution:

$$\text{RCF} = 1.12\text{r}(\text{rpm}/1000)^2$$

In day-to-day use, centrifuge operating speeds are referred to in revolutions per minute, rpm.

Where R = radius in inches: $RCF = 28.38R(rpm/1000)^2$
Where r = radius in centimeters: $RCF = 11.17r(rpm/1000)^2$

Pelleting separates particles of different sedimentation coefficients, the largest particles in the sample arrive at the bottom of the tube first. Pelleting is the most common separation method. The centrifuge tube is filled with a uniform mixture of sample solution. Centrifugation results in two fractions; a pellet that contains the sedimented material, and a supernatant, consisting of the unsedimented material. The pellet is a mixture of all the sedimented components in the sample. The supernatant is easily recovered by decanting.

Differential centrifugation is the successive pelleting of particles from previous supernatants using increasingly higher forces. The supernatant is recentrifuged at higher speeds to obtain further purification, with formation of a new pellet and supernatant.

The relative pelleting efficiency of each rotor is measured by its ***k*** factor, or clearing factor. This factor is useful for estimating the time *t* (in hours) required for pelleting,

$$t = k/s$$

where ***s*** is the sedimentation coefficient of the particle of interest.

The *s* value unit is seconds which is generally expressed in Svedberg units, **S**, where 1 S is equal to 10^{-13} seconds. The usual practice is to use the standard sedimentation coefficient S_{20w}, which is based on sedimentation in water at 20°C. Ribosomal subunits possessing a sedimentation coefficient of 18×10^{-13} seconds are said to be 18 S.

Density gradient centrifugation is a single centrifugation in a gradient or layered medium. This method is slightly more complicated than differential centrifugation, but has compensating advantages. Density gradient centrifugation enables the complete separation of several components in a complex mixture and also permits analytical measurements to be made. It might be advantageous to have a series of density steps rather than a continuous gradient. A component will ultimately band at a particular interface when the next layer is denser than the particle. In banding runs it is often advantageous to place a dense solution at the bottom of the tube to act as a "cushion" to stop material on the cushion that would otherwise become pelleted and be difficult to recover quantitatively.

Isopycnic banding takes place in a continuous gradient when the particle density equals the gradient density. Isopycnic technique separates particles solely on the basis of their density differences, independent of time.

Prior to using a centrifuge for the first time, obtain operating instructions and/or a short demonstration from a veteran lab member and try to find and read the manual for safe handling of centrifuge and rotors. Centrifuges and rotors are expensive and potentially dangerous if used incorrectly.

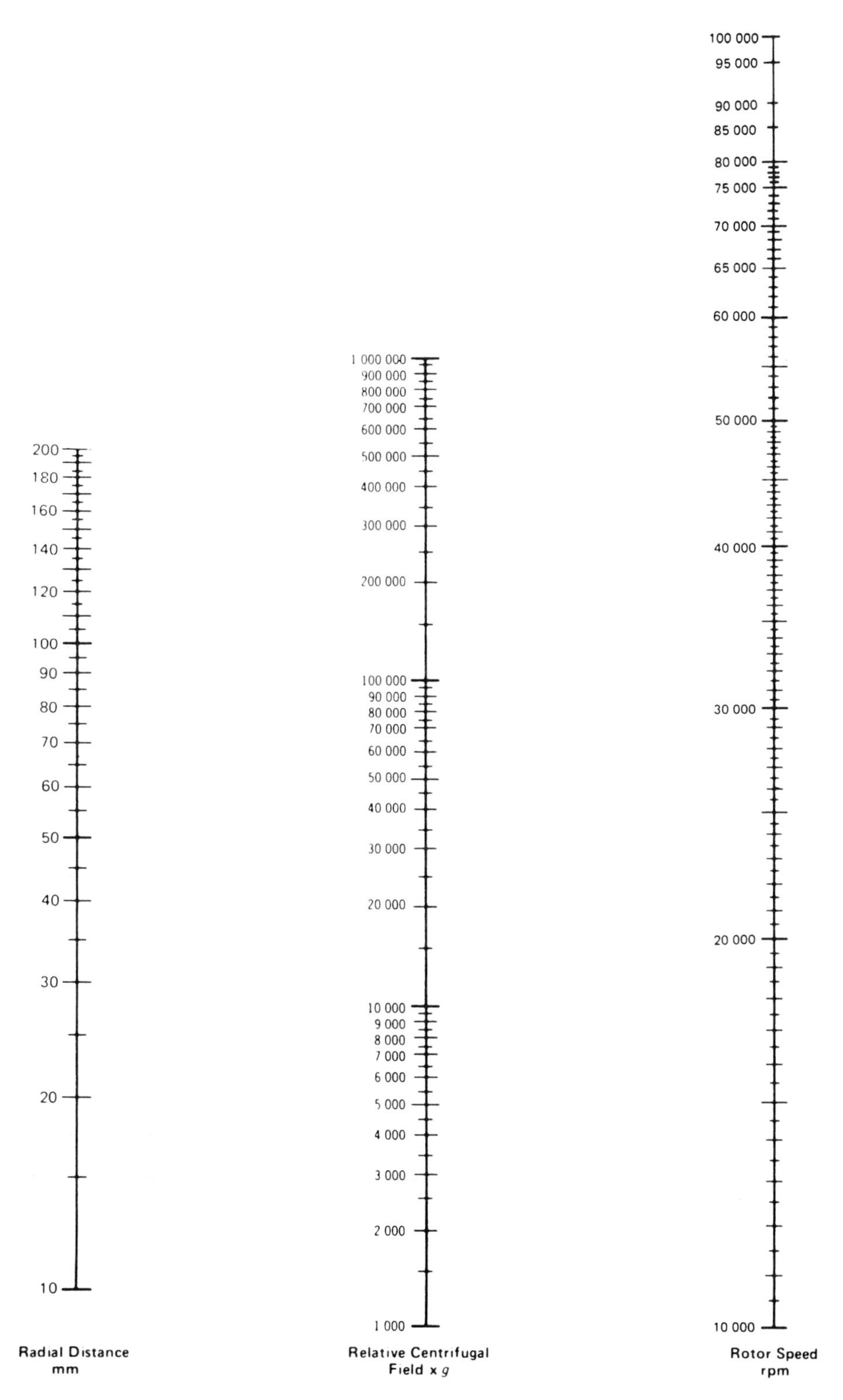

FIGURE F.1 Nomogram. Align a straightedge through known values in two columns; read the desired figure where the straightedge intersects the third column. Reprinted from *Rotors and Tubes for Preparative Ultracentrifuges*, Spinco Business Center of Beckman Instruments, Palo Alto, CA, 1994, by permission of Beckman Instruments, Inc.

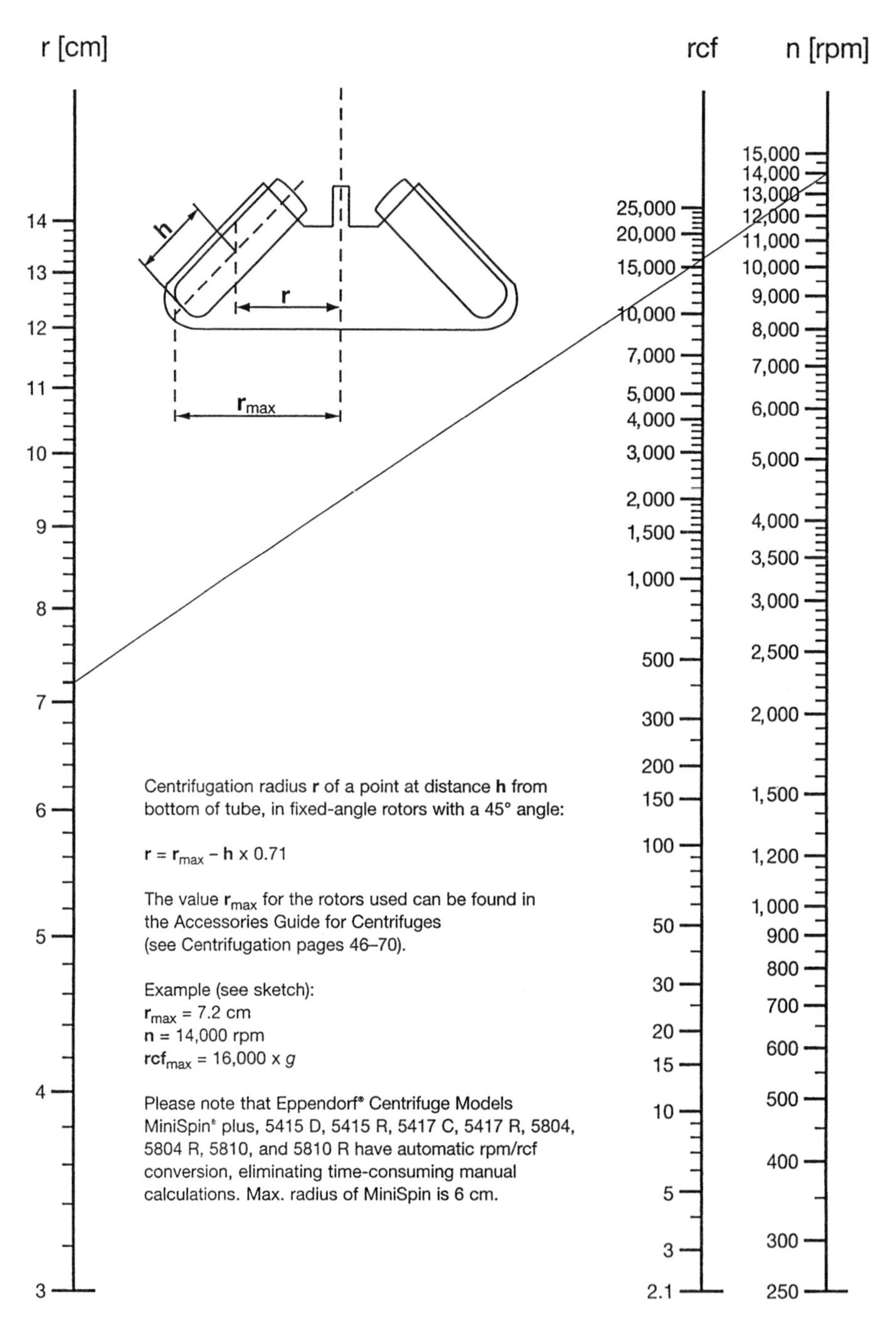

FIGURE F.2 Conversion table for rpm/rcf, lowspeed nomogram.

TABLE F.1 Reprinted from *Relative Centrifugal Fields (× g) at r_{max} for J Rotors in J2 Series Centrifuges*, Spinco Business Center of Beckman Instruments, Palo Alto, CA, 1993, by permission of Beckman Instruments, Inc. Entries in this table are calculated from the equation RCF = 1.12r (RPM/1000)^2 and are rounded to three significant digits. The centrifugal force at a given radius in a rotor is a function of run speed. Comparisons of forces between different rotors are made by comparing the rotors' RCFs. When rotational speed is selected so that identical samples are subjected to the same RCF in two different rotors, the samples have then been subjected to the same maximum force

	ROTOR NAME AND r_{max} IN MILLIMETERS																
RPM	JA-21 102	JA-20 108	JA-20.1 98*	JA-20.1 115*	JCF-Z 81†	JCF-Z 89†	JV-20 93	JA-18 132	JA-18.1 112‡	JA-18.1 116‡	JA-17 123	JA-14 137	JS-13.1 140	JA-12 144	JA-10 158	JS-7.5 165	JS-4.3 204	JE-6B 86**
500	29	30	27	32	23	25	26	37	31	32	34	38	39	40	44	46	57	24
1000	114	121	110	129	91	100	104	148	125	130	138	153	157	161	177	185	228	96
1500	257	272	247	290	204	224	234	333	282	292	310	345	353	363	398	416	514	217
2000	457	484	439	515	363	399	417	591	502	520	551	614	627	645	708	739	910	385
2500	714	756	686	805	567	623	651	924	784	812	861	959	980	1010	1110	1160	1430	602
3000	1030	1090	990	1160	816	897	937	1330	1130	1170	1240	1380	1410	1450	1590	1660	2060	870
3500	1400	1480	1340	1580	1110	1220	1280	1810	1540	1590	1690	1880	1920	1980	2170	2260	2800	1180
4000	1830	1940	1760	2060	1450	1590	1670	2370	2010	2080	2200	2460	2510	2580	2830	2960	3660	1540
4300	2110	2240	2030	2380	1680	1840	1930	2730	2320	2400	2550	2840	2900	2980	3270	3420	4220	1780
4500	2310	2450	2220	2610	1840	2020	2110	2990	2540	2630	2790	3110	3180	3270	3580	3740		1950
5000	2860	3020	2740	3220	2270	2490	2600	3700	3140	3250	3440	3840	3920	4030	4420	4620		2410
5500	3460	3660	3320	3900	2740	3020	3150	4470	3790	3930	4170	4640	4740	4880	5350	5590		2910
6000	4110	4350	3950	4640	3270	3590	3750	5320	4520	4680	4960	5520	5640	5810	6370	6650		3470
6500	4830	5110	4640	5440	3830	4210	4400	6250	5300	5490	5820	6480	6620	6810	7480	7810		
7000	5600	5930	5380	6310	4450	4880	5100	7240	6150	6370	6750	7520	7680	7900	8670	9060		
7500	6430	6800	6170	7250	5100	5610	5860	8320	7060	7310	7750	8630	8820	9070	9950	10400		
8000	7310	7740	7020	8240	5810	6380	6670	9460	8030	8310	8820	9820	10000	10300	11300			
8500	8250	8740	7930	9310	6550	7200	7530	10700	9060	9390	9950	11100	11300	11700	12800			
9000	9250	9800	8890	10400	7350	8070	8440	12000	10200	10500	11200	12400	12700	13100	14300			
9500	10300	10900	9910	11600	8200	9000	9400	13300	11300	11700	12400	13800	14200	14600	16000			

Continued

TABLE F.1 *Continued*

RPM	ROTOR NAME AND r_{max} IN MILLIMETERS																
	JA-21 102	JA-20 108	JA-20.1 98*	JA-20.1 115*	JCF-Z 81†	JCF-Z 89†	JV-20 93	JA-18 132	JA-18.1 112‡	JA-18.1 116‡	JA-17 123	JA-14 137	JS-13.1 140	JA-12 144	JA-10 158	JS-7.5 165	JS-4.3 204	JE-6B 86**
10000	11400	12100	11000	12900	9100	10000	10400	14800	12500	13000	13800	15300	15700	16100	17700			
10500	12600	13300	12100	14200	10000	11000	11500	16300	13800	14300	15200	16900	17300	17800				
11000	13800	14600	13300	15600	11000	12100	12600	17900	15200	15700	16700	18600	19000	19500				
11500	15100	16000	14500	17000	12000	13200	13800	19600	16600	17200	18200	20300	20700	21300				
12000	16500	17400	15800	18600	13100	14400	15000	21300	18100	18700	19800	22100	22600	23200				
12500	17900	18900	17200	20100	14200	15600	16300	23100	19600	20300	21500	24000	24500					
13000	19300	20400	18600	21800	15300	16800	17600	25000	21200	22000	23300	25900	26500					
13500	20800	22000	20000	23500	16500	18200	19000	26900	22900	23700	25100	28000						
14000	22400	23700	21500	25200	17800	19500	20400	29000	24600	25500	27000	30100						
14500	24000	25400	23100	27100	19100	21000	21900	31100	26400	27300	29000							
15000	25700	27200	24700	29000	20400	22400	23400	33300	28200	29200	31000							
15500	27500	29100	26400	30900	21800	23900	25000	35500	30100	31200	33100							
16000	29300	31000	28100	33000	23200	25500	26700	37800	32100	33300	35300							
16500	31100	32900	29900	35100	24700	27100	28400	40200	34200	35400	37500							
17000	33000	35000	31700	37200	26200	28800	30100	42700	36300	37500	39800							
17500	35000	37000	33600	39500	27800	30500	31900	45300	38400	39800								
18000	37000	39200	35600	41700	29400	32300	33700	47900	40600	42100								
18500	39100	41400	37600	44100	31000	34100	35600											
19000	41200	43700	39600	46500	32700	36000	37600											
19500	43400	46000	41700	49000	34500	37900	39600											
20000	45700	48400	43900	51500	36300	39900	41700											
20500	48000																	
21000	50400																	

*Inner and outer rows.

†JCF-Z with small pellet core has an r_{max} of 81 mm; with all other JCF-Z cores r_{max} is 89 mm.

‡The JA-18.1 rotor can spin tubes at two different angles: 45° and 25°. The 45° angle configuration gives an r_{max} of 116 mm. The 25° configuration gives an r_{max} of 112 mm.

**g-force at elutriation boundary.

TABLE F.2 Sorvall®Lowspeed RCF Chart. Reprinted with permission of DU PONT, from SORVALL® Centrifuges.

ROTOR	M & A-384 inner row	M & A-384 outer row	SP/X & A-500	HL-4 omnicarrier	HL-4/50 ml open bucket	HL-4/50 ml sealed bucket	HL-4/100 ml open bucket	H-1000B buckets	H-1000B microplate	F-12/M.18
RADIUS max (in centimeters)	9.13	12.15	12.34	16.39	18.08	17.54	20.43	18.67	13.29	7.54
rpm										
500	25	34	34	46	50	49	57	52	37	21
1000	102	136	138	183	202	196	228	209	148	84
1500	229	305	310	412	454	441	513	469	334	189
2000	408	543	551	732	808	784	913	834	594	337
2500	637	848	861	1144	1262	1225	1426	1303	928	526
3000	918	1221	1241	1435[6]	1818	1763[4]	1543[5]	1877	1336	758
3500	1249	1663	1689	2116[3]	2068[2]	2265[3]	2638[3]	2135[7]	1716[8]	1032
4000	1632	2171	2205		2335[3]					1348
4500	2065	2748	2791							1705
5000	2550	3393	3446							2106
5500	3085	4105	4170							2548
6000	3670	4890	4960							3032
6500										3558
7000										4127
7500										4737
8000										5390
8500										6085
9000										6822
9500										7601
10000										8422

Continued

TABLE F.2 *Continued*

ROTOR	M & A-384 inner row	M & A-384 outer row	SP/X & A-500	HL-4 omnicarrier	HL-4/50 ml open bucket	HL-4/50 ml sealed bucket	HL-4/100 ml open bucket	H-1000B buckets	H-1000B microplate	F-12/M.18
RADIUS max (in centimeters)	9.13	12.15	12.34	16.39	18.08	17.54	20.43	18.67	13.29	7.54
rpm										
10500										9285
11000										10191
11500										11138
12000										12127
500	39	73	39	65	74	58	73	60		
1000	157	291	157	258	295	232	293	241		
1500	352	655	352	581	663	523	659	541		
2000	626	1164	626	1033	1180	929	1172	962		
2500	979	1819	979	1613	1843	1452	1832	1504		
3000	1409	2620	1409	2323	2311[10]	1821[10]	2465[12]	2165		
3500	1918	3566	1918	3162				2620[11]		
4000	2506	4657	2506	4130						
4500	2896	5895	3171	5227						
5000	3000[1]	7280	3915	6453						
5500			4737	7120[9]						
6000			5640							

TABLE F.3 SORVALL® Superspeed RCF Chart. *Relative Centrifugal Forces (in g) for SORVALL® Superspeed rotors in RC-2*, RC-2B*, RC-3B*, RC-3C*, RC-5, RC-5B, RC-5B Plus, RC-5C, RC-5C Plus, RC-24, RC-26 Plus and RC-28S* centrifuges.* **Reprinted with permission of DUPONT, from SORVALL® Centrifuges**

ROTOR RADIUS max (in centimeters) rpm	SE-12 9.33	SM-24 inner 9.10	SM-24 outer 11.07	SS-34** 10.70	SA-600 12.96	SA-300 9.58	GSA 14.57	GS-3 15.13	SH-80 10.16	SLA-1000 11.77	SLA-1500 13.6	SLA-3000 15.14	F20/ Micro 11.43
500	26	25	30	29	36	27	40	42	28	33	38	42	31
1000	104	101	123	119	144	107	162	169	113	131	152	169	127
1500	234	228	278	268	325	241	366	380	255	296	342	381	287
2000	416	406	494	478	579	428	650	676	453	526	608	676	510
2500	651	635	772	746	904	669	1017	1056	709	822	949	1057	797
3000	937	914	1112	1075	1302	963	1464	1521	1021	1183	1367	1522	1149
3500	1276	1245	1514	1464	1773	1311	1993	2070	1390	1611	1861	2072	1563
4000	1667	1626	1978	1912	2316	1712	2603	2704	1815	2104	2431	2706	2042
4500	2110	2058	2503	2420	2931	2167	3295	3422	2298	2662	3076	3425	2585
5000	2605	2541	3091	2987	3619	2675	4068	4225	2837	3287	3798	4228	3191
5500	3152	3074	3740	3615	4379	3237	4923	5112	3432	3977	4595	5116	3862
6000	3751	3659	4451	4302	5211	3852	5858	6084	4085	4733	5469	6088	4596
6500	4403	4294	5224	5049	6116	4521	6876	7140	4794	5555	6418	7145	5394
7000	5106	4980	6058	5856	7093	5243	7974	8281	5560	6442	7444	8287	6255
7500	5862	5717	6955	6722	8142	6019	9154	9506	6383	7395	8545	9513	7181
8000	6669	6505	7913	7649	9264	6849	10415	10816	7263	8414	9722	10823	8171
8500	7529	7343	8933	8635	10459	7731	11758	12210	8199	9499	10976	12218	9224
9000	8441	8233	10015	9681	11725	8668	13182	13689	9192	10649	12305	13698	10341
9500	9405	9173	11159	10786	13064	9658	14687	K = 4211	10242	11865	13710	15263	11522
10000	10421	10164	12365	11951	14476	10701	16274		11348	13147	15191	16911	12767
10500	11489	11206	13632	13176	15960	11798	17942		12511	14495	16748	18645	14075
11000	12610	12299	14961	14461	17516	12948	19692		13731	15908	18381	20463	15448
11500	13782	13442	16352	15806	19144	14152	21523		15008	17387	20090	22365	16884

Continued

TABLE F.3 *Continued*

ROTOR RADIUS max (in centimeters) rpm	SE-12 9.33	SM-24 inner 9.10	SM-24 outer 11.07	SS-34** 10.70	SA-600 12.96	SA-300 9.58	GSA 14.57	GS-3 15.13	SH-80 10.16	SLA-1000 11.77	SLA-1500 13.6	SLA-3000 15.14	F20/ Micro 11.43
12000	15007	14637	17805	17210	20845	15409	23435		16342	18932	21875	24352	18384
12500	16283	15882	19320	18674	22619	16720	25429		17732	20542	23736	K = 2365	19948
13000	17612	17178	20897	20198	24464	18084	27504		19179	22219	25673		21576
13500	18993	18525	22535	21782	26383	19502	K = 2024		20683	23961	27686		23268
14000	20426	19922	24235	23425	28373	20974			22243	25768	29775		25023
14500	21911	21371	25997	25128	30436	22499			23860	27642	31939		26843
15000	23448	22870	27821	26891	32571	24077			25534	29581	34180		28726
15500	25037	24420	29707	28714	34779	25709			27265	31586	K = 1473		30673
16000	26679	26021	31654	30596	37059	27394			29052	33657			32684
16500	28372	27673	33664	32539	39411	29133			30896	35793			34759
17000	30118	29375	35735	34540	41837	30925			32797	K = 1725			36897
17500	31916	31129	37868	36602	K = 747	32771			34755				39099
18000	33766	32933	40063	38724		34671			36769				41366
18500	35667	34788	42319	40905		36624			38840				43696
19000	37622	36694	44638	43146		38630			40968				46089
19500	39628	38651	47018	45447		40690			43153				48547
20000	41686	40658	49460	47807		42803			45394				51069
20500	43796	42717	51965	50228		44970			K = 400				K = 182
21000	45959	K = 593	K = 436	K = 716		47191							
21500	48174					49465							
22000	50441					51792							
22500	52759					54173							
23000	55130					56608							
23500	57553					59095							
24000	60028					61637							
24500	K = 335					64232							
25000						66880							
						K = 606							

*The Superlite™ rotors (SLA-1500, SLA-3000) and the SH-3000 rotor DO NOT run in these centrifuges.

**With or without KSB system.

TABLE F3 *Continued*

ROTOR RADIUS max (in centimeters) rpm	TZ-28 9.52	TZ-28GK 9.52	HB-4 14.61	HS-4 17.22	HB-6 14.63	SH-3000 (Bucket) 18.54	SH-3000 (Microplates) 15.01	SV-80 10.16	SV-288 9.02
500	26	26	40	48	41	52	42	28	25
1000	106	106	163	192	163	207	168	113	100
1500	239	239	367	432	368	466	377	255	226
2000	425	425	652	769	654	828	671	453	403
2500	664	664	1019	1202	1021	1294	1048	709	629
3000	957	957	1468	1731	1471	1864	1509	1021	906
3500	1302	1302	1999	2356	2002	2537	2054	1390	1234
4000	1701	1701	2611	3077	2615	3313	2683	1815	1612
4500	2153	2153	3304	3895	3309	4194	3395	2298	2040
5000	2658	2658	4079	4808	4085	K = 9484		2837	2518
5500	3216	3216	4936	5818	4943			3432	3047
6000	3828	3828	5874	6924	5883			4085	3627
6500	4492	4492	6894	8126	6904			4794	4256
7000	5210	5210	7996	9425	8007			5560	4936
7500	5981	5981	9179	10820	9192			6383	5667
8000	6805	6805	10444	K = 3916	10459			7263	6448
8500	7682	7682	11790		11807			8199	7279
9000	8613	8613	13218		13237			9192	8161
9500	9597	9597	14728		14748			10242	9092
10000	10633	10633	16319		16342			11348	10075
10500	11723	11723	17992		18017			12511	11108
11000	12866	12866	19746		19773			13731	12191
11500	14063	14063	21582		21612			15008	13324
12000	15312	15312	23499		23532			16342	14508
12500	16615	16615	25499		25534			17732	15742

Continued

TABLE F3 *Continued*

ROTOR RADIUS max (in centimeters) rpm	TZ-28 9.52	TZ-28GK 9.52	HB-4 14.61	HS-4 17.22	HB-6 14.63	SH-3000 (Bucket) 18.54	SH-3000 (Microplates) 15.01	SV-80 10.16	SV-288 9.02
13000	17971	17971	27579		27617			19179	17027
13500	19380	19380	K = 1682		K = 1765			20683	18362
14000	20842	20842						22243	19747
14500	22357	22357						23860	21183
15000	23926	23926						25534	22669
15500	25547	25547						27265	24209
16000	27222	27222						29052	25792
16500	28950	28950						30896	27430
17000	30731	30731						32797	29117
17500	32566	32566						34755	30855
18000	34453	34453						36769	32644
18500	36394	36394						38840	34482
19000	38388	38388						40968	36371
19500	40435	K = 672						K = 98	38311
20000	42535								40301
20500	K = 606								K = 210

These tables give the maximum RCF in *g* for all currently available SORVALL® rotors for superspeed applications up to 26,000 rpm.
To find the average or minimum RCF, consult your rotor instruction manual for the average or minimum radius, and use the formula given below.
To calculate the most exact value of RCF at any speed for any given radius, use the RCF equation:

$$RCF = 28.38R\left(\frac{rpm}{1000}\right)^2$$ where R = radius in inches

$$RCF = 11.17r\left(\frac{rpm}{1000}\right)^2$$ where r = radius in centimeters

TABLE F.4 Physical Specifications of Beckman Ultracentrifuge Rotors. Reprinted from *Rotors and Tubes for Preparative Ultracentrifuges,* Spinco Business Center of Beckman Instruments, Palo Alto, CA, 1994, by permission of Beckman Instruments, Inc

Rotor	Maximum Speed[a] (rpm)	Relative Centrifugal Field[b] ($\times g$) at r_{max}	k Factor	Number of Tubes × Nominal Capacity (mL) of Largest Tube	Nominal Rotor Capacity (mL)	For Use in Instruments Classified
Rotors for Centrifuging Extremely Small Particles						
NVT 90	90000	645000	10	8 × 5.1	40.8	H,R,S
Type 90 Ti	90000	694000	25	8 × 13.5	108	F,G,H,R,S
VTi 90	90000	645000	6	8 × 5.1	40.8	H,R,S
Type 80 Ti	80000	602000	28	8 × 13.5	108	F,G,H,R,S
VTi 80	80000	510000	8	8 × 5.1	40.8	H,R,S
Type 75 Ti	75000	502000	35	8 × 13.5	108	F[c],G[d],H,R,S
Type 70.1 Ti	70000	450000	36	12 × 13.5	162	F[c],G[d],H,R,S
NVT 65.2	65000	416000	16	16 × 5.1	81.6	H,R,S
NVT 65	65000	402000	21	8 × 13.5	108	H,R,S
VTi 65.2	65000	416000	10	16 × 5.1	81.6	H,R,S
VTi 65.1	65000	402000	13	8 × 13.5	108	H,R,S
VTi 65	65000	404000	10	8 × 5.1	40.8	H,R,S
Type 65	65000	368000	45	8 × 13.5	108	A,B,C,D,F,G,H,Q,R,S
Type 50 Ti	50000	226000	78	12 × 13.5	162	B,C,D,F,G,H,Q,R,S
Rotors for Centrifuging Small Particles in Volume						
Type 70 Ti	70000	504000	44	8 × 38.5	308	F[c],G[d],H,R,S
Type 60 Ti	60000	362000	63	8 × 38.5	308	B,F[c],G[d],H,R,S
Type 55.2 Ti	55000	340000	64	10 × 38.5	385	G,H,R,S
VC 53	53000	249000	36	8 × 39	312	H,R,S
Type 50.2 Ti	50000	302000	69	12 × 38.5	462	F,G,H,R,S
VAC 50	50000	242000	36	10 × 39	390	H,R,S
VTi 50	50000	242000	36	8 × 39	312	H,R,S
Type 45 Ti	45000	235000	133	6 × 94	564	F,G,H,Q,R,S
Type 42.1	42000	195000	133	8 × 38.5	308	H,R,S
Type 35	35000	143000	225	6 × 94	564	H,R,S
Type 28	28000	94800	393	8 × 40	320	H[e],R,S
Rotors for Differential Flotation						
Type 50.4 Ti	50000	312000[f]	33	44 × 6.5	286	F,G,H,R,S
Type 50.3 Ti	50000	223000	49	18 × 6.5	117	B,C,D,F,G,H,Q,R,S
Type 42.2 Ti	42000	223000	9	72 × 230 μL	16.5	G,H,R,S
Type 25	25000	92500[g]	62	100 × 1	100	C,D,F,G,H,R,S
Rotors for Centrifuging Large Particles						
Type 50	50000	196000	65	10 × 10	100	A,B,C,D,F,G,H,Q,R,S
Type 40	40000	145000	122	12 × 13.5	162	A,B,C,D,F,G,H,Q,R,S
Type 30	30000	106000	213	12 × 38.5	462	H,R,S
Rotors for Centrifuging Large Particles in Volume						
Type 21	21000	60000	402	10 × 94	940	H,R,S
Type 19	19000	53900	951	6 × 250	1500	H,R,S
Type 16	16000	39300	1350	6 × 250	1500	H,R,S
Rotors for Isopycnic and Rate-Zonal Gradients						
SW 65 Ti	65000	421000	46	3 × 5	15	B,C,D,F,G,H,Q,R,S
SW 60 Ti	60000	485000	45	6 × 4.4	26.4	G,H,R,S
SW 55 Ti	55000	368000	48	6 × 5	30	B,C,D,F,G,H,Q,R,S
SW 50.1	50000	300000	59	6 × 5	30	A,B,C,D,F,G,H,Q,R,S

Continued

TABLE F.4 *Continued*

Rotor	Maximum Speed[a] (rpm)	Relative Centrifugal Field[b] (×g) at r_{max}	k Factor	Number of Tubes × Nominal Capacity (mL) of Largest Tube	Nominal Rotor Capacity (mL)	For Use in Instruments Classified
Rotors with Long, Slender Tubes for Rate-Zonal Gradients						
SW 41 Ti	41000	288000	124	6 × 13.2	79.2	C,D,F,G,H,R,S
SW 40 Ti	40000	285000	137	6 × 14	84	G,H,R,S
SW 28.1[h]	28000	150000	276	6 × 17	102	C,D,F,G,H,R,S
Rotors for Larger-Volume Density Gradients						
SW 30.1	30000	124000	138	6 × 8	48	B,C,D,F,G,H,R,S
SW 30	30000	124000	138	6 × 20	120	B,C,D,F,G,H,R,S
SW 28[h]	28000	141000	245	6 × 38.5	231	C,D,F,G,H,R,S
SW 25.1	25000	90400	337	3 × 34	102	A,B,C,D,F,G,H,Q,R,S

[a]Maximum speeds are based on a solution density of 1.2g/mL in all rotors except for the Type 60 Ti, Type 42.1, and the Type 35, which are rated for a density of 1.5g/mL; and the near vertical tube and vertical tube rotors, which are rated for a density of 1.7g/mL.

[b]Relative Centrifugal Field (RCF) is the ratio of the centrifugal acceleration at a specified radius and speed ($r\omega^2$) to the standard acceleration of gravity (g) according to the following formula: RCF = $r\omega^2/g$ where r is the radius in millimeters, ω is the angular velocity in radians per second (2πRPM/60), and g is the standard acceleration of gravity (9807 mm/s^2). After substitution: RCF = 1.12r (RPM/1000)^2.

[c]Class F, Model L2-50 and Model L3's only.

[d]Class G, Model L3's only.

[e]Except L5 and L5B.

[f]Maximum RCF measured at outer row.

[g]Maximum RCF measured at the third (i.e., outermost) row. Radial distances are those of the third row.

[h]SW 28.1M and SW 28M rotors are specially modified versions of the SW 28.1 and SW 28 rotors, and are equipped with a mechanical overspeed system. These rotors are otherwise identical to the SW 28.1 and SW 28 rotors.

TABLE F.5 Density, Refractive Index, and Concentration Data—Cesium Chloride at 25°C, Mol. Wt. 168.37. Reprinted with permission of the Spinco Division of Beckman Instruments, Palo Alto, CA, from *Techniques of Preparative, Zoned, and Continuous Flow Ultracentrifugation*

Density (g/cm^3)*	Refractive Index, η_D	% by Weight	mg/ml of Solution**	Molarity
1.0047	1.3333	1	10.0	0.056
1.0125	1.3340	2	20.2	0.119
1.0204	1.3348	3	30.6	0.182
1.0284	1.3356	4	41.1	0.244
1.0365	1.3364	5	51.8	0.308
1.0447	1.3372	6	62.8	0.373
1.0531	1.3380	7	73.7	0.438
1.0615	1.3388	8	84.9	0.504
1.0700	1.3397	9	96.3	0.572
1.0788	1.3405	10	107.9	0.641
1.0877	1.3414	11	119.6	0.710
1.0967	1.3423	12	131.6	0.782
1.1059	1.3432	13	143.8	0.854
1.1151	1.3441	14	156.1	0.927
1.1245	1.3450	15	168.7	1.002

Continued

TABLE F.5 *Continued*

Density (g/cm^3)*	Refractive Index, η_D	% by Weight	mg/ml of Solution**	Molarity
1.1340	1.3459	16	181.4	1.077
1.1437	1.3468	17	194.4	1.155
1.1536	1.3478	18	207.6	1.233
1.1637	1.3488	19	221.1	1.313
1.1739	1.3498	20	234.8	1.395
1.1843	1.3508	21	248.7	1.477
1.1948	1.3518	22	262.9	1.561
1.2055	1.3529	23	277.3	1.647
1.2164	1.3539	24	291.9	1.734
1.2275	1.3550	25	306.9	1.823
1.2387	1.3561	26	322.1	1.913
1.2502	1.3572	27	337.6	2.005
1.2619	1.3584	28	353.3	2.098
1.2738	1.3596	29	369.4	2.194
1.2858	1.3607	30	385.7	2.291
1.298	1.3619	31	402.4	2.390
1.311	1.3631	32	419.5	2.492
1.324	1.3644	33	436.9	2.595
1.336	1.3657	34	454.2	2.698
1.3496	1.3670	35	472.4	2.806
1.363	1.3683	36	490.7	2.914
1.377	1.3696	37	509.5	3.026
1.391	1.3709	38	528.6	3.140
1.406	1.3722	39	548.3	3.257
1.4196	1.3735	40	567.8	3.372
1.435	1.3750	41	588.4	3.495
1.450	1.3764	42	609.0	3.617
1.465	1.3778	43	630.0	3.742
1.481	1.3792	44	651.6	3.870
1.4969	1.3807	45	673.6	4.001
1.513	1.3822	46	696.0	4.134
1.529	1.3837	47	718.6	4.268
1.546	1.3852	48	742.1	4.408
1.564	1.3868	49	766.4	4.552
1.5825	1.3885	50	791.3	4.700
1.601	1.3903	51	816.5	4.849
1.619	1.3920	52	841.9	5.000
1.638	1.3937	53	868.1	5.156
1.658	1.3955	54	895.3	5.317
1.6778	1.3973	55	922.8	5.481
1.699	1.3992	56	951.4	5.651
1.720	1.4012	57	980.4	5.823
1.741	1.4032	58	1009.8	5.998
1.763	1.4052	59	1040.2	6.178
1.7846	1.4072	60	1070.8	6.360
1.808	1.4093	61	1102.9	6.550
1.831	1.4115	62	1135.8	6.746
1.856	1.4137	63	1167.3	6.945
1.880	1.4160	64	1203.2	7.146
1.9052	1.4183	65	1238.4	7.355

*Computed from the relatioship $\rho^{25} = 10.2402\ \eta_D{}^{25} - 12.6483$ for densities between 1.00 and 1.38, and $\rho^{25} = 10.8601\ \eta_D{}^{25} - 13.4974$ for densities above 1.37 (Bruner and Vinograd, 1965).

**Divide by 10.0 to obtain % w/v.

Density data are from International Critical Tables.

G

Proteases and Proteolytic Enzyme Inhibitors

TABLE G.1 Commonly Used Proteases

Protease	Class	Specificity	pH-optimum	Applications
Acylamino acid peptidase		Exopeptidase	7.5–9.0	Deblocking of peptides for subsequent N-terminal sequence analysis
Aminopeptidase M	Metalloprotease	Peptides and proteins bearing a free α-amino group and containing L-amino acids only	7.0–7.5; up to 9.0 depending on substrate	Study of protein sequences
Bromelain	Thiolprotease	Non-specific endopeptidase	5.0–7.0	Nonspecific cleavage
Carboxypeptidase A	Zn-metalloprotease	C-terminal residues possessing an an L configuration and an unsubstituted α-amino group. Slowly releases Gly, Asp, Glu, Cys Cys, no release of Arg, Pro	~7.5	Sequence analysis of proteins. Successive cleavage of C-terminal amino acids
Carboxypeptidase B	Zn-metalloprotease	Hydrolysis of the basic amino acids L-Lys and L-Arg from the C-terminal position	7.0–9.0	Sequence analysis of proteins. Successive cleavage of basic amino acids from the C-terminus of proteins
Carboxypeptidase P	Serine carboxypeptidase	Hydrolyzes amino acid residues (including Pro) from the C-terminus. Slow release of Ser any Gly	3.7–5.2	Sequence analysis
Carboxypeptidase Y	Serine carboxypeptidase	Hydrolyzes L-amino acids (including Pro) from the C-terminus. High catalysis rate if the penultimate residue has aromatic or aliphatic side chain. Dipeptides completely resistant to cleavage	~5.5 for acidic and 7.0 for basic amino acids	Sequence analysis

TABLE G.1 *Continued*

Protease	Class	Specificity	pH-optimum	Applications
Cathepsin C (Dipeptidyl-transferase)	Thiolprotease	Catalyzes the successive removal of N-terminal dipeptides from from polypeptides. Activity blocked by the presence of N-terminal Lys or Arg, and Pro as the second or third amino acid	7.0–8.0	Its transferase activity makes it suitable for polymerization of dipeptide amides
Chymotrypsin	Serine protease endopeptidase	Cleaves at the C-termini of Phe, Tyr and Trp	7.0–9.0	Peptide mapping, fingerprinting, sequence analysis
Collagenase (from *Achromobacter iophagus)*	Metalloprotease	After X in X-Gly-Ala- slower after X in -*X*-Gly-Pro	7.2	Tissue disruption and cell harvesting
Dispase	Metalloprotease	Non-specific	8.5	Tissue disruption and cell harvesting
Elastase	Serine protease endopetidase	Cleaves at the C-terminal side of Ala, Val, Leu, Ile, Gly, and Ser	8.8	Digestion of elastin, together with collagenase and trypsin for tissue dissociation
Endoproteinase Arg-C (Clostripain from *Clostridium histolyticum*)	Thiolprotease	Cleaves at the C-terminal side of Arg. Reducing agents and Ca + 2 are required for full activity	7.2–8.0	Peptide mapping, fingerprinting and sequence analysis
Endoproteinase Asp-N	Metalloprotease	Catalyzes the hydrolysis of peptide bonds at the N-terminal side of Asp and cysteic acid residues	7.0–8.0	Protein structure and sequence analysis
Endoproteinase Glu-C (Protease V8 from *S. aureus V8*)	Serine protease	Cleaves at the carboxylic side of Glu (in ammonium bicarbonate at pH 7.8 or ammonium acetate buffer pH 4.00 and Glu and Asp (in phosphate buffer, pH 7.8)	4.0 and 7.0	Protein structure and sequence analysis
Endoproteinase Lys-C	Serine protease	Cleaves at the carboxylic side of Lys	8.5–8.8	Protein structure and sequence analysis
Enterokinase	Serine protease	Recognizes the amino acid sequence-(Asp)4-Lys- and cleaves at the carboxylic acid side of Lys	8.0	Processing of fusion proteins at a definite site; activation of trypsinogen to trypsin
Factor Xa	Serine protease	Specifically hydrolyzes peptide bonds at the carboxylic side of arginyl residues within the sequence -I-E-G-R-X	8.3	Processing of fusion proteins at a definite cleavage site, activation of prothrombin to thrombin, coagulation research
Papain	Thiolprotease	Broad specificity	6.0–7.0	Limited hydrolysis of immunoglobulins to yield biologically active fragments, production of glycopeptides from purified proteoglycans

Continued

TABLE G.1 *Continued*

Protease	Class	Specificity	pH-optimum	Applications
Pepsin	Acid protease	Broad specificity endoprotease	1.8–2.2	Non-specific hydrolysis of proteins and peptides in acidic media. For limited hydrolysis of native immunoglobulins to produce biologically active fragments
Pronase (from *Streptomyces griseus*)	Mixture	Non-specific	6.0–7.5	This protease mixture is used for total hydrolysis of proteins
Proteinase K	Serine protease	Broad specificity	7.5–10.5	Used for the total degradation of proteins during the isolation of DNA and RNA. It is used in the specific modification of proteins and glycoproteins on cell surfaces. Active in the presence of SDS and urea.
Subtilisin	Serine protease	Unspecific endopeptidase	7–11	Total hydrolysis of proteins and peptides for sequence studies
Thrombin	Serine protease	Cleaves peptide bonds specifically at the carboxylic side of arginyl residues	8.2–9.0	Coagulation research, processing of fusion proteins
Trypsin	Serine protease	Catalyzes the hydrolysis of proteins and peptides at the carboxlic side of Arg and Lys	8.0	Tryptic mapping, fingerprinting, sequence analysis

Derived from ***Biochemica Information***, compiled and edited by Keesey, 1987. Indianapolis, IN: Boehringer Mannheim.

PROTOCOL G.1 Preparation of Defatted BSA

1. Suspend 5–10g of BSA (fraction V, Sigma) in 100ml of 95% ethanol.
2. Wash the slurry with ethanol using a Buchner funnel until the eluate ceases to be acidic as determined with pH paper.
3. Dissolve the BSA in a minimal volume of 10mM tricine-EDTA, pH 8.0. Dialyze the BSA for 2 days vs 2 × 4 liters of 10mM tricine-EDTA, pH 8.0 and then lyophilize.
4. Prior to use, resuspend the BSA at a concentration of 50mg/ml in the desired buffer and dialyze overnight against 1L of the desired buffer (Shani-Sekler et al,. 1988).

TABLE G.2 Protease Inhibitors. Reprinted from *Protease Inhibitors Technical Guide*, by permission of Boehringer Mannheim Biochemicals

Inhibitor	Specificity of inhibitor	Solubility/stability*	Suggested starting concentration**	Notes
Complete™ Protease Inhibitor Cocktail Tablets	Mixture of several protease inhibitors with broad inhibitory specificity. Inhibits serine, cysteine, and metalloproteases in bacterial, mammalian, yeast, and plant cell extracts. In the acidic pH range, add Pepstatin with Complete tablets to ensure aspartic protease inhibition.	Soluble in H_2O. The tablets are stable at +4°C, stored dry. The stock solution is stable for 1–2 days, stored at +4°C or at –20°C.	For a 25x stock solution, dissolve 1 tablet in 2 ml deion. H_2O or 100 mM phosphate buffer, pH 7.0.	One tablet contains protease inhibitors sufficient for 50 ml cell extract. The efficiency of inhibition can be tested fast with the Universal Protease Substrate
Antipain-dihydrochloride (Papain Inhibitor)	Inhibits papain and trypsin. Plasmin is inhibited to a small extent.	Soluble in: H_2O, methanol, DMSO*** to 20 mg/ml. Sparingly soluble in: ethanol, propanol, butanol. Insoluble in: Benzene, chloroform ($CHCl_3$), hexane, petroleum and ethyl ethers. Dilute solutions should be stored frozen in aliquots at –20°C. Stable approx. 1 month.	50 µg/ml (74 µM) (1 U of papain is inhibited to 49% by 0.9 µg of antipain.)	Molecular Weight: 677.63 Antipain is more specific for papain and trypsin than is leupeptin. The inhibitory potency of antipain is 100-fold higher than that of elastatinal [Ref. 1, 2 (p. 683), 3, 4, 5].
Antithrombin III (Heparin Co-factor) from human plasma	Antithrombin III (AT III) inhibits all serine proteases of the blood coagulation system, including thrombin, plasmin, kallikrein, the protease factors IXa, Xa, XIa, and XIIa. It also inhibits trypsin and chymotrypsin. Does not inhibit cysteine proteases, aspartic proteases, and metalloproteases.	Soluble in H_2O (10 mg/ml) Stable in solution for 1 week at +4°C and pH = 7.0 + 9.0	1 Inh. U/ml Unit definition: One inhibitor unit AT III inactivates 1 U of thrombin (25°C, pH 8.1) in the presence of heparin.	Molecular Weight: 65,000 At III forms an irreversible 1:1 complex with serine proteases. Once formed, the (AT thrombin) complex does not dissociate, even during electrophoresis in the presence of denaturing and reducing agents. AT III from BM is heparin free. The rate of complex formation is accelerated by typically 0.2 units of heparin per unit AT III (Ref. 6,7).

Continued

TABLE G.2 *Continued*

Inhibitor	Specificity of inhibitor	Solubility/stability*	Suggested starting concentration**	Notes
α_1-Antitrypsin (α_1-Protease Inhibitor)	Inhibits most serine proteases, including thrombin, plasmin, trypsin, chymotrypsin, elastase (leukocyte and pancreatic). Collagenase, cathepsin G, and acrosin are also inhibited. Also forms complexes with the zymogens of these enzymes.	Soluble in water and Tris buffer (0.1M, pH 8.4). Stable in solution at +4°C for one week.	10µg/ml (40mU/ml) Unit definition: One inhibitor unit α_1-antitrypsin inactivates 1U of trypsin (25°C, pH 8.4).	Molecular Weight: 54,000 α_1-Antitrypsin is a glycoprotein that does not inhibit most cysteine proteases, aspartic proteases, and metalloproteases. Forms a 1:1 complex with the protease, blocking the access of substrate molecules to the protease active site (Ref. 8.9).
Aprotinin	Serine protease inhibitor. Does not act on thrombin or Factor X. Inhibits plasmin, kallikrein, trypsin, chymotrypsin with high activity.	Freely soluble in H_2O (10mg/ml) or aqueous buffer solution (e.g., Tris, 0.1M, pH 8.0). A solution adjusted to pH 7–8 is stable for approx. 1 week at +4°C. Aliquots stored at –20°C are stable approx. 6 months.	0.06–2.0µg/ml (0.01–0.3µM)	Molecular Weight: 6,512 Avoid repeated freeze-thaws and exposure to strong alkali solutions. Aprotinin is inactive at pH > 12.8 (Ref. 10).
Bestatin	Primarily, if not exclusively, an inhibitor of amino peptidases and other exopeptidases, including aminopeptidases found in wheat germ and reticulocyte *lysate in vitro* translation systems (*e.g.*, amino-peptidase B, leucine aminopeptidase, tripeptide aminopeptidase, and aminopeptidases on the surface of mammalian cells). It does not inhibit carboxypeptidases.	Soluble to 20mg/ml in 1M HCl, 5mg/ml in methanol, and 1mg/ml in 0.15M NaCl. Do not store in HCl. We recommend a stock solution of 2–5mg/ml in methanol. Solutions are stable for 6 months if stored in aliquots at –20°C.	40µg/ml (130µM)	Molecular Weight: 308.4 Bestatin has been found to have antitumor properties and enhances not only blastogenesis and lymphocytes *in vitro*, but also establishes a delayed-type hyper- sensitivity *in vivo* (Ref. 1, 11–14).

Calpain Inhibitor I	Inhibitor of calpains. Calpains are calcium-dependent neutral cysteine proteases. Inhibits activity of Calpain I. ID_{50} for 0.02U platelet Calpain I: 0.05μmol/l. Some inhibitory activity against Calpain II. Inhibits papain to a lesser extent.	Soluble in DMF, ethanol, and methanol to 10mg/ml. For a stock solution, we recommend dissolving 1mg of the inhibitor in 100μl DMF, methanol, or ethanol. Before use, dilute with water or phosphate buffer (0.1M, pH 7.5) to desired concentration. Solutions in DMF, ethanol, or methanol are stable for 2–3 days at +4°C. and approx. 4 weeks at –20°C. Solutions diluted with water or phosphate buffer are stable only a few days at –20°C. We recommend making solutions up fresh before use.	17μg/ml	Molecular Weight: 383.5 Not soluble in water (Ref. 15–18).
Calpain Inhibitor II	Inhibits activity of Calpain II. Inhibits Calpain I (ID_{50} = 0.12μmol/l) and papain to a lesser extent.	(See Calpain Inhibitor I, above)	7μg/ml	Molecular Weight: 401.6 Not soluble in water (Ref. 15, 16, 17, 19).
Chymostatin	Specific inhibitor of α-, β-, γ-, δ-chymotrypsin.	Soluble in: glacial acetic acid, DMSO*** to 20mg/ml. Sparingly soluble in: water, methanol, ethanol. Insoluble in: ethyl acetate, petroleum and ethyl ethers, hexane, chloroform ($CHCl_3$). Dilute solutions should be stored frozen in aliquots at –20°C. Stable approx. 1 month.	6–60μg/ml (10–100μM) Unit definition: One unit chymotrypsin is inhibited to 49% by 1.8μg of chymostatin.	Molecular Weight: 607.71 (Ref. 1, 2 [p. 686], 20)

TABLE G.2 *Continued*

Inhibitor	Specificity of inhibitor	Solubility/stability*	Suggested starting concentration**	Notes
3,4-Dichioro-isocoumarin	Inhibits a large number of serine proteases such as elastase, cathepsin G, and endoproteinase Glu-C (Staph. V-8 protease).	May be dissolved in DMF and stored in aliquots at –20°C.	1–43 μg/ml (5–200 μM)	Molecular Weight: 215.0 Does not inhibit the thiol protease papain, the metalloprotease leucine aminopeptidase or β-lactamase. More sensitive to hydrolysis than pA-PMSF (Ref. 21).
EDTA-Na_2	Inhibits metalloproteases.	Soluble in water to 0.5 M at pH 8–9. Stable at +4°C for at least 6 months.	0.2–0.5 mg/ml (0.5–1.3 mM)	Moleclar Weight: 372.24 The disodium salt of EDTA will not go into solution until the pH of the solution is adjusted to approximately 8.0 by the addition of NaOH (Ref. 22).
E-64	Inhibits papain and other cysteine proteases like cathepsin B and L.	Soluble to 20 mg/ml in a 1:1 (v/v) mixture of ethanol and water. Solutions are stable for 1 month if stored in aliquots at –20°C.	0.5–10 μg/ml (0.14–28.0 μM)	Molecular Weight: 357.4 Stable between pH 2–10. Unstable in strong alkali and strong mineral acids (Ref. 23–25).
Hirudin from *Hirudo medicinalis* (European leeches)	Specifically inhibits thrombin	Soluble in 50% ethanol, water, and commonly used buffers. The lyophilizate is stable at room temperature for approx. 2 years. Solutions can be stored at –20°C for at least 6 months.	150–200 ATU/ml plasma Unit definition: One anti-thrombin unit (ATU) neutralizes one NIH unit of thrombin (fibrinogen assay) at 37°C.	
Leupeptin	Inhibits serine and cysteine proteases such as trypsin, papain, plasmin, and cathepsin B.	Highly soluble in water (1 mg/ml). Stable for at least 1 week at +4°C and 6 months frozen in aliquots at –20°C.	0.5 μg/ml (1 μM)	Molecular Weight: 460 (Ref. 2, 25, 26)

α_2-Macroglobulin	A general endoproteinase inhibitor. Inhibits most endoproteinases, but does not inhibit endoproteinases that are highly specific for one or a limited number of sequences (*e.g.*, tissue kallikrein, urokinase, coagulation factor XIIa, and endoproteinase Lys-C).****	Soluble in water. Stable at least 1 week at room temperature or 3 weeks at +4°C. Can also be frozen in aliquots at –20°C, where it remains stable at least 6 months. Sensitive to acicic pH, denatured below pH 4.0 Ammonia methylamine and hydroxylamine (above pH 7.0) cause irreversible conversion to the inactive form.	Unit definition: One inhibitor unit inhibits 9.1µg of trypsin.	Molecular Weight: 725,000 Do not use α_2-Macroglobulin in presence of DTT. DTT, even at 1mM, causes reversible dissociation into inactive subunits. α_2-Macroglobulin acts by physically entrapping the endoproteinases, usually in a 1:1 ratio (Ref. 27).
α_2-Macroglobulin, carrier-fixed	(See α_2-macroglobulin info)	Can be used in phosphate, Tris, or other buffer systems between pH 6.0 and pH 8.5. The optimum buffer concentration should be >0.05M and <0.1M. Buffer should contain 0.1–0.15M NaCI to avoid non-specific binding to carrier matrix.	Specific binding capacity = 0.3mg trypsin (33U/ml gel). General binding capacity = 13nmol protease/ml gel.	Can be used in batch or column process to remove endoproteinases from solution. Can be recharged with fresh α_2-Macroglobulin. Column stable at +4°C for 1 year. DO NOT FREEZE (Ref. 28).
pA-PMSF (4-Amidino-phenyl)-methane-sulfonyl fluorlde	Specific and irreversible inhibitor of serine proteases (*e.g.*, trypsin, thrombin, factor Xa, plasmin). Unlike PMSF, pA-PMSF does not inhibit chymotrypsin or acetylcholinesterase.	Can be dissolved in water, soluble to 20mg/ml. Solution can be stored frozen at –20°C in aliquots. Stability is pH dependent: $t_{1/2}$: pH 6, 20min; pH 7, 6min; pH 8, 1msec.	0.01–0.04mg/ml (10–20µM)	Molecular Weight: 216.2 Inhibitory action corresponds to that of DFP, but pA-PMSF is not nearly as toxic (Ref. 34).
Pefabloc® SC 4-(2-Aminoethyl)-benzenesulfonyl-flouride, hydro-chloride (AEBSF)	Irreversibly inhibits serine proteases, including trypsin, chymotrypsin, plasmin, plasma kallikrein, and thrombin.	Soluble up to 100mg/ml in aqueous buffers and water. Stable in solution for 1–2 months if stored in aliquots at –20°C. Only slight hydrolysis occurs under weakly basic conditions (pH 8.0–9.0).	0.1–1.0mg/ml (0.4–4mM)	Molecular Weight: 239.5 A safe, stable, and water solublesoluble altemative to PMSF and DFP (Ref. 29–32).

Continued

TABLE G.2 *Continued*

Inhibitor	Specificity of inhibitor	Solubility/stability*	Suggested starting concentration**	Notes
Pepstatin	Inhibits aspartic (acid) proteases such as pepsin, renin, cathepsin D, chymosin, and many microbial acid proteases.	Soluble in methanol to approx. 1 mg/ml. Also soluble to 1 mg/ml in ethanol if allowed to sit over- night, and to 300 μg/ml in 6 N acetic acid. Stable at least 1 week at +4°C, or 1 month if stored in aliquots at –20°C.	0.7 μg/ml (1 μM)	Molecular Weight: 685.9 Insoluble in water (Ref. 33)
PMSF (Phenylmethyl-sulfonyl fluoride)	Inhibits serine proteases (chymotrypsin, trypsin, and thrombin). Also inhibits cysteine proteases such as papain (reversible by DTT treatment).	Soluble to >10 mg/ml in isopropanol, ethanol, methanol, and 1,2 propanediol. Unstable in aqueous solution. In 100% isopropanol, stable at least 9 months at +25°C.	17–170 μg/ml (0.1–1 mM)	Molecular Weight: 174.2 Add fresh PMSF at every isolation/ purification step (from stock solution). Does not inhibit metalloproteases, most thiol proteases, and aspartic proteases (Ref. 35).
Phosphoramldon	Specifically inhibits thermo-lysin, collagenase, and metallo-endoproteinases from various microorganisms (*Bacillus subtilis, Streptomyces griseus and Pseudomonas aeruginosa*)	Salts of phosphoramidon are soluble to 20 mg/ml in water. Also soluble in methanol and DMSO.*** Recommended stock solution 1–20 mg/ml. Stable in solution for 1 month if stored in aliquots at –20°C.	4–330 μg/ml (7–570 μM)	Molecular Weight: 579.6 (Ref. 1, 2, 11, 36)
TLCK · HCL (L-1-Chloro-3-[4-tosylamido]-7-amino-2-heptanone-HCI)	Irreversibly and specifically inhibits trypsin. Also inhibits many other serine and cysteine proteases such as bromelain, ficin, and papain.	Salts of TLCK are soluble to 20 mg/ml in water. We recommend a stock solution of 1 mg/ml in either dilute (1 mM) HCI or buffer, pH ≤6; to ensure stability (see "notes" column).	37–50 μg/ml (100–135 μM)	Molecular Weight: 369.3 Stable at +25°C pH ≤6.0. Rapidly decompose at pH >7.5. For example, at pH 9.0, +25°C, TLCK's half-life is only 5 minutes. Chymotrypsin is not inhibited (Ref. 37)
TPCK (L-1-Chloro-3-[4-tosylamido]-4-phenyl-2-butanone)	Irreversibly inhibits chymotrypsin. Also inhibits many other serine and cysteine proteases such as bromelain, ficin, and papain.	Soluble to 20 mg/ml in ethanol. Recommended stock solution: 3 mg/ml.	70–100 μg/ml (200–284 μM)	Molecular Weight: 351.9 Trypsin is not inhibited (Ref. 38, 39).

Trypsin inhibitors **From chicken egg white** **From soybean**	**Inhibits trypsin. Soybean trypsin inhibitor also inhibits factor Xa, plasmin, and plasma kallikrein. Neither inhibit metallo, cysteine, and aspartic proteases or tissue kallikrein.**	**Both are soluble in water. Recommended stock solution: 1 mg/ml. Store frozen in aliquots at –20°C. Stable at least 6 months.**	**10–100 µg/ml**	**Molecular Weight: (egg white) 28,000 (soybean) 20,100 Egg white inhibitor is stable at acid pH and labile at alkaline pH.** **Soybean inhibitor is sensitive to heat, high pH, and protein-precipitating solutions.**

References

1. Umezawa, H. and Aoyagi, T. (1983) In: *Protease Inhibitors: Medical and Biological Aspects* (Katunuma, N. et al., eds.) pp 3–15, Springer-Verlag, Berlin.
2. Umezawa, H. (1976) Meth. Enzymol. **45**:678.
3. Suda, H. *et al.* (1972) *Journal of Antibiotics* **25**:263.
4. Umezawa, S. *et al.* (1972) *Journal of Antibiotics* **25**:267.
5. Westerich, J.O., and Wolfenden, R. (1972) *J. Biol. Chem.* **247**:8195.
6. Abildgaard, U. (1968) *Scand. J. Clin. Lab. Invest.* **21**:89–91.
7. Rosenberg, R.D. and Damus, P.S. (1973) *J. Biol. Chem.* **248**:6490–6505.
8. Laskowski, M. Jr. and Kato, I. (1980) *Ann. Rev. Biochem.* **49**:593.
9. Carrell, R.W. *et al.* (1982) *Nature* **298**:329.
10. Kassell, B. (1970) *Meth. Enzymol.* **19**:844.
11. Suda, H. *et al.* (1973) *J. Antibiotics* **26**:621.
12. Umezawa, H. (1982) *Am. Rev. Microbiol.* **36**:75.
13. Umezawa, H. *et al.* (1976) *J. Antibiotics* **29**:97.
14. Aoyagi, T. *et al.* (1976) *Biochemistry International* **9**:405.
15. Murachi, T. (1983) *Trends in Biochem. Sci.* **8**:167.
16. Yoshimura *et al.* (1983) *J. Biol. Chem* **258**:8883.
17. Crawford, C. *et al.* (1988) *Biochem. J.* **253**:751.
18. Kajiwara, Y. *et al.* (1987) *Biochemistry International* **15**:935–944.
19. Delbaere, L.T.J., and Brayer, G.D. (1985) *J. Mol. Biol.* **183**:89.
20. Umezawa, H. *et al.* (1970) *Jorunal of Antibiotics* **23**:425.
21. Harper, J.W. *et al.* (1985) *Biochemistry* **24**:1831.
22. Maniatis, T. *et al.* (1982) *Molecular Cloning: A Laboratory Manual*, p 446, Cold Spring Harbor Laboratory, NY.
23. Hanada, K. *et al.* (1978) *Agr. Biol. Chem.* **42**:523.
24. Hanada, K. *et al.* See Ref. #1, p 25.
25. Frommer, W. *et al.* (1979) *J. Med. Plant. Res.* **35**:195.
26. Miyamoto, M. *et al.* (1973) *BBRC* **55**:84.
27. Barrett, Alan J. (1981) *Meth. Enzymol.* **80**:737.
28. Wunderwald, P. *et al.* (1983) *J. Appl. Biochem.* **5**:31.
29. Markwardt, F. *et al.* (1973) *Thromb. Res.* **2**:343.
30. Markwardt. F. *et al.* (1974) *Biochem. Pharmacol.* **23**:2247–2256.
31. Lawson, W.B. *et al.* (1982) *Folia Haematol.*, Leipzig **109**:52–60.
32. Mintz, G.R. (1993) *BioPharm.* **Vol. 6, No. 2**:34.
33. See Ref. #2, p 689.
34. Laura, R. *et al.* (1980) *Biochemistry* **19**:4859.
35. James, G.T. (1978) *Analytical Biochem.* **86**:574.
36. Hanada, K. *et al.* (1983) in *Proteinase Inhibitors: Medical and Biological Aspects* (Katunuma, N. *et al.*, eds.) PP 25–36, Springer Verlag, Berlin.
37. Shaw, *et al.* (1965) *Biochemistry* **4**:2219.
38. Schoellmann, G. and Shaw, E. (1963) *Biochemistry* **2**:252.
39. Kostka, V. and Carpenter, F.H. (1964) *J. Biol. Chem.* **239**:1799.

H

Radioactivity

Amounts of radioactivity are designated in terms of rads, roentgens, Curies, disintegrations per unit time (min) or counts per unit time (cpm). The first two units are used except for measuring animal exposure to ionizing radiation. DPM represents the actual number of β particles emitted per minute and cpm is the number of disintegrations that are detected by a radioactivity detecting device. The specific activity (or specific radioactivity) of a compound is defined as the radioactivity per unit mass of an element or compound containing a radioactive nuclide. It is usually expressed in terms of radioactivity per milligram or per millimole, such as Curies/mole or mCuries/mmole (mCi/mmol).

1 Curie (Ci) = 2.22×10^{12} dpm = 3.7×10^{10} Becquerel (Bq)
1 Ci = 2.22×10^{6} dpm (disintegrations per minute)
1 dpm = cpm/efficiency

TABLE H.1 Physical Properties of some Commonly Used Isotopes

Radioisotope	Half-life	Emission	Energy, max (MeV)	Counting method
^{3}H	12.4 years	β	0.018	With scintillant
^{14}C	5760 years	β	0.155	With scintillant
^{32}P	14.3 days	β	1.71	Dry
^{33}P	25 days	β	0.249	With scintillant
^{35}S	87.4	β	0.167	With scintillant
^{125}I	60 days	γ	0.27–0.035	Dry
^{51}Cr	27.7 days	γ	0.322	Dry
^{45}Ca	162.7	β	0.257	With scintillant

Manual and Machine Film Processing

Autoradiography is used to visualize and quantitate radiolabeled proteins that are resolved by 1-D or 2-D protein gel electrophoresis. During film based autoradiography β particles or γ-rays enter the film and cause the ejection of electrons from silver halide crystals generating local precipitates of silver atoms. Due to the limited dynamic range of X-ray film, it is often necessary to choose different exposure times to allow visualization and quantification of a range of proteins of different abundance. Spots and bands that exceed an absorbance of 1.4 (A_{540}) have saturated the available silver halide crystals and cannot be accurately quantified. Depending on the radioisotope used, the image enhancers employed, and the abundance of the target protein, exposure times can vary from a few hours to several weeks.

X-ray films that are recommended for use in autoradiographic applications can be processed in automated processing machines. Most labs or institutes have automatic processing machines which are convenient, dependable and easy to use. The use of fresh chemicals and strict adherence to the film manufacturer's recommendations should be followed.

Manual processing offers few advantages over the automated method. A significant advantage is that one can control the intensity of the photographic image. For manual processing you will need a darkroom with tanks (or trays large enough to hold your fluids) and metal hangers. When not in use, the tanks should be protected with a cover to decrease the rate of oxidation and prevent contamination from dust and other foreign bodies.

The following manual processing protocol can be used with X-OMAT AR film:

Develop—Submerge the film for 5 min at 20°C in Kodak GBX Developer and Replenisher with intermittent agitation every 5 sec for 60 sec. Check the film periodically. If the band of interest is visible, proceed to the next step. If after 5 min you do not see your band, incubate the film in developer for more time.

Rinse—Submerge the film for 30 sec at 16 to 24°C in Kodak Stop Bath SB-1a, or Kodak Indicator Stop Bath with continuous agitation.

Fix—Submerge the film in the chosen fixer for the times indicated below at 16 to 24°C.

Kodak Fixer	Fixing Times
X-ray Fixer	5–10 min
Rapid Fixer	2–4 min
GBX Fixer and Replenisher	2–4 min

Wash—Place the film in running water for 5 to 10 min at 16 to 24°C.

Dry—Dry the film in a dust-free area at room temperature or in a suitable drying cabinet at a temperature not exceeding 49°C.

Over the past several years the use of phosphorimaging has been replacing film-based autoradiography. The linear dynamic range of phosphor screens which covers five orders of magnitude and fast imaging times, offers greater sensitivity than film as well as simple and accurate quantitation of radioactivity. Phosphorimaging is 10–250 fold more sensitive than film-based autoradiography and also has a greater linear dynamic range than film-based autoradiography. The main drawback to phosphorimaging is the initial expense of the instrument and screens.

I

Tissue Culture

TABLE I.1 Area of Culture Plates for Cell Growth

Size	Growth Area (cm^2)	Relative Area
96 well	0.32	0.2×
24 well	1.99	1×
12 well	3.83	2
6 well	9.4	5
35 mm	8	4.2
60 mm	21	11
100 mm	55	29

TABLE I.2 Height of the Column of Liquid and Volume in the Well (μl)

Height (cm)	384-well (μl)	96-well (μl)
0.25	20	81
0.5	48	171
0.75	77	262
1.0	105	352
1.25	133	443

Transwell Permeable Supports

Permeable supports with micropores that are commercially available in a variety of pore sizes have become a routine method to culture various cell types. Permeable supports treated for cell growth provide significant advantages over solid impermeable cell culture substrates. For epithelial cells and other cell types, the use of permeable supports allows cells to be grown in a polarized fashion producing an environment that resembles the in vivo state. Cellular functions such as transport, adsorption and secretion can be studied since cells grown on permeant supports provide convenient and independent access to apical and basolateral membrane domains.

Transport/Permeability studies

- Macromolecules, ions, water, hormones, growth factors, etc
- Drug transport, effects on permeability
- Invasion and penetration of epithelial barriers

Polarity

- Polarized distribution of ion channels, enzymes, transport proteins, receptors, lipids
- Sorting and targeting
- Polarity development and maintenance
- Synthesis and assembly of tight junctions

Endocytosis

- Protein turnover
- Membrane recycling
- Receptor-ligand interactions

Chemotaxis/Motility studies

- Migration systems
- Phagocytosis
- Metastatic Potential and Invasion
- Invasion inhibitors

Co-culture

- Cell-cell interactions
- Cell-substrate interactions
- Feeder layers

Tissue Remodeling

- Wound healing
- Angiogeneis
- Re-epithelialization
- Inflammation

J

Miscellaneous

PROTOCOL J.1 Siliconizing Glassware

Glassware can be made hydrophobic by coating it with silicon (Schleif and Wensink, 1981). This procedure should be performed in a fume hood as a product of the reaction is HCl vapor.

1. Prepare a 5% (v/v) solution of dichlorodimethyl silane or Sigmacote (Sigma) in either n-hexane or chloroform. This solution can be stored at room temperature for at least one year.
2. Soak the glass vessel in the 5% solution of dichlorodimethyl silane for 5 min at room temperature in a fume hood or well ventilated area.
3. Rinse the glass vessel several times with water.
4. Bake the glassware at 210°C which will remove volatile materials, sterilize the glass and harden the silicon surface.

TABLE J.1 Unit Prefixes

Prefix	Symbol	Factor
exa	E	10^{18}
peta	P	10^{15}
tera	T	10^{12}
giga	G	10^{9}
mega	M	10^{6}
kilo	k	10^{3}
milli	m	10^{-3}
micro	μ	10^{-6}
nano	n	10^{-9}
pico	p	10^{-12}
femto	f	10^{-15}
atto	a	10^{-18}

TABLE J.2 The Greek Alphabet

Α	α	alpha	σ	φ	phi
β	β	beta	Κ	κ	kappa
Ξ	ξ	xi	Λ	λ	lambda
Γ	γ	gamma	Μ	μ	mu
Δ	δ	delta	Ν	ν	nu
Ε	ε	epsilon	Ο	ο	omicron
Ζ	ζ	zeta	Π	π	pi
Σ	σ	sigma	Ρ	ρ	rho
Η	η	eta	Υ	υ	upsilon
Τ	τ	tau	Χ	χ	chi
Θ	θ	theta	Ψ	ψ	psi
Ι	ι	iota	Ω	ω	omega

TABLE J.3 Abbreviations

Ag	antigen
ATP	adenosine triphosphate
A	ampere
Å	angstrom
AA	amino acid
Ag	antigen
AIDS	acquired immune deficiency syndrome
ATP	adenosine triphosphate
AUFS	absorbance units full scale
Bp	basepair
Bq	Becquerel
BSA	Bovine serum albumin
C	complement
C′	activated complement
°C	degree Celsius
cAMP	adenosine 3′,5′-cyclic monophosphate
cDNA	complementary DNA
CFA	complete Freund's adjuvant
CFU	colony forming unit
CHAPS	3-[(3-cholamidoprophyl) dimethyl-ammonio] -1-propanesulfonate
Ci	curie
CMV	cytomegalovirus
ConA	concanavalin A
Cpm	counts per minute
CSF	colony-stimulating factor
CTL	cytotoxic T lymphocyte
Da	Dalton
d	day
DEAE	diethylaminoethyl
DIG	digoxigenin
DMEN	Dulbecco's modified Eagle's medium
DMF	dimethylformamide
DMSO	dimethylsufoxide
DNA	deoxyribonucleic acid
DOC	sodium deoxycholate
dpm	disintegrations per minute
ds	double stranded (i.e. dsDNA)
DTT	dithiothreitol

TABLE J.3 *Continued*

EBV	Epstein-Barr virus
EDTA	ethylenediaminetetraacetic acid
ELISA	enzyme-linked immunosorbent assay
EM	electron microscopy
ER	endoplasmic reticulum
EST	expressed sequence tag
FACS	fluorescence-activated cell sorter
FBS	fetal bovine serum
FCS	fetal calf serum
FITC	fluorescein isothiocyanate
FMLP	formyl-methionyl-leucyl-phenylalanine
Fuc	Fucose
g	gram
g	unit of gravity
GAG	glycosaminoglycan
Gal	Galactose
GalNAc	N-acetyl-D-galactosamine
Glc	Glucose
GlcNAc	N-acetyl-D-glucosamine
h	hour
HAT	hypoxanthine, aminopterine, thymidine
Hb	hemoglobin
HBSS	Hank's balanced salt solution
HEPES	N-2-hydroxyethylpiperazine-N′-2-ethanesulfonic acid
HIV	human immunodeficiency virus
HLA	human histocolmpatibility leukocyte antigens
HPLC	high-performance liquid chromatography
HSV	herpes simplex virus
HTS	high throughput screen
Id	idiotype
IEF	isoelectric focusing
IFA	incomplete Freund's adjuvant
IFN	interferon
Ig	immunoglobulin
IL	interleukin
i.m.	intramuscular
ICAT	isotope-coded affinity tag
IMAC	immobilized metal affinity chromatography
i.p.	intraperitoneal
IPG	immobilized pH gradients
IPTG	isopropyl-1-thio-β-D-galactopyranoside
i.v.	intravenous
kb	kilobase
kDa	kilodalton
L	liter
LD_{50}	50% lethal dose
LPS	lipopolysaccharide
m	meter
M	molar
MAb	monoclonal antibody
MALDI-TOF	matrix-assisted laser desorption ionization time-of-flight
Man	mannose
MCS	multiple cloning site
2-ME	2-mercaptoethanol
MEM	minimal essential media
MHC	major histocompatability complex

Continued

TABLE J.3 *Continued*

ml	milliliter
MLC	mixed lymphocyte culture
MLR	mixed lymphocyte reaction
mol	mole
mol wt	molecular weight
M_r	relative molecular mass
MS	mass spectrometry
M/z	mass to charge ratio
N	normal
NANA	N-acetyl neuraminic acid
ND	not determined
NEPHGE	nonequilibrium pH gradient electrophoresis
NK cell	natural killer cell
NMWC	nominal molecular weight cut off
NP-40	Nonidet P-40
OD	optical density
ORF	open reading frame
osmol	osmole
OVA	ovalbumin
PAGE	polycrylamide gel electrophoresis
PBL	peripheral blood lymphocytes
PBS	phosphate buffered saline
PCR	polymerase chain reaction
PCV	packed cell volume
PG	prostaglandin
PHA	phytohemagglutinin
PIPES	piperazine-N,N′-bis(2-ethane sulfonic acid)
PMA	phorbol myristate acetate
PMSF	phenylmethylsulfonyl fluoride
PPI	protein-protein interaction
PTD	protein transduction domain
PTM	posttranslational modification
PWM	pokeweed mitogen
r	recombinant (e.g., rIFN-γ)
RBC	red bllod cell
RIA	radioimmunoassay
RNA	ribonucleic acid
RT	reverse transcriptase
s	second
s	sedimentation coefficient
S	Svedberg unit of sedimentation
SAS	saturated ammonium sulfate
s.c.	subcutaneous
SCX	strong cation exchanger
SD	standard deviation
SDS	sodium dodecyl sulfate
SE	standard error
SEM	standard error of the mean
SLE	systemic lupus erythematosus
SNP	single nucleotide polymorphism
sp act	specific activity
SRBC	sheep red blood cells
ss	single strande (e.g., ssDNA)
SV40	simian virus 40
$t_{1/2}$	half-life
TCA	trichloroacetic acid

TABLE J.3 *Continued*

TFA	trifluoroacetic acid
TdR	thymidine deoxyribose
TLC	thin-layer chromatography
TNF	tumor necrosis factor
TNP	trinitrophenyl
TRN	trans Golgi network
Tris	tris(hydroxymethyl)aminomethane
U	unit
UV	ultra violet
V	volt
vol	volume
V region	variable region of Ig
w	watt
wt	weight
WWW	world wide web
y	year

TABLE J.4 HPLC Pump Pressure Conversion

	HPLC Pump Pressure Conversion				
	atm	Kg/cm^2	Torr	Bar	Inches Hg
PSI=	0.068	0.0703	51.713	0.06895	2.0359

Example: convert 484 PSI to Kg/cm^2
484 × 0.0703 or 484 PSI = 34.0252 Kg/cm^2

TABLE J.5 Residue Masses of Amino Acids Together with Their Corresponding Immonium Ion Masses[a]

Amino acid	Abbreviations—three letter, single letter	Residue mass[b]	Immonium ion mass[b]
Glycine	Gly (G)	57	30
Alanine	Ala (A)	71	44
Serine	Ser (S)	87	60
Proline	Pro (P)	97	70
Valine	Val (V)	99	72
Threonine	Thr (T)	101	74
Cysteine	Cys (C)	103	76
Isoleucine	Ile (I)	113	86
Leucine	Leu (L)	113	86
Asparagine	Asn (N)	114	87
Aspartate	Asp (D)	115	88
Glutamine	Gln (Q)	128	101
Lysine	Lys (K)	128	129, 101, 84[c]
Glutamate	Glu (E)	129	102
Methionine	Met (M)	131	104
Histidine	His (H)	137	110
Phenylalanine	Phe (F)	147	120
Arginine	Arg (R)	156	129, 112, 100, 87, 70, 43[c]
Tyrosine	Try (Y)	163	136
Tryptophan	Trp (W)	186	159

[a]The values were obtained from Jardine (*20*), and Spengler et al. (*21*).
[b]All masses are given as average integer values.
[c]Arginine and lysine both exhibit multiple immonium ions, and these are listed (they are not of equal intensity).

TABLE J.6 Dipeptide Masses. All Masses are Measured in Daltons

	Gly	Ala	Ser	Pro	Val	Thr	Cys	L/I	Asn	Asp	K/Q	Glu	Met	His	Phe	Arg	CmC	Tyr	Trp
Gly	114																		
Ala	128	142																	
Ser	144	158	174																
Pro	154	168	184	194															
Val	156	170	186	196	198														
Thr	158	172	188	198	200	202													
Cys	160	174	190	200	202	204	206												
L/I	170	184	200	210	212	214	216	226											
Asn	171	185	201	211	213	215	217	227	228										
Asp	172	186	202	212	214	216	218	228	229	230									
K/Q	185	199	215	225	227	229	231	241	242	243	256								
Glu	186	200	216	226	228	230	232	242	243	244	257	258							
Met	188	202	218	228	230	232	234	244	245	246	259	260	262						
His	194	208	224	234	236	238	240	250	251	252	265	266	268	274					
Phe	204	218	234	244	246	248	250	260	261	262	275	276	278	284	294				
Arg	213	227	243	253	255	257	259	269	270	271	284	285	287	293	303	312			
CmC	218	232	248	258	260	262	264	274	275	276	289	290	292	298	308	317	322		
Tyr	220	234	250	260	262	264	266	276	277	278	291	292	294	300	310	319	326	326	
Trp	243	257	273	283	285	287	289	299	300	301	314	315	317	323	333	342	349	349	272

TABLE J.7 Mass Differences Considered in Molecular Weight Analysis of Proteins

Reaction	Protein	Daltons
Amino acid substitution	Lys for Gln	0.04
	Asn for Leu	0.94
	Glu for Lys	0.95
	Asp for Asn	0.99
	Glu for Gln	0.99
	Thr for Val	1.97
	Val for Pro	2.01
	Cys for Thr	2.04
	Met for Glu	2.07
	His for Met	5.95
Amino acid additions	Gly	57.05
	Ala	71.08
	Ser	87.08
	Pro	97.12
	Val	99.13
	Thr	101.10
	Cys	103.14
	Leu, Ile	113.16
	Asn	114.10
	Asp	115.09
	Gln	128.13
	Lys	128.17
	Glu	129.12
	Met	131.19
	His	137.14
	Phe	147.18
	Arg	156.19
	Tyr	163.18
	Trp	186.21
Posttranslational and chemical modifications	SO_3 versus HPO_3	0.08
	—S—S— to —SH HS—	2.02
	CH_2 (e.g., methyl ester)	14.03
	O (e.g., methionine sulfoxide; hydroxyproline; oxidized tyrosine)	15.99
	H_2O (e.g., homoserine lactone/ homoserine after CNBr cleavage; pyroglutamic acid)	18.02
	CO (e.g., N-terminal formylation)	28.01
	CH_2CO (e.g., acetylation)	42.04
	CO_2 (e.g., γ-carboxyglutamic acid)	44.01
	HPO_3	79.98
	SO_3	80.06
	Hexoses	160.14
	N-Acetylaminohexoses	201.19
	Sialic acid	291.26

K

List of Suppliers, Vendors, Manufacturers

This list is not meant to be complete. I apologize in advance to any companies that are left out. Several manufacturers have merged, been bought out, and have gone out of business while new companies have entered the marketplace. More comprehensive lists can be found in issues of *American Biotechnology Laboratory*.

Accurate Chemical and Scientific Corp.
300 Shames Drive
Westbury, NY 11590
Tel: (800) 645-6264
Fax: (516) 997-4948

Agilent Technologies
395 Page Mill Rd.
P.O. Box 10395
Palo Alto, CA 94303
Tel: (650) 752-5000
www.chem.agilent.com

American Type Culture Collection
10801 University Blvd.
Manassas, VA 20110
Tel: (703) 365-2700
www.atcc.org

Amersham Biosciences
800 Centennial Ave.
P.O. Box 1327
Piscataway, NJ 08855
Tel: (732) 457-8000
Fax: (732) 457-0557
www.amershambiosciences.com

Amicon, Inc.
72 Cherry Hill Dr.
Beverly, MA 01915
Tel: (508) 777-3622
Fax: (508) 777-6204

Amresco Inc.
30175 Solon Industrial Pkwy.
Solon, OH 44139
Tel: (440) 349-1313
www.amresco-inc.com

Applied Biosystems
850 Lincoln Center Drive
Foster City, CA 94404
Tel: (800) 327-3002
Fax: (650) 638-5884
www.appliedbiosystems.com

BD Biosciences
2350 Qume Drive
San Jose, CA 95131-1807
Tel: (800) 223-8226
www.bdbiosciences.com

Beckman Coulter
4300 Harbor Blvd. Box 3100
Fullerton, CA 92834-3100
Tel: (800) 742-2345
Fax: (800) 643-4366
www.beckmancoulter.com

Bellco Glass, Inc.
P.O. Box B, 340 Edrudo Rd.
Vineland, NJ 08360
Tel: (800) 257-7043
Fax: (856) 691-3247
www.bellcoglass.com

Biospec Products (Bead-beater)
P.O. Box 788
Bartlesville, OK 74005
Tel: (918) 336-3363
Fax: (918) 336-6060

Bio-Rad Laboratories Main Office
2000 Alfred Nobel Drive
Hercules CA 94547
Tel: (800) 424-6723
Fax: (510) 741-5800
www.bio-rad.com

Biowhittaker Molecular Applications
191 Thomaston Street
Rockland, ME 04841
Tel: (800) 341-1574
www.cambrex.com

BioWhittaker, Inc.
8830 Biggs Ford Road
Walkersville, MD 21793-0127
Tel: 301-898-7025
sales@biowhittaker.com

Branson Ultrasonics Corp.
41 Eagle Rd.
Danbury, CT 06813
Tel: (203) 796-0400
Fax: (203) 796-2240
info@bransoncleaning.com

B. Braun Biotech Intl.
999 Postal Rd.
Allentown, PA 18103
Tel: (610) 266-6262
Fax: (610) 266-9319
www.bbraunbiotech.com

CBS Scientific Co., Inc.
P.O. Box 856
Del Mar, CA 92014
Tel: (800) 243-4959

Calbiochem-Novabiochem Corp.
10394 Pacific Center Ct.
San Diego, CA 92121
Tel: (858) 450-9600
www.calbiochem.com

Cambrex
One Meadowlands Plaza
East Rutherford, NJ 07073
Tel: (201) 804-3000
www.cambrex.com

Chemicon International
28820 Single Oak Drive
Temecula, CA 92590
Tel: (800) 437-7500
www.chemicon.com

Dionex Corp.
P.O. Box 3603
Sunnyvale, CA 94088
Tel: (408)737-0700
www.dionex.com

DuPont NEN®
549 Albany Street
Boston, MA 02118
Tel: (800)551-2121
www.nenlifesci.com

Dynal Biotech Inc.,
5 Delaware Drive
Lake Success, NY 11042
www.dynalbiotech.com

EY Laboratories, Inc.
107 North Amphlett Blvd.
San Mateo, CA 94401
Tel: (800) 821-0044
Fax: (415) 342-2648

Eastman Kodak Co.
Eastman Fine Chemicals
343 State St. B-701
Rochester, NY 14652
Tel: (800) 225-5352
Fax: (800) 879-4979

E. M. Science
P.O. Box 70
480 Democrat Road
Gibbstown, NJ 08027
Tel: (800) 222-0342
Fax: (609) 423-4389

Eppendorf (Brinkmann Instruments Inc)
One Cantiague Rd, PO. Box 1019
Westbury, NY 11590
Tel: (800) 645-3050
Fax: (516) 334-7506
www.brinkmann.com

Genomic Solutions Inc.
4355 Varsity Drive
Ann Arbor, MI 48108
Tel: (877) GENOMIC
www.proteometrics.com

Geno Technology, Inc.
92 Weldon Parkway
St. Louis, MO 63043
Tel: (800) 628-7730
www.genotech.com

Genzyme Corp
One Kendall Square
Cambridge, MA 02139
Tel: (800) 332-1042
Fax: (617) 252-7759

GIBCO BRL
P.O. Box 68
Grand Island, NY 14072
Tel: (800) 828-6686
Fax: (800) 331-2286

Gilson Medical Elec. Inc.
3000 W Beltline Hwy
Middleton, WI 53562
Tel: (800) 445-7661
Fax: (608) 831-4451
www.gilson.com

Glyko, Inc.
11 Pimentel Ct.
Novato, CA 94949
Tel: (800) 33-GLYKO
Fax: (415) 382-3511
www.glyko.com

Grace VYDAC
17434 Mojave St.
P.O. Box 867
Hesperia, CA 92345
Tel: (800) 247-0924
Fax: (619) 244-1984
www.vydac.com

Hamilton Company
P.O. Box 10030
Reno, NV 89520
Tel: (800) 648-5950
www.hamiltonco.com

Hitachi High Technologies America
3100 North First Street
San Jose, CA 95134
Tel: (800) 548-9001
Fax: (408) 432-0704
www.hii-hitachi.com

ICN Biomedicals, Inc.
1263 S. Chillicothe Rd.
Aurora, OH 44202
Tel: (800) 854-0530
Fax: (800) 334-6999
www.icnbiomed.com

International Equipment Co. (IEC)
300 Second Ave.
Needham Heights, MA 02494
Tel: (800) 843-1113
Fax: (781) 444-6743
www.labcentrifuge.com

Invitrogen
3985 B Sorrento Valley Blvd
San Diego, CA 92121
Tel: (800) 955-6288
Fax: (619) 597-6201
www.invitrogen.com

Jule, Inc.
185 Research Drive, #6
Milford, CT 06460
Tel: (800) 648-1772
www.precastgels.com

Kimble/Kontes
P.O. Box 729
Vineland, NJ 08636
Tel: (800) 223-7150
Fax: (609) 692-3242
www.kimble-kontes.com

Molecular Devices Corp.
1311 Orleans Ave

Sunnyvale, CA 94089-1136
Tel: (408) 747-1700
www.moleculardevices.com

Molecular Probes Inc.
P.O. Box 22010
Eugene, OR 97402-0469
Tel: (541) 465-8300
www.probes.com

Millipore Corp., Waters Chromatography
34 Maple St.
Milford, MA 01757
Tel: (508) 478-2000
Fax: (508) 872-1990
www.waters.com

Nalge Nunc International
75 Panorama Creek Dr.
Rochester, NY 14625
www.nalgenunc.com

New England Biolabs, Inc.
32 Tozer Road
Beverly, MA 01915
Tel: (978) 927-5054
www.neb.com

PanVera
501 Charmany Drive
Madison, WI 53719
Tel: (800) 791-1400
www.panvera.com

Pel-Freez Biologicals
205 N. Arkansas St.
P.O. Box 68
Rogers, AR 72757
Tel: (800) 643-3426
Fax: (501) 636-4282
www.pelfreez-bio.com

PerkinElmer Life Sciences, Inc.
549 Albany Street
Boston, MA 02118
Tel: (800) 551-2121
www.perkinelmer.com

PerSeptive Biosystems
38 Sidney St.
Cambridge, MA 02139
Tel: (800) 899-5858

Pfanstiehl Laboratories, Inc.
1219 Glen Rock Ave.
Waukegan, IL 60085
Tel: (800) 383-0126
Fax: (708) 623-9173

Pharmacia Biotech Inc.
800 Centennial Ave.
Piscataway, NJ 08854
Tel: (800) 526-3593
Fax: (908) 457-0557

Pierce
3747 N. Meridan Rd.
P.O. Box 117
Rockford, IL 61105
Tel: (800) 874-3723
Fax: (800) 842-5007

Promega Corp.
2800 Woods Hollow Rd
Madison, WI 53711
Tel: (800) 356-9526
Fax: (800) 356-1970
www.promega.com

Protein Sciences
1000 Research Parkway
Meriden, CT 06450
Tel: (800) 488-7099
www.proteinsciences.com

QIAGEN Inc.
9259 Eton Ave.
Chatsworth, CA 91311
Tech Services: (800) 362-7737
Fax: (818) 718-2056
www.qiagen.com

RepliGen
One Kendall Square Building 700
Cambridge, MA 02139
Tel: (800) 622-2259

Roche Diagnostics Corp
(Formerly Boehringer Mannheim Corporation)
9115 Hague Road
P.O. Box 50414
Indianapolis, IN 46250
Tel: (800) 428-5433
www.bichem.roche.com

Sarstedt, Inc.
1025 St. James Rd.
P.O. Box 468
Newton, NC 28658
Tel: (800) 465-257-5101
Fax: (828) 465-4003
Sarstedt@twave.net

Sartorius Corp.
131 Heartland Blvd.
Edgewood, NY 11717-8358
Te1: 800-368-7178
www.sartorius.com

Savant Instruments, Inc.
110-103 Bi-County Blvd.
Farmingdale, NY 11735
Tel: (800) 634-8886
Fax: (516) 249-4639

Schleicher and Schuell, Inc.
10 Optical Ave.
P.O. Box 2012
Keene, NH 03431
Tel: (800) 245-4024
Fax: (603) 357-3627

Shandon, Inc.
171 Industry Dr.
Pittsburgh, PA 15275
Tel: (412) 788-1133
Fax: (412) 788-1138

Shimadzu
7102 Riverwood Dr.
Columbia, MD 21046
Tel: (800) 477-1227
www.shimadzu.com

Sigma Chemical Co.
P.O. Box 14508
St. Louis, MO 63178
Tel: (800) 325-3010
Fax: (800) 325-5052
www.sigma-aldrich.com

SLM-AMINCO
810 W. Anthony Dr.
Urbana, IL 61801
Tel: (800) 637-7689
Fax: (217) 384-7744

Spectrum Medical Industries Inc.
60916 Terminal Annex
Los Angeles, CA 90060
Tel: (800) 445-7330
Fax: (713) 443-3100

Supelco Inc.
Supelco Park
Bellefonte, PA 16823
Tel: (800) 247-6628
Fax: (415) 359-5459

SynChrom, Inc.
802 Columbia St.
P.O. Box 310
Lafayette, IN 47902
Tel: (800) 283-4752
Fax: (317) 742-2721

Thermo EC
100 Colin Drive
Holbrook, NY 11741-4306
Tel: (800) EC-RANGE
www.thermoec.com

Thermo Finnigan
355 River Oaks Parkway
San Jose, CA 95134-1991
Tel (408) 965-6000
www.thermofinnigan.com

TosoHaas
156 Keystone Drive
Montgomeryville, PA 18936-9637
Tel: (800) 366-4875
www.rohmhaas.com

Vector Laboratories Inc.
30 Ingold Road
Burlingame, CA 94010
Tel: (415) 697-3600
Fax: (415) 697-0339

Wako Chemicals USA, Inc.
1600 Bellwood Rd.
Richmond, VA 23237
Tel: (804) 271-7677

Waters (see Millipore)

Whatman Inc.
9 Bridewell Place

Clifton, NJ 07014
Tel: (973) 773-5800
Fax: (973) 773-0168
www.whatman.com

Wheaton Science Products
1501 N. 10th St.
Millville, NJ 08332
Tel: (800) 225-1437
Fax: (609) 825-1368
www.wheatonsci.com

Worthington Biochemical Corp.
730 Vassar Ave.
Lakewood, NJ 08701
Tel: (732) 942-1660
www.worthington-biochem.com

Zymed Labs., Inc.
561 Eccles Ave.
San Francisco, CA 94080
Tel: (800) 874-4494
Fax: (650) 871-4499
www.zymed.com

References

Aitken A (1990): *Identification of Protein Consensus Sequences. Active Site Motifs, Phosphorylation, and Other Post-translational Modifications.* New York: Ellis Horwood

Altschul SF, Gish W, Miller W, Myers EW, Lipman DJ (1990): Basic local alignment search tool. *J Mol Biol* 215:403–410

Appel RD, Sanchez J-C, Bairoch A, Golaz O, Miu M, Vargas JR, Hochstrasser DF (1993): SWISS-2DPAGE: A database of two-dimensional gel electrophoresis images. *Electrophoresis* 14:1232–1238

Appel RD, Bairoch A, Hochstrasser DF (1994): A new generation of information retrieval tools for biologists: the example of the ExPASy WWW server. *Trends Biochem Sci* 19:258–260

Bairoch A (1993): The PROSITE dictionary of sites and patterns in proteins, its current status. *Nucleic Acids Res* 21:3097–3103

Bairoch A, Boeckmann B (1993): The SWISS-PROT.protein sequence data bank, recent developments. *Nucleic Acid Res* 21:3093–3096

Berners-Lee TJ, Cailliau R, Groff JF, Pollermann B (1992): The World Wide Web, Computer Networks and ISDN Systems. *Electronic Networking: Research, Applications and Policy* 2:52–58

Bourdon MA, Krusius T, Campbell S, Schwartz NB (1987): Identification and synthesis of a recognition signal for the attachment of glycosaminoglycans to proteins. *Proc Natl Acad Sci USA* 84:3194–3198

Chelsky D, Ralph R, Jonak G (1989): Sequence requirements for synthetic peptide-mediated translocation to the nucleus. *Mol Cell Bio* 9:2487–2492

Collawn JF, Stangel M, Kuhn LA, Esekogwu V, Jing S, Trowbridge IS, Tainer JA (1990): Transferrin receptor internalization sequence YXRF implicates a tight turn as a structural recognition motif for endocytosis. *Cell* 63:1061–1072

Cosman D, Lyman SD, Idzerda RL, Beckman MP, Park LS, Goodwin RG, March CJ (1990): A new cytokine receptor superfamily. *Trends Biochem Sci* 15:265–269

Czernik AJ, Pang DT, Greengard P (1987): Amino acid sequences surrounding the cAMP-dependent and calcium/calmodulin-dependent phosphorylation sites in rat and bovine synapsin I. *Proc Natl Acad Sci* USA 84:7518–7522

Dice JF (1990): Peptide sequences that target cytosolic proteins for lysosomal proteolysis. *Trends Biochem Sci* 15:305–309

Dingwall C, Laskey RA (1991): Nuclear targeting sequences-a consensus? *Trends Biochem Sci* 16:478–481

Dreyfuss G, Swanson MS, Pinol-Roma S (1988): Heterogeneous nuclear particles and the pathway of mRNA formation. *Trends Biochem Sci* 13:86–91

Feramisco JR, Glass DB, Krebs EG (1980): Optimal spatial requirements for the location of basic residues in peptide substrates for the cyclic AMP-dependent protein kinase. *J Biol Chem* 255:4240–4245

Gierasch LM (1989): Signal sequences. *Biochemistry* 28:923–930

Gonzalez JM, Saiz-Jimenez C (2003): Optical thermal cycler for use as a fluorimetric plate reader to estimate DNA concentrations. *Bio Techniques* 34:710–712

Good NE, Winget GD, Winter W, Connolly TN, Izawa S, Singh RMM (1966): Hydrogen ion buffers for biological research. *Biochemistry* 5:467–477

Good NE, Izawa S (1968): Hydrogen ion buffers. *Methods Enzymol* 24(Part B):53–68

Gordon JI, Duronio RJ, Rudnick DA, Adams SP, Gokel GW (1991): Protein N-myristoylation. *J Biol Chem* 266:8647–8650

Grantham R, Gautier C, Gouy M, Jacobzone M, Mercier R (1981): Codon catalogue usage is a genome strategy modulated for gene expressivity. *Nucleic Acids Res* 9:r43–r74.

Harlow E, Lane D (1988): *Antibodies*. Cold Springs Harbor, NY: Cold Springs Harbor Laboratory Press

Henikoff S (1993): Sequence analysis by electronic mail server. *Trends Biochem* Sci 18:267–268

Hong W, Tang BL (1993): Protein trafficking along the exocytotic pathway. *Bio Essays* 15:231–238

Joseph CK, Byun H-S, Bittman R, Kolesnick RN (1993): Substrate recognition by ceramide-activated protein kinase. *J Biol Chem* 268:20002–20006

Kehoe B (1992): *Zen and the Art of the Internet: A Beginner's Guide to the Internet*. Upper Saddle River, NJ: Prentice Hall

Kemp BE, Pearson RB (1990): Protein kinase recognition sequence motifs. *Trends Biochem Sci* 15:342–346

Kirshenbaum DM (1976): *Handbook of Biochemistry and Molecular Biology Third Edition*, Vol. 2. GD Fasman, ed. Cleveland, OH: CRC Press

Lathe, R (1985): Synthetic oligonucleotide probes deduced from amino acid sequence data. Theoretical and practical considerations. *J Mol Biol* 183:1–12

Low A, Faulhammer HG, Sprinzl M (1992): Affinity labeling of GTP-binding proteins in cellular extracts. *FEBS Lett* 303:64–68

Magee T, Hanley M (1988): Sticky fingers and CAAX boxes. *Nature* 335:114–115

Moreno S, Nurse P (1990): Substrates for $p34^{cdc2}$: *In vivo* veritas? *Cell* 61:549–551

Munro S, Pelham HRB (1987): A C-terminal signal prevents secretion of luminal ER proteins. *Cell* 48:899–907

Pierschbacher MD, Ruoslahti E (1984): Cell attachment activity of fibronectin can be duplicated by small synthetic fragments of the molecule. *Nature* 309:30–33

Pohlner J, Klauser T, Kuttler E, Halter R (1993): The secretion pathway of IgA protease type proteins in gram negative bacteria. *BioEssays* 15:799–804

Rapoport TA, Jungnickel B, Kutay U (1996): Protein transport across the eukaryotic endoplasmic reticulum and bacterial inner membranes. *Annu Rev Biochem* 65:271–303

Qin N, Pittler SJ, Baehr W (1992): *In vitro* isoprenylation and membrane association of mouse rod photoreceptor cGMP phosphodiesterase α and β subunits expressed in bacteria. *J Biol Chem* 267:8458–8463

Schleif RF, Wensink PC (1981): *Practical Methods in Molecular Biology*. New York: Springer-Verlag

Shin J, Dunbrack RL, Lee S, Strominger JL (1991): Signals for retention of transmembrane proteins in the endoplasmic reticulum studied with CD4 truncation mutants. *Proc Natl Acad Sci USA* 88:1918–1922

Steere NV (1971): *CRC Handbook of Laboratory Safety*. Cleveland, OH: The Chemical Rubber Co.
Towler DA, Gordon JI, Adams SP, Glaser L (1988): The biology and enzymology of eukaryotic protein acylation. *Ann Rev Biochem* 57:69–99
von Heijne G (1990): The signal peptide. *J Membr Biol* 115:195–201
Weber IT, Miller M, Jaskolski M, Leis J, Skalka AM, Wlodawer A (1989): Molecular modeling of the HIV-1 protease and its substrate binding site. *Science* 243:928–931
Wong SH, Hong W (1993): The SXYQRL sequence in the cytoplasmic domain of TGN38 plays a major role in trans-golgi network localization. *J Biol Chem* 268:22853–22862
Yokode M, Pathak RK, Hammer RE, Brown MS, Goldstein JL, Anderson RGW (1992): Cytoplasmic sequence required for basolateral targeting of LDL receptor to livers of transgenic mice. *J Cell Biol* 117:39–46

General References

Beckman (1985): *Rotors and Tubes for Preparative Ultracentrifuges: An Operator's Manual*. Spinco Division of Beckman Instruments, Inc., Palo Alto, CA
Griffith OM (1986): *Techniques of Preparative, Zonal, and Continuous Flow Ultracentrifugation*. Spinco Division of Beckman Instruments, Inc., Palo Alto, CA

Index